Energiya–Buran

The Soviet Space Shuttle

Bart Hendrickx and Bert Vis

Energiya–Buran

The Soviet Space Shuttle

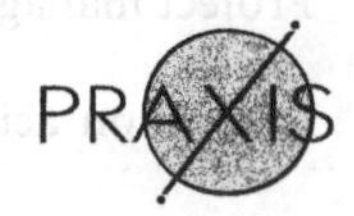

Mr Bart Hendrickx
Russian Space Historian
Mortsel
Belgium

Mr Bert Vis
Spaceflight Historian
Den Haag
The Netherlands

SPRINGER–PRAXIS BOOKS IN SPACE EXPLORATION
SUBJECT *ADVISORY EDITOR*: John Mason, M.Sc., B.Sc., Ph.D.

ISBN 978-0-387-69848-9 Springer Berlin Heidelberg New York

Springer is part of Springer-Science + Business Media (springer.com)

Library of Congress Control Number: 2007929116

Cover design: Jim Wilkie
Project management: Originator Publishing Services Ltd, Gt Yarmouth, Norfolk, UK

Printed on acid-free paper

Contents

Предисловие

«Энергия-Буран» – самый мощный космический корабль, который когда-либо видел мир, и если бы была дана возможность полностью осуществить программу, это было бы огромной пользой не только для Советских людей, но и действительно для всего мира. Это оказалось невозможным из-за зарождающихся неблагоприятных политических и отчасти экономических условий.

Мне выпала честь стать участником программы и ведущим летчиком-испытателем космического корабля «Буран». В этом качестве я двенадцать раз летал на BTS-002, аналоге «Бурана» по программе испытаний, связанных с полетами «Бурана» в атмосфере. Команда Летно-исследовательского института им. М.М. Громова состояла из самых лучших летчиков-испытателей Советского Союза. Два летчика из этой группы – Анатолий Левченко и я – летали в космос на космическом корабле «Союз» в качестве подготовки для испытаний «Бурана» на орбите. Но до пилотируемых полетов «Бурана» дело не дошло. Программа была закрыта после одного автоматического (беспилотного) полета.

«Буран» до сих пор говорит о талантливости людей в России, и многие гордятся тем, что принимали участие в программе, хотя так и не получили результата, хотя бы одного управляемого полета в космос. На космодроме «Байконур» можно встретить модель «Бурана» у главных ворот при выезде из аэропорта. На могильных плитах инженеров и техников, работавших по программе, выгравирован «Буран».

Становится теплее на сердце, когда видишь, что даже вне России «Буран» еще жив. И я рад, что авторам этой книги удалось написать авторитетную историю «Энергии» и «Бурана», используя подлинные советские и российские источники. Я искренне надеюсь, что эта книга и дальше будет способствовать распространению знаний о программе, которая могла бы принести огромную экономическую пользу миру, если была бы возможность ее осуществить.

Игорь Петрович Волк,
Герой Советского Союза,
Заслуженный летчик-испытатель,
Летчик-космонавт Советского Союза.

Foreword

Energiya–Buran is the most powerful space vehicle the world has ever seen, and, had it been given the chance to fully develop, it would have been of great benefit to the people of the Soviet Union and, indeed, the world. It didn't get that chance, but the political and to some extent economical situation were not ideal.

I had the honor of being selected as the lead test-pilot for Buran. As such, I flew Buran's analog BTS-002 on 12 occasions in the program that tested the atmospheric portion of Buran missions. The team from the Flight Research Institute named after M.M. Gromov consisted of some of the best test-pilots in the Soviet Union. Two pilots from this select group, Anatoliy Levchenko and I, flew in space on a Soyuz spacecraft as part of our preparations to test Buran in orbit. But, after one unmanned flight and before we had the chance to fly Buran ourselves, the program was canceled.

Buran still speaks to the imagination of the people in Russia and many take pride to have participated in the program, even though it never resulted in even one manned mission in space. At the Baykonur Cosmodrome, a model of Buran can be seen at the main gate one passes when coming from the airport. Engineers and technicians who worked on the program and have since passed away even have Buran etched on their headstones.

It is heart-warming to see that, even outside Russia, Buran still lives and I am happy to see that the authors of this book have managed to write an authoritative history on Energiya and Buran, using original Soviet and Russian sources. I sincerely hope that this book will further spread the knowledge of a program that might have yielded enormous economical profit to the world, had it been given the chance.

Igor Petrovich Volk
Hero of the Soviet Union
Merited Test Pilot
Pilot-Cosmonaut of the Soviet Union

Authors' preface

This book is about the Energiya–Buran system, the Soviet equivalent of the US Space Shuttle. Originally conceived in 1976, Buran made its one and only flight in November 1988, more than seven years after the inaugural flight of the Space Shuttle. Prudent as the Soviet authorities were, it was conducted in an unmanned mode, a feat not accomplished by NASA in the Space Shuttle program.

Buran was not unique for being a manned spaceflight project that eventually would never carry a man into orbit. There were other Soviet programs that had suffered the same fate, such as the L-1/L-3 lunar program, and the military space station ferry TKS. Unlike these, however, from its conception Buran was a spacecraft without a clearly defined task. It was solely designed and built in response to the Space Shuttle, whose military potential was a source of major concern to the Soviet Union. Unsure what exactly the threat was, the Russians decided to build a vehicle matching the Shuttle's capabilities to have a deterrent in the long run. From the Russian perspective, Buran was just another product of the arms race between the superpowers.

The orbiter resembled its American counterpart to the point that they were aerodynamic twins, but there were important differences between the two systems as well. The most notable one was that Buran did not have main engines and was carried into orbit by a powerful launch vehicle (Energiya) that could be adapted for other missions as well. Despite the copying that unquestionably took place, the Russians still had to develop the technology, the materials, and the infrastructure all by themselves and in doing so often followed their own, unique approach. Building upon the lessons learned from their star-crossed manned lunar program, they brought the project to a state of maturity that allowed them to fly two successful launches of the Energiya rocket and one of the Buran orbiter. This was a remarkable feat, irrespective of whether the expenditures were justified or not.

After the maiden Buran flight in 1988, plans were drawn up for another mission in which the orbiter would again go up and land unmanned, although this time it would be briefly boarded in orbit by a visiting Soyuz crew. Only after the second mission had proven the system to be reliable, would a crew have been allowed to be launched on board the orbiter.

Unfortunately, it would never come to that. As the Cold War drew to a close and the Soviet Union collapsed, the program largely lost its *raison d'être*. In a time where funds allocated to large space undertakings were getting scarcer and scarcer, here was a program that was devouring more and more of that money. Slowly but surely, more and more space program officials began to oppose Buran, emphasizing that all this money was disappearing into a bottomless pit, without anyone being able to give a clear answer to that one question: what do we need Buran for?

Finally, the program died a silent death. It was never officially terminated by a government decree, but those who were involved knew the signs. The cosmonauts who had been training for the manned missions began returning to test flying in their respective institutes, transferred to the Soyuz and Mir program, or tried their luck in private industry.

Hardware was scrapped, stored, or offered for sale. The full-scale test model used for the approach and landing tests was sent to Sydney, where it was put on display. Later it was to be shipped to a museum in Germany, but didn't make it beyond a junkyard in Bahrain, where it still sits at the time of writing.

Another full-scale test model ended up as a tourist attraction in Gorkiy Park in Moscow, while a third has been parked outdoors at Baykonur for several years, where it has been left exposed to the elements. The only Buran orbiter that flew in space was put in storage in the Energiya assembly building, but was totally destroyed when the building's roof collapsed in May 2002.

In spite of the sad fates of these Buran orbiters, the program was a source of great pride for everyone who participated in it, from engineers to prospective cosmonauts. In many places models of the orbiter, or the entire vehicle, were erected, sometimes as monuments, sometimes just to embellish the streets in which they stand. As Buran's lead test pilot Igor Volk says in his foreword and as maybe the ultimate sign of pride, many who were involved in it have the vehicle etched on their gravestones.

Despite cancellation of the project, the technology developed for it has not all disappeared down the drain. The rocket engine of Energiya's strap-on boosters is still being used today by the Zenit rocket and its Sea Launch version and scaled-down versions of the engine currently power the first stage of America's Atlas rockets and will also be employed in a new family of Russian launch vehicles called Angara. The docking hardware originally developed for Buran was used in the Shuttle/Mir program and is now actively used on the International Space Station.

Perhaps Buran was born under an unlucky star, but since the programm ended those who designed and built it have gone to a lot of trouble to make sure that the Soviet/Russian counterpart to the US Space Shuttle will be remembered as a state-of-the-art spaceship that was launched by one of the most powerful launch vehicles the world has ever seen. With this book, we hope we can contribute to that endeavor.

Bart Hendrickx | *Bert Vis* | *April 2007*
Mortsel | Den Haag
Belgium | The Netherlands

Acknowledgments

This book is a cooperative effort by two authors, but would probably not have come about without the initiative of David Shayler, who originally came up with the idea to write the book but in the end could not participate in it due to other commitments.

Of particular help in preparing the book were several people who were either directly or indirectly involved in the Energiya–Buran program. Thanks are extended to Buran lead test pilot Igor Volk for his foreword and also to the numerous other Buran test pilots who granted interviews to Bert Vis during his countless travels to Star City, Zhukovskiy, and other locations. Lida Shkorkina was instrumental in arranging many of those interviews and also acted as interpreter during most of them. Thanks are also due to Emil Popov, a veteran of the Military Industrial Commission, who shared recollections of the meetings and discussions in the early 1970s that eventually led to the decision to go ahead with Buran. Nina Gubanova, the widow of Energiya–Buran chief designer Boris Gubanov, provided an original copy of her husband's hard-to-obtain memoirs.

Our special thanks also go to those who gave the authors access to some rare primary documents, most of them from the archives of the late Ernest Vaskevich, who headed the coordination and planning department of the Departmental Training Complex for Cosmonaut-Testers (OKPKI) in Zhukovskiy, which acted as the Flight Research Institute's own cosmonaut training center. Many of those documents offer unique insight into the training program of the Buran test pilots as well as crewing issues and flight plans.

The authors also wish to thank several researchers who supported them while writing the book. First and foremost among those is Vadim Lukashevich, the web-master of the *www.buran.ru* website and without doubt Russia's leading expert on the history of Buran. Vadim never got tired of answering the authors' frequent and challenging questions and also kindly granted permission to use many of the pictures and illustrations on his website and CD-ROMs. The book would not have been what it is without his dedication, advice, and continued support.

Appreciation is also due to the staff of the unrivaled Russian space magazine *Novosti kosmonavtiki,* whose tireless efforts to unravel the mysteries of Soviet space history were a great source of help and inspiration in writing the book. Asif Siddiqi, the highly respected American authority on Soviet/Russian space history, was always willing to help and share information from his rich archives. Chris van den Berg, who has been patiently monitoring Soviet/Russian space-to-ground communications for over 40 years, assisted the authors in making sense of Buran's communication systems. Peter Pesavento provided valuable information on US intelligence assessments of the Soviet shuttle program. Rex Hall granted access to his archives and helped with his knowledge of the Soviet/Russian space program.

Several people kindly allowed the authors to select pictures from their photo collections, including Igor Afanasyev, Edwin Neal Cameron, Sergey Grachov, Vadim Lukashevich, Igor Marinin, Timofey Prygichev, Asif Siddiqi, Rudolf van Beest, Luc van den Abeelen, and Simon Vaughan. Dennis Hassfeld was kind enough to make several line drawings based on original Russian sketches.

We thank Clive Horwood of Praxis for his continued support and Neil and Bruce Shuttlewood of Originator Publishing Services for copy editing and generation of proofs.

Last but not least, the authors wish to extend a special word of thanks to their relatives, who put up with them during two years of painstaking and time-consuming research.

Figures

1

The roots of Buran

When Buran swooped down to a safe landing on its Baykonur runway on 15 November 1988, it marked the culmination of more than just the 12 years needed to take it to the launch pad since its official approval by a Soviet government and Communist Party decree in February 1976. Even by the start of the Buran program the Soviet Union possessed a rich database on high-speed aeronautics, gradually accumulated through four decades of work on rocket-propelled aircraft, intercontinental cruise missiles and smaller spaceplanes.

THE FATHER OF SOVIET SPACEPLANES

The first man in the Soviet Union to widely advocate the idea of winged spacecraft was Fridrikh Tsander. Born in 1887 in the Latvian capital Riga into an intellectual German family, Tsander became obsessed with the idea of space travel around the age of 20 and was one of the Soviet Union's most prominent popularizers of space exploration in the 1920s (with one of his lectures attended by Lenin himself in December 1920). Although inspired by the work of great spaceflight theoreticians like his compatriot Konstantin E. Tsiolkovskiy and the German Hermann Oberth, Tsander was convinced that the most practical way of reaching other planets was not with powerful and expensive rockets, but with winged vehicles. Tsander outlined his ideas in the journal *Tekhnika i zhizn* in 1924 in an article called "Flights to Other Planets", openly taking issue with the ideas of Oberth and Tsiolkovskiy:

"For flight to the upper layers of the atmosphere and also for landing on planets possessing an atmosphere, it will be advantageous to use an aeroplane as a construction keeping the interplanetary ship in the atmosphere. Aeroplanes, having the capability of conducting a gliding descent in case of an engine

Fridrikh Tsander.

> shutdown, are far superior to parachutes, proposed for the return to Earth by Oberth in his book "Rocket to the Planets".
>
> Parachutes do not offer the possibility of freely choosing a landing site or continuing the flight in case of a temporary engine shutdown, and therefore it would be advisable to use them only for flights without people. The part of the rocket that is operated by a man, should be equipped with an aeroplane. For descending to a planet having sufficient atmosphere, using a rocket, as proposed by K.E. Tsiolkovskiy, will also be less advantageous than using a glider or an aeroplane with an engine, because a rocket consumes much fuel during the descent and its descent will cost, even if there is only one person in the rocket, tens of thousands of rubles, whereas descending with an aeroplane costs only several tens of rubles, and with a glider, nothing at all."

In this and other works Tsander expounded on the design of an interplanetary spaceplane that would reach space by using a combination of propeller, jet, and rocket engines. As the atmosphere got thinner, unneeded metallic components would move into a boiler to be melted into more rocket fuel. For propulsion during the interplanetary cruise, Tsander proposed screens or mirrors driven by solar light, early precursors of today's solar sails.

THE FIRST ROCKET PLANES

The RP-1

Tsander did more than just generate fancy ideas. He set about turning his ideas into practice in the late 1920s with the development of an experimental rocket engine called the OR-1. In the autumn of 1931 Tsander took the initiative to establish an amateur group to study the practical aspects of rocketry and space exploration. Called the Group for the Investigation of Reactive Motion (GIRD), one of its four sections aimed to install rocket engines on gliders and thereby create a high-altitude aircraft, an idea promoted by the young engineer Sergey P. Korolyov. The engine to be used initially would be Tsander's OR-2. Generating 50 kg of thrust, it used gasoline and liquid oxygen as propellants and had sophisticated features such as regenerative cooling of the combustion chamber using gaseous oxygen, a nozzle-cooling system using water, and a pressure feed system using nitrogen.

In early 1932 a decision was made to put the OR-2 on the BICH-11 flying wing glider. The resulting rocket plane, called RP-1, would be a modest machine, capable of developing a speed of 140 km/h, reaching an altitude of 1.5 km, and staying in the air for just about 7 minutes. However, GIRD had plans for more sophisticated rocket planes, including the RP-3, a two-man plane using a combination of piston and rocket engines to reach altitudes of 10–12 km [1].

While development of the engine got underway, Korolyov himself made several unpowered test flights of the BICH-11 to test its flying characteristics. Before tests of the engine got underway, the overworked and frail Tsander was sent to a sanatorium in the Caucasus, but contracted typhoid fever on the way and passed away on 28 March 1933 at the age of 45. His infectious enthusiasm was surely missed by the GIRD team. Korolyov's daughter would later describe Tsander as an "adult child" in everyday affairs, but the "highest authority" in rocket matters [2]. One cannot even begin to imagine what further contributions this man could have made to Soviet rocketry had he not died such an untimely death.

Tests of the OR-2 engine in 1933 proved unsatisfactory and attempts to replace the gasoline by ethanol to facilitate cooling did not produce the expected results either. Modifying the glider to carry a rocket engine also turned out to be more difficult than expected, with one of the requirements being to drop the fuel tanks in flight to increase safety. Before the RP-1 ever had a chance to make a powered flight, GIRD was forced to change direction.

The RP-318-1

Another organization involved in rocket research in the Soviet Union was the Gas Dynamics Laboratory (GDL) in Leningrad. Established in 1921, it was mainly engaged in developing solid-fuel rockets for arming aircraft or assisting aircraft during take-off. In 1929 a small subdivision was added, headed by 20-year-old Valentin P. Glushko, to conduct research on electric and liquid-propellant engines. While the GIRD members were mainly driven by utopian visions of space travel, the GDL team primarily consisted of military-oriented rocketeers and received its modest funding directly from the military.

In 1932 the Red Army Chief of Staff Marshal Mikhail Tukhachevskiy, convinced that the Soviet Union needed modern technology to arm itself against the forces of capitalism, proposed to unite GIRD and GDL into a single institute to develop both solid and liquid-fuel rockets for the military. After many months of negotiations, the new organization, called the Reactive Scientific Research Institute (RNII) was founded in September 1933. Placed in charge of RNII was GDL's Ivan Kleymyonov, with GIRD's Sergey Korolyov acting as his deputy.

The different backgrounds of the two organizations soon led to internal conflicts about the future direction of the new institute. Many of these centered around the types of propellants to be used. While the GDL faction favored solid propellants or storable liquid propellants, the former GIRD team promoted engines burning liquid oxygen. Also, Korolyov was hoping to continue work on rocket planes capable of reaching the stratosphere, but this was of little interest to Kleymyonov, who saw the development of military missiles as the institute's main objective. These and other disagreements caused Korolyov to be demoted to work as a chief engineer in the section for winged missiles in early 1934.

Winged missiles offered several advantages over ballistic missiles in destroying both mobile and stationary targets. Their flight path could be controlled after shutdown of the engines and they could cover much larger distances thanks to the extra lift provided by the wings, thereby compensating for the absence of powerful rocket engines in those days [3]. However, for Korolyov they also provided an opportunity to covertly pursue his dream of achieving manned stratospheric flight.

Ever since the work on the RP-1 rocket plane, Korolyov had become increasingly convinced that it would be difficult to turn existing aircraft or gliders into efficient rocket planes. Neither was the time ripe to put men aboard ballistic missiles. What was needed instead was a new type of winged machine capable of withstanding higher acceleration forces and fitted with low-aspect-ratio wings, a tail section, and a long fuselage to house the propellant tanks. Although the winged missiles tested at RNII in the 1930s were officially seen as precursors to surface-to-air and air-to-surface missiles, Korolyov developed many of them with the goal of manned stratospheric flight in the back of his mind.

At a conference on the use of rockets to explore the stratosphere in March 1935, Korolyov went public with his ideas to build manned winged missiles that could reach altitudes of up to 20–30 km, emphasizing the need to build "flying laboratories" that would pave the way for such vehicles. Apparently, by late 1935 Kleymyonov was impressed enough to include studies of rocket planes in the RNII's plans for 1936. By early 1936 Korolyov had drawn up a step-by-step plan calling for the development of ever more capable piloted rocket planes. The first of these (218, later renamed 318), powered by either a solid-fuel or liquid-fuel rocket engine, would reach an altitude of 25 km and be flown by two pilots wearing pressure suits. The ultimate goal was to push the ceiling to a phenomenal 53 km [4].

The rocket engines needed for such planes were not yet available, but Korolyov got approval to build an experimental rocket plane based on his SK-9 glider, which he had probably built with that idea in mind. The rocket plane was initially called RP-218-1 and later renamed RP-318-1 after a reorganization within RNII. The engine selected to power the plane was Glushko's ORM-65, a nitric acid/kerosene engine capable of generating between 50 and 175 kg of thrust and already under development for the 212 winged missile.

The goals formulated by Korolyov for the rocket plane program in early 1936 were "to achieve a record altitude and speed" and "to obtain the first practical experience in solving the problem of piloted rocket flight" [5]. To him personally, it was probably the first step on the long road to manned space travel, but Korolyov

The RP-318-1 rocket plane.

was well aware that this would not be enough to receive continued support for the program. As he would have to do more than once in his later rocket and space career, he had to justify his efforts by coming up with military applications. In a study requested by Korolyov, the Zhukovskiy Air Force Academy concluded in 1937 that, despite the limited operating time of the rocket engine, rocket planes could play a vital role as fighters [6]. Their main task would be to intercept enemy bombers. With the development of jet engines in an embryonic stage, rocket engines would be the only practical way of significantly increasing speed in the near future.

After an exhaustive series of tests, the ORM-65 was installed in the SK-9 in September 1937 and began a series of integrated test firings in December 1937. Korolyov was intent on piloting the RP-318-1 himself, going down in history as the first man to fly a rocket plane. However, by this time Stalin's purges were beginning to sweep through the ranks of RNII (renamed NII-3 in 1937). Tukhachevskiy, Kleymyonov, and his deputy Langemak were executed in January 1938, and Glushko and Korolyov were arrested on trumped-up charges in March and June 1938, disappearing into the Soviet prison system for the following six years.

Work on the RP-318-1 was not resumed until the end of the year under the leadership of A. Shcherbakov. The ORM-65 was replaced by a somewhat simplified but more reliable version called the RDA-1-150 with a thrust of between 50 and 146 kg, developed by Glushko's successor Leonid Dushkin. After being installed in the plane, it underwent a series of more than 100 test firings between February and October 1939.

In November 1939 the RP-318-1 was transported to an aerodrome in the outskirts of Moscow, where after several more test firings of the rocket engine it made its first historic flight on 28 February 1940. Piloted by Vladimir Fyodorov, the 675 kg and 7.9 m long rocket plane was towed into the air by an R-5 airplane and released at an altitude of 2.8 km. After gliding down to an altitude of 2.6 km, Fyodorov ignited the RDA-1-150 engine, which burned for 110 seconds, accelerating the plane from 80 to 140 km/h and taking it to an altitude of 2.9 km. There were two more flights on 10 and 19 March 1940. If it hadn't been for the delays caused by the repression in the late 1930s, the RP-318-1 might very well have become the world's first aircraft propelled by a liquid-fuel rocket engine. In the event that distinction went to the German Heinkel He-176, which made its maiden flight on 20 June 1939 using a rocket engine fueled by hydrogen peroxide.

WORLD WAR II ROCKET-PROPELLED AIRCRAFT

With the threat of a German invasion looming, there was increasing interest in the use of rocket-propelled aircraft to improve combat efficiency. On the one hand, dedicated rocket-propelled fighters could use such engines to quickly intercept enemy bombers as soon as they appeared over the horizon and then immediately glide back to the runway, completing their mission in a matter of minutes. On the other hand, rocket engines could also be installed on existing aircraft in addition to the traditional piston engines to either assist in take-off or abruptly increase speed during flight to overtake or evade enemy aircraft. Any utopian visions of space travel quickly faded into the background. However, the World War II rocket planes provided further experience in the field of piloted rocket flight and gave the Soviets an opportunity to continue work on rocket engines, no matter how modest their performance was in comparison with the powerful rocket engines concurrently under development in Germany for the A-4 ("V-2") missile.

Short-range interceptors

The only Soviet short-range interceptor that ever made powered flights during the war was the BI (for *blizhniy istrebitel* or "short-range fighter"), developed at the OKB-293 design bureau of Viktor Bolkhovitinov under the leadership of Aleksandr Bereznyak and Aleksey Isayev. It had a fabric-skinned wooden frame, low-mounted straight wings, a "razorback" style canopy and twin 20 mm cannons mounted in front of the canopy. Installed in the aft was a D-1A-1100 nitric acid/kerosene engine developed by Dushkin at NII-3 capable of throttling between 400 kg and 1,100 kg. The originally planned turbopump was replaced by a pressure-fed system designed by Isayev. With a maximum propellant load of 705 kg, the engine could burn for almost two minutes. Maximum take-off mass was 1,650 kg (dry mass 805 kg).

Work on the BI began in late 1940, but was not given priority until after the German invasion of the Soviet Union in June 1941. In July Stalin ordered the OKB-293 team to build the first BI in just 35 days instead of the three to four months that Bereznyak and Isayev had planned. Amazingly, the first BI was duly delivered in mid-September and in the following weeks made 15 unpowered flights, towed by a Pe-2 bomber. In October 1941, with German troops approaching Moscow, Bolkhovitinov's design bureau was evacuated to the Urals near Sverdlovsk, where the BI-1 tests continued with a series of static test firings and take-off runs. The first

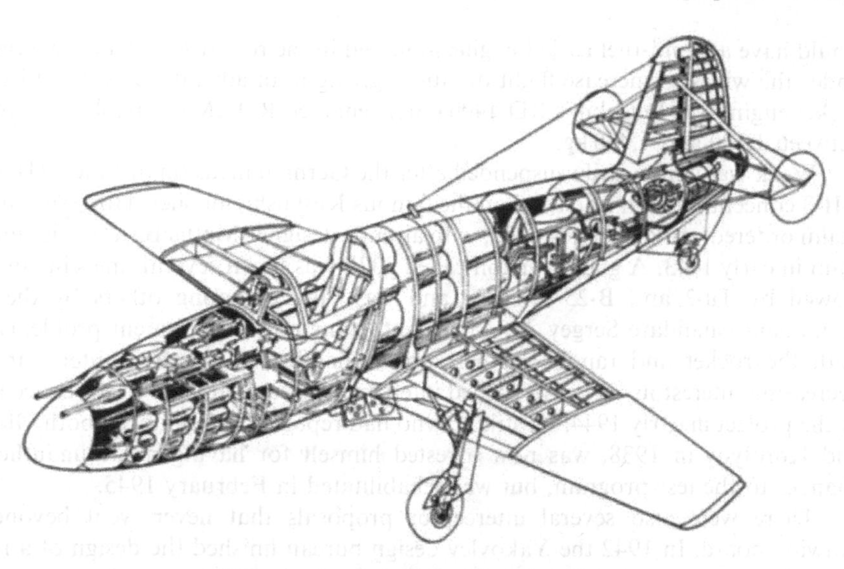

Cutaway drawing of the BI rocket plane.

powered flight took place on 15 May 1942 with Grigoriy Bakhchivandzhi behind the controls. The D-1A-1100 boosted the BI-1 to an altitude of 840 m and a maximum speed of 400 km/h. Just over three minutes after take-off, the BI-1 made a hard landing, seriously damaging its landing gear.

The next test flights were performed with the BI-2 and BI-3, which only differed from the first aircraft in having a skid landing gear. The BI-2 made four flights in January–March 1943 (three by Bakhchivandzhi and one by Konstantin Gruzdev), pushing the maximum speed and altitude to 675 km/h and 4 km, respectively. Bakhchivandzhi was again at the helm for flights 6 and 7 on the BI-3. Unfortunately, the seventh flight on 27 March 1943 ended in tragedy, when the BI-3 went into a dive shortly after engine shutdown and crashed, killing its pilot. The accident had a stifling effect on the rocket plane program, the more so because the cause was never clearly established. Plans for building a batch of 50 production planes (BI-VS) equipped with small bombs were canceled. However, more experimental planes were built with the goal of increasing the very limited flight duration, which was seen as one of the main disadvantages of the plane. One BI version was fitted with two wing-mounted ramjet engines and another one with an improved 1,100 kg thrust engine developed by Isayev (the RD-1). Both made test flights in 1944 and 1945.

NII-3, now headed by Andrey Kostikov, did not only provide the engine for the BI, but also worked on its own short-range interceptor. Originally, the institute had hoped to build upon the success of the RP-381-1 by equipping the plane with an improved 300 kg thrust RD-1-300 engine that would have allowed it to take off on its own power. Also studied was a rocket plane based on the G-14 glider with a liquid oxygen/ethanol RDK-1-150 engine developed by Dushkin. However, in 1940 those plans were abandoned in favor of a short-range interceptor called "302", which would have a liquid-fuel rocket engine mounted in the rear fuselage and two ramjets under the wings to increase flight duration, giving it an advantage over the BI. The rocket engine was Dushkin's RD-1400 (later renamed RD-2M), capable of throttling between 1,100 and 1,400 kg.

Work was temporarily suspended after the German invasion in June 1941, when NII-3 concentrated all its efforts on the famous Katyusha missiles. However, in 1942 Stalin ordered work to be resumed, with aircraft designer Matus Bisnovat joining the team in early 1943. A glider version called 302P was flown several times in late 1943 (towed by Tu-2 and B-25 aircraft) and was piloted among others by the later cosmonaut candidate Sergey Anokhin. Unfortunately, development problems with both the rocket and ramjet engines, the availability of the BI fighters, and the decreasing interest in rocket-propelled interceptors eventually led to the cancellation of the project in early 1944. Kostikov, who had reportedly denounced both Glushko and Korolyov in 1938, was now arrested himself for having made unauthorized changes to the test program, but was rehabilitated in February 1945.

There were also several interceptor proposals that never went beyond the drawing board. In 1942 the Yakovlev design bureau finished the design of a rocket plane based on its YaK-7 aircraft. Called the YaK-7R, it would carry a single Dushkin D-1-A-1100 rocket engine in the tail and two Merkulov DM-4S ramjets under the wings. However, the project stalled due to development problems with the ramjets. In late 1943 the OKB-51 design bureau of Nikolay Polikarpov began work on a rocket plane called Malyutka ("Baby") using an unidentified rocket engine with a thrust of between 1,000 and 1,200 kg. Malyutka was expected to develop a speed of 845 km/h and reach a maximum altitude of 16 km during an 8 to 14 minute flight. The project was canceled after Polikarpov's death in July 1944.

Other proposals were based on the use of a 1,200 kg thrust, four-chamber nitric acid/kerosene engine designed by Valentin Glushko. Equipped with a turbopump, it was called RD-1, not to be confused with its namesake developed by Isayev for the BI. Still a prisoner, Glushko was working at the time for the 28th Special Department of the NKVD secret police in the city of Kazan. This was a so-called *sharaga*, a type of penitentiary for scientists and engineers to work on projects assigned by the Communist Party.

One man displaying interest in the engine was Roberto Bartini, an Italian-born aircraft designer who as a member of the Italian Communist Party was transferred undercover to the USSR in 1923 after the Fascist revolution. Arrested in 1937 because of his ties to Tukhachevskiy, Bartini ended up in the 29th Central Design Bureau (TsKB-29) of the NKVD in Omsk, a *sharaga* that served as the engineering facility for Andrey Tupolev. During 1941–1942 he drew up plans for a rocket plane with swept-back wings called R-114 that would use the RD-1 to reach phenomenal speeds of over 2,000 km/h.

Glushko's RD-1 also figured prominently in rocket plane proposals by Sergey Korolyov, who, after having barely survived the hardships of the Kolyma gulag camps in far eastern Siberia, had been sent back west in 1940 to work under Tupolev at TsKB-29 (first in Moscow, then in Omsk) and was subsequently transferred to Glushko's *sharaga* in Kazan in November 1942. The following month he finished plans for two rocket planes, one weighing 2,150 kg and the other 2,500 kg, with Korolyov preferring the latter version. Equipped with the four-chamber RD-1, the plane would be able to intercept any enemy aircraft at any altitude and even be capable of attacking ground-based targets such as tanks, artillery batteries, and surface-to-air missile installations. It would outperform the BI in every respect, staying in the air for 10.5 minutes (vs. 2 minutes for the BI) at a speed of 800 km/h and carrying 200 kg of weaponry (vs. 50–100 kg for the BI-1). Korolyov also suggested mounting an RD-1 type engine on a 3,500 kg "flying wing" type interceptor that would be able to stay in the air for 30 minutes and reach a maximum altitude of 15 km. A similar flying wing ("RM-1") with a Dushkin RD-2M-3V engine was studied by the OKB-31 design bureau of Aleksandr Moskalyov in 1945–1946.

Advanced as both Bartini's and Korolyov's proposals may have been, they had little resonance. Put forward by men who officially were still "enemies of the people", they stood little chance against official, government-supported projects such as the BI and "302" [7].

Rocket-augmented aircraft

Another application of liquid-fuel rocket engines in aviation was to augment the performance of existing aircraft. In 1932 GDL had worked on a project to install two ORM-52 engines on the I-4 (ANT-5) fighter to improve its combat performance, but those plans were never realized due to the institute's high workload. In 1939–1940 Glushko proposed to use the ORM-65 engines on experimental bombers called S-100 and Stal-7, but he was eventually ordered to develop the four-chamber RD-1. By the time Korolyov arrived in Kazan in late 1942, a 300 kg thrust single-chamber version of the engine was already undergoing tests.

Korolyov proposed to fly this version of the engine on a Pe-2 dive-bomber of aircraft designer Vladimir Petlyakov, not only to augment its performance, but also to speed up the development of the four-chamber version that Korolyov intended to employ on his short-range interceptor. With the engine installed in the rear fuselage of the plane, the RD-1's turbopump would be driven by one of the two M-105 propeller engines mounted under the wings. The combination of the RD-1, the fuel tanks, the turbopump assembly, propellant feed systems, and other components was known as RU-1.

The first rocket-powered flight of the modified Pe-2 bomber (Pe-2RD) took place on 1 October 1943. The first two series of test flights, with the engine either ignited during take-off or at altitudes less than 5 km, revealed problems with the RD-1's electric ignition system, which was therefore replaced by a chemical ignition system. The modified engine (RD-1KhZ) was subsequently flown in a third series of test flights and was ignited at altitudes up to 7 km. In 1943–1945 the Pe-2RD made more than 100 flights. Korolyov was on board for some of the test flights. Plans for further modifications of the rocket-powered Pe-2 were never realized.

The RD-1 and RD-1KhZ were also flown on three Lavochkin planes (La-7R1, La-7R2, and La-120R) in 1944–1946, on the Sukhoy Su-7 in 1945, and on the Yakovlev YaK-3 in 1945. However, partly because of the low reliability of the

RD-1 engine and also because of the emergence of the jet engine, none of these planes ever went into production [8].

POST-WAR ROCKET PLANES

With turbojet development in the Soviet Union slow to take off, there was continued interest in rocket-propelled aircraft in the first post-war years, not only to counter the new threat of US strategic bombers, but also to explore the behavior of aircraft at supersonic speeds.

One project was initiated before the end of the war at NII-1, the new name given to the former NII-3 after it had merged in May 1944 with Bolkhovitinov's OKB-293 (which also included a rocket engine department headed by Isayev). Headed by Ilya Frolov, the new effort was mainly intended to compare the performance of a pressure-fed and a turbopump-fed engine in future rocket fighters. For this purpose NII-1 developed two aerodynamically identical airplanes with straight wings: 4302 nr. 2 with a pressure-fed Isayev RD-1M engine and 4302 nr. 3 with a turbopump-fed Dushkin RD-2M-3. The RD-2M-3 was a two-chamber design with a 1,100 kg thrust main chamber and a 300 kg thrust supplementary chamber. Both chambers would be used for take-off, after which the pilot would shut down the main chamber and use the thrust of the smaller one to search for and engage the target. This technique was more fuel-efficient and allowed the plane to stay in the air longer. A glider version (4302 nr. 1) towed by a Tupolev Tu-2 made 46 flights beginning in 1946. Of the two rocket-powered versions only 4302 nr. 2 was eventually flown, making one single flight in August 1947 before funds were transferred to another rocket plane project initiated by Mikoyan.

In February 1946 the Soviet government ordered both the OKB-301 Lavochkin bureau and the OKB-155 Mikoyan bureau to develop rocket interceptors capable of reaching speeds up to Mach 0.95 and altitudes of up to 18 km. Both planes were to be equipped with modified Dushkin RD-2M-3V dual-chamber engines. Lavochkin's team studied a plane with a radar sight called La-162, but abandoned work on it in late 1946, preferring to fully concentrate their efforts on jet aircraft. The Mikoyan version was called I-270 and was heavily influenced by the German Me-263-V1 rocket fighter, which had been captured by the Red Army and carted off to the Soviet Union. The Me-263-V1 was one in a long line of Messerschmitt rocket fighters, one of which (the "Komet") had been the only rocket fighter ever to be used in combat. Although the I-270 featured a similar cockpit and landing gear arrangement, it was substantially longer than the Me-263 and also had mid-mounted straight wings and a tee tail rather than the swept wings and tailless design of the German interceptor.

Two experimental I-270 planes were built, one designated Zh-1 and the other Zh-2. Towed by a Tu-2 bomber, Zh-1 made 11 unpowered test flights between February and June 1947. Zh-2 performed the first rocket-propelled flight in September 1947, but was irreparably damaged when it landed far off the runway. Zh-1 also suffered damage on its first powered flight in October 1947 when the landing gear failed to deploy, but was repaired for one final flight in May 1948. The test flights were rather conservative, with the planes developing speeds of only about 600 km/h.

In the second half of 1945 the design bureau of Pavel V. Tsybin was tasked with studying various wing configurations for use at near supersonic speeds. For this purpose the bureau developed "Flying Laboratories" (LL) powered by a Kartukhov PRD-1500 solid rocket motor. Two configurations were tested: LL-1 (or Ts-1) with straight wings and LL-3 (or Ts-3) with 30° forward-swept wings. The LL-1 and LL-3 made about 130 flights in 1947–1948, with the latter reaching speeds of up to Mach 0.97. A planned LL-2 with 30° swept-back wings was not flown because that wing configuration had already been tested on the MiG-15 and La-15 fighters.

Meanwhile, the Soviets were also out to break the sound barrier with rocket-propelled airplanes, relying both on a captured German and a derived domestic design. The German rocket plane was the 346, originally developed by the German Institute for Sailplane Flight (DFS) in 1944. After shutting down its engine, the plane was supposed to glide over enemy territory to take reconnaissance photos and then re-ignite the engine to gain enough speed and altitude to glide back to a friendly base in France or Germany. It had a long, slender fuselage reminiscent of a rocket, with 45° swept-back wings. Thrust was to be provided by a dual-chamber HWK 109-509C. The pilot was supposed to lie in a prone position. In case of an emergency the cabin could be separated from the airplane, with the pilot subsequently ejecting for a parachute landing.

Having fallen into Soviet hands at the end of the war, the German team that had been working on the 346 was sent to Podberyozye (some 100 km north of Moscow) in October 1946 to continue development of the rocket plane under the leadership of Hans Rössing in a newly created Soviet–German design bureau called OKB-2. In the second half of 1948 Rössing's team completed work on a glider version of the plane called 346-P, which was flown several times by German test pilot Wolfgang Ziese in 1948–1949. The carrier aircraft for these and subsequent test flights was an American B-29 bomber confiscated by the Russians after having made an emergency landing in Vladivostok in 1944. In 1949 Ziese and Soviet pilot Pyotr Kazmin were behind the controls for three drop tests of a version of the 346 carrying a mock-up rocket engine (346-1). Despite landing problems on all three flights, the team pressed ahead with the development of the first rocket-propelled model called 346-3.

Ziese performed three powered flights in August–September 1951. During the third flight he lost control of the plane after engine shutdown, forcing him to separate the cabin and eject. With only one of the two combustion chambers activated during these test flights, the maximum speed reached was just over 900 km/h. The 346 could have broken the sound barrier with both chambers working, but OKB-2 was reportedly wary of making such an attempt because of aerodynamic insufficiencies in the airplane. Plans for a delta-wing supersonic rocket plane called 486 were scrapped and the German team was eventually repatriated to the GDR in 1953.

Concurrently with the Germans at Podberyozye, a team under Matus Bisnovat worked on an indigenous supersonic rocket plane called Samolyot-5 ("Airplane-5"). Bisnovat had been placed in charge of the OKB-293 design bureau in June 1946,

The 346 rocket plane.

when it separated from NII-1 to once again become an independent entity. With its 45° swept-back wings, Samolyot 5 was outwardly very similar to the 346, also featuring a jettisonable cabin. It would use a twin-chamber Dushkin RD-2M-3F with a total thrust of 1,600 kg to reach speeds of up to Mach 1.1.

Bisnovat's team initially developed a scale model called "Model 6" that was dropped on four occasions from a Tu-2 bomber between September and November 1947. Claims that speeds of Mach 1.28 and Mach 1.11 were attained on the two final flights are hard to verify because the speedometers were not retrieved intact. With the Dushkin engine not yet ready, a glider version designated 5-1 performed three drop tests from a Pe-8 bomber between July and September 1948, but it was damaged beyond repair in a landing accident on the final mission. Eight or nine more drop tests were conducted with airplane 5-2 between January and June 1949, but that same year financing for Samolyot-5 was discontinued even though the Dushkin engine had been test-fired and installed on the 5-2 [9].

By the end of the 1940s the era of the rocket planes was drawing to a close. Because of their limited flight times, they had little military value, among other things because the pilot had a hard time gliding back to a safe landing. Another safety issue was the use of toxic nitric acid in all the rocket engines developed for the Soviet rocket planes. A much cheaper, safer, and more efficient way of shooting down enemy bombers was the use of surface-to-air missiles, something that the Russians realized shortly after the end of the war when they stumbled on advanced German SAM missiles such as the Wasserfall.

After the war, rocket engines were a handy way of testing aircraft performance at supersonic speeds, but their development was gradually being overtaken by that of the jet engine, even though the Russians mainly relied on copies of German and British engines. While Chuck Yeager had become the first man to break the sound barrier on the X-1 rocket plane on 14 October 1947, the Russians achieved the same feat a year later, but not on one of their rocket-propelled aircraft. On 26 December 1948 Oleg Sokolovskiy became the first Soviet pilot to exceed Mach 1, flying a

The Samolyot 5 rocket plane.

Lavochkin La-176 jet plane. Despite the simultaneous development of the 346 and Samolyot-5, no Russian pilot ever flew faster than the speed of sound on a rocket-propelled aircraft. In the first half of 1950 a special commission set up under the Ministry of the Aviation Industry concluded that rocket engines had little future in aviation and a government decree in June 1950 closed down all further work on rocket engines for aircraft at NII-1.

INTERCONTINENTAL BOMBERS

Ever since the reörganization of NII-1 under the Ministry of the Aviation Industry in 1944, one of its main goals had been to incorporate rocket and ramjet technology into aviation. On 29 November 1946 the new head of NII-1 became Mstislav Keldysh. One of his first assignments was to study a German hypersonic winged trans-continental bombardment aircraft known as the Silbervogel ("Silver Bird") or the "antipodal bomber". This was the brainchild of Dr. Eugene Sänger and mathematician-cum-wife Irene Brendt, who proposed it in a document in August 1944, one copy of which was found by the Russians in Germany after the war. Launched horizontally by a rocket-powered sled, it would use its own rocket engines to boost itself to orbital altitude and subsequently ricochet off the Earth's atmosphere, dropping a bomb over the desired target during one of the dips. Stalin was impressed enough by the Silbervogel to dispatch an Air Force officer named Grigoriy Tokaty-Tokayev to kidnap Sänger in France in 1948, but Tokaty-Tokayev took the opportunity to defect to the West instead [10].

Actually, by this time Keldysh had already come to the conclusion that the Silbervogel was an unrealistic design, among other things because of the high specific impulse required for the engines and the high propellant load. In 1947 he had come up with an alternative intercontinental bomber that would use a combination of supersonic ramjets (scramjets) and rocket engines to perform a mission very similar

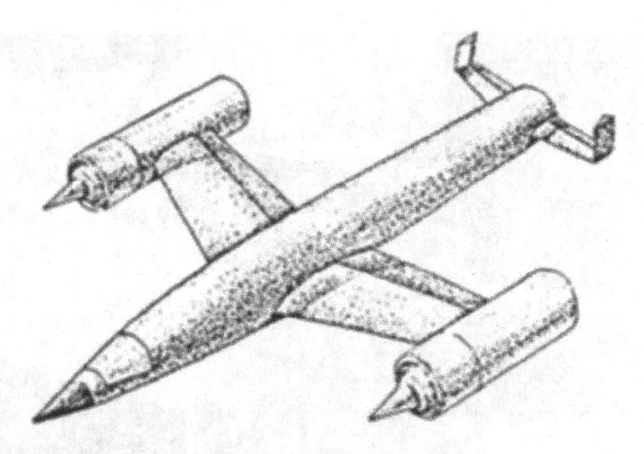

Soviet antipodal bomber.

to the Silbervogel. In what amounted to the first serious Soviet spaceplane proposal, Keldysh's single-seater bomber would be horizontally rail-launched as the Sänger/Bredt bomber, but then switch to two wing-mounted scramjets to reach an altitude of 20 km. Subsequently, the scramjets would be jettisoned, after which a 100-ton thrust rocket engine would kick in to send the vehicle to the upper reaches of the atmosphere, where it would fly several "dip-and-skip" trajectories to reach its final target. The scramjet and rocket engines would share several systems such as a common kerosene tank and also a common hydrogen peroxide tank to drive the turbopumps [11].

INTERCONTINENTAL CRUISE MISSILES

As the Cold War shifted into higher gear in the late 1940s, the Russians began looking at more realistic ways of delivering nuclear warheads over intercontinental distances. Unlike the US, the USSR did not have the luxury of having bases along the enemy borderline, making rockets a more convenient way of transporting nuclear bombs than strategic bombers. The leading rocket research institute was NII-88, set up in 1946. Sergey Korolyov, released from the *sharaga* in 1945 to study V-2 missiles in Germany, headed one of its departments. By the end of the decade NII-88 was investigating both ballistic and cruise missiles as a means of delivering nuclear bombs over long distances. In the winged arena, the emphasis now shifted from a horizontally launched piloted bomber with rocket and scramjet engines to a vertically launched unmanned two-stage cruise missile carrying rocket engines in the first stage and supersonic ramjets in a winged second stage. Unlike the ballistic missiles, the cruise missiles would remain within the boundaries of the Earth's atmosphere, developing top speeds of around Mach 3.

A Soviet government decree issued on 4 December 1950 approved a new rocket research program, one part of which ("theme N-3") focused on intercontinental missiles. The conclusion of the N-3 studies was that the development of a cruise missile with a range of 8,000 km was feasible, but needed to be preceded by further research into scramjets and navigation systems. A government decree on 13 February 1953 gave the go-ahead for the simultaneous development of both intercontinental ballistic and cruise missiles, assigning both tasks to OKB-1, which was the name given to Korolyov's reorganized department within NII-88 in 1950. Since the cruise missile required a significant leap in technology, OKB-1 would first design an intermediate Experimental Cruise Missile (EKR). With a targeted range of 730 km, it would consist of an R-11 missile as the first stage and a winged second stage with a Bondaryuk RD-40 scramjet.

By the end of 1953 preliminary ground-based testing of EKR components had given the Russians enough confidence to skip this intermediate step and move directly to an intercontinental cruise missile with a range of 8,000 km. Since Korolyov's OKB-1 was too heavily preoccupied with its R-7 ICBM, responsibility for the cruise missiles was entrusted to the aviation industry by a government decree released on 20 May 1954. Three aviation design bureaus were tapped to build cruise missiles with different missions:

- OKB-49 (Georgiy Beriyev): a missile called Burevestnik ("Petrel") or "P-100" to be used for long-range reconnaissance (fulfilling the same role as the American U-2 reconnaissance aircraft) and also to deliver small 1.2-ton nuclear warheads.
- OKB-301 (Semyon Lavochkin): a missile called Burya ("Storm") or La-350 to transport 2.18-ton atomic bombs.
- OKB-23 (Vladimir Myasishchev): a missile called Buran ("Blizzard") or M-40 to transport 3.4-ton hydrogen bombs.

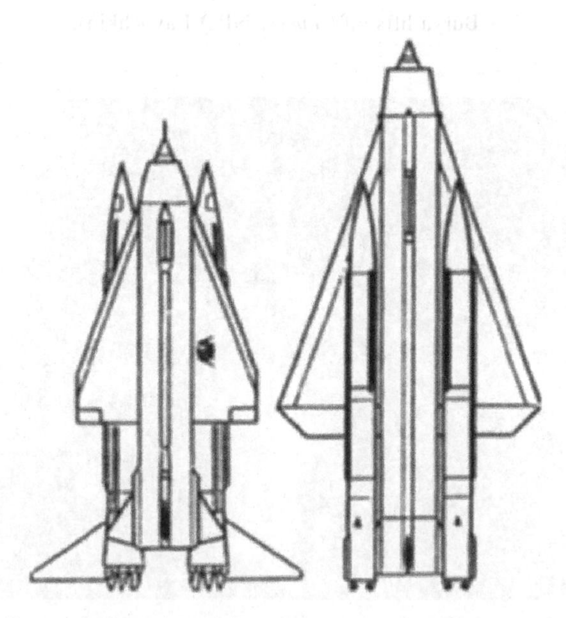

The Burya (left) and Buran cruise missiles (*source*: *www.buran.ru*)

Burya lifts off (*source*: NPO Lavochkin).

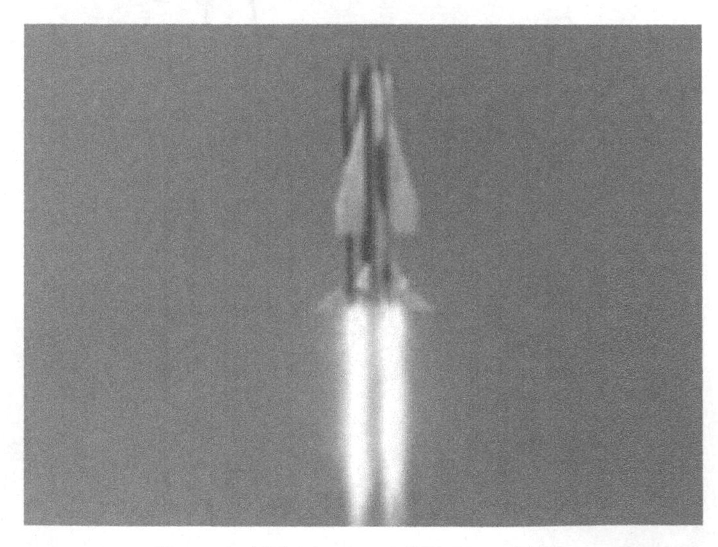

Burya in flight (*source*: NPO Lavochkin).

Keldysh's NII-1, which had once again become independent in 1952 after having been a branch of the Central Institute for Aviation Materials (TsIAM) since 1948, had overall scientific supervision of the cruise missile effort, relying on its earlier experience in high-speed and high-altitude aeronautics obtained during the antipodal bomber projects.

All cruise missiles consisted of a "core stage" with air-breathing scramjet engines, flanked by rocket-powered "strap-on boosters" (two for Burevestnik and Burya and four for Buran), giving them an appearançe somewhat reminiscent of the Space Shuttle. Burevestnik seems to have been a very short-lived program that never came close to flying. Buran featured four first-stage boosters with Glushko RD-212 kerosene/nitric acid engines and a core stage with a Bondaryuk RD-018 scramjet. In August 1956 OKB-301 started work on an improved version (Buran-A or M-40A) with upgraded RD-213 engines, increasing payload capacity to 5 tons. Interestingly, Myasishchev at one point considered equipping Buran with a small crew cabin, from which the pilot was to eject prior to impact. One of the objectives of this plan was to see if a man could endure the psychological and physical rigors of hypersonic flight. Buran was scrapped in November 1957 before making its first flight.

Burya's two strap-ons had Isayev S2.1100 rocket engines (later replaced by the S2.1150) and the core stage was powered by a Bondaryuk RD-012 scramjet. The missile underwent a series of 17 test flights from the Vladimirovka test range in the Volgograd region between September 1957 and December 1960. The maximum range achieved was 6,425 km, less than the prescribed 8,000 km, because the scramjet had a tendency of igniting too early. However, even as Burya was overcoming its teething problems, its fate had already been sealed by the successful test flights of OKB-1's R-7 ICBM, the first of which took place in August 1957. Cruise missiles were very vulnerable to defensive measures due to their low flight altitudes (around 17–25 km) and took more than two hours to reach their targets, whereas ICBMs could deliver their deadly cargo in a matter of minutes. The US Air Force had come to the same conclusion, closing down its Navaho intercontinental cruise missile project in July 1957. The Soviet government followed suit by releasing a decree on 5 February 1960 that canceled all further work on Burya, although it allowed some of the remaining missiles to be tested in flight [12].

SPACEPLANES OF THE 1950s AND 1960s

For many years official histories of the Soviet space program created the impression that Vostok had been the only Soviet piloted space project in the late 1950s/early 1960s. Not until the days of *glasnost* in the late 1980s/early 1990s did it emerge that just like the United States the Soviet Union had considered winged spacecraft as an alternative to ballistic capsules in the early years of the space program. Surprisingly, it turned out that this option was studied in no fewer than five design bureaus.

In the US winged spacecraft were long seen as the logical culmination of research into high-speed aeronautics conducted since the mid-1940s with air-launched rocket-propelled X-planes. The first phase had seen aircraft such as the X-1, X-2, and

Skyrocket gradually push the envelope from Mach 1 to Mach 3 between 1947 and 1956. Phase 2 had been initiated in late 1954 with the decision to press ahead with the development of the X-15 high-altitude hypersonic research aircraft, which eventually performed a largely successful test program between 1958 and 1969. Ultimately, suborbital and orbital capability would be achieved using the "boost–glide" principle, where a spaceplane would be launched vertically with the help of a conventional rocket and eventually glide back down to the runway like an ordinary aircraft. In late 1957, responding to Sputnik, the Air Force consolidated three "boost–glide" feasibility studies (Hywards, Brass Bell, and Rocket Bomber) into a single program called "Dyna-Soar" or X-20. Unlike the X-15, however, Dyna-Soar was not seen as an experimental system, but an operational weapon system capable of orbital nuclear bombardment, reconnaissance, and satellite identification and neutralization [13].

During 1958 the exigencies of the Cold War and the fledgling space race with the Soviet Union gradually pushed the ballistic capsule approach to the foreground, especially after the formation of NASA in October of that year. Having lost face after the early Sputnik successes, the United States was intent on restoring its reputation by putting the first man into orbit and capsules were a more efficient and quicker way of achieving that goal than winged spacecraft. The Air Force continued work on Dyna-Soar against the backdrop of NASA's Project Mercury, but in December 1963, with the first flight an estimated three years away, Secretary of Defense Robert McNamara canceled the program. X-20 funds were reappropriated to a military space station called the Manned Orbiting Laboratory (MOL). After that the United States did not have another officially sanctioned spaceplane project until the approval of the Space Shuttle by President Nixon in early 1972.

The winged approach to piloted spaceflight was probably less central in Soviet thinking than it was in the US, at least when it came to building the *first* manned spacecraft. For one, the Soviet aviation industry and the Air Force were far removed from missiles and space-related matters after the Ministry of the Aviation Industry had declined offers in 1945–1946 to bear responsibility for long-range missile programs. Instead, the assignment went to the Ministry of Armaments, which had developed artillery during the Second World War. This had far-reaching implications for the Soviet space program (essentially an offshoot of the missile program), which until the break-up of the USSR remained tightly in the grip of the "artillery" camp. Moreover, missiles were soon favored over strategic bombers to deliver nuclear warheads to US territory and there was little incentive for research into high-speed, high-altitude aircraft, reflected in the absence of high-altitude "X-type" airplane research programs in the Soviet Union. On top of that, Soviet leader Nikita Khrushchov had become particularly enamored with missiles in the mid-1950s, curtailing contracts for the aviation industry and even dissolving several aviation design bureaus towards the end of the decade.

The earliest plans for piloted missions beyond the atmosphere revolved around the use of converted R-2 missiles to send people on vertical trajectories to altitudes of up to 200 km. Although one common cabin design was planned, different methods were studied for returning the capsule to Earth. One option presented by Korolyov

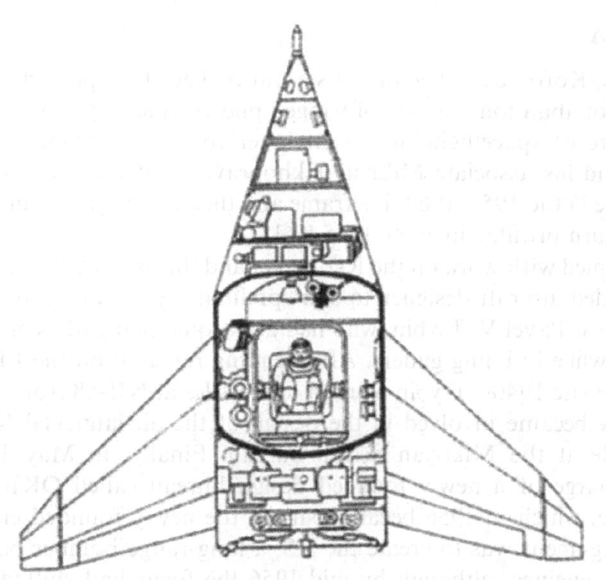

Winged capsule for suborbital mission (reproduced from Peter Stache, *Sowjetischer Raketen*, Berlin, 1987).

during a speech in September 1955 was to equip such a capsule with wings, allowing it to make a long ballistic suborbital flight rather than a short vertical hop [14].

Research on piloted spaceflight began in earnest in the spring of 1957 with the establishment within OKB-1 of Department 9, which was to focus exclusively on the development of lunar probes and piloted spaceships, signaling the beginning of the bureau's reorientation from missiles to spaceflight. Between September 1957 and January 1958 OKB-1 and the NII-1 research institute carried out a comparative analysis of various basic shapes for piloted spaceships, paying particular attention to thermal protection requirements and the g-forces exerted on the crew. The conclusion was that the heat-resistant alloys available at the time were not up to the task of protecting winged vehicles with high lift-to-drag ratios against the severe thermal stresses of re-entry. Instead, the recommendation was that the first piloted spaceship should have a lift-to-drag ratio between just 0.5 and 0, depending on the g-forces that were deemed acceptable for the crew. The ship would preferably be shaped as a blunt cone with a rounded nose and a spherical base, with the pilot being ejected from the descent capsule before touchdown.

In April 1958 one of the main obstacles to manned ballistic flight was eliminated when a key meeting of leading experts in the field of aviation medicine came to the conclusion that people could withstand forces of up to $10g$ as long as the body was properly positioned inside the capsule. All this would lead later that year to preliminary designs for the manned vehicle that eventually became Vostok, redesigned in early 1959 to serve the dual function of carrying people into space and performing unmanned photoreconnaissance missions [15].

Tsybin's PKA

Nevertheless, Korolyov, a veteran of several rocket plane projects in the 1930s and 1940s, did not abandon the idea of winged piloted spaceflight. Outlining their ideas on the future of spaceflight in a joint letter to the government on 5 July 1958, Korolyov and his associate Mikhail Tikhonravov called for developing a manned space capsule in the 1958–1960 timeframe and then to design a manned vehicle "with a gliding return profile" in 1959–1965 [16].

Preoccupied with work on the R-7 rocket and the first satellites, Korolyov turned to a befriended aircraft designer to start preliminary research on a manned spaceplane. This was Pavel V. Tsybin, who had got acquainted with Korolyov back in the early 1930s while building gliders. After leading research on the LL "flying laboratories" in the late 1940s, Tsybin worked on missiles at NII-88 from 1949 to 1951 and subsequently became involved in the design of the air-launched Kometa anti-ship cruise missile at the Mikoyan design bureau. Finally, in May 1955 Tsybin was placed in charge of a newly founded design bureau called OKB-256, situated in Podberyozye, which in 1956 became part of the newly founded city of Dubna. Its primary assignment was to create the RS, a long-range bomber powered by supersonic ramjet engines, although by mid-1956 the focus had shifted to a supersonic reconnaissance aircraft named RSR.

Sometime later, presumably in 1958, Korolyov proposed Tsybin to design a small winged spaceship that could be orbited by an R-7 based rocket. Tsybin's team readily set to work, assisted by specialists from OKB-1. What they came up with was a vehicle called PKA (for "Gliding Space Apparatus"), which because of its shape was also nicknamed Lapotok ("little bast shoe").

Having a launch mass of 3.5 tons, the one-man spaceplane was to be placed into a circular 300 km orbit by a Vostok rocket for missions lasting up to 24–27 hours. Built into the fuselage was a small pressurized cabin with a control panel, life support systems, and three windows, one of them for an astronavigation system. In case of a launch abort, the pilot could eject from the cabin up to an altitude of 10 km and in an emergency at higher altitudes the entire spaceplane would be separated from the rocket. Located behind the cabin was a pressurized instrument compartment with on-

Pavel Tsybin.

orbit and re-entry support systems. The spaceplane also had a detachable engine compartment with two 2,350 kg thrust nitric acid/kerosene engines, one for on-orbit maneuvers and the other for the deorbit burn. Also on this compartment were an infrared vertical sensor and a thermal control system using radiators. The dry mass of the engine unit was 350 kg and the propellant mass at launch was 430 kg. For orientation in orbit and during the early stages of re-entry the ship used small hydrogen peroxide thrusters.

The deorbit, re-entry, and landing phase was to last up to 90 minutes. After the deorbit burn the engine compartment was to be separated at an altitude of 90 km. During re-entry the spaceplane's steel fuselage was protected from the high temperatures by a heat shield consisting of a 100 mm thick organic silicon layer and a 70 mm thick fibre layer as well as by special air ducts to cool the outside structure. Places with maximum heat exposure such as the nose of the heat shield and the leading edges of the two elevons and the tail were to be cooled with the help of liquid lithium. During maximum heating the angle of attack was 55 to 60°. At an altitude of 20 km, having reduced its speed to 500–600 m/s, the PKA would deploy two wings with a span of 7.5 m and an area of 8.7 m^2, which until then had remained folded back to protect them against the highest temperatures during re-entry. The spaceplane was to land on a dirt runway using a skid landing gear. Landing speed was 180–200 km/h and landing mass was 2.6 tons.

The preliminary design ("draft plan" in Russian terminology) for the PKA was officially approved by Tsybin on 17 May 1959 and the following day Korolyov sent a letter to the State Committee of Defense Technology (GKOT, the former Ministry of Armaments) with the request to include the spaceplane in its long-range plans and assign OKB-256 to the project as the lead organization [17]. However, wind tunnel tests conducted at the Central Aerohydrodynamics Institute (TsAGI) showed that the PKA would be exposed to much higher temperatures than expected (up to 1,500°C), requiring significant changes to the heat shield. Moreover, it turned out

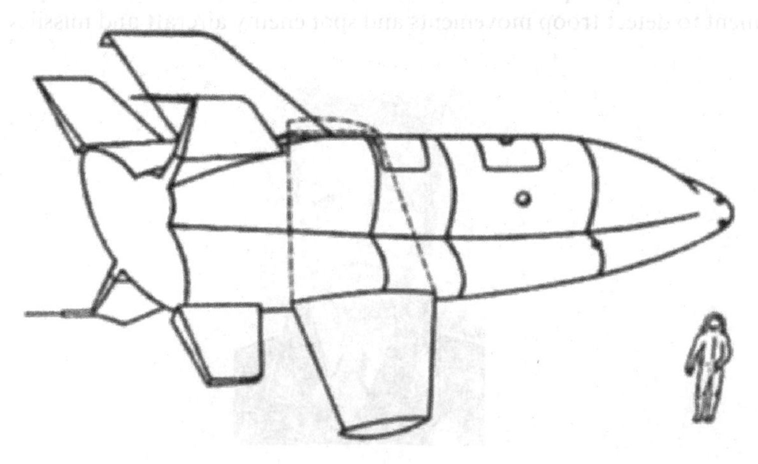

The PKA spaceplane (*source*: Igor Afanasyev).

that the use of liquid lithium to cool the hottest parts of the fuselage would make the design much heavier and more complex than anticipated [18].

Tsybin invited specialists of the All-Union Institute of Aviation Materials (VIAM) to deal with these issues, but by the end of 1959 clouds were gathering not only over the PKA, but over Tsybin's design bureau as well. The RS supersonic strategic bomber had been canceled in the wake of the Soviet Union's early ICBM successes and in October 1959 OKB-256 was absorbed by Myasishchev's OKB-23. When OKB-23 in turn became a branch of Vladimir Chelomey's OKB-52 in late 1960, Tsybin returned to Korolyov's OKB-1, where he would eventually go on to play an important role in the Energiya–Buran program and later in the design of single-stage-to-orbit spaceplanes [19].

Myasishchev's Projects 46 and 48

Vladimir Myasishchev's OKB-23 (situated in the Moscow suburb of Fili) was mainly engaged in the development of long-range strategic bombers, but branched out into cruise missiles with the M-40/Buran project in 1954–1957 and also did considerable research on spaceplanes even *before* Tsybin had started his PKA project. Unfortunately, most of the archival materials related to Myasishchev's spaceplane projects have not been preserved, making it difficult to piece together their history. According to Russian historians Myasishchev, inspired by plans for the X-15 and US boost–glide concepts, began spaceplane research "on his own initiative" as early as 1956 under a program named Project 46. Also involved in the research were the NII-1 and NII-4 research institutes.

By 1957 he came to the conclusion it would be feasible in the short run to develop a reusable vehicle called a "satelloid" or "intercontinental rocket plane". Its primary goal would be to conduct strategic reconnaissance over enemy territory without the risk of being shot down by anti-aircraft defense means. Such missions would last 3 to 4 hours, with the spaceplane using radar and both optical and infrared photographic equipment to detect troop movements and spot enemy aircraft and missiles. Included

Vladimir Myasishchev.

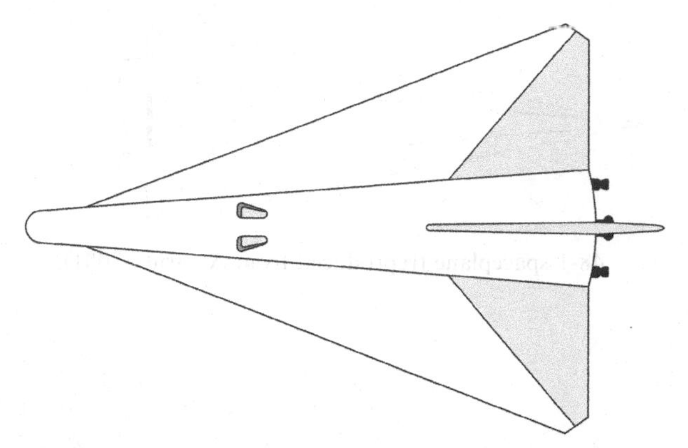

Project 46 spaceplane (reproduced from A. Bruk, 2001).

in the early warning network would be high-orbiting relay satellites. Later goals were to send vehicles of this type on bombing missions or to destroy enemy missiles and satellites. A reconnaissance version was expected to be ready by 1963 and a combined reconnaissance/bombing version was planned for 1964–1965. Myasishchev is said to have presented his ideas for spaceplanes during a visit to OKB-23 by Khrushchov in August 1958, but the Soviet leader was unimpressed, telling Myasishchev to stick to the field of aviation and leave rocket-related matters to others.

Undeterred by Khrushchov's scepticism, OKB-23 pressed on with its spaceplane research. By April 1959 the bureau had worked out plans for a 10-ton rocket plane flying between altitudes of 80 and 150 km and capable of increasing orbital altitude by 100 km (to a maximum of 250 km) and changing orbital inclination by 3°. As Dyna-Soar, it was envisaged as a "boost–glide" system, being launched into orbit by a conventional ballistic rocket and then gliding back to a horizontal runway landing. The launch vehicle was to be an upgraded three-stage version of Korolyov's R-7 missile. The third stage apparently consisted of four "boost engines" drawing propellant from four jettisonable tanks mounted on the spaceplane itself. In April 1960 Myasishchev revised his plans and was now aiming for a 6-ton vehicle flying in 600 km orbits and capable of performing inclination-changing maneuvers of as much as 6°.

Meanwhile, OKB-23 was tasked with the development of another manned space vehicle by a government and party decree (nr. 1388-618) issued on 10 December 1959. This decree, considered to be the first macro-policy statement on the Soviet space program, encompassed a wide range of space projects. Myasishchev's bureau in particular was assigned to develop a manned vehicle capable of ensuring "a reliable link" between the ground and "heavy satellites". Known as Project 48, this appears to have been an early version of a transportation system for space stations, although it was supposed to solve defense-related tasks as well. It was only the second piloted space project to be officially approved by a party/government decree after Vostok. Work on the project got underway after orders from the State Committee of Aviation Technology (GKAT) on 7 January and 4 March 1960.

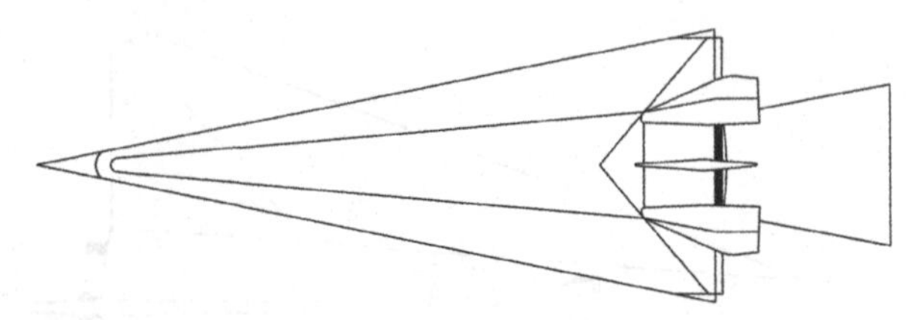

48-1 spaceplane (reproduced from A. Bruk, 2001).

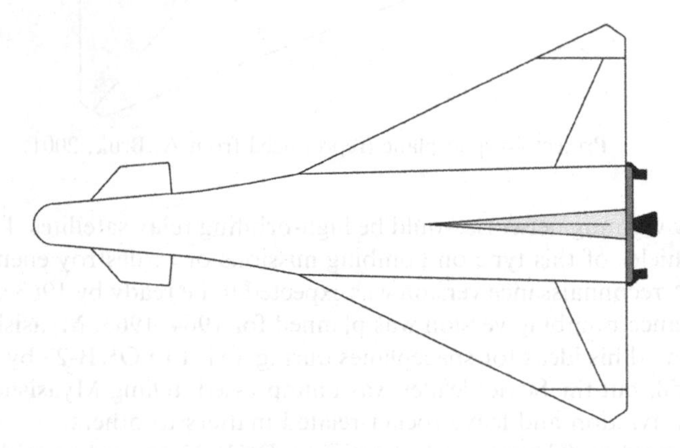

48-2 spaceplane (reproduced from A. Bruk, 2001).

Weighing no more than 4.5 tons, the spacecraft was to be launched into a circular 400 km orbit by an R-7 based launch vehicle and stay in orbit anywhere from 5 to 27 hours. Re-entry through the atmosphere was to consist of a ballistic and a "controlled gliding" phase, reducing deceleration forces to no more than 3–4*g*. This required an aerodynamic shape providing at least some lift and ruled out a Vostok-type spherical design. Thermal protection was to be provided by ceramic tiles and/or by super-cold liquid metals circulating under the spacecraft's skin.

Myasishchev's team came up with four possible designs to meet these requirements, each capable of carrying two men. Vehicle 48-1 (launch mass 4.5 tons) had a cone-shaped fuselage with highly swept delta wings (79°) and fins on the wings and fuselage to provide braking during re-entry. The crew cabin was located in the back. Both the fins and the glider's engine compartment were to be jettisoned when the spaceplane had decelerated to a speed of Mach 5. Vehicle 48-2 (launch mass 4.3 tons) had a cylindrical fuselage with delta wings (leading edge sweepback of 65°) and small canards in the front. There were vertical tails both on top of and under the fuselage. The crew cabin was situated in the middle and the spaceplane was outfitted with a non-jettisonable engine compartment. The two other schemes envisaged a Mercury/Gemini look-alike inverted cone with a rotor for a helicopter-type landing (48-3) and a conically shaped spacecraft for a parachute landing (48-4). Missions of the two-man ship were to be preceded by test flights of a single-seater spaceplane to demonstrate

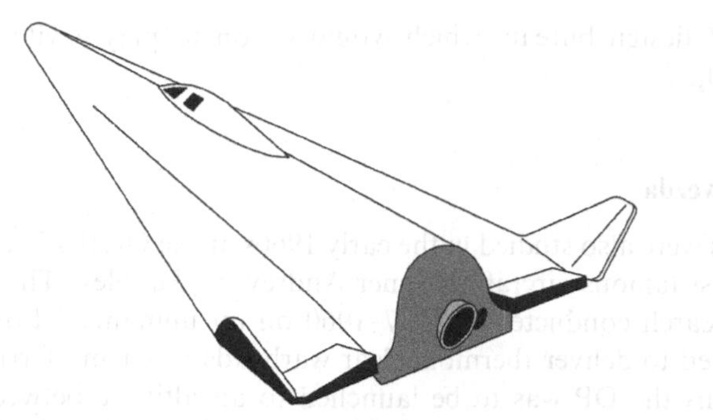

One version of the VKA-23 spaceplane (reproduced from A. Bruk, 2001).

the functioning of life support systems and test the "gliding re-entry" technique. The proposals were reviewed at a meeting of leading aviation specialists on 8 April 1960, but no consensus was reached on the way to go forward.

There was yet another OKB-23 proposal for a single-seater spaceplane, which Myasishchev historians also link to Project 48, although it does not appear to have been the aforementioned one-man demonstration vehicle. It has been referred to as VKA-23 (VKA standing for "Aerospace Apparatus" and "23" referring to the name of the design bureau) and was the brainchild of OKB-23 designers L. Selyakov and G. Dermichov, who had originally presented it to NII-1 chief Mstislav Keldysh. Two versions of the delta-wing VKA-23 were studied between March and September 1960, one with a single fin at the rear (launch mass between 3.5 and 4.1 tons, length 9.4 m) and one with two fins at the tips of the wings (launch mass between 3.6 and 4.5 tons, length 9.0 m).

The VKA-23 was to be launched either by an R-7 based rocket or a much more powerful rocket developed in-house under the so-called Project 47. In a launch emergency, the pilot could eject from the vehicle up to an altitude of 11 km, higher than that the entire vehicle would be separated from the rocket. The VKA-23 was supposed to borrow some elements from the Vostok spacecraft such as the Chayka orientation system and the Zarya communication system. Thermal protection was provided by ultra lightweight ceramic foam tiles very similar in shape to the ones later used by the US Space Shuttle and Buran. The leading edges of the wings were protected by a thick layer of siliconized graphite. A small turbojet engine was to give the ship extra maneuverability during re-entry. Just like the Vostok cosmonauts, the pilot was not supposed to land inside the ship, but eject at an altitude of about 8 km, with the spacecraft itself making an automatic landing on skids.

Although Project 48 had received the official nod with the party/government decree of December 1959, it was no longer mentioned in an even bigger space decree released on 23 June 1960. Actually, OKB-23 was counting its final days, falling victim to Khrushchov's policy of downsizing aviation in favor of missiles. In October 1960 Myasishchev's design bureau became Branch Nr. 1 of the OKB-52 design bureau of Vladimir Chelomey and was assigned to various missile, rocket, and spacecraft projects. Myasishchev was named head of TsAGI, but in 1967 was placed in charge of the EMZ design bureau, which would go on to play a vital role in the Buran program [20].

Tupolev's Zvezda

Spaceplanes were also studied in the early 1960s at the OKB-156 bureau of the Soviet Union's most famous aircraft designer Andrey N. Tupolev. These studies had their roots in research conducted in 1957–1960 on an unmanned Long-Distance Glider (DP) intended to deliver thermonuclear warheads to enemy territory. According to original plans the DP was to be launched to an altitude between 50–100 km by a missile, either the R-5 or R-12, or a booster built at the Tupolev bureau itself. After separation from the rocket, it would gradually glide to its target, located up to 4,000 km from the launch pad. An on-board altimeter would then detonate the thermonuclear bomb at the required altitude.

Scale models of the DP were launched to speeds of up to Mach 2 with small solid rocket motors from Tu-16LL aircraft. OKB-156 also developed an experimental prototype of the DP called 130 or Tu-130. Weighing 2.5 tons, the tailless glider was 8.8 m long and 2.2 m high with a wingspan of 2.8 m. However, on 5 February 1960, just as the first Tu-130 was being readied for launch on a modified R-12 missile, the Soviet government issued a decree to cancel the DP project, now considered useless in the wake of the early ICBM successes. By this time OKB-156 had been aiming to launch the DP with a three-stage rocket built in-house, enabling the glider to cover distances of 9,000 to 12,000 km and carry a thermonuclear warhead weighing 3 to 5 tons.

The experience gained during the research on the DP came in handy for Tupolev's spaceplane project, presumably started around 1960 under the names Aircraft 136, Tu-136, or Zvezda ("Star") (Tu-136 is also the name of a recently developed regional cargo/passenger plane). The ultimate goal was to build a 10- to

Andrey Tupolev.

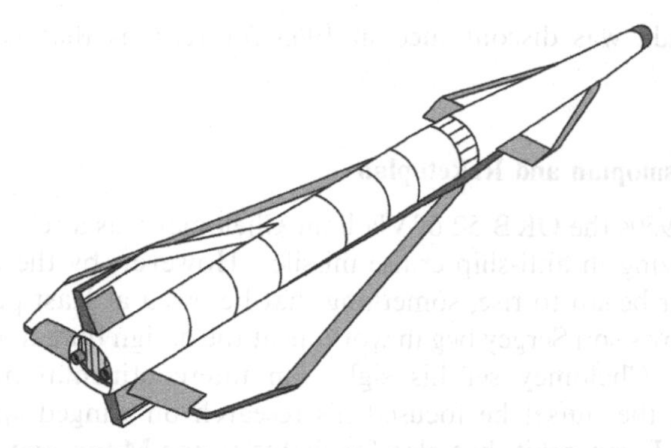

The Zvezda spaceplane (reproduced from V. Rigmant, 2001).

20-ton spaceplane to be orbited by a newly developed launch vehicle. Several aerodynamic shapes were studied, one closely resembling that of the 130 glider and another that of Dyna-Soar. In the end the designers opted for a canard configuration. If the experimental version was successful, it would serve as the basis for a whole series of rocket planes to be used for reconnaissance, bombing, and anti-satellite missions. Tupolev envisaged a grueling two-phase test program to verify the design at hypersonic speeds in the lower and upper atmosphere and to try out re-entry and landing techniques.

The first phase would see scale models of Zvezda being launched from Tu-16 aircraft and with the help of R-5 and R-14 missiles. The air-launched version would have a built-in solid rocket motor to reach an altitude of up to 40 km and a speed of 9,000 km/h. Models launched by the R-5 and R-14 would climb to 45 km and 90 km, respectively, and develop speeds of 14,000 km/h and 23,000–28,000 km/h.

The second phase involved the use of three manned test vehicles. One was a scaled-down version of Zvezda known as 136-1, air-launched from a Tu-95K. Having reached a peak altitude of 10 km and a top speed of 1,000 km/h, it would land at a speed of about 300 km/h, just like the real Zvezda. The next step was to use the Tu-95K as a launch platform for a hypersonic vehicle designated "139". The Soviet equivalent of the X-15, it was to fly as high as 200 km and develop a speed of 8,000 km/h. The third vehicle was dubbed 136-2, an improved version of the 136-1 with an additional rocket engine to reach speeds of up to 12,000 km/h and a maximum altitude of about 100 km.

After this the stage would be set for the first launches of the actual Zvezda vehicle, which would fly between altitudes of 50 and 100 km and therefore be limited to single-orbit missions. There were also plans for an unmanned version called 137, Tu-137, or Sputnik, capable of performing multi-orbit missions. The only launch vehicle capable of launching Zvezda was Chelomey's UR-500/Proton, but this was only in the very early stages of development when Zvezda was conceived. Therefore OKB-156 worked out plans for its own two- or three-stage rocket to launch the spaceplane. Also considered was a scheme in which the spaceplane would be launched with a missile from the back of a strategic supersonic plane (the Tu-135 or Tu-139).

Work on Zvezda was discontinued in 1963 for reasons that have not been disclosed [21].

Chelomey's Kosmoplan and Raketoplan

Until the late 1950s the OKB-52 of Vladimir Chelomey was a relatively minor design bureau specializing in anti-ship cruise missiles. However, by the end of the decade Chelomey's star began to rise, something that he owed at least partially to the fact that Khrushchov's son Sergey began working at the design bureau in 1958. Brimming with ambition, Chelomey set his sights on intercontinental missiles and space projects. From the outset he focused his research on winged spacecraft, not just for missions in Earth orbit, but also for flights to the Moon and planets.

In 1958–1959 OKB-52 began working on two projects called Kosmoplan and Raketoplan. Kosmoplan was a rather futuristically looking family of spacecraft primarily designed to fly to the Moon, Mars, and Venus and then return to Earth. During re-entry the winged landing vehicle would be protected from thermal stresses by a jettisonable container shaped somewhat like a furled umbrella and it would land on a conventional runway using turbojet engines. One early version of the Kosmoplan was also intended for military reconnaissance missions in low Earth orbit. Initial Kosmoplan missions would be automated, with the eventual goal being to switch to piloted flights, first for the Earth-orbital version and later for the deep-space versions.

Raketoplan was initially conceived as a suborbital vehicle to carry passengers and cargo over intercontinental distances and, more importantly, to perform bombing missions. Launched by a conventional rocket or a winged fly-back booster, it would perform suborbital ballistic flights with aerodynamic braking, maneuvering and landing on a runway using turbojets. Two versions were studied, one for a range of 8,000 km and the other for a range of 40,000 km.

Chelomey received official support for the projects with a government and party decree of 23 June 1960, which saw a clear shift in emphasis from civilian to military space projects compared with the space plan outlined in the December 1959 decree. It

Vladimir Chelomey.

called for the development of two unmanned deep-space versions of the Kosmoplan ("Object K") by 1965–1966, one with a mass of 10–12 tons and the other with a mass of 25 tons. The vehicles were to be launched by a new Chelomey rocket with a launch mass of 600 tons. Raketoplan ("Object R") was now eyed as an *orbital* spaceplane with a mass of 10–12 tons. An unmanned version would be ready in 1960–1961, a piloted variant in 1963–1965, and an anti-satellite version in 1962–1964 [22].

However, these goals turned out to be overly ambitious and another government decree on 13 May 1961 ordered OKB-52 to limit this work to a piloted version of the Raketoplan for military missions in Earth orbit and for deep-space missions. By 1963 engineers had completed the preliminary design for four variants of such a vehicle: two single-seat Earth-orbital versions for anti-satellite and bombing missions, a two-seat scientific spacecraft for circumlunar flight, and a seven-seat passenger ballistic spacecraft for intercontinental ranges. The first three were to be launched by the Chelomey bureau's UR-500/Proton, the fourth by the UR-200. Despite the name Raketoplan, the circumlunar spacecraft appears to have been a wingless vehicle for a ballistic re-entry from lunar distances, one that would later evolve into a vehicle called LK-1 that had a shape reminiscent of the US Gemini capsule.

By early 1964 the Raketoplan project was left with only military goals, namely orbital reconnaissance and anti-satellite missions. At this time OKB-52 was planning two versions, the unmanned R-1 and the manned R-2, both weighing 6.3 tons. The R-1 was a model of the piloted version designed to test all essential systems in orbit. The R-2, manned by a single pilot, would fly 24-hour missions in a nominal orbit of 160 × 290 km.

Two test vehicles were developed in the framework of the Raketoplan project to test heat shield materials, flight control systems, and maneuvering characteristics at hypersonic speeds. One was a 1,750 kg model called MP-1, a cone-shaped vehicle with two graphite rudders and a set of speed brakes at the base resembling an unfurled umbrella. The MP-1 was launched by an R-12 missile from the Vladimirovka test site near Kapustin Yar (Volgograd region) on 27 December 1961. Having reached a

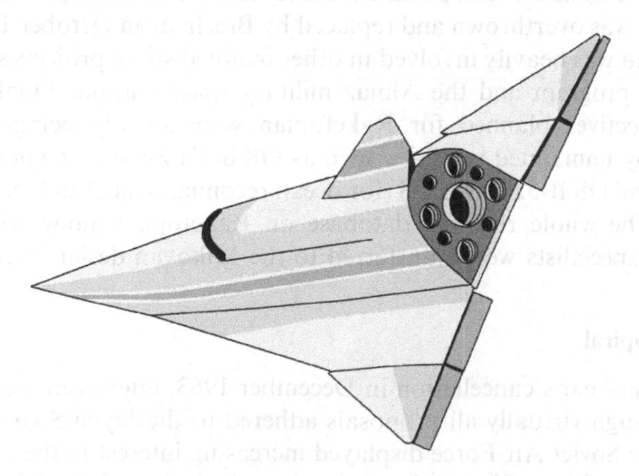

The R-2 Raketoplan (*source*: Dennis Hassfeld).

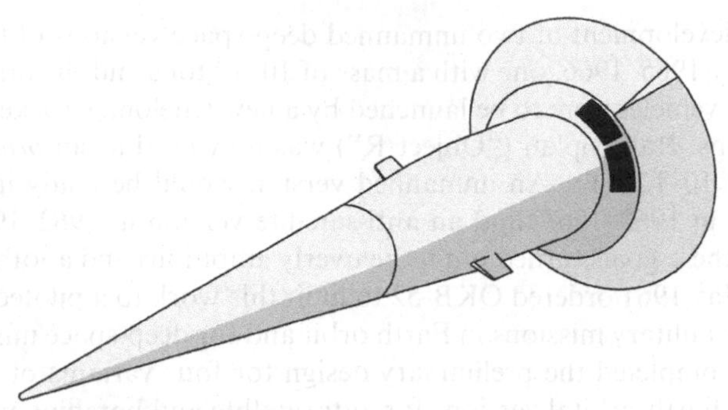

The MP-1 (reproduced from G. Yefremov, 2004).

maximum altitude of 405 km, it successfully re-entered the atmosphere at a speed of 3,800 m/s and safely landed on three parachutes 1,880 km downrange. This marked the first ever re-entry test of an aerodynamically controlled vehicle. It came about two years before the US Air Force began similar flights under the so-called START program.

The other vehicle was named M-12 and looked quite similar to its predecessor, except that the umbrella-shaped braking panels were replaced by four titanium rudders. Using the same missile and launch site as the MP-1, the 1,700 kg M-12 was launched on 21 March 1963, but was lost during re-entry, probably because of a problem with its heat shield. The data obtained during the tests were also applicable to OKB-52's research on maneuverable warheads. This was particularly the case for the M-12, which was seen as a subscale model of the AB-200 warhead. The MP-1 and M-12 were significant in that they were the only hardware ever launched in support of the multitude of Soviet spaceplane projects conceived in the late 1950s and early 1960s.

The Raketoplan project was discontinued in 1964–1965. There appear to have been several reasons for this. First, Chelomey lost much of his political support when Khrushchov was overthrown and replaced by Brezhnev in October 1964. Second, the design bureau was heavily involved in other manned space projects such as the LK-1 circumlunar program and the Almaz military space station. Finally, many of the military objectives planned for Raketoplan were already being or about to be performed by unmanned satellites such as OKB-1's Zenit (for photographic reconnaissance) and OKB-52's own US (for ocean reconnaissance) and IS (for anti-satellite missions). The whole research database on Raketoplan along with a number of Chelomey's specialists were transferred to the Mikoyan design bureau [23].

Mikoyan's Spiral

Despite Dyna-Soar's cancellation in December 1963, interest in spaceplanes did not abate. Although virtually all proposals adhered to the Dyna-Soar type boost–glide principle, the Soviet Air Force displayed increasing interest in the early 1960s in air-launched spaceplanes. Unlike the rocket-launched spaceplanes, these would not be

Artyom Mikoyan.

tied to specific launch sites and could be launched from virtually any place in the world into a wide variety of orbital inclinations. This made the system less vulnerable to attack and gave it far more flexibility in fulfilling key military objectives such as timely reconnaissance of ground-based enemy targets and inspection and neutralization of enemy satellites. Air-launched systems were promoted at an Air Force conference at the Monino Air Force Academy in January 1962 [24]. Studies conducted in 1964–1965 by the Soviet Air Force research institute TsNII-30 also concluded that an air-launched vehicle would best meet the military requirements formulated for spaceplanes.

On 30 July 1965 the Ministry of the Aviation Industry (MAP) assigned the task of building such a system to the OKB-155 design bureau of Artyom Mikoyan. Renamed MMZ Zenit in 1966, the bureau was most renowned for its MiG fighter jets, but at the same time was no stranger to air-launched systems. Back in the 1950s Mikoyan had been involved in the development of the air-launched Kometa anti-ship cruise missile and in the early 1960s he had briefly worked on an air-to-space missile to be launched from a MiG-25 to destroy enemy missiles in flight [25]. In a newspaper article in January 1962, coinciding with the Air Force conference in Monino, Mikoyan had even publicly proclaimed the need for what he called a *kosmolyot* (a compound of *kosmicheskiy samolyot*, "spaceplane") to provide the Soviet Air Force with an operational capability in space [26].

Mikoyan's team wasted no time in getting down to business and by July 1966 had completed a preliminary design for the air-launched spaceplane system, called Spiral. Placed in charge of the project was 55-year-old Gleb Yevgenyevich Lozino-Lozinskiy, a deputy of Mikoyan who had worked at the bureau since 1941 and had played a crucial role in the development of propulsion systems (especially afterburners) for numerous MiG jets. As a sign of his dedication to Spiral, Mikoyan set up a special space branch of his design bureau in the town of Dubna in April 1967. This was located on the same premises where Pavel Tsybin's OKB-256 had worked on the PKA spaceplane a decade earlier. The chief of the branch was Pyotr A. Shuster and

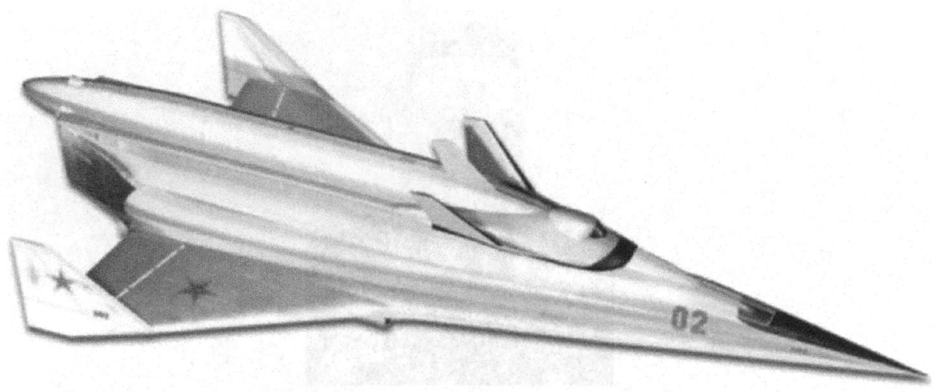

The Spiral system (*source*: *www.buran.ru*).

the head of its design bureau Yuriy D. Blokhin. Lozino-Lozinskiy's deputy was Gennadiy P. Dementyev, the son of Minister of the Aviation Industry Pyotr Dementyev.

Spiral was a 115-ton system consisting of a Hypersonic Boost Aircraft (GSR or "Product 50-50"), an Orbital Plane (OS), and a two-stage rocket to place the OS into orbit. The GSR, probably supposed to be built by the Tupolev bureau, was a 38 m long aircraft with a wingspan of 16.5 m and four air-breathing turbojet engines fixed under the main fuselage. An early version would burn kerosene and the final one hydrogen, with hydrogen gas being used to drive the turbine that in turn rotated the turbojet compressor. The two-stage rocket mounted on the back of the GSR was to be propelled by liquid oxygen/liquid hydrogen engines, but the designers ultimately wanted to replace the oxygen by fluorine. Although this is a highly toxic substance, it provided a higher specific impulse and would require smaller tanks than the LOX version. The hydrogen/fluorine engines were to be developed by Glushko's Energomash design bureau, which by that time had already acquired extensive experience with testing fluorine-based engines.

When picking the shape of the spaceplane, Mikoyan's engineers may at least partially have been inspired by flight tests of suborbital and atmospheric lifting bodies in the United States in the early 1960s, but in the end they came up with their own, unique design. The spaceplane proper was an 8 m long flat-bottomed lifting body with a large upturned nose and wings that could be rotated to vertical position during launch and the initial portion of re-entry. The vehicle's aerodynamic design was such that thermal stresses during re-entry were minimized. The spacecraft's reusable heat shield was not solid, but was composed of a set of sheets, much like a fish's scales. Suspended on ceramic bearings, these sheets could move relative to the vehicle's body as the temperatures on various parts of the ships changed during re-entry. The plates were made of a niobium alloy with a molybdenum disilicide coating and could withstand temperatures up to about +1,500°C.

Situated in the front was the single pilot's cockpit, which in case of an emergency could be ejected from the spaceplane and land by parachute. The headlight-shaped

Spiral spaceplane in orbit (*source*: *www.buran.ru*).

capsule even had a small engine and a heat shield to deorbit and re-enter independently if an emergency arose in orbit. The power plant, located in the back, consisted of a single main engine for changing orbital inclination and deorbiting, two back-up deorbit engines, 16 attitude control thrusters, and a turbojet engine for subsonic propulsion and landing. The landing gear was made up of four skids mounted on the sides of the spaceplane.

In between the cockpit and the engine compartment was a $2\,m^3$ payload section stowed full with reconnaissance equipment or weapons, depending on the mission. There were two reconnaissance versions of the spaceplane, one with optical cameras with a resolution of up to 1.2 m for detailed photography and another with an externally mounted radar antenna with a resolution of 20–30 m for spotting large objects such as aircraft carriers. An attack version of the OS was designed to destroy sea-based targets with a 1,700 kg nuclear-tipped space-to-surface missile, which required an additional $2\,m^3$ of volume in the mid-section of the spaceplane (at the expense of fuel).

Finally, there were two interceptor versions of the spaceplane. One was supposed to catch up with targets in orbit for close inspection and had six 25 kg homing missiles on board for destroying them (if necessary) from a maximum range of 30 km. The other was a long-range interceptor outfitted with 170 kg homing missiles to neutralize targets from a maximum distance of 350 km. Both interceptor versions had enough fuel on board to destroy two targets orbiting at altitudes of up to 1,000 km. The OS weighed 8.8 tons in all configurations, carrying 500 kg of payload for reconnaissance and interception missions, and 2,000 kg in its attack configuration.

A typical Spiral mission would begin with the GSR taking off at a speed of 380–400 km/h using a "launch truck". Having accelerated the system to a hypersonic speed of Mach 6, the carrier aircraft would release the OS/booster combination at an altitude of 28–30 km and return to its home base. Subsequently, the two-stage rocket would place the spaceplane into a low orbit of approximately 130 × 150 km with inclinations varying between 45 and 135° (if launched from the territory of the USSR). If equipped with a main engine burning liquid fluorine (F_2) and amidol (50% N_2O_4, 50% $BH_3N_2H_4$), the reconnaissance and interception versions could change their inclination by 17° for a second target run and the attack version by 7–8°. The interception version could also simultaneously change inclination by 12° and ascend to an altitude of up to 1,000 km.

After a mission of maximum three orbits, the spaceplane would fire its deorbit engine and dive into the atmosphere at a 45–60° angle of attack with the wings folded to near-vertical position, allowing the air stream to flow from the body down to the wings, rather than to the wing leading edges. Cross-range capability was between 1,100 and 1,500 km, offering the pilot much flexibility in choosing landing sites. After unfolding the wings to a near-horizontal position and igniting the turbojet engine, the pilot would land the spaceplane on a dirt runway at a speed of no more than 250 km/h.

According to plans formulated in the preliminary design in 1966 the Spiral program was to be conducted in four stages. The first step was to build three suborbital prototypes and launch them from the back of a Tu-95KM aircraft, the same type of plane Tupolev had intended to use for his own spaceplane tests. Subsonic flights were to begin in 1967, followed by X-15 type supersonic and hypersonic flights in 1968 to altitudes of 120 km and speeds of Mach 6–8. In the second stage Soyuz rockets would be used to launch full-scale versions of the OS ("EPOS") into orbit on both unmanned and manned missions in 1969 and 1970, with one of the mission objectives being to perform an 8° plane-changing burn. Stage 3 would see test flights of a kerosene-fueled version of the GSR in 1970, with the hydrogen-fueled version being introduced in 1972. That same year the fourth stage was to begin with an all-up test of the Spiral complex using a kerosene-fueled GSR and a LOX/liquid hydrogen rocket booster. In 1973 the hydrogen-fueled GSR would be used for a manned test of the Spiral system. Later steps were the introduction of fluorine-based engines for both the rocket and the spaceplane and the replacement of the expendable rocket by a reusable rocket with hypersonic scramjet engines burning liquid hydrogen.

Spiral was by far the largest-scale Soviet spaceplane program of the 1960s, although the amount of money invested in it must still have been dwarfed by what the US Air Force spent on Dyna-Soar. It was also the first for which cosmonauts began training. A Spiral training group was set up at Star City in 1966 and existed until 1973. The Air Force cosmonauts known to have been involved in Spiral at one time or another are Gherman Titov, Anatoliy Kuklin, Vasiliy Lazarev, Anatoliy Filipchenko, Leonid Kizim, Anatoliy Berezovoy, Vladimir Dzhanibekov, Vladimir Kozelskiy, Vladimir Lyakhov, Yuriy Malyshev, Aleksandr Petrushenko, and Yuriy Romanenko. The training mainly involved flying a variety of aircraft from an Air Force test site in Akhtubinsk (Volgograd region) to acquire the status of test pilot.

By the end of the 1960s the Spiral project had still not been officially sanctioned by a party/government decree. One man who tried to change that situation was

Nikolay Kamanin, the Air Force Commander-in-Chief's Aide for Space Matters, who had been pushing for spaceplanes since the early 1960s, seeing them as a logical extension of military aircraft. Sometime in late 1969 Kamanin and his entourage worked out a draft for such a decree to be sent to the Council of Ministers and the Central Committee. The draft was supposed to be signed by seven ministers and high-ranking military officials, but, as Kamanin recounts in his diaries, by April 1970 only four had done so. The delay was caused at least partly by a conflict that had arisen over the missions of future spaceplanes between Sergey Afanasyev, who headed the Ministry of General Machine Building (MOM, the "space and missile ministry"), and Minister of the Aviation Industry Pyotr Dementyev. Afanasyev had signed the draft with the remark that besides military spaceplanes there should also be winged spacecraft adapted as transportation systems. Subsequently, Dementyev refused to put his signature under it, fearing that the organizations under his ministry would become overloaded with space-related work, which was not their primary line of business.

In the middle of 1970 Defense Minister Andrey Grechko sent a letter to Central Committee Secretary for Defense Matters Dmitriy Ustinov, in which he justified the need to build spaceplanes and asked him to order several ministries to reach a consensus on a draft government decree on Spiral. Three months later that was apparently achieved, but by that time Grechko, who was not at all space-minded, seems to have had a change of heart on the issue. When the moment came for him to sign the draft himself, he vetoed it, writing on the document in question that Spiral was "a fantasy" and that money should be spent on more realistic things. Grechko's rejection of the draft sounded the death knell for Spiral. Kamanin asked Air Force Commander-in-Chief Pavel S. Kutakhov (assigned to the post in March 1969) to try and change Grechko's mind, but Kutakhov himself seems to have shown little enthusiasm for the project [27]. To make matters even worse, Mikoyan, backing the program with his authority, died in December 1970, and Lozino-Lozinskiy was forced to divert his attention from space matters after having been assigned chief designer of the new MiG-31 interceptor in 1971.

All this meant that by the turn of the decade the prospects for Spiral were very bleak indeed. Aside from interdepartmental squabbling and budgetary issues, other reasons for the lukewarm interest in Spiral may have been the challenges involved in mastering advanced technologies such as the hypersonic carrier aircraft and the reusable heat shield. In addition to that, it was probably realized by now that at least some of the missions planned for Spiral could just as well be performed by unmanned satellites.

Remarkably enough, the program continued on what appears to have been a semi-legal basis and eventually did see some hardware make it off the ground, even after the Buran program was approved in 1976. Why Spiral wasn't canceled outright is a fact that remains to be satisfactorily explained. In the mid-1970s designers looked at enlarged versions of the Spiral spaceplane to be launched by the Proton rocket or the massive Energiya booster (see Chapter 2), but these appear to have been short-lived paper studies not enough to justify the continuation of a test flight program.

Recent evidence indicates MAP may eventually have seen Spiral as no more than a trump card in a seemingly mundane tug-of-war with MOM over subordinate organizations. One source of acrimony between the two ministries was that many factories and research institutes of MAP had been transferred to MOM after the latter's establishment in 1965. Around the mid-1970s Lozino-Lozinskiy and MAP deputy minister Aleksey Minayev reportedly convinced MAP minister Dementyev that by demonstrating its ability to fly Spiral hardware, MAP would eventually muster the political support required to have some of those organizations transferred back to its ranks. Another factor enabling the continuation of Spiral may have been the death in April 1976 of Defense Minister Grechko, one of the program's most vigorous opponents [28].

Whatever the real motives for keeping Spiral alive, several test flights were conducted in support of the program between 1969 and 1978. The Gromov Flight Research Institute (LII) built several scale models of the spaceplane known as BOR-1, 2 and 3 ("Unmanned Orbital Rocket Plane"). These were launched on suborbital trajectories by R-12 missiles from Kapustin Yar between 1969 and 1974. A full-scale prototype for subsonic flights (105.11, nicknamed "Lapot" or "Bast Shoe" and sometimes also called EPOS, like the orbital test bed) was ready for test flights by the mid-1970s. Staged from the Air Force site in Akhtubinsk, they began in December 1975 with a series of taxi runs and brief flights (first in June 1976) in which the plane took off on its own power. At the helm for those tests were civilian test pilots Aviard Fastovets, Valeriy Menitskiy, Vasiliy Uryadov, Igor Volk, and Aleksandr Fedotov. After a number of "captive–carry" tests, the 105.11, piloted by Fastovets, was dropped from the belly of a Tu-95K from an altitude of 5 km for the

The 105.11 atmospheric test bed (B. Hendrickx).

first time in October 1977. Five more drop tests followed the following months, three performed by Fastovets, one by Pyotr Ostapenko, and one by Uryadov. The final one in September 1978 ended with the plane making a hard landing to the right of the runway, causing some damage to the landing gear. The 105.11 was never refurbished for another flight. It can still be seen today at the Monino Air Force museum outside Moscow. A model for supersonic tests (105.12) was built but never flown and a model for hypersonic tests (105.13) was partially built.

Apparently, Spiral died a silent death in the late 1970s as work on Buran got underway in earnest. It did have at least one important legacy for the Buran program. A subscale model (BOR-4) originally intended for orbital test flights of the Spiral spaceplane was eventually launched on single-orbit missions in 1982–1984 to test materials for Buran's thermal protection system (see Chapter 6). The work on Spiral also served as the basis for studies of new air-launched spaceplanes in the 1980s and 1990s, particularly the MAKS project (see Chapter 9) [29].

SPACEPLANE STUDIES AT THE ZHUKOVSKIY ACADEMY

Winged spacecraft were not only studied at spacecraft and aviation design bureaus, they were also the subject of academic studies by the original Air Force cosmonauts during the 1960s. In September 1961 most of the cosmonauts of the original "Gagarin group" began studying at the prestigious Zhukovskiy Academy in Moscow to improve their engineering skills in preparation for future space missions. In 1964 they were joined by the five women who had been selected for cosmonaut training in 1962. The culmination of the studies would be a thesis in a chosen field of specialization that the candidates would defend before their tutors in written and oral sessions at the end of their course.

Rather than pick a completely different subject for each candidate, it was decided that all would work on one general theme. In 1965 Sergey Korolyov recommended that the cosmonauts should study a practical design for a winged reusable spacecraft, a suggestion that was accepted by the cosmonauts' supervisor Professor Sergey Belotserkovskiy. Fifteen cosmonauts were involved in the thesis work: Yuriy Gagarin, Gherman Titov, Andrian Nikolayev, Pavel Popovich, Valeriy Bykovskiy, Aleksey Leonov, Boris Volynov, Yevgeniy Khrunov, Georgiy Shonin, Viktor Gorbatko, Dmitriy Zaikin, Valentina Tereshkova, Irina Solovyova, Tatyana Kuznetsova, and Zhanna Yorkina. Each of them was given the liberty of choosing from a series of topics suggested by their tutors. Gagarin, for instance, decided to focus on aerodynamics during approach and landing, Nikolayev on aerodynamics at hypersonic and supersonic speeds as well as on thermal protection, Titov on emergency escape systems, Popovich on engine design, Khrunov on orientation systems, Bykovskiy on fuel supply, etc.

By mid-1966 the cosmonauts had picked a lifting body shape somewhat reminiscent of the M2F1 "flying bathtub" that NASA had been testing at Edwards Air Force Base since 1963. The vehicle would have small wings that would only be

Lifting body studied at the Zhukovskiy Academy (*source*: *www.buran.ru*).

Gagarin practicing landings in a simulator.

Gagarin and fellow cosmonauts examining an unidentified spaceplane model.

unfolded for the final approach and landing. In order to improve stability at supersonic speeds, the cosmonauts decided to add small lattice wings to the nose section similar to the ones used in the emergency escape system of the Soyuz launch vehicle.

Part of the work was to test wooden scale models of the spaceplane in wind tunnels and also to practice landings on a crude simulator using primitive analog computers. The tutors followed the cosmonauts' work with due scrutiny and their critical remarks were not always easily accepted by national heroes like Gagarin. In the autumn of 1967 Gagarin's thesis failed to pass a critical review because the vehicle had poor gliding characteristics during the final phase of the flight. Gagarin's solution to the problem, namely to land the spaceplane by parachute, was deemed unacceptable. It took Gagarin several more weeks of theoretical and simulator work to refine the design such that the ship could make an unpowered runway landing.

Most of the cosmonauts defended their thesis projects in January 1968. Gagarin's turn came on 17 February 1968, only weeks before he died in a plane crash. All of them graduated from the Academy with the diploma of "Pilot–Engineer–Cosmonaut". Only Gorbatko's thesis got the result "good" rather than "excellent".

As it turned out later, this was not because his thesis was worse than the others', but simply because it was felt not all of them should have the same result. Gorbatko, as one of the unflown cosmonauts at the time, had the misfortune of being picked as the "victim" [30].

The cosmonauts' spaceplane studies were considered top secret, as was *all* diploma work at the Zhukovskiy Academy for that matter. Professor Belotserkovskiy was not even allowed to take snapshots of his pupils, but used a hidden camera nevertheless to record their activities [31]. After having been safely hidden in a safe for almost twenty years, many of them were eventually published in 1986 in a book called "Gagarin's Thesis" [32]. However, even that provided little solid information on the diploma work and did not contain a single proper picture of the spaceplane. This had to wait until another publication by Belotserkovskiy in 1992, where the spaceplane was nicknamed "Buran-68" [33]. There is also a famous picture released in the 1970s showing Gagarin and several other cosmonauts examining a model of a delta-wing spaceplane, but that is not "Buran-68". Belotserkovskiy claims it is Dyna-Soar, but it clearly is not that either. Some have questioned the authenticity of the picture, but the model in question has recently been seen at the Zhukovskiy Academy.

Research on the diploma project coincided with early work on the Spiral system, but as yet there is no evidence of any interaction between Mikoyan's team in Dubna and the Air Force cosmonauts, although Titov began training for the Spiral program in 1966–1967 and must have been aware of the project's details. Although the finished thesis projects were sent to Lozino-Lozinkiy's team, there are no indications that they in any way influenced the design of Spiral or future spaceplanes [34].

REFERENCES

[1] G. Vetrov, *S.P. Korolyov i kosmonavtika: pervye shagi*, Moscow: Nauka, 1994, pp. 56, 72.

[2] N. Korolyova, *Otets (kniga pervaya)*, Moscow: Nauka, 2001, p. 262.

[3] N. Korolyova, *op. cit.*, p. 293.

[4] G. Vetrov, *op. cit.*, pp. 110–111; N. Korolyova, *op. cit.*, p. 316.

[5] N. Korolyova, *op. cit.*, p. 316.

[6] G. Vetrov, *op. cit.*, p. 111.

[7] A. Nikolayev, "This is how liquid-fuel rocket engines and liquid-fueled rockets began" (in Russian), *Dvigateli*, 4/2004, pp. 44–47, 5/2004, pp. 26–30; M. Yeftivyev, *Ognyonnye krylya: istoriya sozdaniya reaktivnoy aviatsii SSSR (1930–1946)*, Moscow: Veche, 2005, pp. 28–87; G. Vetrov, B. Raushenbakh, *S.P. Korolyov i ego delo*, Moscow: Nauka, 1998, pp. 538, 540; A. Moskalyov, *Golubaya spiral*, Voronezh, 1999.

[8] M. Yevtifyev, *op. cit.*; G. Vetrov, B. Raushenbakh, *op. cit.*, pp. 538–539.

[9] M. Yevtifyev, *op. cit.*; on-line aviation encyclopedia *Ugolok neba* at *http://www.airwar.ru/* The latter source claims there were an additional four flights in November–December 1949.

[10] A. Siddiqi, *Challenge to Apollo*, Washington, D.C.: NASA, 2000, p. 53.

[11] M. Yevstafyev, *Dolgiy put k Bure*, Moscow: Vuzovskaya kniga, 1999, pp. 34–41.

[12] M. Yevstafyev, *op. cit.*, pp. 45–82; A. Siddiqi, *op. cit.*; A. Bruk *et al.*, *Illyustrirovannaya entsiklopediya samolyotov OKB V.M. Myasishcheva (tom 2, chast 1)*, Moscow: Aviko Press, 2001, pp. 54–59, 93–116.

[13] R. Houchin, "Why the Air Force Proposed the Dyna-Soar X-20 Program", *Quest*, Winter 1994, pp. 4–12.

[14] G. Vetrov, B. Raushenbakh, *op. cit.*, pp. 190–200.

[15] Y. Semyonov (ed.), *Raketno-kosmicheskaya korporatsiya Energiya 1946-1996*, Moscow: RKK Energiya, 1996, p. 105. Winged spacecraft are known to have been studied at Department 9 by Konstantin S. Shustin, presumably in the 1957–1958 timeframe. His conclusion was also that the challenges posed by thermal protection were too daunting. See K. Feoktistov, *Trayektoriya zhizni*, Moscow: Vagrius, 2000, p. 47.

[16] M. Keldysh (ed.), *Tvorcheskoye naslediye akademika Sergeya Pavlovicha Korolyova*, Moscow: Nauka, 1980, p. 406.

[17] G. Vetrov, B. Raushenbakh, *op. cit.*, pp. 271–272.

[18] Korolyov is known to have tasked TsAGI in September 1959 with studying various shapes for returnable spacecraft, including winged vehicles, but it is unclear if the PKA was part of those studies. See G. Vetrov, B. Raushenbakh, *op. cit.*, pp. 272–276.

[19] V. Bobkov, "Little bast shoe in space" (in Russian), *Krylya Rodiny*, 11/1991, p. 25; I. Afanasyev, "Unknown ships" (in Russian), *Kosmonavtika, astronomiya (Znaniye)*, 12/1991, pp. 5–7; M. Rebrov, "PKA or simply 'little bast shoe'" (in Russian), *Krasnaya Zvezda*, 17 June 1995.

[20] V. Petrakov, M. Chernyshov, "Unkown Buran" (in Russian), *Sovetskaya Rossiya*, 10 April 1991, p. 4; V. Petrakov, "Two Projects of V.M. Myasishchev", *Journal of the British Interplanetary Society*, September 1994, pp. 347–354; E. Kulaga "The mini-shuttle of V.M. Myasishchev" (in Russian), *Tekhnika Vozdushnogo Flota*, 5/1997; A. Bruk, *op. cit.*, pp. 69–70, 74–76, 135–158. According to the latter source there was yet another spaceplane proposal called "Project 49", but information on this is very sketchy.

[21] V. Rigmant, "Under the signs 'ANT' and 'Tu'" (in Russian), *Aviatsiya i kosmonavtika*, 11/1999; V. Rigmant, *Samolyoty OKB A.N. Tupoleva*, Moscow: Rusavia, 2001.

[22] G. Vetrov, B. Raushenbakh, *op. cit.*, pp. 298–300.

[23] I. Afanasyev, "Kosmoplan" (in Russian), *Krasnaya zvezda*, 26 August 1995; A. Siddiqi, *op. cit.*, pp. 230–232, 241, 305–312, 441–442; G. Yefremov (ed.), *60 let samootverzhennogo truda vo imya mira*, Moscow: Oruzhiye i tekhnologii, 2004, pp. 97–100.

[24] N. Kamanin, *Skrytyy kosmos (kniga 1)*, Moscow: Infortekst, 1995, p. 87.

[25] Y. Kazarov, "The child mortality of Buran" (in Russian), *Nezavisimaya gazeta/Nauka*, 13 November 2003. In 1960 Mikoyan had also been involved together with Korolyov and chief designer of anti-missile systems Grigoriy Kisunko in a proposal to develop an anti-satellite system launched by an R-7 derived rocket. See G. Kisunko, *Sekretnaya zona*, Moscow: Sovremennik, 1996, p. 393.

[26] A. Mikoyan, "Future of aerospace technologies" (in Russian), *Krasnaya zvezda*, 9 January 1962.

[27] N. Kamanin, *Skrytyy kosmos (kniga 4)*, Moscow: Novosti kosmonavtiki, 2001, pp. 144–145, 222–223, 240.

[28] TV documentary *"Udarnaya sila"* , shown on the Russian ORT television channel, 20 March 2007; B. Hendrickx correspondence with Vadim Lukashevich, 7 April 2007.

[29] A detailed history of the Spiral program can be found in a series of articles by V. Lukashevich, V. Trufakin, S. Mikoyan in the journal *Aerokosmicheskoye obozreniye*, 3/2005, 4/2005, 5/2005, 6/2005, 1/2006, 2/2006 and in *Aviatsiya i kosmonavtika*, 10/2006,

11/2006, 12/2006, 1/2007, 2/2007; An English summary can be found in: V. Lukashevich, "Predecessor of Shuttle and Buran: Spiral Orbital Aircraft Programme, *Air Fleet*, 4/2004.

[30] S. Belotserkovskiy, *Pervoprokhodtsy vselennoy*, Moscow: Mashinostroyeniye, 1997, pp. 73–99.

[31] J. Doran, P. Bizony, *Starman: The Truth Behind the Legend of Yuriy Gagarin*, London: Bloomsbury, 1999, pp. 171–174.

[32] S. Belotserkovskiy, *Diplom Gagarina*, Moscow: Molodaya gvardiya, 1986.

[33] S. Belotserkovskiy, *Gibel Gagarina*, Moscow: Mashinostroyeniye, 1992.

[34] TV documentary *Udarnaya sila, op. cit.*

2

The birth of Buran

Although there was significant spaceplane research in the Soviet Union in the 1960s, it was still dwarfed by the effort the country put into its mainstream manned space program, the one that was visible to the outside world. In terms of successes, there were two distinct periods in the Soviet piloted space program in the 1960s. The first part of the decade was marked by amazing triumphs that stunned the whole world. There was the pioneering flight of Yuriy Gagarin in 1961, the first flight into space by a woman (Vostok-6 in 1963), the first three-man flight (Voskhod in 1964), and the first spacewalk (Voskhod-2 in 1965). Then things started going downhill in spectacular fashion. First, there was the death in January 1966 of chief designer Sergey Korolyov, the mastermind behind the Soviet Union's early space triumphs. After a two-year gap in piloted space missions, the maiden manned flight of the Soyuz capsule ended in disaster with the death of cosmonaut Vladimir Komarov in April 1967.

Meanwhile, in August 1964 the Soviet Union had secretly decided to send men to the Moon in response to the Apollo program, kicked off three years earlier by President Kennedy's announcement in May 1961. The Soviet piloted Moon program was to be carried out in two stages, beginning with manned circumlunar flights (using the L-1 capsules and the Proton rocket) and culminating in manned landings on the lunar surface (using the L-3 complex and the massive N-1 rocket). While the Russians came relatively close to beating America in the circumlunar race, they never stood a chance of upstaging the United States in putting a man on the Moon. Already months behind schedule, the L-3 lunar-landing program was thrown into complete disarray by the catastrophic failure of the first two test flights of the N-1 rocket in February and July 1969.

At the same time, the Soyuz program continued as an independent effort, with a couple of missions flown in 1968 and 1969 (albeit with mixed success). While Soyuz shared many features with the manned lunar craft, the Soyuz program, essentially a remnant of a canceled circumlunar project of the early 1960s, lacked a clear sense of direction.

Realizing that the ailing Soviet manned space program needed a fresh impetus, a small group of engineers within the Korolyov design bureau started working out plans in mid-1969 for an Earth-orbiting space station (Long-term Orbital Station or DOS) that could be built relatively quickly using available technology and would use Soyuz as a ferry vehicle. By early 1970 they saw their plans approved with the release of a key government decree that would determine the course of the Soviet Union's piloted space activities for the remainder of the century. After a herculean effort lasting just over one year, the space station, officially dubbed Salyut, rocketed into orbit in April 1971. Unfortunately, the three cosmonauts who boarded the station two months later died during the return to Earth.

THE ORIGINS OF THE SPACE SHUTTLE

Meanwhile, even as NASA was still preparing to land the first Apollo astronauts on the Moon, the space agency was drawing up plans for the post-Apollo era. In January 1969 NASA appointed four aerospace companies to study possible configurations for what it called an "Integrated Launch and Re-entry Vehicle" (ILRV), what would eventually become the Space Shuttle. As these studies got underway, a Space Task Group (STG) headed by Vice President Spiro Agnew recommended that America embark on a manned flight to Mars and devised three options to achieve this goal, each of which would need the logistical support of a reusable spacecraft shuttling back and forth between Earth and low orbit. Also part of the space infrastructure would be a low-orbiting space station, a space tug, a lunar base, and a nuclear propulsion system for interplanetary missions. The most modest of the three options was to develop only a shuttle and a space station and defer a decision on a manned Mars flight until after 1990.

However, when the STG released its report in September 1969, waning public interest in the space program and the escalating cost of the Vietnam War were about to take their toll on America's ambitious space plans. NASA's budget was drastically cut back, dashing any hopes of turning the STG's plans into reality anytime soon. It turned out that even the cheapest of the three options (requiring $5 billion per year until 1980) cost more than the nation could afford. When President Nixon officially reacted to the STG report in March 1970, all he left standing of the STG plan was a shuttle vehicle "designed so that it will be suitable for a wide range of scientific, defense and commercial uses [and] help us realize important economies in all aspects of our space program."

If the Shuttle was going to be turned into a satellite-carrying truck, it would only be economically effective if it achieved an extremely high launch rate and placed all government, commercial, and military payloads into orbit. In other words, it had to replace all existing expendable launch vehicles. Therefore, it was crucial for NASA to gain agreement from the military community to use the Space Shuttle to launch all military and intelligence payloads, which were projected to constitute one-third of all

future space traffic. For the military this was not a bad deal, because they would acquire a launch vehicle built at NASA's expense. Their only major investment in the Shuttle would be the construction of a launch pad at Vandenberg Air Force Base in California to enable launches of military payloads into polar orbits. At the same time, Defense Department (DoD) requirements also had a very serious impact on the Shuttle's design.

Until then the favored option within NASA had been to develop a completely reusable vertically launched system consisting of a relatively small spaceplane and a flyback booster, mated either belly-to-belly or piggyback. The flyback booster (either manned or unmanned) would act as the first stage, carrying the shuttle to a significant altitude before separating and returning to the launch site to make a horizontal runway landing. The shuttle would then use its on-board fuel supply to complete the trip to orbit. The preferred design for the spaceplane was a vehicle with stubby straight wings. This was designed to re-enter the Earth's atmosphere at a high angle of attack, which would reduce frictional heating. It would make only minor hypersonic maneuvers and have excellent subsonic glide characteristics.

The Defense Department requirements, first of all, changed the dimensions of the orbiter. The DoD needed an orbiter that could handle payloads up to 18 m long and launch 18 tons into polar orbit from Vandenberg and over 27 tons into a due-east orbit from Cape Canaveral. This was significantly more than what NASA had asked for in its original request for proposals in 1969. Even more significantly, the DoD required a much higher cross-range capability, the ability to maneuver to either side of the vehicle's ground track during re-entry. The Air Force wanted a cross-range capability of about 2,000 km, which would allow the Shuttle to quickly return to its secure launch site runway at Vandenberg after a single revolution while the Earth rotates to the east under it. However, this requirement dictated a delta-wing vehicle with a much higher hypersonic lift-to-drag ratio as well as a much more robust thermal protection system. This is because most of the cross-range maneuvering is performed at extremely high speeds, exposing large portions of the airframe to the thermal effects of re-entry. Also, the delta-shaped wings entail a much worse performance at subsonic speeds, with the orbiter making a very steep descent and coming down at a much higher speed.

The net result was that the orbiter was going to be much bigger and heavier than originally anticipated, making it impossible to retain the spaceplane/flyback booster concept. Instead, the orbiter's propellant would now have to be carried in an expendable external fuel tank and the flyback booster was replaced by two solid rocket boosters, which is the Space Shuttle configuration as we know it today. On 5 January 1972 President Nixon gave his final go-ahead for the development of the Space Shuttle, but it would take another two years for the design to be frozen. One of the last changes was the deletion of an air-breathing propulsion system in early 1974 [1].

A SLOW RESPONSE

Early work on the Space Shuttle in the late 1960s did not spark an immediate response from the Soviet side for a number of reasons. The only two design bureaus capable of building manned spacecraft had more pressing concerns. TsKBEM, the former Korolyov design bureau (now headed by Vasiliy Mishin), was preoccupied with Soyuz, the civilian DOS space station, and the N-1/L-3 manned lunar program. TsKBM, the Chelomey design bureau, was busy working on the military Almaz space station and its TKS transport ship, not to mention a variety of unmanned military satellites and anti-ship missile projects.

Not only would the development of a large reusable spacecraft place an extra burden on the already overtaxed design bureaus, there simply was no clear need for such a system in the near future. The Soyuz and TKS spacecraft could perfectly handle transportation tasks for the DOS and Almaz stations and the Soviet Union had a varied fleet of expendable rockets to satisfy satellite launch requirements for many years to come.

Looking at the more distant future, TsKBEM was studying a so-called Multipurpose Orbital Complex (MOK), an entire orbital infrastructure aimed at lowering space transportation costs. Even here there was no immediate need for a large reusable shuttle system. The centerpiece of the MOK was to be a giant N-1 launched space station called MKBS (Multipurpose Space Base Station) that would serve as an orbiting garage. The idea was that satellites in the constellation would be serviced either at the MKBS or regularly be visited by MKBS-based crews flying light versions of the Soyuz outfitted with a manipulator arm. The satellites themselves would be orbited by expendable rockets or partially reusable rockets based on the N-1. In April 1972 Mishin and Chelomey got approval for a joint proposal to turn Almaz into a combined civilian/military space station serviced by Soyuz spacecraft, allowing TsKBEM to focus on the more distant goal of creating the MOK. The two chief designers agreed that Chelomey's TKS would be the MOK's key transportation system during the program's experimental phase. Reusable transportation systems *were* part of the MOK plans, but only at a later stage [2].

There were also other obstacles to the initiation of a Space Shuttle type project. Requiring a blend of aviation, rocket, and space technology, it would be an organizational nightmare. The leading missile and space design bureaus, including TsKBEM and TsKBM, came under the "space and missile industry" known as the Ministry of General Machine Building (MOM) (headed by Sergey Afanasyev), while the leading aviation design bureaus were under the Ministry of the Aviation Industry (MAP) (headed by Pyotr Dementyev). Although both were willing to participate in such an effort, neither was eager to take on prime responsibility for it, considering it to be "the other ministry's field of business".

Finally, the atmosphere around the turn of the decade may not have been conducive to the start of a totally new program. It was a period marked by many spectacular failures in the Soviet space program, both launch vehicle mishaps (notably the Proton and the N-1) and spacecraft malfunctions (notably lunar and deep-space probes). The string of failures even led to the creation of an investigative commission, which concluded that one of the root causes for the numerous setbacks was the lack of proper ground-testing facilities such as engine test stands, vacuum chambers, and the like. Embarking on a completely new, costly, and technologically advanced project under such conditions would not have been a logical course of action.

However, while any final decision on a Soviet shuttle was still years away, some in the Soviet space community did think it was time to begin preliminary research on such a system. The initiative seems to have come from the Military Industrial Commission (VPK), a body under the Council of Ministers (the Soviet government) that oversaw all defense-related ministries (including MOM and MAP). Among its tasks was to formulate new proposals for military and space projects (with the necessary input from the design bureaus and the military community), which could then be officially approved in the form of joint decrees of the Central Committee of the Communist Party and the Council of Ministers. These decrees would set rough timelines for projects, outline their major goals, and also assign the main organizations that would be involved. It was then again up to the VPK to implement those decrees by dividing the work among the design bureaus, setting concrete timetables, and convening meetings of the people in charge.

In a draft proposal for the Soviet Union's next five-year space program dated 27 November 1970, the VPK suggested that both MOM and MAP as well as other organizations should work out a so-called "draft plan" for a "unified reusable transport ship" in 1972. This is the first known written evidence of the Soviet Union's intention to respond to the Space Shuttle. Essentially, it was an order to produce nothing more than paperwork. The "draft plan" is just one of the preliminary stages that Soviet space projects went through before metal was actually cut.

Indications are that the phrase about the reusable transportation system was not included in the final government and party decree describing the country's goals in space for the next five years. Clearly, the time was not quite ripe enough even for preliminary research on a shuttle system. There may have been opposition from

MOM and MAP but, perhaps more importantly, there was no urgent need to begin this work because the US Space Shuttle had not even been officially approved.

Even President Nixon's go-ahead for the Space Shuttle project in January 1972 did not set in motion a concerted effort to develop a reusable spacecraft. The first high-level meeting in response to Nixon's January 1972 announcement was organized by the VPK on 31 March 1972. It was attended by both industry and military officials, more particularly representatives of TsNIIMash (MOM's leading space test and research facility), the TsNII-30 and TsNII-50 military research institutes, the Chief Directorate of Space Assets or GUKOS (the "space branch" of the Strategic Rocket Forces) and the Air Force, but no consensus was reached on the need for a response. At this stage the VPK once again formulated a draft proposal asking MOM, MAP, and other organizations to develop a draft plan for a shuttle system, but it met with stiff opposition from MOM minister Afanasyev and was not accepted.

In late April 1972 another meeting took place at TsNIIMash, attended by some of the chief designers (Mishin, Chelomey, Glushko), officials of MOM and TsNII-50. Their conclusion was that a reusable space transportation system was a less efficient and less cost-effective way of delivering payloads to orbit than expendable boosters. Also, they did not see an immediate need for using such a system to return satellites or other hardware back to Earth, certainly not after Mishin and Chelomey had received approval for the MOK/TKS plan that same month. Moreover, at this point the US Space Shuttle was not considered a military threat to the Soviet Union [3].

PRELIMINARY STUDIES

Meanwhile, six Soviet military and civilian research institutes were tasked with performing a study of the Soviet Union's future space transportation needs to help determine the need for a response to the Space Shuttle. These were TsNIIMash and NIITP (the Scientific Research Institute for Thermal Processes, the current Keldysh Research Centre) under MOM, TsAGI (the Central Aerohydrodynamics Institute) under MAP, TsNII-30 and TsNII-50 under the Ministry of Defense, and IKI (the Institute of Space Studies) under the Academy of Sciences. TsNIIMash was given the lead role.

Actually, the studies centered not solely on shuttles, but on a wide array of expendable and reusable launch vehicles that would provide the most economical access to space in the future. They also extended to various reusable space tugs and expendable upper stages with either liquid or nuclear rocket engines for interorbital maneuvers and deep-space missions. Four future directions were considered for the Soviet Union's space transportation program:

- the continued use of expendable launch vehicles and spacecraft until the year 2000;
- the continued use of expendable launch vehicles, but with standardized satellites;
- the use of a reusable space transportation system capable of returning spacecraft back to Earth for servicing and subsequent reuse;

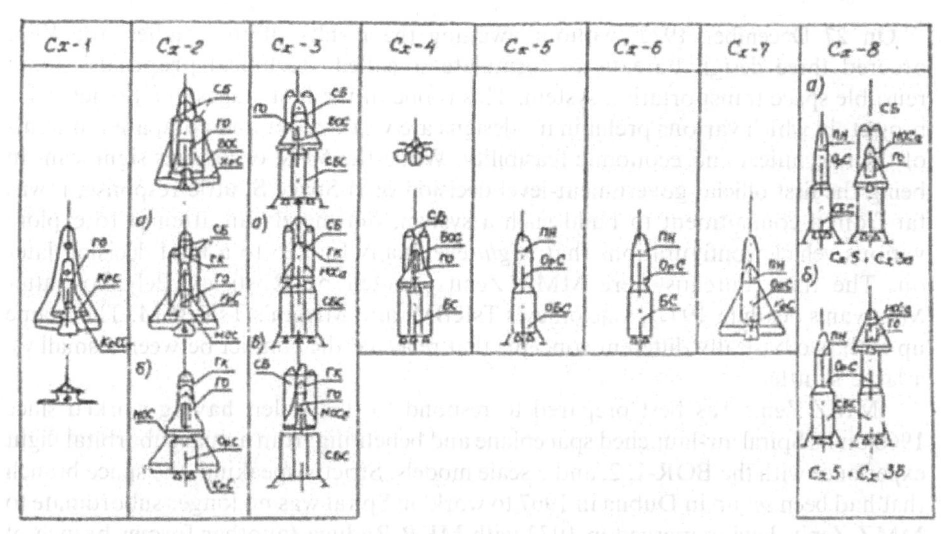

Reusable space transportation systems studied at Soviet research institutes in the early 1970s (*source*: Ts. Solovyov).

- the use of a reusable space transportation system capable of servicing and repairing satellites in orbit.

As for reusable systems, the institutes explored two vehicle sizes, one able to accommodate payloads of 30-40 tons (like the Space Shuttle) and another for payloads weighing 3-5 tons. Furthermore, two ways were studied of recovering the first stage, one involving the use of standard recovery techniques, the other requiring the use of reusable flyback boosters.

The six institutes presented their joint findings in June 1974. They concluded that the development of a reusable launch vehicle was only economically justified if the launch rate was very high, more particularly if the annual amount of cargo delivered to orbit would exceed 10,000 tons. However, it was stressed that much also depended on the vehicle's capability of servicing satellites in orbit or returning them to Earth for repairs. The size of the spaceplane in itself would not determine its effectiveness and would have to depend on the mass and size of the payloads that needed to be launched or returned. Finally, it was recommended to perfect future reusable systems by developing first stages with air-breathing engines and eventually to introduce high-thrust nuclear engines for single-stage-to-orbit spaceplanes.

Basically, the conclusion was that a Space Shuttle type transportation system would not provide any major cost savings even if a relatively high launch rate was achieved and should be seen as nothing more than a first step towards developing more efficient transportation systems. However, the consensus was that if a reusable system were developed, preference should be given to a big shuttle akin to the American one [4].

On 27 December 1973, without awaiting the results of the studies, the VPK ordered three design bureaus to formulate so-called "technical proposals" for a reusable space transportation system. This is one of the first stages in a Soviet space project, in which various preliminary designs are worked out and compared in terms of their technical and economic feasibility. While the VPK order was significant in being the first official government-level decision on a Space Shuttle response, it was far from a commitment to build such a system, but merely an attempt to explore various vehicle configurations that *might* eventually lead to to a final decision later on. The three bureaus were MMZ Zenit (headed by Rostislav Belyakov after Mikoyan's death in 1971), Chelomey's TsKBM, and Mishin's TsKBEM. They came up with two basically different concepts that reflected the conflict between a small vs. a large shuttle.

MMZ Zenit was best prepared to respond to this order, having worked since 1966 on its Spiral air-launched spaceplane and benefiting from actual suborbital flight experience with the BOR-1, 2, and 3 scale models. Strictly speaking, the space branch that had been set up in Dubna in 1967 to work on Spiral was no longer subordinate to MMZ Zenit, having merged in 1972 with MKB Raduga (another former branch of Mikoyan's bureau) to form DPKO Raduga. The VPK order must have been a major morale booster for Lozino-Lozinskiy's Spiral team, which because of a lack of government and military support had been forced to do its work on an almost semi-legal basis. It did require the team to divert its attention from the small air-launched Spiral, which had been primarily designed for reconnaissance, inspection, and combat missions. With the focus now shifting to transportation tasks, a larger version of Spiral with a higher payload capacity was needed. Although the details are sketchy, Lozino-Lozinskiy's team seems to have studied an enlarged 20-ton version of the Spiral spaceplane launched by the Proton rocket.

TsKBM also set its sights on a 20-ton spaceplane to be orbited by the Proton, which itself was a product of Chelomey's bureau. However, the spaceplane project seems to have been low on Chelomey's list of priorities at this stage. Indications are that the bulk of spaceplane research at TsKBM was done in the late 1970s, by which time Buran had already been approved (see Chapter 9) [5].

TsKBEM was to focus on a Space Shuttle sized vehicle to be orbited by the N-1 rocket, but it appears that little, if any, work was done on this. The research was to be done by a small team headed by Valeriy Burdakov, but as Burdakov later recalled, the team's work was limited to studying the possibility of reusing the first stage of the N-1 and keeping track of foreign literature on reusable space systems [6]. However, big changes were ahead at TsKBEM that would turn these plans upside down.

GLUSHKO TAKES CHARGE

Ever since the death of Korolyov in 1966, TsKBEM had been run by Vasiliy Mishin, a long-time associate of Korolyov. Eight years on, his position had been significantly weakened by the deadly accidents in the Soyuz and Salyut programs and the repeated failures of the N-1 rocket. Although Mishin was a talented engineer and the seeds for many of those failures had been sown under Korolyov's leadership, he clearly lacked the authority and the managerial qualities of his predecessor. His predilection for alcohol had not done his reputation any good either. On 17 May 1974 the Soviet leadership decided it was time to act. Mishin was sacked as general designer of TsKBEM and replaced by Valentin P. Glushko, who until that time had headed the KB Energomash organization, the main design bureau for Soviet rocket engines. With Glushko's arrival, Energomash was absorbed by TsKBEM, which was renamed NPO Energiya.

One of Glushko's first orders was to suspend all work on the N-1 rocket and the associated L-3 and MOK projects. After the back-to-back failures in 1969, two more N-1 flights in June 1971 and November 1972 had ended in first-stage failures. However, Glushko's move had as much to do with past rivalries as with sound engineering reasons. Energomash had been excluded from taking part in the N-1's development because of a major disagreement between Korolyov and Glushko in the early 1960s over the types of engines and propellants to be used and the relations between the two men had remained strained until Korolyov's death. Therefore, Glushko's assignment to the top job at Korolyov's former bastion was an ironic twist of fate, to say the least.

After his arrival at NPO Energiya, Glushko started a one-year effort to map out a future course for the Soviet manned space program. Apart from the ongoing Soyuz/Salyut effort, there was no consensus on what that future should be. For this purpose NPO Energiya was reorganized into five departments. Aside from the Apollo–Soyuz and Salyut departments, headed by Konstantin Bushuyev and Yuriy Semyonov, respectively, Glushko established a department that would study various concepts for a reusable space transportation system akin to the American Space Shuttle. This was headed by Igor Sadovskiy. A fourth department, overseen by Ivan Prudnikov, would focus on the establishment of a lunar base, and a fifth department, led by Yakob Kolyako, would study a new generation of heavy-lift launch vehicles called RLA to replace the N-1. The new structure was approved on 28 June 1974 by MOM minister Sergey Afanasyev [7].

Valentin Glushko.

A LUNAR BASE OR A SHUTTLE?

Although studies of reusable space transportation systems were underway by the time TsKBEM was reorganized into NPO Energiya in May 1974, at that time they were certainly not considered the main priority. Glushko himself was opposed to the development of a Space Shuttle equivalent and in the first week following his appointment had even disbanded Burdakov's shuttle team, only to reinstate it on the insistence of Igor Sadovskiy, who in turn was placed in charge of the shuttle department [8].

Glushko feared the shuttle program would jeapordize plans to establish a permanent base on the Moon. Despite the suspension of the N-1 project, the manned lunar program was not dead. Not only was a lunar base considered an appropriate response to the short-duration Apollo flights, for Glushko personally it would be a sweet revenge on Korolyov's star-crossed N-1/L-3 effort. Initially at least, Glushko had the support of Dmitriy Ustinov, who in his capacity of Communist Party Secretary for Defense Matters served as the *de facto* head of the Soviet space program from 1965 until 1976. A long-time ally of Glushko, Ustinov may very well have been instrumental in getting him the top job at NPO Energiya. Opening a top meeting at NPO Energiya on 13 August 1974, Ustinov said:

> "In recent days the Politburo has held serious discussions on our space problems ... It was said at the Politburo that, taking into account the successful landings of the Americans, the task of conquering the Moon remains especially important for us. Whatever task we carry out, this will remain our main general task, but in a new [form]." [9]

Even though work on a lunar base *was* already underway at the KBOM design bureau of Vladimir Barmin, Prudnikov's department at NPO Energiya set to work. By the end of 1974 it completed preliminary plans for a permanently manned lunar base called Zvezda ("Star") that would see three-man crews working on the surface of the Moon for up to a year before being changed out. The plan included a Lunar Expedition Ship (LEK) to transport the crews to the Moon and a lunar base consisting of a Lab-Hab Module, a Lab-Factory Module, a manned lunar rover, and a small nuclear power plant to provide power to the various elements of the base. The scheme required multiple launches of a massive rocket in the RLA family capable of putting 230 tons into low Earth orbit, 60 tons into lunar orbit, and 22 tons on the lunar surface [10].

By the first half of 1975 NPO Energiya had devised a so-called "Integrated Rocket and Space Programme", which included the plans for the RLA rocket family, the Zvezda lunar base, and reusable spacecraft. It was submitted for approval to the Ministry of General Machine Building and the Ministry of Defense. Apparently, the hope was that all these elements would be approved. However, by this time Zvezda, Glushko's pet project, stood little chance of surviving. Not surprisingly, the lunar base received no support whatsoever from the military. Neither did it receive the blessing of the Academy of Sciences (in the person of Keldysh) and Ustinov's initial support had dwindled for a variety of reasons [11]. Clearly, many had been sobered up by the fact that the estimated price tag for the project was 100 billion rubles [12]. In fact, very few people apart from Glushko himself seem to have believed in the ambitious plans he outlined after becoming chief of NPO Energiya.

The Space Shuttle's military threat

By mid-1975 the reusable spacecraft had moved to the foreground as the next major step in the Soviet manned space program. Based on the evidence currently available, it would seem this decision was made mainly under pressure from the Soviet military community, which was becoming increasingly worried about the military potential of the US Space Shuttle. These concerns seem to have been triggered by several studies made at Soviet research institutes, including TsNIIMash. TsNIIMash specialists came to the conclusion that the Space Shuttle would never become economically viable if it was only used for the goals that NASA officially announced. As TsNIIMash director Yuriy Mozzhorin later said:

> "[The Space Shuttle] was announced as a national program, aimed at 60 launches per year ... All this was very unusual: the mass they had been putting into orbit with their expendable rockets hadn't even reached 150 tons per year, and now they were planning to launch 1,770 tons per year. Nothing was being returned from space and now they were planning to bring down 820 tons per year. This was not simply a program to develop some space system ... to lower transportation costs (they promised they would lower those costs tenfold, but the studies done at our institute showed that in actual fact there would be no cost savings at all). It clearly had a focused military goal."

In their opinion the Shuttle's 30-ton payload-to-orbit capacity and, more significantly, its 15-ton payload return capacity, were a clear indication that one of its main objectives would be to place massive experimental laser weapons into orbit that could destroy enemy missiles from a distance of several thousands of kilometers. Their reasoning was that such weapons could only be effectively tested in actual space conditions and that in order to cut their development time and save costs it would be necessary to regularly bring them back to Earth for modifications and fine-tuning [13].

A study often cited with respect to the origins of the Soviet shuttle program was performed at the Academy of Sciences' Institute of Applied Mathematics (IPM). Headed since 1953 by Mstislav Keldysh (President of the Academy of Sciences from 1961 to 1975), this institute had been involved in mission modeling and ballistics computations since the early days of the space program. The IPM studies were conducted under the leadership of Yuriy Sikharulidze and Dmitriy Okhotsimskiy, two of its leading scientists.

The IPM studies focused on the Shuttle's possible use as a bomber, more particularly its capability to launch a nuclear first strike against the United States. Efraim Akin, one of the institute's scientists, later recalled:

Space Shuttle Enterprise during pad tests at Vandenberg.

"When the US Shuttle was announced we started investigating the logic of that approach. Very early our calculations showed that the cost figures being used by NASA were unrealistic. It would be better to use a series of expendable launch vehicles. Then, when we learned of the decision to build a Shuttle launch facility at Vandenberg for military purposes, we noted that the trajectories from Vandenberg allowed an overflight of the main centers of the USSR on the first orbit. So our hypothesis was that the development of the Shuttle was mainly for military purposes. Because of our suspicion and distrust we decided to replicate the Shuttle without a full understanding of its mission.

When we analysed the trajectories from Vandenberg we saw that it was possible for any military payload to re-enter from orbit in three and a half minutes to the main centers of the USSR, a much shorter time than [a submarine-launched ballistic missile] could make possible (ten minutes from off the coast). You might feel that this is ridiculous but you must understand how our leadership, provided with that information, would react. Scientists have a different psychology than the military. The military, very sensitive to the variety of possible means of delivering the first strike, suspecting that a first-strike capability might be the Vandenberg Shuttle's objective, and knowing that a first strike would be decisive in a war, responded predictably" [14].

The report produced by the IPM scientists has never been made public, leaving unanswered many questions about the technical details of such a mission. Apparently, the Russians believed the Shuttle could drop bombs on Soviet territory while re-entering from a single-orbit mission from Vandenberg or by briefly "diving" into the atmosphere and then returning to orbit. As Energiya–Buran chief designer Boris Gubanov writes in his memoirs:

> "The studies ... showed that the Space Shuttle could carry out a return maneuver from a half or single orbit ..., approach Moscow and Leningrad from the south, and then, performing ... a "dive", drop in this region a nuclear charge, and in combination with other means paralyze the military command system of the Soviet Union." [15]

What lent this scenario particular credibility from the Russians' perspective was the Shuttle's 2,000 km cross-range capability, demanded by the Air Force to enable the Shuttle to return to Vandenberg after a single orbit around the Earth. However, such single-orbit missions from Vandenberg were not considered for a nuclear strike against the Soviet Union, but to quickly service polar-orbiting US spy satellites or even pluck enemy satellites from orbit, barely giving the Russians a chance to detect such operations with their space-tracking means [16]. Leaving aside the question whether such missions were feasible, the capability to return to Vandenberg after a single revolution was needed anyway to allow the Shuttle to perform a so-called "Abort Once Around" in case it ended up in an unacceptably low orbit after a main engine failure.

One can only guess what led the Russians to believe that the Shuttle had a nuclear first-strike capability. Possibly, they were "inspired" by their own plans for a so-called Fractional Orbit Bombardment System, an orbital nuclear weapons system designed to attack the US via the South Pole rather than passing through the net of radar systems at the northern approach corridor. The Soviet Union worked on three such systems in the 1960s, one of which (using Yangel's R-36 missile) actually reached operational status by the end of the decade.

Even though it bordered on paranoia, IPM's assessment of the Shuttle's first-strike capability is said by many to have been a decisive factor in convincing the Soviet leadership of the need to build an equivalent system (although a more rational reaction would probably have been to upgrade anti-missile defense systems). Gubanov writes:

> "On the basis of the results of the analysis, M.V. Keldysh sent a report to the Central Committee of the Communist Party, as a result of which L.I. Brezhnev, actively supported by D.F. Ustinov, took the decision to work out a set of alternative measures to guarantee the safety of the country." [17]

However, new evidence shows that Keldysh put his signature under the IPM report on 26 March 1976, which was more than a month *after* the official party and government decree that sanctioned the Soviet shuttle program [18]. Still, it cannot be ruled out that the studies began well *before* that time and that preliminary results did play some role in the Soviet decision to move forward with a Space Shuttle equivalent.

Mstislav Keldysh.

Keldysh, an influential figure in the Soviet space program until his death in 1978, seems to have been a major supporter of a Soviet shuttle system, which may seem surprising given his background as a scientist. However, rather than being a stereotypical armchair scientist, Keldysh had always been keen on putting his mathematical talents to practical use, making major contributions to Soviet strategic programs in his capacity as head of NII-1 (1946–1955) and IPM (1953–1978). His appointment as President of the Academy of Sciences in 1961 was seen as symbolizing the marriage of the Academy as the headquarters of fundamental science to the military–industrial complex. One joke that reportedly circulated among scientists was that "instead of representing the Academy in the Central Committee, Keldysh represented the Central Committee in the Academy" [19]. Roald Sagdeyev, the former head of the Academy's Institute of Space Sciences, recalls how Keldysh reacted when an Academy workshop was asked to formulate an opinion on the need to develop a shuttle:

> "Though we tried very hard, the workshop was unable to find even one single scenario in which the shuttle could provide a comparative advantage. Finally, I drafted a negative response to the government's request for the Academy of Sciences' opinion, in which I stated that the Academy did not see any sensible way to use this Russian version of the shuttle. Cautious Keldysh, however, did not want to get into conflict with the military, so he modified my wording, saying: 'We do not see any sensible scenario that would support the shuttle for *scientific uses* [author's stress]'." [20]

Moving towards approval

Whatever the motives, by the middle of 1975 a number of joint meetings between officials of the Ministry of General Machine Building and the Ministry of Defense resulted in a Soviet shuttle taking center stage in future plans for the country's piloted space program. There seems to have been particular pressure from GUKOS, headed at the time by Andrey Karas. The consensus by now was also that the vehicle should be similar in size to the Space Shuttle in order to respond to whatever threat the

American vehicle would eventually pose. It was also felt that the time needed to develop a big or small shuttle wouldn't be too different anyway.

From an economic and operational viewpoint, there was clearly no immediate need for the Soviet Union to build a shuttle, but in times of almost limitless budgets for defense-related programs any such considerations were easily outweighed by military arguments. Still, there was much division in the industry, mainly within NPO Energiya, on the need to press ahead. Therefore, GUKOS ordered the TsNII-50 research institute to perform a study of the military potential of such a system. Strangely enough, TsNII-50 head Gennadiy Melnikov, wishing to satisfy both camps, ordered preparation of *two* reports, one confirming the need to build a Shuttle equivalent, and the other demonstrating there was no need for such a system. The negative report was sent to the opponents of the reusable spacecraft and the positive report to the proponents. Eventually, however, both reports landed on the desk of Dmitriy Ustinov, who was dismayed to learn that two contradictory reports had been prepared by one and the same institute. Ustinov subsequently summoned Glushko to his office to clarify the situation, but Glushko, still not enthusiastic about a shuttle program, instead decided to send Valeriy Burdakov.

Burdakov, an avid shuttle supporter, had headed the shuttle team under Mishin, but after Glushko's arrival had been demoted to a position under shuttle chief designer Sadovskiy. Glushko's decision not to go himself and not even send Sadovskiy was his way of showing his lack of interest in the program, but it apparently had a boomerang effect. Burdakov and Ustinov talked at length about reusable spacecraft, with Ustinov showing particular interest in the military applications of such systems. Asked about the goals of the US Space Shuttle, Burdakov told Ustinov among other things about its capability to place giant laser complexes into orbit. The two agreed that much of the N-1 infrastructure at Baykonur (mainly the giant N-1 assembly building and the two launch pads) could be modified for use by a reusable spacecraft. The conversation ended with Ustinov ordering Sadovskiy's department to draw up a detailed report outlining the possible designs, missions, and operational aspects of a Soviet reusable space system [21].

Given Ustinov's influence, this order was more than a trivial matter and a considerable step on the road to final approval of a Soviet shuttle system. In

Dmitriy Ustinov.

September 1975 Ustinov convened a meeting at NPO Energiya, where it was agreed to speed up the release of a government and party decree on such a system, seen as the official endorsement of the program and the go-ahead to actually design and build the hardware [22]. In a letter dated 21 December 1975, KGB chief Yuriy Andropov once again reminded Ustinov of the Space Shuttle's military capabilities, emphasizing that its 30-ton payload capacity allowed it to orbit big spy satellites and space-to-ground weapons [23]. Roald Sagdeyev confirms Ustinov's role in the final decision to build a Space Shuttle equivalent:

> "I heard that [Buran] was adopted mainly due to insistence from Ustinov, who had made the following argument: if our scientists and engineers do not see any specific use of this technology now, we should not forget that the Americans are very pragmatic and very smart. Since they have invested a tremendous amount of money in such a project, they can obviously see some useful scenarios that are still unseen from Soviet eyes. The Soviet Union should develop such a technology, so that it won't be taken by surprise in the future" [24].

THE OFFICIAL GO-AHEAD

The decree (nr. 123-51) was finally issued by the Central Committee of the Soviet Communist Party and the Council of Ministers of the USSR on 17 February 1976 and called "On the Development of a Reusable Space System and Future Space Complexes". In the typical style of those days, the official go-ahead for the Soviet shuttle was literally worded as follows:

> "The Central Committee of the Soviet Communist Party and the Council of Ministers, attaching special importance to increasing the defense capabilities of the country and strengthening the work to create future space complexes for solving military, economic, and scientific tasks, has decided: (1) to accept the proposals of the Ministry of General Machine Building, the Ministry of Defense of the USSR, and the Academy of Sciences of the USSR to create a Reusable Space System consisting of a rocket boost stage, an orbital plane, an interorbital tug, a complex to control the system, launch, landing and repair complexes, and other ground-based means to launch into northeasterly orbits with an altitude of 200 kilometers payloads weighing up to 30 tons and return to the launch and landing complex payloads weighing up to 20 tons and with the purpose of:
>
> – counteracting the measures taken by the likely adversary to expand the use of space for military purposes;
> – solving purposeful tasks in the interests of defense, the national economy, and science;
> – carrying out military and applications research and experiments in space to support the development of space battle systems using weapons based on known and new physical principles;

Подлежит возврату в течение 24-х часов
в группу № 1
Особой части Управления Делами
Совета Министров СССР

ОСОБОЙ ВАЖНОСТИ

РАССЕКРЕЧЕНО

Центральный Комитет КПСС и Совет Министров СССР

ПОСТАНОВЛЕНИЕ

от 17 февраля 1976 г. № 132-51

Москва, Кремль

О создании многоразовой космической системы
и перспективных космических комплексов

Центральный Комитет КПСС и Совет Министров СССР, придавая особое значение повышению обороноспособности страны и усилению работ по созданию перспективных космических комплексов для решения военных, народнохозяйственных и научных задач, ПОСТАНОВЛЯЮТ:

1. Принять предложения Министерства общего машиностроения, Министерства обороны СССР и Академии наук СССР:

а) о создании многоразовой космической системы (МКС) в составе ракетной разгонной ступени, орбитального самолета, межорбитального корабля-буксира, комплекса управления системой, стартово-посадочного и ремонтно-восстановительного комплексов и других наземных средств, обеспечивающей выведение на северо-восточные орбиты высотой 200 километров полезных грузов весом до 30 т и возвращение на стартово-посадочный комплекс полезных грузов весом до 20 т и предназначенной для:

комплексного противодействия мероприятиям вероятного противника по расширению использования космического пространства в военных целях;

решения целевых задач в интересах Министерства обороны СССР, народного хозяйства и науки;

Экз. № 13

Секретарь
Центрального Комитета КПСС
Л.Брежнев

Председатель
Совета Министров СССР
А.Косыгин

The historic February 1976 government and party decree on Buran (*source*: OmV Luch/Russian Space Agency).

– putting into near-Earth orbits, servicing in these orbits, and returning to Earth space vehicles for different purposes, delivering to space stations cosmonauts and cargo and returning them back to Earth ...” [25].

Underscoring the military motives for developing the Soviet shuttle, the decree placed the Ministry of Defense in charge of determining the system's specifications. MOM was assigned as the lead ministry to develop the shuttle and associated space weaponry. MAP was tasked with developing the orbiter's airframe as well as building the runway and its associated equipment and the carrier aircraft to ferry the orbiter to the launch site.

Not surprisingly, Glushko's NPO Energiya became the lead organization for the shuttle under MOM, performing a role similar to that of a “prime contractor” in the West. Igor Sadovskiy remained the system's chief designer. The choice of an organization within MAP was less obvious. Probably, the most logical choice would have been MMZ Zenit's space branch in Dubna (now part of DPKO Raduga), which had been working on the Spiral project for a decade. However, MAP minister Pyotr Dementyev decided to set up a new organization called NPO Molniya, which was an amalgam of three existing design bureaus: MKB Molniya, MKB Burevestnik, and Myasishchev's Experimental Machine Building Factory (EMZ). Only EMZ had earlier experience with spaceplane projects, having proposed a futuristic single-stage-to-orbit spaceplane called M-19 in response to the Space Shuttle (see Chapter 9). Given the limited background of the three organizations in space-related work, some leading specialists of MMZ Zenit and the Dubna space branch were transferred to NPO Molniya to occupy the leading positions. These included Spiral chief designer Gleb Lozino-Lozinskiy, who was placed in charge of the new organization, and his deputy Gennadiy Dementyev, the son of the MAP minister. NPO Molniya was officially created on the basis of MAP orders dated 24 February and 15 March 1976 (see Chapter 4) [26].

The decree was not just restricted to the Soviet shuttle, it also sanctioned the development of multimodular space stations (what would eventually become Mir), a new type of Soyuz ship to transport cosmonauts to those stations (what later became Soyuz-T) as well as a system of geostationary data relay satellites called GKKRS (Global Space Command-Relay System) (what would eventually become Geyzer and Luch/Altair). Surprisingly, it also ordered NPO Energiya to work out a preliminary design in 1976–1978 for a “Lunar Expeditionary Complex” in 1976–1978, essentially a continuation of the Zvezda work begun in the middle of 1974. However, the project does not seem to have received much support and was closed down by a commission headed by Keldysh in 1978 [27]. In a final attempt to keep his lunar aspirations alive, Glushko tabled a proposal for a more modest manned lunar project using the Energiya rocket, but this never saw the light of day either [28].

However, if any Soviet cosmonauts were going to the Moon anytime soon, it was certainly not going to be on the N-1. With work on the ill-fated Moon rocket already suspended in 1974, the decree now officially terminated all work on the N-1/L-3 project, although it did call for using the N-1's cosmodrome infrastructure to the maximum extent possible.

Official approval of the Soviet shuttle came more than four years after President Nixon's Space Shuttle decision. In some ways this slow response was reminiscent of the Soviets' 1964 decision to go to the Moon, which was made more than three years after President Kennedy's announcement of the Apollo program. The official history of NPO Energiya gives both political and strategic motives for the decision:

> "... on the one hand [the Reusable Space System] was to consolidate the leading position of the USSR in the exploration of space and on the other hand [it] was to exclude the possible technical and military [advantage], connected with the appearance among the potential enemy of the ... Space Shuttle, a principally new technical means of delivering to near-Earth orbit and returning to Earth payloads of significant masses" [29].

While national prestige certainly played a role in the Buran decision, it was not as dominant as it had been in the Moon race. For one, there was no intention to upstage the US Space Shuttle. The maiden flight of the Soviet shuttle was planned for no earlier than 1983, which was four years later than the expected launch date of the first Space Shuttle. Clearly, the driving force behind Buran was the urge to maintain strategic parity with the United States. As far as the Russians were concerned, Buran was just another part of the Cold War. Another government/party decree in 1976 ordered NPO Energiya to begin studies of space-based weapons "for combat operations in and from space", in which the new heavy-lift launch vehicles and the shuttle would play a crucial role (see Chapter 6).

This is not to say there was unanimous support for the project among the military. The payloads for the shuttle and the super-heavy boosters derived from it were not clearly defined. With the benefit of hindsight, the official history of the Military Space Forces says:

> "There was no well-founded need for the USSR Ministry of Defense [to develop] such a system. Buran's main characteristics were close to those of the Space Shuttle and it had [the same] shortcomings, and moreover it was even less economical" [30].

Despite all the similarities, there was also a basic difference with the American Shuttle philosophy. The Space Shuttle was advocated as a system that would replace all existing expendable launch vehicles and launch all types of payloads (both government and commercial), a decision for which NASA had to pay dearly after the Challenger disaster in 1986. The Soviet shuttle was never intended to be a substitute for expendable launch vehicles, but a system that would be used exclusively for tasks that could not be handled by conventional rockets, such as the launch of heavy payloads and the maintenance and retrieval of satellites in orbit.

THE RLA ROCKET FAMILY

Even as officials were still pondering over the need to respond to the Space Shuttle, specialists were already busy figuring out what the Soviet equivalent should look like. Glushko had not come to NPO Energiya empty-handed. He and his engineers at Energomash had devised plans for a new family of heavy-lift launch vehicles called RLA, which stood for "Rocket Flying Apparatus". This was the same term that Glushko had used for some experimental liquid-fuel boosters he had developed way back in the early 1930s while working for the Gas Dynamics Laboratory in Leningrad. In Glushko's original vision, the Soviet shuttle was going to be just one payload for the RLA family.

With Glushko's background in engine development, it was logical that his initial efforts at NPO Energiya focused mainly on launch vehicles. Until then he had only concentrated on designing and building the rocket engines themselves, with other design bureaus (Korolyov, Chelomey, Yangel) being responsible for building the rockets that were powered by those engines. Now, with the merger of Energomash and TsKBEM, Glushko received the infrastructure and workforce to design not only engines, but also the rockets themselves.

The RLA plan revolved around two key concepts. First, it required the development of a new generation of powerful rocket engines using liquid oxygen (LOX) as oxidizer, and kerosene and hydrogen as fuel. Second, it envisaged the use of standardized rocket stages that could be assembled into different configurations tailored to the specific payloads to be placed into orbit.

Back to kerosene

The RLA's LOX/kerosene engines were to be the first such engines developed under Glushko in almost 15 years. In the mid to late 1950s Glushko had supervised the development of the RD-107/RD-108 engines for the R-7 missile and derived launch vehicles (sea-level thrust around 80 tons) and the RD-111 for the R-9 ICBM (sea-level thrust 144 tons). All of these were four-chamber LOX/kerosene engines using an open combustion cycle, in which the gases used to drive the turbopumps are vented overboard. This system is also known in Russian terminology as "liquid–liquid", because both the fuel and the oxidizer are injected into the combustion chamber in a liquid state. However, the development of the RD-111 was plagued by serious problems, including high-frequency oscillations in the combustion chamber, intermittent combustion, and the need to protect the chambers and nozzle walls from overheating.

In the early 1960s Glushko turned his attention to closed-cycle engines, in which the gases used for driving the turbines are routed to the combustion chamber to take part in the combustion process. This, together with the increased chamber pressure, produced much higher specific impulses than had been obtained earlier. One of the propellants entered the combustion chamber in a liquid form and the other in a gaseous form (which is why this system is also called the "gas–liquid" system by the Russians). Given the painful experience with the RD-111, Glushko was wary of using LOX/kerosene for these even more powerful engines. Instead, he decided to con-

The RD-107 LOX/kerosene engine (B. Hendrickx).

centrate on storable propellants based on unsymmetrical dimethyl hydrazine (UDMH), which he had already mastered while developing open-cycle engines for the R-12, R-14, and R-16 missiles. In fact, Glushko's preference for storable over cryogenic propellants can be traced back all the way to his years as a rocket pioneer at the GDL and RNII rocket research institutes in the 1930s.

All this had dire implications for the N-1 program. Glushko's reluctance to build closed-cycle LOX/kerosene engines and Korolyov's refusal to use the highly toxic storable propellants for the rocket effectively ended the cooperation between the two chief designers. It forced Korolyov to rely on LOX/kerosene engines of the much less experienced OKB-276 Kuznetsov design bureau in Kuybyshev (which actually *were* of the closed-cycle type).

For the remainder of the 1960s, Glushko was mainly engaged in building closed-cycle engines with storable propellants for a variety of missiles and launch vehicles of the Chelomey and Yangel bureaus. Except for the R-7 derived rockets, all Soviet space launch vehicles that were operational around the turn of the decade (Kosmos, Tsiklon, Proton) were powered by such engines. They had been derived from nuclear missiles, which traditionally use storable propellants to enable them to be launched at short notice.

Energomash didn't end its boycott on LOX/kerosene engines until the late 1960s, by which time enough experience had been gained with the closed-cycle combustion principle for engineers to feel confident enough to apply it in powerful LOX/kerosene engines. An opportunity to build such an engine arose in 1969, when the Chelomey bureau drew up plans for a mammoth rocket called UR-700M, intended to send Soviet cosmonauts to Mars. One version of the rocket that Chelomey looked into would have 600-ton thrust LOX/kerosene engines in the first and second stages. In 1970 Glushko's engineers worked out plans for such an engine called RD-116 or 11D120, which presumably was a modified LOX/kerosene version of the single-chamber RD-270, a hypergolic engine earlier planned for Chelomey's (unflown) UR-700 Moon rocket [31]. Although the UR-700M remained no more than a fantasy, Energomash was reportedly also ordered to investigate the possibility of using the same engine on the first stage of the N-1, which had suffered two launch failures in 1969. A small cluster of RD-116 engines would be enough to replace the N-1's thirty NK-15 first-stage engines [32].

In the end the idea was dropped because it would also have implied a radical redesign of the N-1 rocket. However, it does seem to have whetted Glushko's appetite to continue studies of such engines, the more so because a new policy was emerging in the early 1970s to abandon storable propellants in favor of cryogenic and hydrocarbon propellants in new, dedicated space launch vehicles.

As a result, work on high-thrust LOX/kerosene engines at Energomash resumed in earnest in 1973. The studies focused not only on standard kerosene, but also an advanced synthetic hydrocarbon fuel known as *tsiklin* or *sintin*. Based on furfural and propylene, it had a higher specific impulse than ordinary kerosene, but was also much more expensive.

In the course of 1973 proposals were presented for single-chamber, two-chamber, and four-chamber versions of a 500+ ton thrust LOX/kerosene engine. There was serious debate between the proponents of the single and four-chamber versions, which both had their advantages and drawbacks. A key meeting at Energomash in the second half of 1973 opted for the four-chamber version. After all, Energomash had had experience with multi-chamber engines since the 1950s. Furthermore, there had been numerous problems with the development of the 640-ton single-chamber RD-270 for the UR-700. Finally, by using four smaller combustion chambers it would be easier to test them by modifying test models of existing combustion chambers for storable propellants. The meeting also approved a so-called "modular design" for the engines, making it possible to use them in a standardized fleet of rockets [33].

Still, all these were no more than internal decisions within Energomash that didn't stand much chance of being implemented until the bureau merged with TsKBEM to form NPO Energiya in May 1974 and Glushko got the opportunity to advance his RLA idea. But even at this stage there was no consensus what the LOX/kerosene engines should look like. Some of the disagreements centered around such things as the pressure in the combustion chamber and the type of combustion cycle. Some claimed the pressure in the combustion chamber shouldn't exceed 200 atmospheres, making the engine more reliable. However, a lower pressure translates into bigger combustion chambers and less payload, and the compromise reached was to have a pressure of 250 atmospheres. Others felt the engine should use a fuel-rich combustion cycle, lowering the risk of turbopump burn-throughs. That was countered by the argument that an oxidizer-rich preburner engine is more efficient and easier to reuse because it leaves behind less soot residue [34].

There was also more fundamental debate over the thrust of the engine. Some felt that the task of building a four-chamber engine with a single, powerful turbopump assembly was too challenging and instead preferred single-chamber engines in the 150-ton thrust range with smaller, individual turbopumps. In other words, rather than having a handful of very powerful engines, it would be better to install a large number of low-thrust engines [35]. One concern with the high-thrust engines was that they would expose the rocket to serious vibrations in case of a sudden emergency shutdown, making it necessary to strengthen the rocket's structure and lower its payload capacity [36].

Bearing in mind these two schools of thought, two design departments at Energomash got down to studying engines in two thrust classes. Department 729 focused on engines ranging in thrust from 112.5 to 263.5 tons: the RD-128, RD-129, and RD-124 for the first stage of the RLA family and the RD-125, RD-126, and RD-127 for the second and third stages. Department 728 initially concentrated on an engine with a phenomenal thrust of 1,003 tons (the RD-150), but then scaled back its ambitions to a 600-ton thrust engine called RD-123 [37]. This is the engine that finally got selected in 1975 for use in the first stage of the Soviet space shuttle stack and the progenitor of the eventually developed RD-170. A determining factor in this choice must have been the negative experience of flying many low-thrust engines on the first stage of the N-1. Moreover, Glushko must have feared that if the choice did fall on the low-thrust engines, there would have been attempts to de-mothball the Kuznetsov bureau's N-1 engines rather than introduce his new LOX/kerosene engines. However, the debate would flare up again in the early 1980s when the RD-170 was plagued by serious development problems (see Chapter 6).

Introducing liquid hydrogen

Glushko's initial position was to use only hydrocarbon fuels in the RLA family and introduce liquid hydrogen (LH_2) at a later stage, when the technology was ripe.

Glushko had always disliked liquid hydrogen. In the 1960s he had opposed the use of liquid hydrogen on the upper stages of the N-1 rocket, arguing that the low density of hydrogen required large tanks and worsened the rocket's mass characteristics. At an August 1974 meeting where Glushko outlined his plans for the RLA rocket family, several participants urged him to move to liquid hydrogen straightaway, but Glushko remained adamant [38]. At another meeting he reportedly said:

> "The person who can find a way of building a rocket suited for the orbiter but with the use of oxygen–kerosene will become my deputy" [39].

However, by the end of the year Glushko had to yield to the pressure. On 30 November 1974 MOM minister Sergey Afanasyev signed an order to start the development of powerful cryogenic engines [40].

Despite Glushko's wariness, Energomash had already performed some initial research on LOX/LH_2 engines. In 1967 Glushko had tabled a proposal for a 200 to 250-ton cryogenic engine for the N-1 and a similar proposal had come from the Kuznetsov bureau [41]. Then there were studies at Energomash of two cryogenic engines for the RLA family, namely the RD-130 (200-ton vacuum thrust) in 1973 and the RD-135 (250-ton vacuum thrust) in 1974 [42]. Actually, the original idea was that Energomash would go on to build the engine, but the bureau was too preoccupied with the development of the powerful LOX/kerosene engines. Therefore, the task was entrusted to the Chemical Automatics Design Bureau (KB Khimavtomatiki or KBKhA) in Voronezh (the former "Kosberg bureau"). The deal was that KBKhA in turn would hand over to Energomash the development of a 85-ton thrust LOX/kerosene engine for the second stage of the medium-lift 11K77 ("Zenit") rocket [43].

KBKhA was not the most obvious choice. First, the only space-related engines developed by KBKhA before this had been LOX/kerosene upper stages for R-7 derived launch vehicles and engines burning storable propellants for the second and third stages of the Proton rocket. Second, there were two design bureaus in the Soviet Union that had already pushed research on LOX/LH_2 engines beyond the drawing board. These were KB Khimmash (the "Isayev bureau") and KB Saturn (the "Lyulka bureau"), both of which had developed cryogenic engines for the upper stages of the N-1 (the 7.5-ton thrust 11D56 of KB Khimmash and the 40-ton thrust 11D54 and 11D57 of KB Saturn). One can only speculate that Glushko had second thoughts about relying on design bureaus that had been involved in the N-1, a rocket he wanted to erase from history.

Not only was KBKhA a newcomer to the field of LOX/LH_2, it was now supposed to build from scratch a cryogenic engine several times more powerful than any developed in the Soviet Union before. With an anticipated vacuum thrust of 250 tons, the engine (called RD-0120) would even outperform the Space Shuttle Main Engine, which was related to the fact that the Russians had to compensate for the higher latitude of the Baykonur cosmodrome. Not surprisingly, KBKhA engineers began their work on the RD-0120 by consulting specialists from KB Khimmash and KB Saturn. They also extensively analysed the data available on the Space Shuttle Main Engines [44]. They almost certainly also benefited from the preliminary

The 11D56 engine (*source*: Igor Afanasyev).

research done by Energomash on the RD-135, which had exactly the same performance characteristics as the original version of the RD-0120.

A test bed for the RLA first stage

The RLA concept fitted well in a new philosophy to replace the existing fleet of Soviet launch vehicles by a new generation of rockets. By the early 1970s the Soviet Union was operating five basic types of fundamentally different launch vehicles each derived from a specific intermediate or intercontinental ballistic missile: the Kosmos and Tsiklon rockets (based on the R-12, R-14, and R-36 missiles of the Yangel bureau), the Vostok/Soyuz family (based on the R-7 missile of the Korolyov bureau) and the Proton (based on the UR-500 missile of the Chelomey bureau). While the R-7 based rockets used LOX/kerosene as propellants, all the others relied on storable hypergolic propellants.

Around the turn of the decade plans were being drawn up for new generations of satellites with more complex on-board equipment and longer lifetimes, requiring the use of heavier, more capable launch vehicles. By early 1973 a research program called Poisk ("Search"), conducted by the Ministry of Defense's main space R&D institute (TsNII-50), had concluded that future satellites should be divided into four classes: "light" satellites up to 3 tons, "medium-weight" satellites up to 10–12 tons, "heavy" satellites (up to 30–35 tons), and "super-heavy" satellites (not specified). The last three classes were not served by the existing launch vehicles.

The new family of launch vehicles was to have two key characteristics. First, in order to cut costs to the maximum extent possible, it would use unified rocket stages and engines. Second, it would rely on non-toxic, ecologically clean propellants, with preference being given to liquid oxygen and kerosene. The reasoning behind this reportedly was that "the number of launches of [space rockets] would be much higher than the number of test flights of nuclear missiles with [storable] propellants." What also may have played a role were a series of low-altitude Proton failures that had contaminated wide stretches of land at or near the Baykonur cosmodrome. The basic conclusions of the study were approved on 3 November 1973 at a meeting of GUKOS [45]. Although not stated specifically, the long-range goal of this effort seems to have been to phase out all or most of the existing missile-derived launch vehicles.

Initially, the primary focus apparently was on unifying the light to heavy class of rockets, because at the time these decisions were made super-heavy rockets and reusable shuttles were still a distant and vague goal. It would seem that three design bureaus were ordered to come up with proposals for such a family of launch vehicles under a competition called Podyom ("Lift"). Both Branch nr. 3 of TsKBEM in Kuybyshev (which became the independent TsSKB in July 1974) and Chelomey's TsKBM put forward plans to refit their respective Soyuz and Proton launch vehicles with the Kuznetsov bureau's LOX/kerosene NK engines originally built for the N-1 rocket. The former Yangel bureau in Dnepropetrovsk (now called KB Yuzhnoye and headed by Vladimir Utkin) also weighed the idea of using NK engines (which after all were around and had been tested), but in the end favored the new generation of LOX/ hydrocarbon engines being designed by Energomash [46].

Actually, by this time Yuzhnoye had been working for several years on a new medium-lift launch vehicle (11K77) burning *storable* propellants. In December 1969 it had been ordered by GUKOS to develop a new rocket capable of placing 8 tons into low polar orbits and 2 tons into highly elliptical Molniya-type orbits. The initial idea in 1970 was to build a three-stage rocket with 3.6 m diameter modules. The following year attention turned to a rocket derived from the bureau's R-36M, a new ICBM that had been approved by a government decree in September 1969. By retaining the 3.0 m diameter of the R-36M's rocket stages, no fundamentally new manufacturing techniques would be required. The 11K77 would now consist of a pair of R-36M first and second stages stacked on top of one another plus a newly developed third stage. In 1972 Yuzhnoye also designed a slightly less powerful rocket designated 11K66, essentially a two-stage R-36M with increased propellant load capable of orbiting 5.9 tons.

When Yuzhnoye made the switch to LOX/kerosene under the Podyom program in 1974, it opted for a vehicle maintaining the 3.0 m diameter of the rocket stages, but employing parallel staging. This version of the 11K77 would fly the single-chamber LOX/kerosene engines then being designed at Energomash. The second stage, acting as the core, would have a 130-ton thrust RD-125 engine, and the two first-stage modules flanking the core would each have three 113-ton thrust RD-124 engines. These engines had combustion chambers of roughly the same size and pressure as those of the R-36M first-stage engine. Payload capacity to low orbit was now 12 tons.

By early 1975 Energomash's Department 728 had made enough progress on the powerful 600-ton thrust RD-123 for Yuzhnoye to incorporate it into the 11K77 design. This made it possible to replace the two modules of the 11K77's first stage by a single module, although its diameter would have to be increased to 3.9 m, which was the maximum that could be transported by rail. The second stage would now be placed on top the first stage, turning the 11K77 into what the Russians call a "mono-block" booster. After a convoluted development path, the rocket had now acquired the configuration in which it later became known to the world as Zenit. It could also serve as the basis for a lighter rocket (11K55) and a heavier version (11K37), with the three covering the whole payload range from "light" to "heavy" (see Chapter 8) [47].

Being in the heavy to super-heavy class, Glushko's RLA family was not part of the Podyom competition, but as plans for a large Soviet shuttle gained more support in 1975, so did the idea of unifying the first stage of the RLA rockets with that of the future medium-lift rocket. It is unclear if this consideration played a significant role early on in the Podyom competition. If it did, TsSKB and TsKBM had betted on the wrong horse by anticipating that whatever followed the N-1 would also carry NK engines, while Yuzhnoye had made the right choice by picking the new Energomash engines. However, the history of the 11K77/RLA unification is a complicated chicken-and-egg story, with sources differing on whether the initiative to unify the first stages came from Glushko or Utkin. At any rate, Yuzhnoye emerged as the winner of the Podyom competition in 1975, no doubt because its medium-lift rocket could now act as a test bed for the first stage of the rocket that would power the Soviet space shuttle to orbit.

The 11K77 is the only thing that ever came out of Podyom. Yuzhnoye's proposed light-class and heavy-class rockets never flew and the whole idea of a standardized fleet of dedicated, environmentally clean space launch vehicles in the light to heavy class remained a distant dream, probably because it infringed too much on design bureau interests. In fact, most of the IRBM and ICBM-derived launch vehicles conceived in the 1960s continue to fly today, although another attempt is being made to develop a standardized rocket fleet under the Angara program (see Chapter 8).

RLA variants

The basic configuration of the RLA rockets was a common core stage complemented by a different number of standard first-stage strap-on boosters, depending on the mass of the payload. Very little has been revealed about the RLA launch vehicles studied in 1974–1975 and various sources have also given different designators for rockets with similar capabilities. Apparently, the design evolved significantly even during that short period, dictated by the progress made in the concurrent research on kerosene/hydrocarbon and hydrogen engines. It would appear that original plans for large clusters of low-thrust engines eventually gave way to small clusters of high-thrust engines as the confidence in the latter grew.

Glushko is known to have presented plans for three RLA rockets (RLA-120, 135, and 150) during a meeting on 13 August 1974, which was attended by most of the chief designers and also by Dmitriy Ustinov. This was Glushko's *third* RLA proposal in the barely three months he had been in office at NPO Energiya. The August 1974 RLA plans revolved around the exclusive use of kerosene and *sintin*, the 1,003-ton thrust RD-150 engine, and massive 6 m diameter rocket modules. The RLA-120, expected to be ready in 1979, would have a payload capacity of about 30 tons and among other things launch modules of a permanent space station. The RLA-135 had a 100-ton payload capacity and could be used to orbit a reusable space shuttle or elements of a lunar base and was expected to make its debut in 1980. Finally, there was the massive RLA-150, capable of placing up to 250 tons into low orbit and seen by Glushko as the rocket that would eventually send Soviet cosmonauts to Mars. Its first flight was anticipated in 1982 [48].

Other sources have identified three rockets known as RLA-110 or Groza ("Thunderstorm"), RLA-120 or Grom ("Thunder"), and RLA-130 or Vulkan ("Volcano"). The RLA-110, equipped with two boosters, would have a payload capacity "higher than the Proton rocket". The RLA-120, using four boosters, would have about the same payload capacity as the N-1. Finally, the RLA-130, toting eight boosters, would play a key role in establishing a Soviet lunar base [49].

While Glushko's RLA plan may have looked very appealing on paper, upon closer analysis it did raise the necessary questions among fellow designers. Some warned that because of the unification not all the launch vehicles in the series would be the most efficient in their particular payload class. The design of the common core stage had to be tailored to the heaviest 250-ton class booster, which was the one expected to fly least. The implication was that the core stage was oversize for the 30-ton class RLA, exactly the one that would probably be launched most frequently [50]. This is probably the very reason preference was eventually given to Yuzhnoye's 11K37 "heavy Zenit" to fill this niche in the payload spectrum. Some felt that a better way of developing a standardized rocket fleet was to first fly a light booster, then use its first stage as the second stage for a heavier booster, subsequently turn that second stage into a third stage for an even heavier rocket, etc. [51]. This was the approach that Korolyov and Chelomey had suggested for their respective N-I/N-II/N-III and UR-200/UR-500/UR-700 families in the 1960s. One big disadvantage, however, was that each vehicle in the fleet would require its own launch pad.

In the end, the only rocket that emerged from the RLA plans was the 100-ton class booster that later became known as Energiya. Proposals were later tabled for Energiya derivatives such as Energiya-M (30-ton class), Groza (60-ton class), and Vulkan (200-ton class), but these never made it off the ground (see Chapter 8). And so the dream of a standardized rocket fleet in the heavy to super-heavy class never materialized either, although the main reason here was the absence of payloads to justify its existence.

CONFLICTING CONFIGURATIONS

By the middle of 1975 two competing designs had emerged within NPO Energiya for a Soviet response to the Space Shuttle. In Glushko's vision the orbiter would be just one of the payloads for his RLA rockets, mounted on top of the rocket as a conventional payload. If a winged orbiter was going to be mounted atop an RLA booster, it would place very high loads on the core stage, especially during the phase of maximum aerodynamic pressure. Therefore, the core stage would have to be strengthened, making it even heavier than it already was and decreasing the rocket's payload capacity [52]. Therefore, the top-mounted orbiter would have to be a wingless, vertical-landing lifting body. This configuration was backed by NPO Energiya luminaries such as Boris Chertok, Yuriy Semyonov, and Konstantin Bushuyev, who were convinced the USSR was not capable of building a reusable space transportation system akin to the Space Shuttle [53]. It would also eliminate some thorny organizational problems, requiring minimal involvement from the Ministry of the Aviation Industry.

Another option under consideration was to mimic the Space Shuttle as closely as possible, namely to build a winged orbiter with main engines which would be strapped to an external fuel tank with strap-on boosters. Known as OS-120, it would enable the Russians to benefit from research and development done in the US and thereby minimize risk. While backed by Igor Sadovskiy, it was Glushko's nightmare, since this design left no room for the family of launch vehicles he had been dreaming of for many years. The philosophy behind the OS-120 was that the Soviet Union would solely be able to match the US Shuttle's capability to place 30 tons into orbit and return 20 tons back to Earth and nothing more. After all, the 100 to 200-ton payload capacity of the heavy RLA rockets, mainly needed for establishing lunar bases and staging manned interplanetary missions, was of little interest to the Soviet military.

The MTKVP lifting body

Called the Reusable Vertical Landing Transport Ship (MTKVP), the lifting body was a 34 m long vehicle consisting of three main sections: a front section with the crew cabin, a mid-section containing a huge payload bay, and an aft section with orbital maneuvering engines. After using its limited aerodynamic characteristics during the hypersonic stage of re-entry, the vehicle would deploy a series of parachutes at an

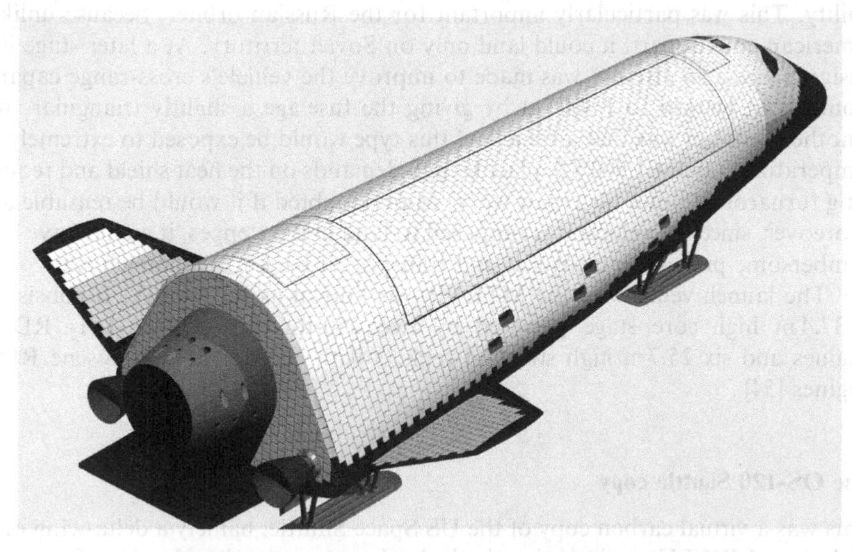

The MTKVP lifting body (*source*: *www.buran.ru*).

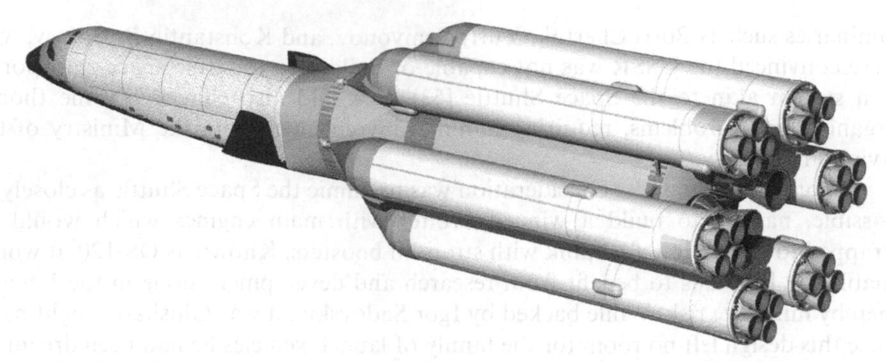

The MTKVP sitting atop the RLA-130V launch vehicle (*source*: *www.buran.ru*).

altitude of 12 km and a speed of 250 m/s. Vertical landing speed would be dampened with small soft-landing engines and horizontal speed with a ski landing gear.

One of the big advantages of this design was that the ship did not need expensive runways, although some of the plans did envisage landing on a prepared dirt surface. The absence of wings, which are dead weight for most of the flight anyway, also saved a lot of mass. The MTKVP would weigh 88 tons and have a payload capacity of 30 tons to a low 50.7° inclination orbit and a return capacity of 20 tons. Moreover, the MTKVP could rely on proven technologies such as other aerodynamically shaped objects (in particular, the Soyuz descent capsule and nuclear warheads) and parachute and soft-landing systems that had been used for some years by airborne troops to safely land heavy cargos. The idea also was to retrieve the RLA strap-on boosters in a similar fashion, leading to additional cost savings.

However, a major disadvantage of the MTKVP was its low cross-range capability. This was particularly important for the Russian orbiter, because unlike its American counterpart, it could land only on Soviet territory. At a later stage in the design process an attempt was made to improve the vehicle's cross-range capability from about 800 km to 1,800 km by giving the fuselage a slightly triangular shape. Another problem was that a vehicle of this type would be exposed to extremely high temperatures (about 1,900°C), placing high demands on the heat shield and requiring long turnaround times for repair work. Many doubted if it would be reusable at all. Moreover, since the vehicle was supposed to land in the steppes, it would have been a cumbersome process to recover it and transport it back to the launch site.

The launch vehicle for the MTKVP was known as RLA-130V. It consisted of a 37.4 m high core stage powered by two 250-ton thrust LOX/LH_2 RD-0120 engines and six 25.7 m high strap-on rockets with 600-ton LOX/kerosene RD-123 engines [54].

The OS-120 Shuttle copy

This was a virtual carbon copy of the US Space Shuttle, namely a delta-wing orbiter with three LOX/LH_2 main engines in the back and strapped to the side of an external

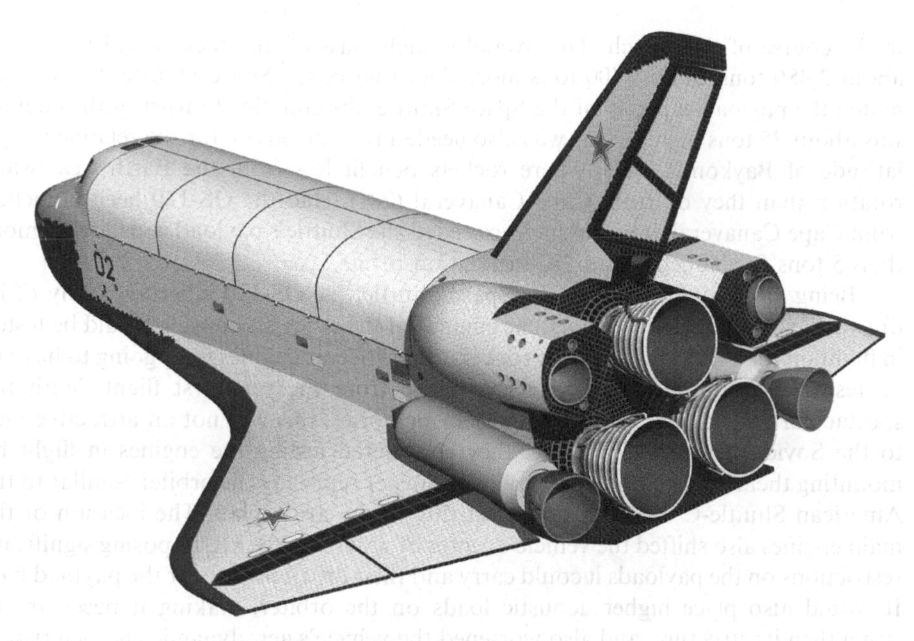

The OS-120 orbiter (*source*: *www.buran.ru*).

fuel tank. Sadovskiy's team even went as far as studying the use of large solid-fuel rockets. Sadovskiy was no newcomer to solids, having earlier headed the development of two large solid-fuel nuclear missiles (the RT-1 and RT-2), the only such rockets ever built at the Korolyov bureau. This may even have been the very reason he was placed in charge of shuttle development at NPO Energiya.

However, the idea to use solids was abandoned because of the absence of the necessary industrial basis for the development of large solid-fuel rockets and the equipment to transport the loaded boosters to the launch site. Also, it would have been difficult to operate the boosters in the temperature extremes of Baykonur. Still, the idea to use solids was briefly reconsidered in the early 1980s as the RD-170 suffered serious development problems (see Chapter 6).

Instead, it was decided to use four 40.75 m high LOX/kerosene strap-ons, each powered by a single RD-123 engine. The orbiter itself would be fitted with three RD-0120 LOX/LH_2 engines with a vacuum thrust of 250 tons. The orbital maneuvering system engines and reaction control system thrusters were arranged almost identically as on the US Orbiter (in two aft pods and a forward module) and would also use hypergolic propellants, the only difference being that the fuel on the Soviet vehicle would be dimethyl hydrazine rather than monomethyl hydrazine. The most striking novelty on the Soviet vehicle was the presence of two jettisonable 350-ton thrust solid-fuel escape motors on the aft fuselage that would have allowed it to instantly separate from the external tank in case of a launch accident.

The OS-120 orbiter owed its name to its 120-ton mass, which was the launch mass with a full 30-ton payload and minus the 35-ton emergency escape system, jettisoned in the course of the launch. The overall launch mass of the stack would have been about 2,380 tons, almost 400 tons more than that of the Space Shuttle. In order to match the payload capacity of the Space Shuttle, the combined thrust of the engines was about 75 tons higher. This was also needed to compensate for the relatively high latitude of Baykonur (45°), where rockets benefit less from the Earth's eastward rotation than they do from Cape Canaveral (28°). Had the OS-120 been launched from Cape Canaveral, it would have exceeded the Shuttle's payload capacity by more than 5 tons for launches into 28° inclination orbits.

Being virtually identical to the Space Shuttle, the OS-120 inherited many of its drawbacks. While the LOX/kerosene engines of the strap-on boosters could be tested in flight on the 11K77 medium-lift rocket, the cryogenic engines were going to have to be tested with the priceless orbiter in place from the very first flight. With the spectacular N-1 failures still fresh in their memories, this was not an attractive idea to the Soviet planners. Therefore, they considered testing the engines in flight by mounting them on an unmanned payload canister replacing the orbiter (similar to the American Shuttle-C configuration), but this was a costly plan. The location of the main engines also shifted the vehicle's center of gravity to the aft, imposing significant restrictions on the payloads it could carry and their distribution over the payload bay. It would also place higher acoustic loads on the orbiter, making it necessary to strengthen its structure, and also worsened the vehicle's aerodynamic characteristics.

The OS-120 design also posed problems specific to the Soviet situation. If the engines were going to be on the orbiter, they would have to be made reusable, making their development even more challenging than it already was for an industry with very limited experience in cryogenic engine technology. Perhaps even more significantly, the orbiter would be so heavy that it could not be transported by any of the aircraft available at the time. The only aircraft capable of doing so, the Antonov design bureau's An-124 "Ruslan", was only on the drawing boards and years away from its first flight. Such an aircraft would also be needed for atmospheric drop tests, similar to the ones performed by NASA with Enterprise and a modified Boeing 747 [55].

Seeking a compromise

Eventually, Sadovskiy's Department 16, then numbering fewer than 80 people, got down to working out a compromise plan that would satisfy all players. After several weeks of work, they came up with a Space Shuttle type configuration with a side-strapped winged orbiter (OK-92), but with the engines mounted on the "external tank" rather than on the orbiter itself. This turned the external tank and the strap-on boosters into a universal launch vehicle capable of flying not only the orbiter, but other payloads as well. Moreover, the number of strap-ons could be varied to match the required payload. On the one hand, the plan made it possible to build an orbiter very similar to the American one (and thereby benefit from American R&D) and, on the other hand, it allowed Glushko to retain his beloved family of launch vehicles.

The report prepared by Sadovskiy's team was called "Reusable Space System With The Orbital Ship OK-92" and contained a comparative analysis with the OS-

120 and MTKVP. Before being sent to Ustinov, it needed to be endorsed by Glushko. Even though the new design went a long way to accommodate his wishes, Glushko realized his signature would probably be the death warrant for his lunar program. Finally, Burdakov was able to talk him around, arguing among other things that Glushko would still go down in history as the man having built the most powerful rocket engine in the world. On 9 January 1976 Glushko signed the report, albeit with mixed feelings. He even called it the "Bloody Sunday" of Soviet cosmonautics, referring to an incident where unarmed, peaceful demonstrators marching to present a petition to Tsar Nicholas II were gunned down by Imperial guards in St. Petersburg on 9 January 1905, exactly 71 years earlier [56].

Perhaps another reason Glushko came around was that the new plan enabled him to deal one final deadly blow to the N-1. Even though work on the N-1 had been suspended in 1974, the project had not yet been officially terminated. Boris Dorofeyev, the chief designer of the N-1, had even prepared an address to the 25th Congress of the Soviet Communist Party (to be held in February 1976) calling for the N-1 program to be resurrected. With the most likely payload for the N-1 now being a shuttle-type vehicle, the very same arguments against mounting a winged orbiter on top of the RLA could now be used against the N-1 as well [57]. Later the Russians would justifiably describe the absence of main engines on the orbiter as one of their system's main advantages, although they rarely or never pointed out that this design had not been a foregone conclusion from the beginning.

The OK-92

The new launch vehicle was called RLA-130. While there were still four strap-on boosters with 600-ton RD-123 engines, the three 250-ton thrust RD-0120 engines were now on the "external tank" rather than on the orbiter. The overall launch mass was the same as that of the OS-120 (2,380 tons), but the orbiter now weighed just 116.5 tons at launch as compared with 155.35 tons for the OS-120.

Now devoid of main engines, the OK-92 orbiter was to be equipped instead with two kerosene-fueled D-30KP turbojet engines mounted in external pods on either side of the aft fuselage. Widely used on the Ilyushin-62 passenger airliner, these would have to give the orbiter more flexibility in reaching the runway. Also, with a shorter landing roll-out (2.5–3 km vs. 4 km for the OS-120), the vehicle could land on many ordinary runways in the USSR. The engine inlets and outlets would be protected from the vacuum of space and the heat of re-entry by jettisonable covers. The engines were to be activated at an altitude of between 5 and 8 km.

The orbital maneuvering engines and aft reaction control thrusters were now installed in pods adjacent to those housing the D-30KP engines. The nozzles of the forward reaction control thrusters were protected during launch and re-entry by a special cover. As on the OS-120, the propellants to be used by the engines would be hypergolic. The orbital maneuvering engines were to be derived from the 15D619 engine used on the second stage of KB Yuzhnoye's UR-100 ICBM. The ultimate goal, however, was to replace the toxic propellants by a combination of hydrogen peroxide (H_2O_2) and kerosene, the latter of which could then be used both by the

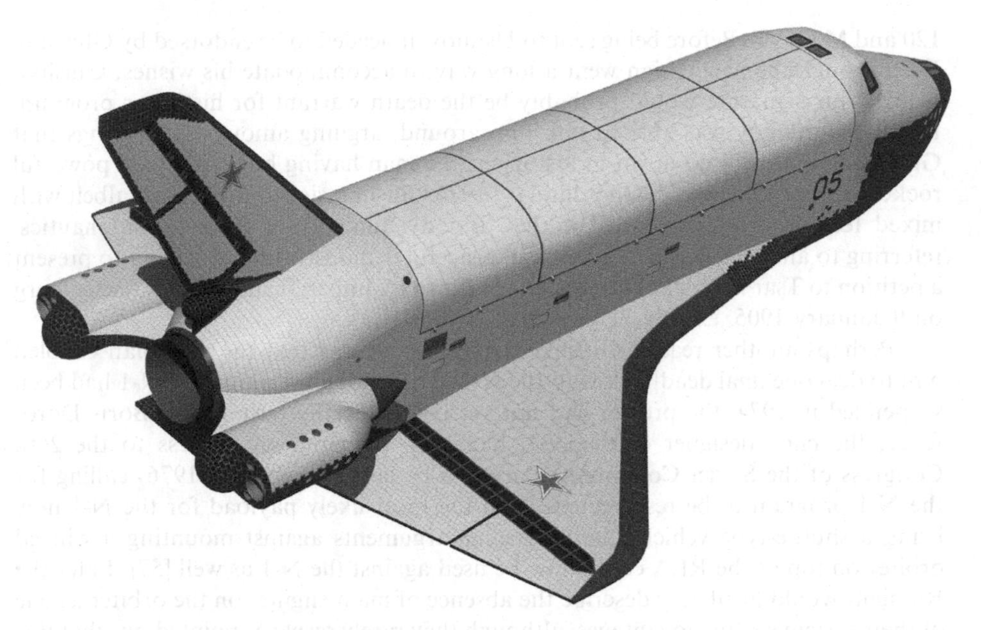

The OK-92 orbiter (*source*: *www.buran.ru*).

D-30KP turbojet engines and the on-orbit propulsion system. For that combination engineers would draw on the experience gained with the RD-510, an H_2O_2/kerosene engine developed at Energomash for the lunar module of the canceled N-1/L-3M manned lunar project.

OK-92 retained a solid-fuel emergency escape system for early launch aborts, but it now consisted of a single engine installed under the vertical stabilizer. This would be jettisoned 56 seconds into the launch, after which the vehicle would have gained enough speed and altitude to reach the runway with its turbojet engines after an emergency separation from the rocket. In case they were needed in a launch abort, the D-30KP engines could be activated in about 30–50 seconds.

Another feature which set OK-92 apart from the US Orbiter was the use of two remote manipulator arms to deploy payloads from the cargo bay. It was also planned to use the arms in docking operations, pretty much like the Space Shuttle Endeavour used its mechanical arm to dock the Unity module with Zarya during the first Shuttle ISS assembly mission (STS-88) in 1998.

Prior to the maiden orbital mission, the Russians were planning to carry out an extensive series of atmospheric approach and landing tests. In the first stage many on-board systems (such as the propulsion and emergency escape engines, various cabin systems, the remote manipulator arms, etc.) were to be replaced by mock-ups and the vehicle could take off either on its own power or (if its mass was reduced to 60–80 tons) on the back of the An-22 "Antey" aircraft, which would then release it at an altitude of about 2 km. If the mass was reduced to 51–60 tons, the An-22 could also be used to transport the orbiter over a distance of about 2,000 km at an altitude of 2 km. Both for the drop tests and the ferry flights the OK-92's D-30KP turbojet engines could be used to assist during take-off and the climb to cruise altitude.

For the second stage of the landing tests the OK-92 would be outfitted with most of the systems needed for orbital flight, including the propulsion system and the solid-fuel escape motor. Now too heavy to be carried by the An-22, the OK-92 would fly all the remaining atmospheric test flights on its own power. Plans called for using the combined thrust of the turbojet engines, orbital maneuvering engines, and the solid-fuel motor to take the vehicle to an altitude of 21 km and a speed of 1,800 km/h (Mach 1.5) to simulate the final portion of its mission. This was far higher and faster than NASA had been able to do with Enterprise. At this point in time the Russians did not plan a dedicated atmospheric test vehicle, but one that would later be modified into a spaceworthy orbiter. At the time, NASA was planning to do exactly the same with Enterprise, until it was realized in late 1977 that it would be cheaper to turn Structural Test Article STA-099 into the second flight article (what became OV-099 Challenger).

The turbojet engines would also have allowed the OK-92 to fly on its own power to the Baykonur cosmodrome either from the manufacturer or from back-up runways. The range would have been 1,600 km at a cruise altitude of 3 km. However, that could have been increased to 3,000 km by increasing the fuel supply for the engines and using the emergency solid fuel motor as an afterburner on take-off [58].

FREEZING THE DESIGN

Although the decree of 17 February 1976 constituted the formal approval of the Soviet shuttle, it did not stipulate what type of design should be chosen. It merely endorsed the basic requirements for the system laid down earlier by the military (30 tons up, 20 tons down). By the time the decree was passed, the OS-120 and MTKVP concepts had been pretty much abandoned and engineers had settled on the January 1976 OK-92 plan, namely a winged orbiter strapped to the side of a massive launch vehicle consisting of a core stage with three RD-0120 cryogenic engines and four strap-on boosters with one RD-123 LOX/kerosene engine each.

A change made soon afterwards was to increase the number of main engines on the RLA-130 core stage to four and reduce their vacuum thrust from 250 to 190 tons. The additional engine provided extra redundancy in case of a main engine failure during the climb to orbit [59]. At the same time, the sea-level thrust of the LOX/kerosene engines in the strap-on boosters was increased from 600 to 740 tons, resulting in an improved engine called the RD-170. The RLA-130 had now almost acquired the configuration that would eventually become known as Energiya.

Orbiter: a Space Shuttle twin or a big Spiral?

Although NPO Energiya's favored design was by now a delta-wing vehicle virtually identical in shape to the US Space Shuttle Orbiter, the Mikoyan engineers that had been transferred to NPO Molniya had their own ideas. In February 1976 Yuriy

Blokhin, the head of DPKO Raduga's space design bureau in Dubna, had already written a report for the Central Committee stating that the 75 million rubles invested in Spiral were the only practical basis in the USSR for the creation of a reusable space transportation system [60]. However, by this time the designers were bound by the payload requirements and mission goals for the shuttle spelled out by the February 1976 party/government decree and the small air-launched Spiral was way below specifications. Therefore, the former Mikoyan engineers hatched a plan to build a much enlarged version of the Spiral lifting body capable of carrying the required amount of payload and launch that with the new heavy-lift rocket [61].

In the weeks after the decree was issued the "big Spiral" (code-named "305-1") and NPO Energiya's delta-wing orbiter ("305-2") were the subject of a comparative analysis carried out by NPO Energiya, NPO Molniya, TsAGI, and TsNIIMash. There seems to have been division within the newly created NPO Molniya itself, with the former Mikoyan people strongly lobbying for the Spiral-based system and the Myasishchev branch supporting the delta-wing orbiter. In fact, NPO Energiya designers had been consulting with Myasishchev's specialists on the delta-wing design since 1974. The former Mikoyan engineers, backed by TsAGI, pointed to the significant experience accumulated over the past ten years in research on Spiral (BOR missions, wind tunnel tests, etc.), while the NPO Energiya people argued that copying the shape of the American vehicle could save at least two years, reasoning it made no sense "to re-invent the wheel". The whole matter turned into a fierce debate between Lozino-Lozinskiy and Sadovskiy, which was eventually put to rest by MAP minister Dementyev, who left the final decision to Glushko. Being no expert in aerodynamics, Glushko in turn delegated the decision to the Council of Chief Designers, which by a simple majority of votes selected NPO Energiya's delta-wing vehicle on 11 June 1976 [62]. However, one source claims the final decision was not made until late 1978 [63].

The January 1976 OK-92 plan was taken as the basis for the orbiter's design, but a couple more changes were made. The solid-fuel emergency separation motor was removed from the orbiter and the hypergolic propellants for the orbital propulsion system were replaced by LOX/*sintin*, with all the engines drawing their propellant from common tanks. The two D-30KP turbojet engines were replaced by a pair of Lyulka AL-31 turbojet engines, already under development at the time for use on the Sukhoy Su-27 fighter. They were mounted in special niches on either side of the vertical stabilizer and covered with thermal protection material. The engines were eventually installed on a full-scale test model of Buran (BTS-002) used for approach and landing tests in 1985–1988, which also had two additional afterburner-equipped versions of the engine to take off on its own power. However, only months before the maiden space mission of Buran, it was decided not to install the AL-31 engines (see Chapter 7).

Boosters: marrying medium-lift and heavy-lift

By 1976 the design of the RLA-130 had evolved such that it was becoming an ever more daunting task to unify the design of the strap-on boosters and the first stage of

Soviet shuttle evolution. From left to right : OS-120, OK-92, and Buran (*source*: *www.buran.ru*).

the 11K77 medium-lift launch vehicle. For one, the strap-ons would now carry the 740-ton thrust RD-170, providing much more muscle than what was needed for the 11K77 first stage and shifting the latter's impact zone in Kazakhstan [64]. However, the biggest problem was that because of their location in the RLA-130 stack the strap-ons experienced high bending loads, which made it necessary to make the tank walls tougher than was necessary for the 11K77. NPO Energiya proposed to construct the tanks out of a lightweight, but strong 1201 aluminum alloy that was also used for the core stage tanks, but KB Yuzhnoye's manufacturing facility did not possess the welding technology needed to build such tanks [65].

At one point the differences seemed so irreconcilable that KB Yuzhnoye chief designer Vladimir Utkin was on the verge of ending his co-operation with NPO Energiya. His bureau was simply not up to the task of building two fundamentally different rocket stages. To Glushko the idea of using the 11K77 first stage as a test bed for the strap-ons was so critical that Yuzhnoye's withdrawal jeopardized the very existence of the RLA-130. There simply was no other organization that could build the strap-ons. Fortunately, Glushko and Utkin were able to hammer out a compromise during an exhausting two-day meeting at KB Yuzhnoye in Dnepropetrovsk [66]. The tank walls would be made of an aluminum–magnesium alloy with a waffle-grid structure and would be somewhat thinner on the 11K77. Other than that, the 11K77 first stage would use virtually the same engines and propellant feed systems. In the end, the commonality between the stages was about 70–75 percent.

On 16 March 1976 the Soviet Communist Party and government issued another decree, which gave KB Yuzhnoye the final go-ahead for the development of the 11K77 rocket. Coming just about a month after the decree on the Soviet shuttle, it seems to have formalized agreements that had been made in the previous months on unifying the 11K77 and RLA-130 designs. That same month Yuzhnoye gave Energomash the required parameters for the first and second-stage engines. The first-stage engine, the RD-171, was almost identical to the RD-170 of the RLA-130 with the exception that it could be gimballed in only one axis rather than two.

The second stage was to be powered by the single-chamber RD-120 LOX/kerosene engine. Its development was assigned to Energomash rather than KBKhA

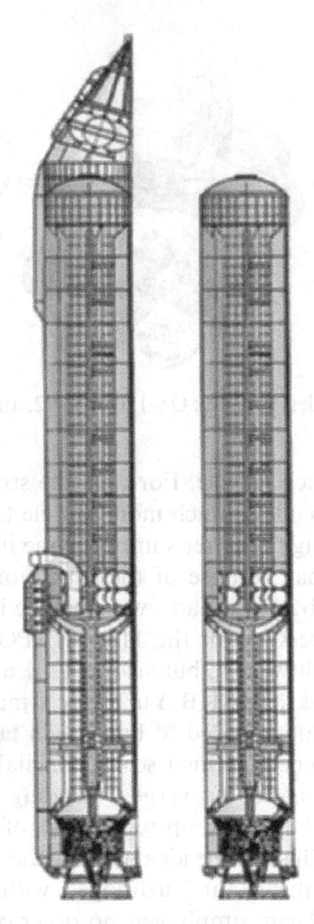

Energiya strap-on booster (left) and Zenit first stage (*source*: *www.buran.ru*).

in Voronezh, with the latter getting the RD-0120 in return. One of the reasons was that the development of this engine should get underway as soon as possible so that engineers might learn the necessary lessons for the larger RD-170/171. In support of its test program for the heavy-thrust first-stage engine, Energomash had already developed a prototype LOX/kerosene engine that had characteristics very similar to those required for the Zenit second stage. This was based on the RD-268 nitrogen tetroxide/UDMH engine, which was being serially produced for the first stage of Yuzhnoye's MR-UR-100 ICBM. Finally, the second stage was also to carry a four-chamber RD-8 vernier engine for thrust vector control, developed in-house at KB Yuzhnoye [67].

The preliminary design ("draft plan") for the 11K77 was finished in February 1977. The March 1976 decree had called for a maiden flight in the second quarter of 1979, but numerous problems (mainly with the first-stage engine) would eventually delay that until April 1985 (see Chapter 6).

Core stage: one or two sections?

On 30 July 1976 Dmitriy Ustinov officially appointed GUKOS as the military organization in charge of the program. Ustinov had been promoted to Minister of Defense in April 1976, but continued to serve as Central Committee Secretary for Defense Matters until October 1976, when he was replaced in that capacity by Yakov Ryabov. In co-operation with NPO Energiya, GUKOS worked out updated technical requirements for the Soviet shuttle system throughout the year. These were approved by Ustinov himself on 8 November 1976, reportedly the first time this had been done at such a high level and again underscoring the military objectives of the Soviet shuttle program. Nothing changed to the basic requirement mentioned in the February 1976 decree, namely the capability to place 30-ton payloads into 200 km orbits inclined 51.6° to the equator and return 20-ton payloads back to Earth. Added to the requirements was a payload capacity of at least 16 tons for missions into 97° inclination orbits, roughly matching the payload capacity of the Space Shuttle from Vandenberg. The orbiter was supposed to be able to fly a total of 100 missions, another requirement identical to that for the US orbiter. The strap-on boosters were expected to fly at least ten times each, creating an inventory of boosters that would help the system achieve a wildly ambitious minimum turnaround time of 20 days [68].

On 12 December 1976 Glushko placed his signature under the "draft plan" for the Soviet shuttle system, but he did not have the final say. The draft plan was reviewed by an Interdepartmental Expert Commission chaired by TsNIIMash director Yuriy Mozzhorin. Even at that point, almost a year after the February 1976 decree, there was no consensus on the need to build a heavy-lift shuttle. Several members of the commission spoke out in favor of Chelomey's 20-ton spaceplane, which could solve practical tasks like servicing space stations. Others called for the development of *both* a small and a large shuttle. In the end though, the commission's recommendation was to press ahead with NPO Energiya's big orbiter, mainly in order to have a deterrent to the US system in the long run [69].

The findings of Mozzhorin's commission were discussed at a joint meeting of the Ministries of Defense, the Aviation Industry, and General Machine Building in March 1977, which recommended making some amendments to the draft plan. These were approved by the Council of Chief Designers in July 1977 and called for some significant changes to the core stage. In the original design the core stage was a single element with one LOX and one LH_2 tank and a diameter of 8.2 m (comparable with the Space Shuttle External Tank's 8.4 m diameter). Now its diameter was reduced to 7.7 m and it was lengthened by 7.9 m. More importantly, the core stage was now to consist of two separate sections, each having its own LOX and LH_2 tanks (four tanks in all). This was mainly dictated by the fact that the carrier aircraft being studied at the time (a modified Myasishchev bomber) could not transport the stage in one piece. It was also supposed to improve the stability of the rocket by keeping its center of gravity as high up as possible. During the initial stages of ascent the engines would consume the propellants in the lower section and as the tanks emptied they would gradually be refilled with propellants from the upper section through a special cross-feed system. Once the upper section ran out of propellant, it would be jettisoned, as a result of which the rocket shed a significant amount of dead weight during the final phase of the launch [70].

The amendments formed the basis for a new government/party decree (nr. 1006-323) on 21 November 1977, which gave the go-ahead for the next step in the design phase, namely the completion of the so-called "technical plan" in the first quarter of 1978. This was to be followed by the release of "design documentation" for the rocket in 1978 and for the orbiter in 1980, usually the last step before the construction of actual flight hardware begins. The first flight remained optimistically targeted for 1983.

In 1978 it was decided to return the core stage to its original configuration, although it retained the 7.7 m diameter and was now even a bit shorter than the version originally proposed in 1976. Disadvantages of the dual-element design had been the need to develop a complex propellant cross-feed system and the requirement to find safe impact zones for the upper section, which imposed further restrictions on the rocket's possible trajectories. However, the return to the single-element design did not really solve the transportation problem. The core stage still had to be flown to the cosmodrome in two sections, with the LH_2 and LOX tanks being ferried separately before being joined together at the launch site. The later An-225 Mriya carrier aircraft *was* capable of carrying the core stage (and the strap-ons) in one piece, although it was never used in that capacity. The final amendments to the "technical plan" were completed in June 1979, which can be considered the date that the design of the Energiya–Buran system was frozen [71].

A SPACE SHUTTLE COPY?

The unavoidable impression one gets when comparing drawings and pictures of the Space Shuttle and Energiya–Buran is that the Soviet vehicle is in many ways a copy of the American one. There *were* of course basic differences with the Space Shuttle, the most notable ones being the use of liquid vs. solid-fuel boosters, the placement of the cryogenic engines on the external tank rather than the orbiter, the use of cryogenic rather than hypergolic propellants for the orbiter's orbital maneuvering and reaction control systems and Buran's higher degree of automation.

However, there is no denying the fact that other differences between the two systems were in details rather than in fundamental design. The similarities far outnumbered the differences. The tank section of the core stage was a virtual carbon copy of the Shuttle's External Tank and the orbiter was almost identical in layout, dimensions, and shape to its US counterpart. The similar dimensions were a logical result of the requirement to match the payload capacity of the Space Shuttle. As for the strikingly similar shape, when asked about this, Soviet officials usually responded along the lines that the laws of aerodynamics left little room for other designs. However, the dozens of orbiter outlines studied by NASA in the late 1960s and early 1970s and by the Russians themselves disprove this claim. In the end, Buran's shape was largely determined by the very same Defense Department requirements that

Space Shuttle and Energiya–Buran compared (*source*: *www.buran.ru*).

had forced NASA into the design for its Space Shuttle Orbiter (high cross-range capability and ability to transport large payloads). As one veteran admits:

> "The deciding factor was not aerodynamics. We were in a position of having to play catch-up [with the Americans] . . . This is where the, unfortunately, classical opinion in our defense industry surfaced: the Americans aren't dumber, do it the way they do!" [72].

Indeed, Buran was not the only example of following Western designs. Similar examples can be found in other branches of the Soviet industry as well, particularly in aviation. Among the more striking ones were the Tu-4 bomber, a clone of the B-29, and the Tu-144 "Konkordski", the Soviet equivalent of Concorde. In some instances this was simply the fastest and most practical way of achieving parity, with the Russians apparently being not all too concerned about losing face in the process.

It should be pointed out though that Buran was the *only* obvious case of copying in the Soviet space program. While several Soviet manned and unmanned space projects were a response to American programs and intended to match their capabilities, the Russians usually came up with their own design solutions (e.g., Spiral vs. Dyna Soar and Almaz vs. the Manned Orbiting Laboratory). In the N-1/L-3 manned lunar-landing project, a program comparable in scale with Energiya–Buran, the Russians adopted the same Lunar Orbit Rendezvous technique as the Americans, but built a rocket that was fundamentally different from the Saturn V. Perhaps the negative experience with that project was one of the reasons that led them to more closely mimic the US design when the next program of comparable proportions came along. If things went wrong again, managers and designers would at least not be held accountable for "having done it differently than the Americans".

More fundamentally though, this time around the Russians were not sure what the ultimate objectives of the American program where. Whereas the goal of Apollo unequivocally had been to put a man on the Moon, the motives behind the development of the Space Shuttle were much more nebulous from the Russian perspective. Fearing the military capabilities of the Shuttle, they felt it necessary to build an equivalent system, but, unsure of what *exactly* the threat was, they had little choice but to stick closely to the American design to make sure they would be able to respond to whatever strategic missions the Shuttle would eventually perform. Buran was built not out of some fundamental need in the Soviet space program, but as an answer to *potential* military and other applications of the Space Shuttle. This would eventually become the root cause of its downfall in the early 1990s.

By copying many aspects of the Space Shuttle design, the Russians could take advantage of the American experience, saving them a lot of research and development time. Although there is little evidence to support this, there can be little doubt that in designing their vehicle the Russians made use of the literature openly available on the Space Shuttle Orbiter. By the time the Energiya–Buran project got underway in 1976, the Shuttle's design had been frozen for nearly two years and the first Orbiters were under construction. On the other hand, there is probably nothing fundamental that the Russians changed in their design based on actual US flight experience, because by the time Columbia flew STS-1 in April 1981 Buran's own design had been finalized.

Still, the copying that irrefutably took place should not be used as an argument to belittle the Soviet accomplishment. No matter how much the Russians relied on Shuttle literature and blueprints, they still needed to develop the technology, the materials, and the infrastructure and do the testing all by themselves. Considering their overall less mature state of technology and the country's smaller economic potential, this was a remarkable feat, irrespective of whether the expenditures were justified or not.

WHAT'S IN A NAME?

The system

The term used for the Soviet shuttle program in the February 1976 party/government decree was Reusable Space System (*Mnogorazovaya Kosmicheskaya Sistema* or MKS). This covered not only the rocket and orbiter, but extended to the interorbital space tug mentioned in the decree as well as the cosmodrome infrastructure needed to prepare, launch, and land the vehicle. MKS is probably the closest equivalent to "(National) Space Transportation System" ((N)STS) in the United States (officially changed into "Space Shuttle Program" in 1990).

On 27 May 1976 chief designer Igor Sadovskiy approved an MKS structure consisting of 13 elements, each of which got its own Ministry of Defense designator beginning with the number 11. Others were added later in the program. The flight hardware was given the following designators:

- the orbiter: 11F35;
- the core stage: 11K25Ts;
- the strap-on boosters: 11K25A;
- the interorbital space tug: 11F45.

The MKS itself received the designator 1K11K25. Combinations of individual elements got their own designators. The most important ones were [73]:

- core stage + strap-on boosters: 11K25;
- core stage + strap-on boosters + orbiter: 11F36.

Another term used later in the program was Universal Rocket and Space Transportation System (*Universalnaya Raketno-Kosmicheskaya Transportnaya Sistema* or URKTS), which seems to have referred more specifically to the rocket family, reflecting the fact that Energiya could also fly with two or eight strap-on boosters and carry other payloads than the orbiter. The word "reusable" was reportedly not included because the reusability of the strap-on boosters had not yet been demonstrated, nor would it ever be [74].

The combination of orbiter and rocket was also known as the Reusable Rocket Space Complex (*Mnogorazovyy Raketno-Kosmicheskiy Kompleks* or MRKK). A general word for the spaceplane, comparable with Orbiter in the US, was "Orbital Ship" (*Orbitalnyy Korabl* or OK). The core stage was called "Central Block" (*Blok-Ts* in Russian spelling) and the strap-on boosters "Block-A" (*Blok-A*). "Block" is a commonly used word in Russian to designate rocket stages. It also appears in the names of famous upper stages such as the Blok-D and Blok-L.

Of course, the orbiter and rocket became known to the world in the late 1980s by the less prosaic names Buran and Energiya. While the name Energiya was specifically invented for public consumption late in the program, the name Buran was used in internal documentation from the very beginning, long before the program entered the public domain.

The word *buran*, imported into Russian from the Turkish language family, refers to a violent, cold northeast wind in the Central Asian steppes that lifts snow from the ground, usually during the winter. The same wind also occurs, but less frequently, in summer, when it darkens the skies by raising dust clouds and is then called *karaburan* ("black buran"). Although usually translated simply as "snowstorm", *buran* is not the general Russian word for a snowstorm, but a much more specific term that could better be defined as "a blizzard in the steppes".

The name Buran had already been applied to a canceled cruise missile designed by the OKB-23 Myasishchev bureau in the 1950s (see Chapter 1). Of course, Myasishchev later became closely involved in the shuttle program as head of the Experimental Machine Building Factory (EMZ) and one might speculate that the suggestion to recycle the name came from him.

The first use of the name Buran in connection with the Soviet shuttle seems to have come in NPO Energiya's proposals for the OS-120 design in 1975. Since this was the Space Shuttle type integrated configuration with an external tank and the main engines on the orbiter, it referred to the whole stack, not the orbiter individually. Even after the final decision had been made to turn the external tank into the second stage, the name Buran continued to be used for the combination of the now engineless orbiter and its launch vehicle (and therefore denoted the same as "11F36" and "MRKK"). For some reason, NPO Energiya's internal RLA-130 designator for the rocket did not become established. The configuration in which the orbiter was replaced by an unmanned cargo canister was known as Buran-T. The common name for orbiter and rocket caused quite some confusion in the space community and was sometimes conveniently misused by opponents to criticize the system as a whole while reacting to problems with one of the two elements [75].

The name Energiya was not coined until May 1987, when the Russians needed to make a public announcement about the rocket's first launch. In contrast to earlier plans this was not flown with a shuttle, but with a quickly improvised payload called Polyus. When Soviet leader Mikhail Gorbachov visited the Baykonur cosmodrome in the final days prior to the launch, Glushko proposed the name Energiya, mainly because this was one of the buzzwords of Gorbachov's policy of *perestroyka*. The fact that it was also the name of Glushko's design bureau was probably less convincing to Gorbachov [76]. By giving the rocket an individual name, the Russians also underlined that this was a launch vehicle in its own right, capable of launching not only shuttles, but other heavy payloads as well [77]. Of course, not enough time was left to paint the new name on the rocket as there was for the second launch.

Orbiter names and mission designators

The name Buran was first publicly applied to the orbiter individually when the TASS news agency announced the launch date for the first mission on 23 October 1988. Actually, the name originally painted on the first flight vehicle had been "Baykal" (after the famous Siberian lake), but this was later erased. Strictly speaking, Buran had now become the name of the vehicle that made the one and only Soviet shuttle flight on 15 November 1988, placing it on an equal footing with NASA's Shuttle Orbiter names Enterprise, Columbia, Challenger, Discovery, Atlantis, and Endeavour. However, since Buran was the only vehicle ever flown, the name later also began to be used for Soviet orbiters in general, as it will be in this book.

It is not known what official names the other vehicles would have been given had they ever flown. The only other ship that came close to flying was sometimes referred to in the press as "Buran-2", but it is unclear if this would have become its official name. A persistent myth is that it was called Ptichka ("Birdie"), which actually was a general nickname for Soviet orbiters that somehow got misinterpreted by Western journalists as being the name of the second orbiter. There is some speculation that it was to be dubbed Burya ("Storm"), continuing a tradition of naming orbiters and some heavy-lift launch vehicles and their upper stages after violent natural phenomena. Burya, incidentally, had also been the name of the Lavochkin bureau's cruise missile that won the competition from Myasishchev's Buran back in the 1950s. Presumably, the Russians would have given this matter serious thought only if the second orbiter had entered final launch preparations, which it never did. No name was ever painted on this vehicle and therefore it can be said that it was never officially named.

The individual vehicles did have designators comparable with the OV designators of the US Shuttle Orbiters (OV-099, 101, 102, 103, 104, 105). Buran was 1K, the second orbiter was 2K, the third one 3K, etc. These designators also appeared in the mission designations. The first flight of orbiter 1K was 1K1, the (planned) first flight of orbiter 2K was 2K1, etc. In documentation these numbers were also used to refer to the vehicles themselves, so "vehicle 1K1" would be "flight vehicle 1 as configured for its first mission". Some Western publications claimed the first flight was designated "VKK-1", but this is not true. VKK (*Vozduzhno-Kosmicheskiy Korabl*) literally means "aerospace ship", a general term for winged spacecraft, although it is most often used for single-stage-to-orbit spaceplanes. Within NPO Molniya the airframes of the flight articles had designators such as 1.01, 1.02 (for the first two orbiters) and 2.01 (for the third orbiter).

Even though the word "Buran" has been used to refer to different things at different times and the name "Energiya" was not introduced until 1987, for the sake of clarity the two names will be used further in this book to refer to the orbiter and the rocket, respectively, irrespective of when the events discussed took place.

REFERENCES

[1] D. Jenkins, *The History of the National Space Transportation System*, Hinckley: Midland Publishing, 2001.

[2] Y. Semyonov, *Raketno-kosmicheskaya korporatsiya Energiya imeni S.P. Korolyova 1946–1996*, Moscow: RKK Energiya, 1996, pp. 278–279, 294–296.

[3] B. Hendrickx interviews with VPK veteran Emil Popov, July 2005; V. Lukashevich, "A Soviet copy of the Shuttle: the orbital ship OS-120" (in Russian), *Novosti kosmonavtiki*, 8/2006, pp. 64–67.

[4] Ts. Solovyov, "Work in the USSR on reusable space transportation systems (1972–1975) (in Russian), paper presented at the Tsiolkovskiy readings in Kaluga, September 1992; V. Burdakov *et al.*: "Reusable space transportation systems: History and future prospects" (in Russian), paper presented at the *International Symposium on the History of Aviation and Cosmonautics in Moscow, June 1995*; V. Favorskiy, I. Meshcheryakov, *Voyenno-kosmicheskiye sily, kniga 2*, Moscow: Izdatelstvo Sankt-Peterburgskoy tipografii, 1998, pp. 54, 292.

[5] A. Kuznetsov, *Mnogorazovaya kosmicheskaya sistema Energiya-Buran*, Moscow: OmV-Luch, 2004, p. 18; V. Lukashevich, *op. cit.*

[6] B. Hendrickx correspondence with V. Lukashevich, 14 April 2006.

[7] Y. Semyonov, *op. cit.*, pp. 40–41.

[8] V. Lukashevich, "The OK-92 that became Buran (part 1)" (in Russian), *Novosti kosmonavtiki*, 3/2006, pp. 56–58.

[9] B. Chertok, *Rakety i lyudi: lunnaya gonka*, Moscow: Mashinostroyeniye, 1999, p. 473.

[10] Y. Semyonov, *op. cit.*, pp. 281–286.

[11] Y. Semyonov, *op. cit.*, p. 288; R. Sagdeyev, *The Making of a Soviet Scientist*, New York: Wiley, 1994, pp. 183–184.

[12] G. Nazarov, "You cannot paper space with rubles" (in Russian), *Molodaya Gvardiya*, April 1990, pp. 192–207.

[13] S. Aleksandrov, "Who needs it, this Buran?" (in Russian), *Propeller/Apogey*, June 1995, p. 1; Y. Mozzhorin, *Tak eto bylo*, Moscow: ZAO Mezhdunarodnaya programma obrazovaniya, 2000, pp. 340–341.

[14] J. Harford, *Korolev*, New York: Wiley, 1997, p. 314.

[15] B. Gubanov, *Triumf i tragediya Energii (tom 3: Energiya-Buran)*, Nizhniy Novgorod: Izdatelstvo Nizhegorodskogo instituta ekonomicheskogo razvitiya, 1998, p. 33.

[16] D. Jenkins, *op. cit.*, p. 101; D. Payson, "The Buran decision" (in Russian), *Rossiyskiy kosmos*, 6/2006, pp. 40–43.

[17] B. Gubanov, *op. cit.*, p. 33.

[18] B. Hendrickx interviews with Emil Popov.

[19] R. Sagdeyev, *op. cit.*, p. 147.

[20] *Ibid.*, p. 213.

[21] V. Lukashevich, "The OK-92 that became Buran (part 1)", *op. cit.*

[22] V. Favorskiy, I. Meshcheryakov, *op. cit.*, p. 55.

[23] TV documentary *"Udarnaya sila"* , shown on the Russian ORT television channel, 20 March 2007.

[24] R. Sagdeyev, *op. cit.*, p. 213.

[25] The text of the decree is almost completely reproduced in A. Kuznetsov, *op. cit.*, pp. 6–10.

[26] A. Bruk, *Illyustrirovannaya entsiklopediya samolyotov EMZ im. V.M. Myasishcheva (tom 3, chast 1)*, Moscow: Aviko Press, 1999, p. 28.

[27] B. Chertok, *op. cit.*, p. 523.

[28] R. Sagdeyev, *op. cit.*, p. 184.

[29] Y. Semyonov, *op. cit.*, p. 362.

[30] V. Favorskiy, I. Meshcheryakov, *Voyenno-kosmicheskiye sily, kniga 1*, Moscow: Izdatelstvo Sankt-Peterburgskoy tipografii, 1997, p. 240.

[31] A. Siddiqi, *Challenge to Apollo*, Washington, D.C.: NASA, 2000, p. 752; B. Katorgin, *NPO Energomash imeni akademika V.P. Glushko. Put v raketnoy tekhnike*, Moscow: Mashinostroyeniye/Polyot, 2004, p. 427.
[32] V. Rakhmanin, *Odnazhdy i navsegda: kniga o Valentine Petroviche Glushko*, Moscow: Mashinostroyeniye, 1998, p. 602; V. Trofimov, *Osushchestvleniye mechty*, Moscow: Mashinostroyeniye/Polyot, 2001, p. 45.
[33] V. Rakhmanin, *op. cit.*, p. 603; V. Trofimov, *op. cit.*, p. 49.
[34] V. Trofimov, *op. cit.*, pp. 70–71.
[35] *Ibid.*
[36] V. Gladkiy, "How the Energiya-Buran project was born" (in Russian), *Aviatsiya i kosmonavtika*, 1/2002.
[37] B. Gubanov, *op. cit.*, p. 92; B. Katorgin, *op. cit.*, pp. 202, 208, 428.
[38] B. Chertok, *op. cit.*, p. 476.
[39] V. Filin, *Put k Energii*, Moscow: Logos, 2001, p. 39.
[40] *KB Khimavtomatiki, stranitsy istorii, tom 1*, Voronezh: KBKhA, 1995, p. 125.
[41] A. Siddiqi, *op. cit.*, p. 649.
[42] B. Katorgin, *op. cit.*, p. 427.
[43] V. Trofimov, *op. cit.*, p. 140; S. Konyukhov, *Prizvany vremenem: rakety i kosmicheskiye apparaty konstruktorskogo byuro Yuzhnoye*, Dnepropetrovsk: Art Press, 2004.
[44] *KB Khimavtomatika, stranitsy istorii, op. cit.*, p. 72.
[45] V. Favorskiy, I. Meshcheryakov, *Voyenno-kosmicheskiye sily, kniga 1, op. cit.*, pp. 237–239; B. Gubanov, *op. cit.*, pp. 55–56.
[46] N. Anfimov, *Generalnyy konstruktor: kniga o Vladimire Fyodoroviche Utkine*, Korolyov: TsNIIMash, 2003, p. 47; V. Favorskiy, I. Meshcheryakov, *Kosmonavtika i raketno-kosmicheskaya promyshlennost: razvitiye otrasli (1976–1992)*, Moscow: Mashinostroyeniye, 2003, p. 54; D. Vorontsov, "The compromise N-1" (in Russian), *Novosti kosmonavtiki*, 1/2006, p. 69.
[47] B. Gubanov, *op. cit.*, pp. 55–59; S. Konyukhov, *op. cit.*
[48] B. Chertok, *op. cit.*, pp. 475–476.
[49] G. Vetrov, "Development of Heavy Launch Vehicles in the USSR", paper presented at the *10th International Symposium on the History of Astronautics and Aeronautics in Moscow, June 1995*; V. Gladkiy, *op. cit.*
[50] V. Gladkiy, *op. cit.*
[51] V. Lukashevich, "The OK-92 that became Buran" (part 1), *op. cit.*
[52] V. Gladkiy, *op. cit.*
[53] V. Lukashevich, "The OK-92 that became Buran" (part 1), *op. cit.*
[54] I. Afanasyev, "Unknown ships" (in Russian), *Astronomiya, kosmonavtika (Znaniye)*, 12/1991, pp. 57–60; I. Afanasyev, "The reusable ship with vertical landing" (in Russian), *Novosti kosmonavtiki*, 6/2004, pp. 71–72; V. Lukashevich, "The OK-92 that became Buran" (part 2) (in Russian), *Novosti kosmonavtiki*, 4/2006, pp. 58–62; V. Lukashevich, "The history of the reusable orbital ship Buran" (in Russian), on-line at *www.buran.ru*
[55] B. Gubanov, *op. cit.*, pp. 38–39; V. Lukashevich, "A Soviet copy of the shuttle: The orbital ship OS-120", *op. cit.*
[56] V. Lukashevich, "The OK-92 that became Buran" (part 1), *op. cit.*
[57] V. Gladkiy, *op. cit.*
[58] V. Lukashevich, "The OK-92 that became Buran" (part 2), *op. cit.*
[59] *KB Khimavtomatiki, stranitsy istorii, op. cit.*, pp. 72, 125; B. Gubanov, *op. cit.*, p. 39.
[60] V. Kazmin, "The quiet tragedy of EPOS" (in Russian), *Krylya rodiny*, January 1991, pp. 4–5.

[61] Interview with G. Lozino-Lozinskiy on the *www.buran.ru* website.
[62] Y. Semyonov, *op. cit.*, p. 380; V. Gladkiy, *op. cit.*; A. Bruk, *op. cit.*, pp. 28–30, 209.
[63] Y. Trufakin, *Orbitalnyy korabl Buran i proyektno-nauchnyy otdel "Dinamika polyota" v NPO Molniya,* 2002, p. 10.
[64] B. Gubanov, *op. cit.*, p. 58.
[65] S. Konyukhov, *op. cit.*
[66] V. Gladkiy, *op. cit.*
[67] V. Rakhmanin, *op. cit.*, pp. 610–615; B. Katorgin, *op. cit.*, p. 126.
[68] V. Favorskiy, I. Meshcheryakov, *Voyenno-kosmicheskiye sily, kniga 2, op. cit.*, pp. 55–56; B. Gubanov, *op. cit.*, pp. 40–41.
[69] S. Aleksandrov, *op. cit.*; Y. Mozzhorin, *op. cit.*, pp. 342–343.
[70] Y. Semyonov, *op. cit.*, pp. 362–363; B. Gubanov, *op. cit.*, p. 43.
[71] Y. Semyonov, *op. cit.*, p. 363; V. Favorskiy, I. Meshcheryakov, *Voyenno-kosmicheskiye sily, kniga 2, op. cit.*, pp. 57; B. Gubanov, *op. cit.*, p. 44.
[72] V. Filin on website *www.buran.ru*
[73] V. Bugrov, "15th anniversary of Buran's flight" (in Russian), *Novosti kosmonavtiki*, 1/2004, pp. 70–72.
[74] V. Gudilin, "Take-off and landing of Buran" (in Russian), on-line at *www.buran.ru/htm/gudilin.htm*
[75] B. Gubanov, *op. cit.*, pp. 41–42.
[76] B. Gubanov, *op. cit.*, p. 297.
[77] A. Yuskovets, "On the launch pad: A conversation with Yuriy Pavlovich Semyonov" (in Russian), *Leninskoye znamya*, 16 February 1989.

3

Systems and scenarios

ENERGIYA CORE STAGE

The core stage was designed jointly by NPO Energiya in Kaliningrad and its Volga Branch in Kuybyshev, with manufacturing taking place at the Progress factory in Kuybyshev. With a length of 58.7 m and a maximum diameter of 7.75 m, it was the backbone of the Energiya–Buran stack, providing structural support for attachment with the strap-on boosters and orbiter. It was very similar in design to the Space Shuttle's External Tank (ET), with the exception of a tail section housing the engine compartment. The core stage was made up of an upper liquid oxygen (LOX) tank, an unpressurized intertank, a lower liquid hydrogen (LH_2) tank, and a tail section containing the four RD-0120 engines. The wet mass was 776 tons.

Both the LOX and LH_2 tanks were made of a 1201 aluminum alloy. The 552 m^3 LOX tank could hold about 600 tons of oxidizer. It consisted of a forward ogive section (itself made up of three sections), a cylindrical section (itself made up of two sections), and a spherical aft dome. All sections were welded together. The tank had anti-slosh baffles to dampen any motions of the LOX that might throw the rocket off course. The LOX feed line exited the LOX tank at a 7° angle to the longitudinal axis of the tank to facilitate oxidizer supply during the final moments of the launch. It ran to the tail section right through the LH_2 tank. This is a major difference with the Space Shuttle External Tank's LOX feed line, which emerges from the ET's intertank area to convey the oxidizer to the aft right-hand ET–Orbiter disconnect umbilical. The LOX feed system had gas accumulators to dampen longitudinal oscillations ("pogo"). These were located in the lower part of the LOX feed line in the bottom of the LH_2 tank and also in the engines' turbopump inlet ducts.

The intertank was the structural connection joining the liquid hydrogen and oxygen tanks. Flanges were affixed at the bottom and top of the intertank so the two tanks could be attached to it. Also installed in the intertank was the instrumentation for the core stage's flight control system. Prior to launch the intertank was

The Energiya core stage (*source*: *www.buran.ru*).

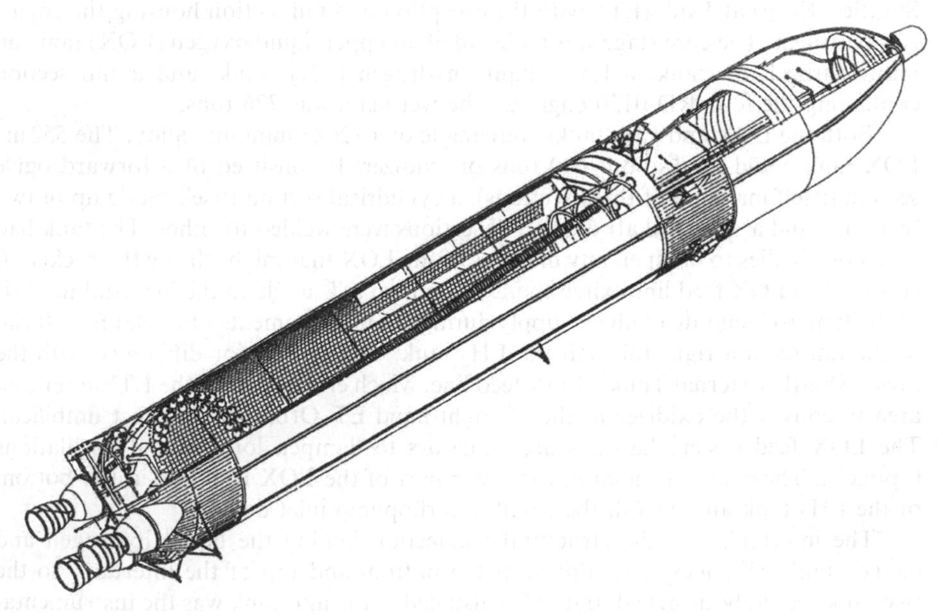

Cutaway drawing of the Energiya core stage (*source*: Boris Gubanov).

purged with nitrogen gas to prevent the build-up of moisture and explosive mixtures of hydrogen and oxygen gas.

The 1,523 m^3 LH_2 tank, which could hold about 100 tons of liquid hydrogen, consisted of spherical aft and forward domes and a large cylindrical section. The tank walls were machined in a waffle-grid pattern, something not employed in the ET hydrogen tank until the introduction of the Super Lightweight External Tank in 1998. Although LH_2 is so light that sloshing does not induce significant forces, Energiya's LH_2 tank, unlike that of the ET, did have anti-slosh baffles.

Just like the Shuttle's ET, the Energiya core stage was covered with a combination of polyurethane spray-on foam insulation (Ripor-2N, PPU-17) and ablative material (PPU-306) for thermal insulation and thermal protection. This reduced boil-off losses during the countdown, maintained the propellants at the proper temperatures for normal engine operation, limited ice formation on the outer surface, and protected the core stage against the flames from the strap-on boosters' separation motors. The original plan *not* to apply thermal insulation to the upper part of the LOX tank was abandoned due to fears that ice might break off from that part of the core stage and damage the orbiter's fragile heat shield. Various non-destructive methods were used to test these materials after they were applied: electric methods to check their thickness, radioisotope techniques for density, and acoustic methods to detect debonding.

Shedding of tank insulation became a big issue in the US after the February 2003 Columbia accident, caused by a piece of foam insulation breaking off the tank and inflicting lethal damage to one of the Reinforced Carbon–Carbon panels on the Shuttle's left wing. Russian sources do not mention whether Energiya's foam insulation was less or more prone to shedding simply because this was not a matter of major concern in the pre-Columbia days. Moreover, any foam loss that might have occurred on the two Energiya launches in 1987 and 1988 would have been virtually impossible to photographically document because the first launch took place in darkness and the second in poor weather conditions. Tile damage suffered by Buran on its sole mission has usually been attributed to ice falling off the core stage and the launch pad and not to foam impacts.

Electric power for the core stage was provided by four simultaneously operating turbogenerators driven by air, nitrogen, hydrogen, and helium gas. In order to simplify the design and reduce mass, common plumbing was used for all four gases. Each generator weighed 330 kg and provided 24 kWt of power.

The cryogenic propellants were loaded at lower temperatures than on the Space Shuttle (−255°C vs. −253°C for the liquid hydrogen and −195°C vs. −182°C for the liquid oxygen). This made the propellant denser and also significantly reduced boil-off losses. Techniques for subcooling liquid oxygen were pioneered by the Russians with the R-9 missile in the early 1960s, but Energiya marked the first use of subcooled liquid hydrogen. The liquid hydrogen was subcooled by passing it through two double-walled cooling devices in which tubular heat exchangers were immersed in a bath of liquid hydrogen boiling at reduced pressure.

Loading of the core stage began several hours before launch with a slow-fill mode to condition the tanks. When the core stage was 2 percent full, the fueling process was sped up to 19,000 liters per minute for the liquid oxygen and 45,000 liters per minute for the liquid hydrogen. This fast-fill mode continued until 98 percent of the propellant was loaded. Topping off continued until $T - 3m02s$ for the LOX tank and $T - 1m52s$ for the LH_2 tank.

After the start of fueling, electrically powered pumps in the main engines began to circulate the liquid hydrogen in the fuel tank through the four engines and back to the tank to chill down the liquid hydrogen lines, ensuring that the path was free of any gaseous hydrogen bubbles and was at the proper temperature for engine start. The LH_2 was recirculated to the tank rather than returned to ground facilities because it loses a lot of pressure during the circulation process. The engines' LOX lines were also thermally preconditioned, but the LOX used for this purpose was dumped overboard.

In the final minutes of the countdown the tanks were pressurized to maintain their structural integrity during launch, minimize the build-up of volatiles in the tanks, and to prevent cavitation of the main engine low-pressure boost pumps. Pre-launch pressurization was performed with ground-supplied helium and began at $T - 2m23s$ for the LOX tank and $T - 1m20s$ for the LH_2 tank. After lift-off the LOX tank was pressurized with hot gaseous oxygen produced by heat exchangers in the main engines and the LH_2 tank with gaseous hydrogen tapped from the turbines of the main engine LH_2 boost pumps. At lift-off the LOX tank was pressurized to 2.6 atmospheres and the LH_2 tank to 3.1 atmospheres. During launch the pressure in the LOX and LH_2 tanks was maintained between 1.41–1.55 atmospheres and 2.25–2.39 atmospheres, respectively.

Each tank had a dual-function vent and relief valve at its forward end. It could be opened by ground-supplied helium before launch for venting or by excessive tank pressure for relief during launch. Excess hydrogen gas left the core stage via the intertank area, while excess oxygen gas was directly vented overboard [1].

THE RD-0120 ENGINE

The RD-0120 (also labelled 11D122), developed at the Chemical Automatics Design Bureau (KBKhA) in Voronezh, was a LOX/LH_2 engine with a vacuum thrust of 190 tons and a vacuum specific impulse of 454 s. It was a staged combustion cycle engine in which the gases from the gas generator are cycled back into the main combustion chamber for complete combustion. The propellants first passed through low-pressure turbopumps ("boost pumps") that boosted the pressure significantly to prevent cavitation of the main turbopump assembly. The low-pressure hydrogen pump used a gas turbine driven by gaseous hydrogen from the main chamber cooling loop. The low-pressure oxygen pump had a hydraulic turbine powered by liquid oxygen.

Each RD-0120 had a single-shaft turbopump consisting of a two-stage axial turbine, a three-stage hydrogen pump, and two oxygen pumps. One of the oxidizer pumps was intended to feed the main combustion chamber and the other to feed the gas generator and the hydraulic turbine of the low-pressure oxygen pump. The 32,500 rpm turbopump was driven by a single fuel-rich preburner operating at 530°C.

The RD-0120 engine (*source*: KBKhA).

During ascent the RD-0120 could be gimbaled plus or minus 11 degrees in pitch and yaw to help steer the rocket. Each engine was gimbaled with the help of two hydraulic servoactuators developed by KB Saturn (the "Lyulka bureau"), with hydraulic pressure being provided by pumps driven by high-pressure hydrogen gas from the engine itself. The engine could be throttled over a range of 45 to 100 percent, a significantly higher throttlability than the Space Shuttle Main Engine (SSME) (67–104%). The pneumatic control system included pressure helium bottles, pneumatic and electro-pneumatic valves, and system piping.

Nominal burn duration was between 450 and 500 seconds, although this could be significantly extended in case one engine had to compensate for the loss of another. The RD-0120 engines for the first three flightworthy Energiya rockets (6SL, 1L, and 2L) were certified for a total burn time of 1,670 seconds (230 seconds for test firings, 480 seconds for the launch, and 960 seconds back-up capability). This was increased to 2,000 seconds for flight vehicle 3L. Theoretically, this meant the engine could be reused for about three to four missions, although they were of course destroyed on re-entry together with the core stage. However, there were plans to certify the engines for 10 to 20 missions for reusable versions of Energiya and for possible use on foreign reusable launch vehicles. It was also planned to gradually uprate the engine, increasing the vacuum thrust to 230 tons and the vacuum specific impulse to 460.5 s. One of the modifications would have been the inclusion of an extendable nozzle to prevent loss of specific impulse in vacuum conditions.

Although the RD-0120 was built to the same overall performance specifications as the SSME, it certainly was not a copy of the SSME, differing from it in several important aspects. Also, the Russians could draw on their extensive experience with staged-combustion cycle engines used on the Proton rocket and various intercontinental missiles. While the RD-0120 had a single turbopump assembly both for liquid oxygen and hydrogen, the SSME has separate turbopumps for each propellant. Soviet engineers did consider a similar scheme, but opted for the single turbopump system because it simplified the computer control system and the ignition sequence. The RD-0120 had a channel-wall nozzle with fewer parts and welds than the SSME nozzle and was therefore easier to manufacture. In the late 1990s NASA even considered building a similar nozzle for the SSME, eliminating the tubular construction as a potential source of nozzle leaks. The new nozzle was also expected to have a higher degree of reusability [2].

ENERGIYA STRAP-ON BOOSTERS

The Energiya "Blok-A" strap-on boosters were largely designed and built by KB Yuzhnoye and its associated production facility in Dnepropetrovsk, although major elements were also supplied by NPO Energiya in Kaliningrad. Each strap-on booster was 39.4 m high and had a maximum diameter of 3.9 m, dictated by railway transportation requirements. The wet mass was about 372 tons. The booster consisted of a nose section, an upper LOX tank, an intertank structure, a lower kerosene tank, and

Nose section of strap-on booster (B. Vis).

Tail section of strap-on booster (B. Vis).

a tail section with the gimbal-mounted LOX/kerosene RD-170 engine. The four boosters were numbered 10A, 20A, 30A, and 40A.

The nose section housed the avionics bay, which among other things contained an M4M computer that interacted with the core stage's M6M central computer. Also installed in the nose section were flight data recorders, mounted in a special casing that protected them from the force of impact. Two of the four Blok-A boosters were equipped with radio beacons, enabling ground controllers to follow their trajectory after separation from the core stage.

The propellant tanks were made of an aluminum–magnesium alloy, with the walls being about 30 mm thick. The LOX and kerosene tanks had useful volumes of 208 m^3 and 106 m^3, respectively, holding approximately 220 tons of LOX and 80 tons of kerosene. The upper LOX tank fitted in a concave depression at the top of the kerosene tank and the LOX feed line passed right through the middle of the lower kerosene tank. There were vent and relief valves in the forward domes of both tanks. The kerosene fill and drain valves were in the aft dome of the kerosene tank, and the LOX fill and drain valves in the lower part of the LOX feed line. The LOX feed line also had a damper to suppress "pogo" oscillations.

Embedded in the lower part of the LOX tank were two rows of helium tanks for in-flight pressurization of both the LOX and kerosene tanks. The helium for the kerosene tank was supplied directly, while that for the LOX tank first passed through a heat exchanger in the lower engine compartment. Before launch the tanks were pressurized with ground-supplied helium. Electrical power for the boosters was provided by batteries.

There was a pair of strap-ons on either side of the core stage. Each pair was mechanically linked and jointly separated from the core stage, with the two boosters not splitting until about half a minute later. This was done in order to minimize the risk of any of the boosters hitting the orbiter after separation. The boosters were pyrotechnically separated from the core stage and subsequently activated 11 small solid-fuel motors (seven on the nose section and four on the tail section) to ensure safe separation from the core stage and payload. They came down some 425 km downrange from the launch site.

Although they were mechanically linked, the strap-ons operated independently from one another. The only electrical connections were with the core stage (one interface with 408 contacts for each booster). There were also twelve electrical, pneumatic, and hydraulic connections between each strap-on and the launch pad (eight for propellant, fluid, and gas supply and four electrical connections). The connections were between the nose section and the launch tower and the tail section and the launch table.

From the very beginning of the Energiya–Buran program the idea was that the strap-on boosters would be reusable. The degree of reusability mainly depended on the robustness of the RD-170 engine, which was certified to fly at least 10 missions. After having studied several schemes, designers opted for a horizontal landing system using parachutes, soft-landing engines, and a set of shock absorbers. Parachutes would be deployed from the forward and aft ends of each booster, orienting it to a horizontal attitude for descent. The plan had much in common with the landing

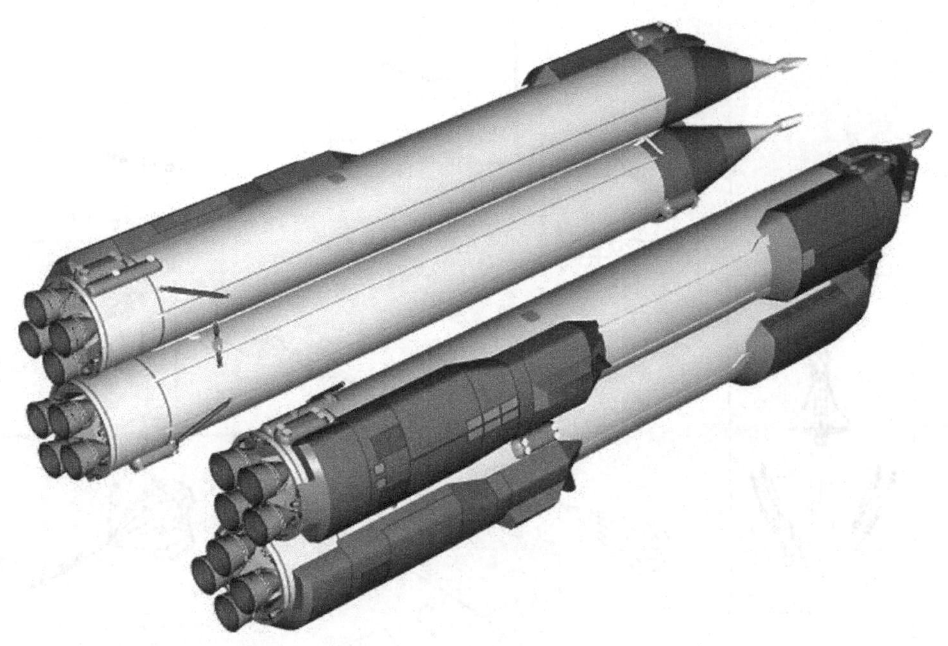

Two pairs of strap-on boosters (*source*: *www.buran.ru*).

technique for the giant MTKVP lifting body that had been studied before the delta-wing concept of Buran was picked.

The strap-ons were designed from the beginning with special containers in their nose and tail sections to house the parachutes, other recovery systems, and control equipment. The containers were installed on the strap-ons flown on the two Energiya missions in 1987 and 1988 (6SL and 1L), but were loaded with instrumentation rather than recovery equipment. However, there were plans to demonstrate the recovery technique on the 2L mission with the GK-199 payload (see Chapter 8).

The boosters shared about 70 percent of their systems with the first stage of the Zenit rocket. The part of the booster that was largely similar to the Zenit first stage was known as the "modular part" (or 11S25) and included the propellant tanks, pressurization systems, and the engine compartment. The main differences with the Zenit first stage were in the gimbal axes of the engines and also in that the tank walls were slightly thicker because of the bending loads imposed by the strap-on configuration. Elements unique to the strap-ons included the aft skirt surrounding the engine compartment (which needed to be compatible with the Energiya core stage), the entire nose section, and the parachute containers.

The original requirement was for the Zenit first stage to be reusable as well, but, if it was to use the same recovery systems and recovery zones as the Blok-A, the Zenit would need to have the same speed at the moment of first-stage separation as Energiya (1,800 m/s instead of 2,500 m/s). This would have made it necessary to make the second stage 38 tons heavier, reducing the rocket's payload capacity to

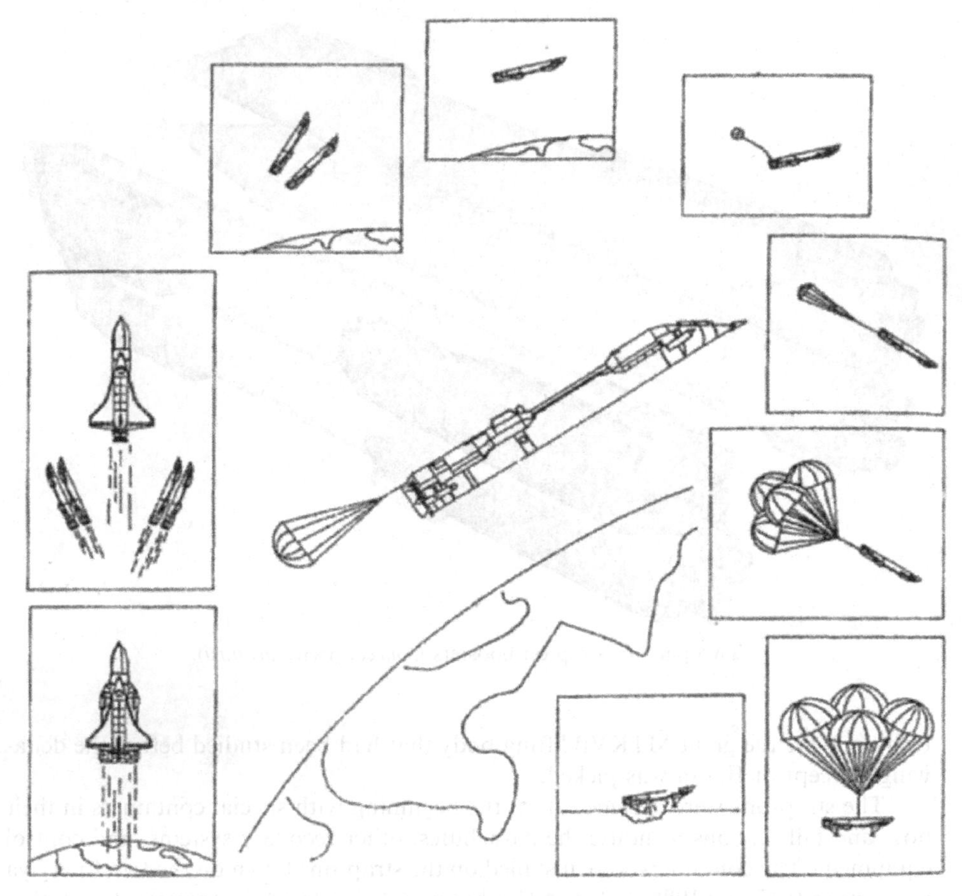

Booster separation and landing sequence (*source*: Boris Gubanov).

an unacceptable 7 tons. A later idea to recover only the tail section with the engine was dropped as well because this was expected to produce a return on investment after only about 500 launches [3].

THE RD-170 ENGINE

The RD-170 (also known as 11D521), designed and manufactured by KB Energomash in Khimki near Moscow, was a LOX/kerosene engine employing the staged combustion cycle. Providing 740 tons of thrust and a specific impulse of 308.5 s at ground level, it remains not only the most powerful LOX/kerosene engine built to date, but also the highest-thrust liquid-fuel engine flown on any launch vehicle in the world.

The RD-170 engine (*source*: *www.buran.ru*).

Although Energomash had gained significant experience with staged-combustion cycle engines burning *hypergolic* propellants, the RD-170 marked the bureau's first foray into closed-cycle LOX/kerosene engines. The only other closed-cycle LOX/ kerosene engines built in the Soviet Union until then had been much less powerful single-chamber engines such as the ones used on the Blok-L and Blok-D upper stages (built by the OKB-1 Korolyov bureau) and the NK engines for the first three stages of the N-1 rocket (developed by the Kuznetsov bureau in Kuybyshev). The United

States has *never* built a staged-combustion cycle LOX/kerosene engine. The only powerful LOX/kerosene engine ever flown by the United States was the F-1, five of which powered the first stage of the Saturn V. This was an open-cycle engine inferior in most aspects to the RD-170.

The RD-170 consisted of four combustion chambers, one turbopump assembly, and two gas generators. The turbopump assembly incorporated a single-stage active axial-flow turbine, an oxidizer pump, and a two-stage fuel pump. Connected to the assembly were low-pressure oxidizer and fuel pumps to increase the pressure of the propellant and thereby prevent cavitation of the turbopump assembly. The turbopump was driven by two oxidizer-rich gas generators. Originally, it was planned to have a single gas generator consuming 1.5 tons of propellant per second, but this would have been too big. In the RD-170 the entire oxidizer supply and just a small fraction of the kerosene (6% of the overall propellant mass) passed through the gas generators. The turbopump produced about 257,000 horsepower, which the Russians like to compare with the combined horsepower of three of their heavy nuclear icebreakers.

The RD-170 could be throttled down to 50 percent of rated thrust and could be gimbaled about 8° with the help of hydraulic actuators. The engine could be gimbaled in two axes, whereas the Zenit's RD-171 had only single-axis gimbal capability. Therefore, each RD-170 required a total of eight hydraulic actuators, two for each combustion chamber. Unlike the RD-171 nozzles, those of the RD-170 entered the air stream impinging on the rocket when they were swiveled, requiring the use of more powerful actuators to counter the aerodynamic pressures.

With a nominal flight burn time of 140–150 seconds, the engine was designed to be used at least ten times, a capability confirmed during bench tests. Although the RD-170 was used only for the two Energiya missions in 1987 and 1988, its nearly identical twin (the RD-171) continues to fly today on the two-stage Zenit launch vehicle and its three-stage Sea Launch version. A derived version with just two combustion chambers (the RD-180) now powers America's Atlas-5 rockets and a single-chamber version (the RD-191) is expected to become the power plant of Russia's Angara family of launch vehicles (see Chapter 8). [4]

ENERGIYA GUIDANCE AND CONTROL

With Buran being only one of many possible payloads of Energiya, flight control functions were divided between the rocket and the orbiter, each using their own set of computers. This is very different from the integrated US Space Shuttle system, where the Orbiter's General Purpose Computers are in control of all flight events. Being the most complex rocket ever built by the Russians, flight control proved to be a daunting task, facing designers with many unprecedented problems.

Originally, the flight control systems for both Energiya and Buran were to be built at NPO AP (Scientific Production Association of Automatics and Instrument Building), a Moscow-based organization headed between 1948 and 1982 by Nikolay A. Pilyugin. However, in 1978 the development of Energiya's control system was entrusted to NPO Elektropribor, an organization based in the Ukrainian city of Kharkov and originally founded as OKB-692 in 1959 (now called NPO Khartron). Since the early days it had been headed by Vladimir Sergeyev, replaced in 1986 by A.G. Andryushchenko. The chief designer of the Energiya control system was Andrey S. Gonchar and a leading role in its development was also played by Yakov E. Ayzenberg, who would go on to lead the organization in 1990. Production of the hardware took place at the Kiev Radio Factory. NPO AP built the orbiter flight control system and remained in overall charge of the Energiya–Buran flight control system [5].

The core stage had a primary computer (called M6M) and a computer charged with continuously monitoring the operation of all Energiya's engines and shutting any one of them down if needed. Each Blok-A strap-on booster had a M4M computer in its nose section. There was continuous interaction between the computers in the core stage and the strap-on boosters. Crucial commands such as nominal or emergency shutdown of both core stage and Blok-A engines and separation of the boosters were issued by the core stage computers [6].

Each booster had a single inertial guidance platform (17L27) built by NPO Elektromekhanika in Miass (Chelyabinsk region). The core stage's intertank area housed three inertial guidance platforms (KI21-36) developed by NPO Rotor in Moscow and based on similar systems built for the 15A35 (SS-19 "Stiletto") and 15A18 (SS-18 "Satan") missiles. Pre-launch alignment of the booster platforms took place with an optical system (17Sh14) (precision 7′) and that of the core stage platforms with an automatic system (17Sh15) (precision 45″) The automatic system consisted of three instruments mounted on a black plate outside the intertank area of the core stage. The plate was detached from the core stage and retracted to the launch tower with less than a minute to go in the countdown after the final pre-launch alignment. Failure of the plate to properly disengage led to the abort of the first Buran launch attempt on 29 October 1988 [7].

With the N-1 failures fresh in their memories, Soviet designers went to great lengths to protect the rocket against the consequences of leaks and engine failures. There was a so-called Fire and Explosion Warning System, consisting of gas and fire detectors and a system to purge the tail sections of the core stage and boosters with nitrogen and extinguish fires with freon. This was activated both during the countdown and launch. Installed on the pad was a hydrogen burnoff system to eliminate hydrogen vapors exhausted into the RD-0120 engine nozzles during the start sequence. This differed from the hydrogen igniters on the Space Shuttle launch pads in that the hydrogen was burnt off well away from the engine nozzles [8].

Energiya was also equipped with a so-called Engine Emergency Protection System, comprising a wide range of sensors in the engine compartments to monitor pressures, temperatures, turbine rates, etc. In case an anomaly was detected, any of the engines could be shut down immediately before failing catastrophically. Depending on the moment when the shutdown took place and the type of payload carried (an orbiter or unmanned payload canister), the flight control system could then decide on a further course of action. This could involve shutting down the diametrically opposed engine to continue controlled flight, increasing the burn time of the remaining engines to deploy the payload in a lower or even nominal orbit, initiating a return to launch site maneuver, guiding the rocket to a safe impact area, etc. This would not only ensure the safety of the crew, but also facilitate post-flight analysis of the failure. The system was designed to deal with over 500 types of anomalies and was said to be a major improvement over the analogous "KORD" system on the N-1 rocket.

The safety systems were not only used during launch countdowns and ascent, but also during bench tests of the RD-170 and RD-0120 engines and test firings of the core stage and strap-ons. The bench tests, especially those of the RD-170, showed that the Engine Emergency Protection System could not always respond to rapidly escalating problems such as turbopump burn-throughs or cracks in the turbopump rotors, a problem that had not been fully solved by the time Energiya made its two missions [9].

THE BURAN AIRFRAME

Buran was a double-delta winged spacecraft capable of putting people and cargo into low Earth orbits and returning them to a controlled gliding landing. Aerodynamically, it was a near-copy of the US Space Shuttle Orbiter. The Soviet orbiter was 36.37 m long and had a maximum diameter of 5.50 m. Buran's airframe consisted of the crew module shell, the forward fuselage, the mid fuselage, the aft fuselage, a body flap, delta wings with elevons, and a vertical tail. The airframe was largely made of aluminum alloys such as D16 (also widely used in aircraft) and 1201 (specifically developed for Buran), designed to withstand temperatures between about −130°C and +150°C. Other materials used were various titanium alloys for areas experiencing higher stresses as well as a variety of high-temperature and tensile steels. Playing a key role in the development of these materials was the All-Union Institute of Aviation Materials (VIAM). All elements of the airframe were covered by reusable thermal insulation to protect the structure against the wide range of temperatures experienced during ascent, in orbit, and during re-entry.

The *forward fuselage* (Russian acronym NChF) was 9 m long, 5.5 m wide, and 6 m high. It housed the pressurized crew module and forward reaction control system thrusters. Structurally, it consisted of the nosecap, the forward thruster module, and an upper and lower section. The latter two were manufactured separately to allow the pressurized crew module to be inserted during final assembly.

The *mid fuselage* (SChF) was 18.5 m long, 6 m wide and 5.5 m high and contained the payload bay, the nose landing gear, the electricity-generating fuel cells, and their fuel tanks, wiring for the power system and flight control system, various tanks for the environmental control system and thermal control system, and also propellant lines connecting the forward and aft thruster sections. Twenty-six frame assemblies provided stabilization of the mid fuselage structure. Longerons on either side absorbed the bending loads of the vehicle and contained the hinges of the payload bay doors. Mounted in the side walls were several doors to vent the vehicle's unpressurized compartments and to service the fuel cells. The front part of the mid fuselage housed the nose gear wheel well, nose landing gear, and nose gear

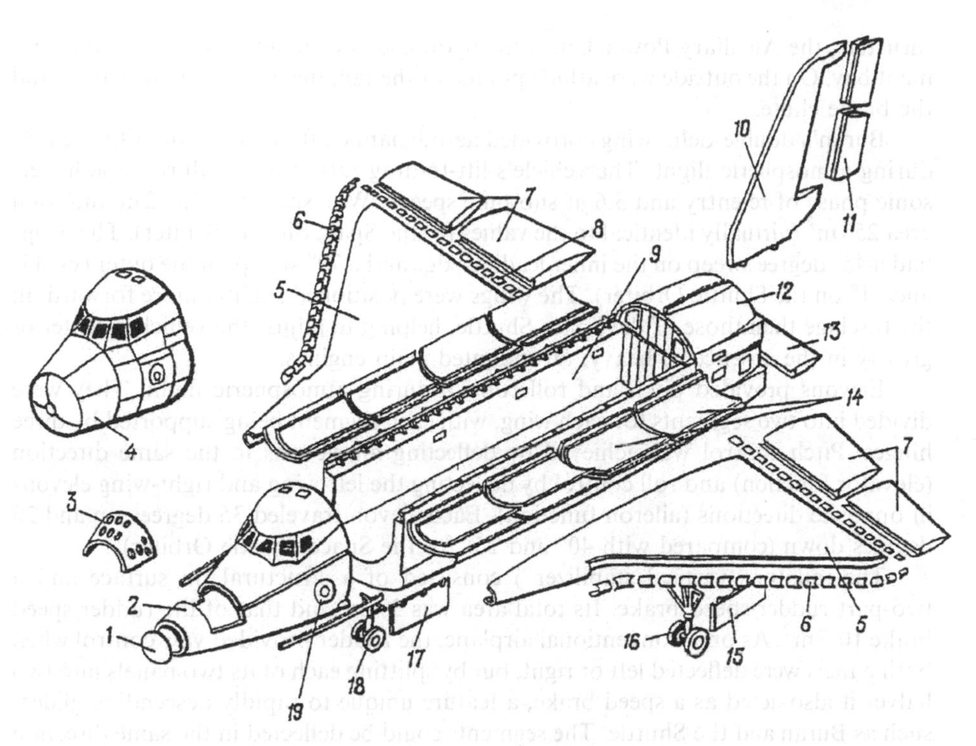

Main elements of Buran's airframe: 1, nosecap; 2, forward fuselage; 3, forward thrusters; 4, crew module; 5, wing; 6, reinforced carbon–carbon panels; 7, elevons; 8, elevon hinges; 9, mid fuselage; 10, tail; 11, rudder/speed brake; 12, aft fuselage; 13, body flap; 14, payload bay doors with radiators; 15, main landing gear door; 16, main landing gear; 17, nose gear door; 18, nose gear; 19, entry hatch (*source*: Yuriy Semyonov/Mashinostroyeniye).

doors. The nose gear was situated farther to the back than the Orbiter's, where it is part of the forward fuselage.

The *payload bay doors* consisted of port and starboard doors hinged at each side of the mid fuselage. They were 18.5 m long and had a gross area of 144 m^2. Each door was made up of four segments, each of them resting on 12 hinges. The doors were held closed by a total of 33 latches, consisting of 16 bulkhead latches (eight forward and eight aft) and 17 payload bay door centerline latches. The doors were composed of a lightweight graphite–epoxy composite material (KMU-4E) that was much lighter than the D-16 aluminum alloy used in most of Buran's airframe. At a later stage it was supposed to have been replaced by an even lighter material called KMU-8. The total mass of Buran's doors was 1,625 kg, which was 620 kg lighter than the mock-up aluminum doors built for the BTS-002 vehicle that flew the atmospheric approach and landing tests. The doors also served as a strongback for the radiator panels that allowed heat rejection in orbit.

The *aft fuselage* (KhChF) was 3.6 m long, 5.5 m wide, and 6 m high. It housed the Combined Engine Installation with its orbital maneuvering engines and steering thrusters, the Auxiliary Power Units, the hydraulic system, and a pressurized equipment bay. On the outside were attach points for the tail, the body flap, the wings, and the brake chute.

Buran's double-delta wings provided aerodynamic lift and control of the vehicle during atmospheric flight. The vehicle's lift-to-drag ratio was 1.3 during the hypersonic phase of re-entry and 5.6 at subsonic speeds. Wingspan was 23.92 m and total area 250 m^2 (virtually identical to the values for the Space Shuttle Orbiter). The wings had a 45° degree sweep on the inner leading edge and a 79° sweep on the outer (vs. 45° and 81° on the Shuttle Orbiter). The wings were positioned slightly more forward on the fuselage than those of the Space Shuttle, helping to adjust the vehicle's center of gravity in the absence of heavy, aft-mounted main engines.

Elevons provided pitch and roll control during atmospheric flight. They were divided into two segments for each wing, with each segment being supported by three hinges. Pitch control was achieved by deflecting all elevons in the same direction (elevator function) and roll control by deflecting the left-wing and right-wing elevons in opposite directions (aileron function). Each elevon traveled 35 degrees up and 20 degrees down (compared with 40° and 25° for the Space Shuttle Orbiter).

The tail (or "vertical stabilizer") consisted of a structural fin surface and a two-part rudder/speed brake. Its total area was 39 m^2 and that of the rudder/speed brake 10.5 m^2. As on a conventional airplane, the rudder provided yaw control when both panels were deflected left or right, but by splitting each of its two panels into two halves it also acted as a speed brake, a feature unique to rapidly descending gliders such as Buran and the Shuttle. The segments could be deflected in the same direction for rudder control of plus or minus 23° (27° on the Orbiter) and the halves could be moved in opposite directions for speed brake control for a maximum of about 43.5° each (49.3° on the Orbiter).

An aerodynamic surface not seen on conventional airplanes is the body flap, attached to the bottom rear of the aft fuselage. With a maximum deflection angle of 30°, it provided pitch control trim to reduce elevon deflections [10].

LANDING GEAR AND DRAG CHUTES

Buran's landing gear was arranged conventionally, consisting of a nose landing gear and left and right main gear. All three gear wells were covered by one door each (as opposed to the two doors on the Orbiter's nose gear). Each gear was actuated by a single hydraulic cylinder. If the hydraulic systems failed, there was a back-up procedure to deploy the gears pyrotechnically. The wheels, two on each gear, were about twice as light as similarly loaded wheels on aircraft thanks to the use of tubeless tires made from natural rubber and beryllium brake disks. Because of the heat that accumulates in the brakes during roll-out, the main landing gear wheels were cooled with nitrogen gas right after the completion of the landing roll-out. During long missions the landing gear was maintained at the proper temperature by electric heaters and also by circulating hydraulic fluid through it.

BTS-002 atmospheric test model with drag chutes deployed (*source*: *www.buran.ru*).

Buran as well as the BTS-002 atmospheric test vehicle were equipped with drag chutes to relieve the stress on the brakes and reduce the landing roll-out distance by 500 m. Stored in a container under the vertical stabilizer, the drag chutes were automatically activated by a pyrotechnic system as soon as the main landing gear touched the runway. The three chutes (each having an area of 25 m^2) were extracted by three small pilot chutes and then jettisoned once the speed had been reduced to 50 km/h. Heaters and thermal protection ensured that the temperatures inside the parachute compartment did not drop below −50°C in orbit and did not exceed +100°C during re-entry [11]. NASA originally also planned to have drag chutes for the Orbiter flight tests, but deleted them in 1974 because it was reasoned that the lakebed runways at Edwards Air Force Base were more than long enough. However, they were eventually introduced on Endeavour in 1992 and later installed on the other Orbiters as well.

HYDRAULICS

Buran's hydraulic system provided hydraulic pressure for positioning actuators needed to move the aerodynamic surfaces (elevons, body flap, rudder/speed brake), deploy the landing gear, operate the main landing gear brakes, and conduct nose wheel steering. Three independent hydraulic circuits were available to provide the necessary redundancy, with one being enough to safely land Buran. A four-circuit system was considered (as it was for the Space Shuttle Orbiter), but rejected due to weight considerations. Each circuit had a hydraulic pump and reservoir, containing a hydraulic fluid. The hydraulic system was designed to operate in temperatures ranging from −60°C to +175°C. In order to keep the system warm enough in orbit, the hydraulic fluid was circulated periodically by an electric-motor-driven circulation pump to absorb heat from heat exchangers in each hydraulic circuit. To prevent the system from overheating during re-entry, each circuit was equipped with a water spray boiler.

Whereas airplanes use their engines to power the hydraulic pumps, gliders such as the Shuttle and Buran need Auxiliary Power Units (APUs) to perform the same function. Just like the Shuttle, Buran had three Auxiliary Power Units (Russian acronym VSU) in the aft fuselage. The VSUs were developed and built by NPO Molniya. They were fueled by hydrazine, which was decomposed in a gas turbine to produce a hot gas that powered a turbine that in turn ran a hydraulic pump. Engineers looked at several fuel combinations (*tsiklin* + an oxide, ammonia + nitrous oxide, hydrogen peroxide + hydrazine), but in the end settled for a hydrazine monopropellant system, as on the Orbiter. The Russians probably made this decision before NASA realized that hydrazine-fueled APUs were not the best of choices. Aside from being a toxic fluid that requires special handling provisions, hydrazine is also a highly flammable chemical. This became all too apparent on the STS-9 mission in 1983, when a hydrazine leak caused a potentially catastrophic fire in Columbia's aft fuselage only minutes before landing. The replacement of a hydrazine-fueled APU by an electric APU was high on NASA's priority list of Shuttle upgrades before the 2003 Columbia accident.

Having a dry mass of 235 kg, each VSU consisted of a fuel unit, the power unit itself, and a system controller. The fuel unit and power unit were built as one integrated system, two located on the left inner wall of the aft fuselage and one on the right inner wall. The system controllers were installed in an equipment bay at the base of the aft fuselage.

The fuel unit contained a single tank with 180 kg of hydrazine and several gaseous nitrogen tanks. Nitrogen was stored in these tanks at a pressure of 32 megapascals (MPa) and first passed through a pressure regulator where the pressure was reduced to 3.5 MPa before it entered the fuel tank to push the hydrazine to the power unit. Each fuel unit was hermetically sealed to prevent any hydrazine leakage into the aft fuselage of Buran. To minimize the fire hazard, the enclosure was purged with nitrogen during re-entry beginning at an altitude of 30 km.

The main elements of the power unit itself were the gas generator, the turbine, and an oil tank. In the gas generator the hydrazine passed over a catalyst bed, which decomposed it into a hot gas that drove a single-stage turbine. While the gas was vented overboard via an exhaust duct, a double-reduction gear reduced the turbine speed from 55,000 rpm to 4,500 rpm before the mechanical drive was imparted to the hydraulic pump. Oil was circulated through the system to lubricate and cool the gear-box and the turbine bearings.

The main difference between the Shuttle's APUs and Buran's VSUs is that the latter used a pressure-fed system rather than a pump to deliver the hydrazine to the gas generator. While the pressure-fed system consumes a slightly larger amount of fuel, it is less prone to fires and other serious malfunctions. Also, in the Shuttle the

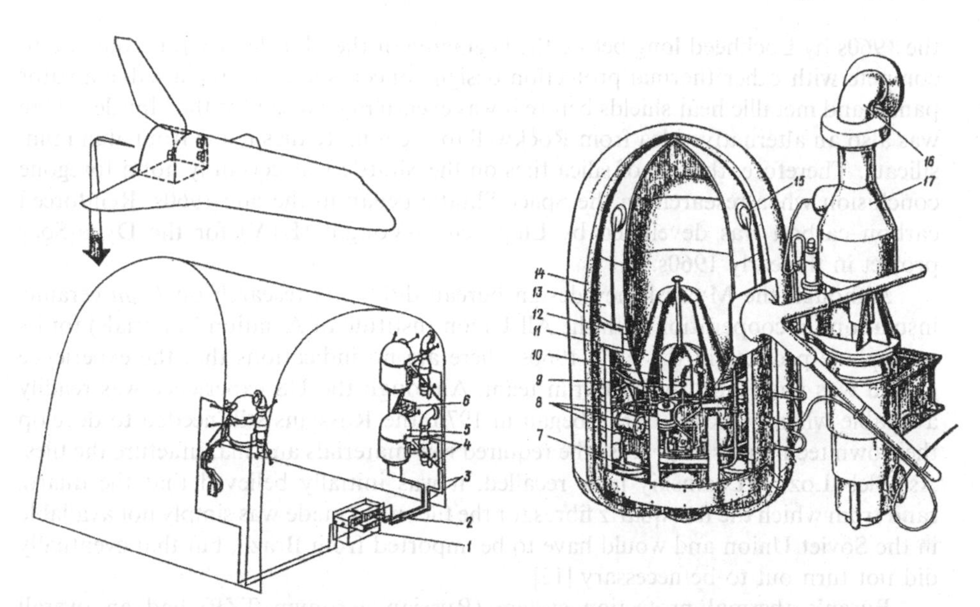

Buran Auxiliary Power Units: 1, instrument compartment; 2, system controller; 3, electric cables; 4, fuel unit attachment system; 5, fuel unit; 6, access panel; 7, nitrogen tanks; 8, fuel unit automatic systems; 9, inertial fuel inlet; 10, negative-*g* section; 11, fuel inlet system; 12, fuel tank primary structure; 13, fuel tank; 14, thermal insulation casing; 15, electric heater; 16, exhaust duct; 17, power unit (*source*: Yuriy Semyonov/Mashinostroyeniye).

fuel tanks are in different locations than the power units, complicating the plumbing and increasing the fire hazard.

The VSUs were designed to operate continuously for a maximum of 75 minutes. Because of the absence of main engines on Buran, the VSUs were not needed for gimbaling the main engine nozzles as was the case for the Shuttle's APUs. However, the VSUs were still started shortly before launch to enable the vehicle to make an emergency landing in certain abort scenarios. They were shut down about 200 seconds into the launch because at that point Buran had enough energy to reach orbit if one or more core stage engines failed. After an in-orbit check-out the VSUs were not reactivated until after the deorbit burn at an altitude of about 100 km. On Shuttle missions one of the three APUs is activated before the deorbit burn, with the other two following afterwards [12].

THERMAL PROTECTION

Thermal protection was definitely one area where the Russians heavily benefited from US experience. Like the Shuttle Orbiter, Buran was largely covered with black and white silica tiles, with reinforced carbon–carbon (RCC) protecting the nosecap and the leading edges of the wings. Ceramic reusable surface insulation was developed in the 1960s by Lockheed long before the beginning of the Shuttle program, but had to compete with other thermal protection design concepts such as replaceable ablator panels and metallic heat shields before it was eventually chosen for the Shuttle. There was also an alternative idea from Rockwell to use mullite tiles made from aluminum silicate. Therefore, the use of silica tiles on the Shuttle was certainly not a foregone conclusion when research on the Space Shuttle began in the late 1960s. Reinforced carbon–carbon was developed by Ling-Temco-Vought (LTV) for the Dyna-Soar project in the early 1960s.

Although the Myasishchev design bureau did some research on *foam* ceramic insulation (in cooperation with the All-Union Institute of Aviation Materials) for its spaceplane projects in the early 1960s, there are no indications that the experience gained was passed on to the Buran team. Although the US experience was readily available when work on Buran began in 1976, the Russians still needed to develop their own techniques to process the required raw materials and manufacture the tiles. As Gleb Lozino-Lozinskiy later recalled, it was initially believed that the quartz sand from which the fine quartz fibres for the tiles were made was simply not available in the Soviet Union and would have to be imported from Brazil, but that eventually did not turn out to be necessary [13].

Buran's thermal protection system (Russian acronym TZP) had an overall mass of 9 tons and was designed to protect the vehicle's aluminum skin from the very low temperatures in Earth's shadow (down to −150°C) to the extremely high temperatures encountered during re-entry (up to +1,600°C). The temperature of the aluminum skin was not allowed to be lower than −120°C or exceed +160°C and shouldn't be any higher than +50°C prior to the beginning of re-entry. These conditions needed to be met in order for Buran to make a total of 100 missions. Buran had five types of thermal protection: white tiles, black tiles, felt material, carbon–carbon, and thermal barriers.

Tiles

Buran was covered with approximately 38,800 heat-resistant tiles (compared with nearly 31,000 tiles on Columbia for STS-1). Each tile consisted of a substrate and a coating. The substrate came in two types with different densities. One was called TZMK-10 (with a density of 0.15 g/cm^3) and the other TZMK-25 (density 0.25 g/cm^3). These were more or less comparable in characteristics and performance with the two basic types of Shuttle tile substrate (Li-900 and Li-2200). They were used in regions where Buran was exposed to temperatures of anywhere between 700°C and 1,250°C. The tiles were made of high-purity 98–99 percent amorphous silica fibres derived from common sand (SiO_2—silica) with minimum amounts of natrium, potassium, and calcium oxides to lower the melting point of the fibers. The thickness of the tiles depended on where they were attached to the aluminum skin and the temperatures and aerodynamic stresses that any particular part of Buran was exposed to.

Both the TZMK-10 and 25 had special 0.3 mm thick glass coatings to reject heat and protect the tiles against wind loads and moisture penetration. This was very

Post-flight picture of black and white tiles near Buran's entry hatch. Note "smearing" of some tiles (B. Vis).

similar to the Reaction-Cured Glass (RCG) coating on the Shuttle's tiles. Chemicals were added to the coating to give the tiles different colors and heat rejection capabilities. Black coating (both for TZMK-10 and 25) was mainly needed to protect the underside of Buran against the high temperatures of re-entry, with the higher-density TZMK-25 only being used in regions exposed to the highest stresses. The black-coated tiles on the belly could not be permanently exposed to sunlight for more than 6 hours. White coating (only applied to TZMK-10) mainly served the purpose of protecting the upper surfaces of the vehicle against solar radiation in orbit.

Although the coating provided some protection against moisture penetration, any cracks in the coating would easily let moisture through. Therefore, additional measures had to be taken to make the tiles waterproof. During manufacture the tiles were treated with a special silicon polymer solution, but that burned out during the first flight in all areas where temperatures exceeded +450°C. Therefore, the tiles would have needed to be rewaterproofed for any subsequent missions (had they been flown). For that purpose the Russians developed a varnish-like coating as well as a technique to permeate the tile with a substance known as hexamethyl disilazane. NASA uses a similar substance (dimethylethoxysilane) for rewaterproofing Shuttle tiles, but injects the material into the tiles, whereas the Russians planned to use a gas diffusion technique.

Since the fragile tiles could not withstand structural deflections and expansions of the aluminum skin, they were not attached directly to the skin, but to 4 mm thick felt

Buran sitting atop Mriya at the Paris Air Show in 1989. Square-shaped ATM-19PKP panels are visible on the mid fuselage (surrounding the name "Buran") and on the upper portions of the payload bay doors (*source*: Luc van den Abeelen).

pads, which then in turn were bonded to the actual skin. Similar to the Shuttle's Strain Isolation Pads, they were attached to the tiles as well as to the skin of Buran with an adhesive based on silicon rubber, ensuring a reliable bond in a temperature range of −130°C to +300°C.

Since the tiles thermally expanded or contracted very little, small gaps were left between them to permit relative motion and allow for the deformation of the aluminum structure under them due to thermal effects. The gaps were filled with a special felt-type material based on organic fibers and capable of withstanding temperatures of up to 430°C.

Tests showed that, if a tile was lost but the underlying felt pad remained in place, the temperature of the aluminum skin would not reach its 500°C melting point, even in areas where temperatures reached 1,250°C. If the felt pad was also lost, there could be damage to the skin, but only in regions close to where carbon–carbon panels were used.

About 28,000 of the tiles were trapezoidal in shape with sizes ranging from about 150 × 150 mm to 200 × 200 mm. Approximately 6,000 tiles were irregular and formed complex patterns on the hatches, around the nozzles of the engines, and on certain edges. Approximately 4,800 tiles had even more complex shapes. Although the distribution of black and white tiles over Buran's surface was very similar to that on the Orbiter, there were different layout patterns. A fan-type pattern was used on the nose section, elevons, and the vertical stabilizer to avoid the use of triangular and sharply angular tiles of low strength.

Felt reusable surface insulation

For regions exposed to temperatures of up to 370° Buran had multiple-layer, square-shaped panels of flexible insulation, similar to the Felt Reusable Surface Insulation (FRSI) employed by the Shuttle. Known as ATM-19PKP, the material was similar to that used for the felt pads under the tiles and was applied to the upper payload bay doors, portions of the upper wing surfaces, and portions of the mid fuselage.

Carbon–carbon

The areas where Buran incurred the highest heating during re-entry (up to 1,650°C) were the nosecap and the leading edges of the wings. As on the Orbiter, these parts were covered with a reinforced carbon–carbon (RCC) material. Until 1978 efforts focused on an RCC material known as KUPVM-BS, but despite its high thermal resistance and strength, it turned out to be too difficult to use. Eventually, the choice fell on a material called GRAVIMOL, an acronym reflecting the names of the three organizations that developed it (NII Grafit, VIAM, and NPO Molniya). There were some small differences in the composition of the RCC material used in the nosecap and the wing leading edges (GRAVIMOL-B in the wing leading edges). The material's density was 1.85 g/cm^3. The RCC had a coating of molybdenum disilicide to prevent oxidation. As on the Shuttle Orbiter, each wing leading edge was covered with 22 RCC panels.

Thermal barriers

Flexible thermal seals protected the vehicle in between certain types of thermal protection material and also in areas containing movable parts. Brush-type seals covered small gaps between sections of the payload bay doors and also in the vertical stabilizer, body flap, and elevons. Seals made of quartz fibers protected areas between the thermal protection system and various doors and hatches. Seals composed of silicon carbide fibers were used in areas exposed to extremely high temperatures such as the gaps between the RCC panels on the wing leading edges and areas where the RCC material bordered on the tiles. An ablative material capable of withstanding temperatures up to +1,800°C covered the gaps between the elevons [14].

VENTILATION

Buran had a so-called Airframe Pressurization and Ventilation System (SNVP). Similar to the Orbiter's Purge, Vent, and Drain System, it served several purposes:

GRAVIMOL material covers Buran's nosecap (B. Vis files).

to maintain proper temperature and moisture levels in the vehicle's unpressurized compartments on the ground, to cool the aluminum skin after landing, to vent the unpressurized compartments during ascent and re-entry, and to prevent big differences in pressure between the mid and aft fuselage. The SNVP consisted of fourteen inward opening 510 × 200 mm vent doors, six on either side of the mid fuselage and one on either side of the aft fuselage, and a series of air ducts with non-return valves. Half of the vent doors were equipped with filters.

Thermal control of Buran's unpressurized compartments was particularly important in the harsh climate of the Baykonur cosmodrome. The SNVP was used for this purpose whenever the vehicle was not in the hangar, whether it be on the pad or during transportation from the assembly building to the pad or from the runway back to the hangar. After circulating through the vehicle the air was released via the vent doors. Another task of the SNVP on the ground was to prevent accumulation of hazardous gases inside the vehicle.

By cooling Buran's aluminum skin after landing, the SNVP played an important role in ensuring the ship's reusability. The aluminum alloy from which the bulk of Buran's airframe was made (D16) could not be repeatedly exposed to temperatures higher than +150/160°, even though Buran would face the same kind of heating during re-entry as the Shuttle (whose skin can withstand +175°C). Therefore, it was necessary to extensively ventilate the vehicle with cool air (no warmer than +10°C)

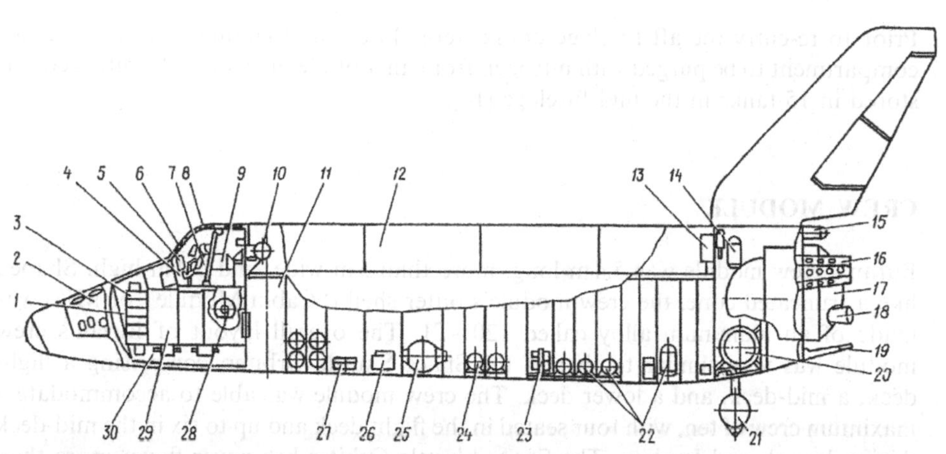

Buran's main systems: 1, crew module; 2, forward thrusters; 3, instrument compartment; 4, flight deck; 5, RM-1 and RM-2 workstations; 6, windows; 7, ejection seats; 8, ejection seat escape hatches; 9, RM-3 workstation; 10, radio altimeter; 11, payload bay; 12, payload bay doors; 13, upper narrow-beam antenna (ONA-I) (stowed); 14, Auxiliary Power Units; 15, drag chute compartment; 16, aft thrusters; 17, propulsion system "base module"; 18, orbital maneuvering engines; 19, body flap; 20, pressurized instrument compartment; 21, lower narrow-beam antenna (ONA-II) (deployed); 22, gas, water, and ammonia tanks; 23, equipment units; 24, tanks of fire suppression system; 25, fuel cell tanks; 26, fuel cells; 27, electric power system instrument module; 28, entry hatch; 29, mid-deck; 30, lower deck (*source*: Yuriy Semyonov/Mashinostroyeniye).

after landing. Ground equipment was hooked up to Buran's SNVP for this purpose within 8 minutes after touchdown.

The SNVP's vent doors were primarily used to equalize inside and outside pressure during launch and landing. During launch the doors were opened between altitudes of 200 m and 35 km. In orbit the ventilation doors situated in the mid fuselage were again briefly opened prior to payload bay door opening to dump any residual pressure that might affect the operation of the payload bay door latches. Those same vent doors were opened during the return phase at an altitude of 22.5 km. The ones lacking filters were closed again at 400 m to prevent dust contamination of Buran's interior. The doors in the aft fuselage remained open throughout the orbital phase of the mission. By creating a near-perfect vacuum in the aft fuselage, it became easier to maintain the liquid oxygen tank of Buran's propulsion system at cryogenic temperatures.

The SNVP also allowed Buran's internal compartments to be purged with nitrogen to minimize the fire hazard in both the mid and aft fuselage during the early stages of launch and the final phases of landing. At $T-40$ minutes the airflow was stopped, after which the vehicle's interior was purged with ground-supplied nitrogen until $T-5$ minutes. Subsequently, the vent doors were closed to ensure that enough nitrogen remained inside the vehicle during the early phase of launch.

Prior to re-entry the aft fuselage doors were closed until landing, allowing the aft compartment to be purged with nitrogen from an altitude of 30 km. The nitrogen was stored in 15 tanks in the mid fuselage [15].

CREW MODULE

Buran's crew module was 5.4 m long, more than 5 m wide, and 5.4 m high. Shaped like a truncated cone, the crew module's outer shell ("Cabin Module" or MK) was made of an aluminum alloy called 1201-T1. The overall layout of Buran's crew module was very similar to that of the Space Shuttle Orbiter, comprising a flight deck, a mid-deck, and a lower deck. The crew module was able to accommodate a maximum crew of ten, with four seated in the flight deck and up to six in the mid-deck during launch and landing. The Space Shuttle Orbiter has never flown more than eight astronauts (one single time on STS-61A in 1985), but could theoretically carry two more if the bunk sleep stations in the mid-deck are removed. The overall volume was 73 m^3.

Flight deck ("Command Compartment" or KO)

The flight deck provided seating for four crew members. The commander and co-pilot occupied the left and right front seats, respectively. Directly behind them would have been a so-called "specialist" (middle position) and a flight engineer (right position). This was a different seating arrangement than in the Space Shuttle Orbiter, where the mission specialist acting as flight engineer during launch and re-entry is seated in the middle behind the commander and co-pilot, looking over their shoulders to check vital instruments. On Buran the flight engineer would have had individual displays on the right-hand side of the cockpit.

There were six forward windows, two overhead windows, one aft window looking out into the payload bay (vs. two on the Shuttle), and a smaller aft-looking porthole permanently occupied by a crew visual navigation instrument. In front of the overhead windows were two jettisonable panels that would have allowed the commander and co-pilot to escape from the vehicle with ejection seats in case of an emergency during launch or landing.

The crew workstations were quite reminiscent of the Space Shuttle's original Multifunction CRT Display System, using the traditional cathode ray tubes rather than the full-color liquid-crystal multifunction display units of the Shuttle's "glass cockpit" introduced in the late 1990s.

The crew had six workstations (RM) at their disposal:

– RM-1 and RM-2 (front left and front right): the commander and co-pilot workstations, used during launch, re-entry, and also some orbital operations. RM-2 duplicated many of RM-1's systems. The instrument boards and control panels of RM-1 and RM-2 were known together as Vega-1 or 17M27. There were three CRTs, controlled by a display processor known as

Adonis. There was one keyboard for interaction with the vehicle's on-board computers.

- RM-3 (middle right): the flight engineer's workstation, used to control vital systems during launch, in orbit, and during re-entry. The console (Vega-2/17M28) featured two CRTs controlled by two US3-DISK display processors and a keypad for interaction with the on-board computer system. It was felt by some that the RM-3 unnecessarily duplicated the functions of other

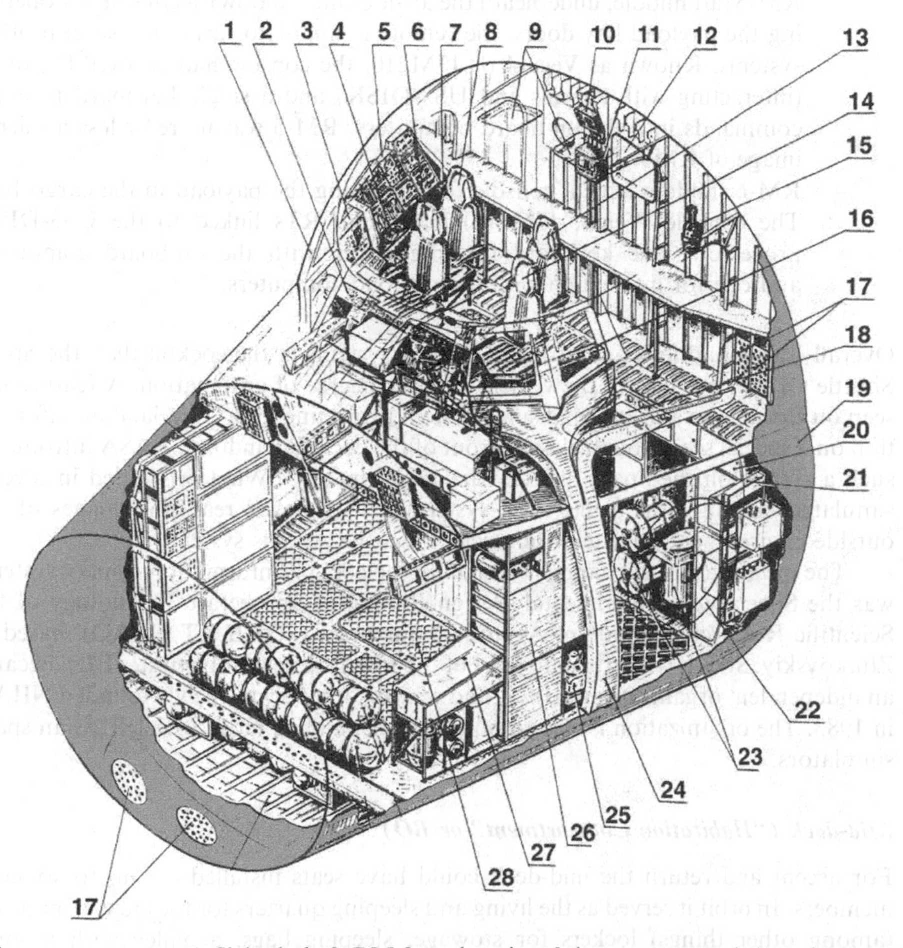

Crew compartment: 1, flight deck; 2, RM-2 workstation; 3, instrument panel; 4, equipment; 5, RM-3 workstation; 6, co-pilot seat; 7, depressurization valve; 8, flight engineer seat; 9, overhead windows; 10, instrument panel; 11, instrument panel; 12, aft window; 13, passenger seat; 14, commander seat; 15, fire extinguisher; 16, RM-1 workstation; 17, feed-through plates; 18, interdeck opening; 19, mid-deck; 20, lockers; 21, instrument bay; 22, entry hatch; 23, toilet; 24, air duct; 25, cooling/drying device; 26, galley; 27, instrument bay; 28, access panel to lower deck (*source*: NPO Molniya/Moscow Aviation Institute).

workstations and that most or all of those functions could eventually be transferred to RM-1/2 and RM-4/5, thereby saving mass.
- RM-4 (aft middle, underneath the porthole with the navigational instrument): a workstation used for orbital operations such as rendezvous and docking, orbit corrections, and navigational measurements and corrections. The console (Vega-3 or 17M29) had two CRT displays linked to the Adonis and US3-DISK processors and a single keypad that interfaced with the ship's computer complex.
- RM-5 (aft middle, underneath the aft-looking window): a console for operating the payload bay doors, the remote manipulator arm, and several other systems. Known as Vega-4 or 17M210, the console had two CRT screens (interacting with Adonis and US3-DISK) and a single keyboard to enter commands into the on-board computers. RM-5 was more or less a mirror image of RM-4.
- RM-6 (middle left): a console for operating the payload in the cargo bay. The console (Vega-5/17M211) had two CRTs linked to the US3-DISK processors, one keyboard for interaction with the on-board computers and one for interaction with the payload computers.

Overall Buran had fewer control and display systems in the cockpit than the Space Shuttle Orbiter because of the vehicle's higher degree of automation. A feature not seen on Buran was a heads-up display system projecting important landing information on a special see-through glass in front of the cockpit windows. NASA introduced such a system on the Space Shuttle Challenger in 1983. What *was* tested in several simulated landings was a television system that displayed real-time images of the outside environment on a television screen via the Adonis system.

The organization in charge of designing the cockpit information display systems was the Specialized Experimental Design Bureau of Spacecraft Technology of the Scientific Research Institute of Aviation Equipment (SOKB KT NIIAO), based in Zhukovskiy. SOKB was originally part of the Flight Research Institute (LII), became an independent organization in 1971, and was absorbed by the newly founded NIIAO in 1983. The organization is also responsible for building most Soviet/Russian space simulators.

Mid-deck ("Habitation Compartment" or BO)

For ascent and return the mid-deck could have seats installed for up to six crew members. In orbit it served as the living and sleeping quarters for the crew, containing (among other things) lockers for stowage, sleeping bags, a galley with a small reclining table, washing facilities, and a toilet. In the aft of the mid-deck there was room for an internal airlock to conduct spacewalks during non-docking missions. For docking missions Buran would have carried a combined docking system/airlock installed in the cargo bay just behind the crew compartment.

The mid-deck also housed three small equipment bays with radio equipment and thermal control systems that could be accessed by the crew via panels. There was a

Buran's cockpit for maiden flight (*source*: Yuriy Semyonov/Mashinostroyeniye).

hatch on the port side of the mid-deck for normal crew ingress and egress, which could be opened very quickly by the crew in emergency situations. As on the Orbiter, there was a small porthole in the middle of the side hatch. Access to the flight deck was via two interdeck openings (left and right), although only the left one was supposed to be used in flight. The mid-deck had its own instrument panel (17M212) with among other things an on-board clock and an emergency warning system. There were also separate instrument panels for the airlock (17M213) and the docking adapter (17M214).

Lower deck ("Aggregate Compartment" or AO)

The lower deck contained life support systems such as air ducts, condensate collectors, oxygen tanks, regenerators, the toilet's waste collection system, and a fire extinguisher bottle. Also installed here were elements of the vehicle's thermal control and power supply systems. The lower deck could be reached by crew members via panels in the floor of the mid-deck.

It should be noted that the fully outfitted crew module as described above was never flown. Since the one and only mission performed by a Buran orbiter was unmanned, the cabin was stripped of much of the equipment essential to support a crew [16].

LIFE SUPPORT AND ENVIRONMENTAL CONTROL

Air supply

Like all earlier Soviet manned spacecraft, Buran used a mixed oxygen/nitrogen atmosphere very similar in composition and pressure to what we breathe on Earth. NASA did not introduce the oxygen/nitrogen mix until the early 1970s on Skylab, having used 100% oxygen atmospheres on Mercury, Gemini, and Apollo. A 100 percent oxygen atmosphere allows for the construction of lighter vehicles and obviates the need for spacewalk pre-breathing, but, on the other hand, significantly increases the fire hazard, as vividly demonstrated by the Apollo-1 fire in 1967.

Buran had three subsystems to provide the crew with breathable air both in standard and emergency situations. These were the Pressurization and Depressurization System (SNiR), the Gas Composition System (SGS), and the Personal Life Support System (ISZhO). The SNiR maintained cabin absolute pressure between 93.3 and 107.3 kilopascals (kPa), supplying oxygen and nitrogen to the cabin from tanks situated in the mid fuselage. As on the Shuttle Orbiter, the oxygen was stored cryogenically in the tanks of the fuel cell system. The SNiR would pump up to 1.5 kg of air into the crew module per day to compensate for routine loss of cabin air and would also repressurize the airlock after spacewalks. The system was automatically activated whenever cabin pressure sank to 98.7 kPa and would then repressurize it to a level of 101.3 kPa. It could also be operated via manually controlled valves. The SNiR was also designed to respond to various emergencies. It could replace the cabin air after a fire or a malfunction of the carbon dioxide removal system and in case of cabin depressurization due to a micrometeorite or space debris impact would blow air into the cabin to give the cosmonauts more time to don pressure suits. If Buran was to have re-entered with a depressurized cabin, the SNiR would have opened a valve to allow outside air to stream into the cabin and minimize pressure differences.

The SGS maintained oxygen partial pressure between 18.7 and 29.3 kPa, making sure that oxygen levels never exceeded 40 percent to limit the fire hazard. The system kept carbon dioxide partial pressure below 1.07 kPa. This was accomplished with regenerators in which CO_2 reacted with potassium superoxide to produce oxygen, which was then recirculated to the cabin air. The ratio of absorbed CO_2 to regenerated oxygen was roughly the same as the respiratory quotient of a human being—that is, the ratio of the volume of carbon dioxide released to the volume of oxygen consumed by the body. The regenerators also had filters to remove trace contaminants from the cabin atmosphere. Similar CO_2 removal systems had also flown on earlier Soviet piloted spacecraft.

Depending on crew size and mission duration, Buran would have needed to carry 6 to 18 regenerators on a single flight. The crew's responsibility was to regularly rehook flexible hoses between cabin ventilators and the regenerators as the potassium superoxide ran out. The Shuttle Orbiter has usually relied on non-regenerative lithium hydroxide canisters for CO_2 removal, as many as 30 of which may be needed on a single flight. NASA did install a regenerative carbon dioxide removal system on the Orbiters Columbia and Endeavour for Extended Duration Orbiter missions, but it did not produce oxygen as a byproduct of the chemical reaction.

The ISZhO was primarily designed to provide life support functions to a full pressure suit that the crew was supposed to wear during critical mission operations such as launch, docking, undocking, and re-entry. Called Strizh ("Swift"—the bird), the suit was derived from the Sokol ("Falcon") pressure suits worn by Soyuz cosmonauts and adapted to be used in conjunction with ejection seats. The system could operate either in an open-cycle or closed-cycle mode. With the loop open, the suit was ventilated with cabin air, which was then released back into the cabin via the helmet (if that was open) or through pressure regulators (if the helmet was closed). With the loop closed, oxygen was supplied to the suit from the fuel cell liquid-oxygen tanks or (if that didn't work) from back-up gaseous oxygen tanks. There were also small portable oxygen containers that could sustain a crew member for 20 minutes. After having passed through the suit, the air moved through a contamination control assembly to remove carbon dioxide and other gases and through a unit that cooled the air and removed the moisture. Finally, the gas was enriched with oxygen and recirculated through the suit. The main operating pressure of the suit was 440 hectopascals (hPa), but could be manually reduced to 270 hPa. A single ISZhO unit formed a ventilation loop for two suits.

The system automatically switched from open loop to closed loop in the event of cabin depressurization or when smoke or other harmful substances were detected in the crew module. The closed-loop mode could also be manually activated by the crew. If the crew members were in shirtsleeves during cabin depressurization, they were able to individually don the Strizh within five minutes, with the SNiR supplying enough air to the cabin to keep them alive during that time (assuming the leak wasn't *too* big). Since as many as 12 hours could elapse between depressurization and an emergency landing, the Strizh also had a waste collection and water supply system. The suits were put to the test in 1990–1991 at a vacuum chamber of the Air Force Scientific Test Institute in Akhtubinsk, when test engineers wore the suits for up to 18 hours, including 12 hours in a mode simulating a depressurized cabin. Unlike the Strizh suits, the pressure suits worn by Space Shuttle astronauts only provide protection during launch and entry, not during in-orbit emergencies.

An additional task of the ISZhO was to support a cosmonaut clad in an Orlan spacesuit during pre and post-spacewalk operations in the airlock, thereby increasing the resources of the suit during the spacewalk itself. More particularly, the system was used to feed oxygen to the suits, to cleanse and cool the air circulating in the suit, and provide water to the cooling garment. The ISZhO was also used to dry the spacesuits in preparation for the next spacewalk.

For unmanned missions the oxygen content in the cabin atmosphere was supposed to be lower to reduce the fire hazard. For instance, Buran had a 90 percent nitrogen/10 percent oxygen atmosphere on its one and only mission in 1988.

The SNiR and SGS were developed by the NPO Nauka organization in Moscow, while the ISZhO and associated pressure and spacesuits were products of the Zvezda organization in Tumilino just outside Moscow [17].

Water supply

Buran had a Water Supply System (SVO) that consisted of the Potable Water System (SPV), designed to provide drinking water to the crew, and the Process Water System (STV), intended to supply water to the thermal control and hydraulic systems.

The bulk of the potable water on Buran was to be produced as a byproduct of the fuel cells, which use oxygen and hydrogen to generate electrical power. Before ending up in one of two reservoirs inside the crew compartment (one prime, one back-up), the near-distilled water passed through a cleansing unit filled with hydrogen gas and then through another unit where it was enriched with silver ions. The water was extracted from the reservoirs via a manually operated pump and then passed through a cooling device or a heater. It could be used either for drinking or for preparing food. A 10-liter back-up supply of potable water was to be pumped into Buran before launch.

Process water was needed for the flash evaporators of the Thermal Control System and the hydraulic system. It was stored in four separate units containing four 25 liter tanks each, giving a total capacity of 400 liters. Since Buran needed this water for cooling during launch, some 370 liters were pumped into the tanks on the ground, with the fuel cells being capable of supplying an additional 30 liters during the final countdown and ascent. During on-orbit operations the STV collected additional water from the fuel cells, dumping overboard any excess supplies. The process water was distilled and saturated with silver ions. It was pushed out of the tanks with compressed air [18].

Thermal control

The Thermal Control System (STR) had two internal and two external loops. Each loop operated completely independently, with pumps circulating cooling agents through it. The cooling agents were substances known as "Antifreeze-20" for the internal loops and "PMS-1.5" for the external loops. The internal loops maintained proper temperature (18–28°C) and humidity (30–70%) in the crew compartment, collected excess heat from equipment inside and just outside the crew compartment, and then transferred that heat to the external loops via liquid-to-liquid heat exchangers. The external loops removed heat from systems in the unpressurized part of Buran (including the fuel cells, the hydraulic system, the payload, the maneuvering and attitude control engines) and finally delivered the excess heat to three types of "heat sinks": the radiator panels on the payload bay doors, flash evaporators, and ammonia boilers.

The eight radiator panels (one on each payload bay door) were the primary means of heat rejection in orbit. Just as on the Shuttle Orbiter, the two forward panels on each side could be unlatched and tilted to allow heat to be radiated from both sides of the panel. The fixed aft panels only dissipated heat from the outer side. When the payload bay doors were closed, heat loads from the external coolant loops were rejected by the flash evaporators or ammonia boilers, which cooled the loops by evaporating water and ammonia, respectively, and venting the resulting gases overboard. Water for this purpose was produced by the fuel cells and stored in the tanks of the Process Water System, whereas the ammonia was loaded in two small tanks prior to launch. The flash evaporators were apparently used during launch and the initial part of re-entry, but since water evaporation becomes ineffective under higher atmospheric pressure, the ammonia boilers took over at an altitude of 35 km [19].

POWER SUPPLY

The Electric Power System (SEP) supplied power to Buran's systems during the final countdown, the mission itself, and during initial post-landing servicing. As on the Shuttle Orbiter, electricity was to be generated with the help of fuel cells ("electrochemical generators" in Russian terminology) using cryogenically stored oxygen and hydrogen reactants. Whereas NASA introduced fuel cells back in the Gemini program, the Russians had always used battery systems and/or solar panels on Vostok, Voskhod, and Soyuz. They *did* develop a fuel cell system called Volga-20 for the Soyuz-based LOK lunar orbiting ship to be used in the N-1/L-3 manned lunar-landing program, but the only LOK ever flown was lost in the fourth and final launch failure of the N-1.

The SEP consisted of the Oxygen/Hydrogen Cryostats, a Power Module, an Instrument Module, and the Distribution and Commutation System. The first three subsystems were situated in the mid fuselage under the front section of the payload bay, so that receding fuel levels in the cryogenic tanks would not affect Buran's center of gravity. Although the oxygen and hydrogen were delivered to the fuel cells in gaseous form at a temperature of about 10°C, they were stored cryogenically to save mass. Buran could accommodate two oxygen and two hydrogen tanks, which needed to be filled in the final days before launch via 500 × 600 mm doors in the mid fuselage.

The oxygen and hydrogen were fed to the Power Module, which contained the actual fuel cells. There were a total of four fuel cell units (as compared with three on the Shuttle Orbiter), each consisting of eight 32-cell stacks connected in parallel and with an active electrode area of 176 m^2. The alkaline fuel cells used a potassium hydroxide electrolyte immobilized in an asbestos matrix and had an oxygen electrode (cathode) and a hydrogen electrode (anode). Each fuel cell unit provided 10 kW continuous and 25 kW peak at between 29 and 34 volts of direct current. Only three

Buran fuel cells (*source*: ESA).

sets of fuel cells were needed for a nominal mission and two for an emergency landing.

The Instrument Module turned the fuel cells on and off and automatically controlled all processes taking place in the system. In case it detected an anomaly, the crew was notified of this on the control panels in the cockpit and with a master alarm. Although the fuel cells were designed to operate entirely automatically, they could also be controlled by the crew or from the ground. Power was distributed to all parts of the vehicle by the Distribution and Commutation System, which consisted of two redundant subsystems, one running along the starboard side, and the other along the port side.

The fuel cells produced water as a byproduct (more than 100 kg per day) for consumption by the crew and also for use in the flash evaporators of the Thermal Control System and the hydraulic system. The liquid oxygen stored in the SEP tanks could also be turned into gaseous oxygen for the crew compartment.

For extended missions, Buran could carry a "cryo kit" located near the middle of the payload bay and equipped with up to six liquid hydrogen tanks. During a long mission the fuel cells would first use the hydrogen supply from the cryo kit before switching to the standard LH_2 tanks under the payload bay. Extra oxygen would be drawn from the LOX tank of Buran's propulsion system situated in the aft fuselage. Buran's cryo kit was comparable with that developed for the Shuttle's Extended Duration Orbiter missions, although that was to be mounted in the *aft* payload bay and had *both* liquid hydrogen and liquid oxygen tanks (given the use of storable rather than cyrogenic propellants in the on-orbit propulsion system).

In addition to the fuel cells, Buran also had battery packs that were charged by the fuel cells and fed electricity to various power-hungry systems, mainly in the aft fuselage. For the first multi-day test flights it was also planned to fly battery packs operating *independently* from the fuel cells to give one day of back-up power in case of a fuel cell failure, enough to make an emergency return to Earth. Since Buran's one and only mission lasted just several hours, the fuel cells were *not* installed, with the vehicle's systems drawing power from batteries in the BDP payload stowed in the payload bay (see Chapter 7). Fuel cells *were* installed on the second vehicle and underwent loading tests at the launch pad.

Called Foton ("Photon"), Buran's fuel cells were developed jointly by NPO Energiya and the Ural Electrochemical Integrated Factory in Verkh-Neyvinsk (Sverdlovsk region), which had also developed the Volga-20 fuel cells for the LOK back in the early 1970s. Although never actually flown in space, the Buran fuel cells attracted the interest of the European Space Agency, which tested a Buran flight-model fuel cell in 1993 at the facilities of ESTEC in Noordwijk, Holland as part of studies to incorporate foreign technology into the Hermes spaceplane. A modified version of Foton powered the first Russian fuel cell car, the Niva, presented at a Moscow auto show in 2001. RKK Energiya is also considering a Foton-derived fuel cell system for its new Kliper spacecraft [20].

PROPULSION

Although Buran lacked main engines for ascent, it did have engines and thrusters for on-orbit maneuvers and attitude control functions. Buran's propulsion system was known as the Combined Engine Installation (ODU or 17D11) and consisted of an integrated set of orbital maneuvering engines, primary thrusters, vernier thrusters, and associated plumbing.

While the overall number and general location of these engines were similar to those of the Space Shuttle Orbiter's Orbital Maneuvering System (OMS) and Reaction Control System (RCS), there were some fundamental differences between the two vehicles, notably the types of propellant used. Orbital maneuvering and attitude control engines on manned spacececraft have traditionally used hypergolic propellants or hydrogen peroxide, which can be stored for long periods of time and do not require complex ignition and turbopump systems. The Space Shuttle Orbiter uses a hypergolic mix of nitrogen tetroxide and dimethyl hydrazine for both its OMS and RCS engines. Although Soviet designers also planned to use hypergolic propellants in their original orbiter concepts (OS-120 and OK-92), they eventually opted for a combination of liquid oxygen and a synthetic hydrocarbon fuel known as *sintin*. This marked the first time that such propellants were used in *any* type of orbital maneuvering and attitude control system. Next to the absence of main engines, this was probably the most significant difference between Buran and the Space Shuttle Orbiter.

Cryogenic propellants offered a number of advantages. They gave the orbital maneuvering engines a better performance than those of the Shuttle (although

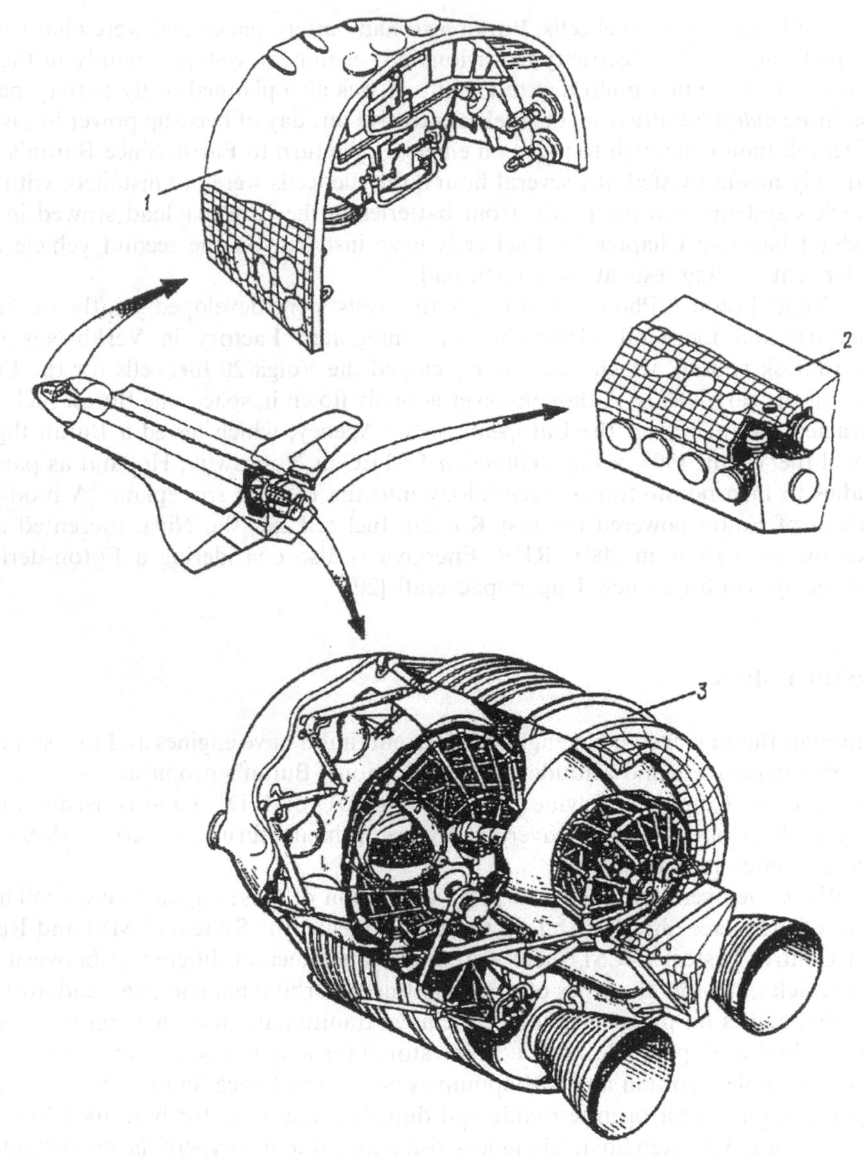

Buran propulsion system: 1, forward thruster module; 2, aft thruster module; 3, base unit (*source*: Yuriy Semyonov/Mashinostroyeniye).

thruster performance was virtually identical) and were safer to handle by ground personnel because of their non-toxicity. Moreover, the LOX could be cross-fed to the storage tanks of Buran's electricity-generating fuel cells, providing extra redundancy to the power system and, indirectly, to the life support system, which drew oxygen and water from the fuel cell system. The drawbacks were that the plumbing was more complex, making the ODU 1,100 kg heavier than the Shuttle's RCS/OMS system. Also, the mix did not ignite spontaneously on contact, such as was the case with hypergolic propellants, but required an electric ignition source. In addition to that, extra measures needed to be taken to prevent the cryogenic oxidizer from boiling off during long missions.

It is interesting to note that in the 1990s US Shuttle engineers considered a cryogenic OMS/RCS as a long-term Shuttle upgrade. This would have used a combination of LOX and ethanol and would have enabled the forward RCS, aft RCS, and OMS engines to draw propellant from common tanks, just as on Buran. It is not clear if this upgrade was in any way inspired by the design of Buran's ODU.

Orbital maneuvering engines

The two orbital maneuvering engines (Russian acronym DOM, also referred to as 17D12) were a further development of NPO Energiya's 11D58 engine used in the Blok-D, an upper stage for the Proton rocket and later also employed by Sea Launch's Zenit-3SL. Each having a vacuum thrust of 8.8 tons and a specific impulse of 362 s, they performed final orbit insertion, orbit circularization, orbit corrections, and the deorbit burn, and were also supposed to be activated in certain launch abort scenarios to burn excess propellant. A long-term objective was to use the DOM engines to provide additional thrust during a nominal launch, a technique that NASA introduced with the Shuttle's OMS engines on STS-90 in 1998.

Usually, only one of the two was required for any given standard burn, with the other acting as a back-up. Simultaneous ignition of both engines was only required in launch emergencies. The DOM ignition process began with a 20–25 second burn of two primary thrusters to force the LOX and *sintin* out of their tanks. The engine used a closed-cycle scheme, re-routing the gases used to drive the turbopump to the combustion chamber. During each burn the propellant tanks were pressurized with gaseous helium. In order to save helium, gaseous oxygen was used to pressurize the LOX tank for the deorbit burn. The engine nozzles could be gimbaled up to 6 degrees in two axes (pitch and yaw) for thrust vector control. Each DOM was designed to be ignited up to 15 times during a single mission.

Thrusters and verniers

Buran had 38 primary thrusters ("Control Engines" or UD) (exactly the same number as on the Space Shuttle Orbiter) and eight verniers ("Orientation Engines" or DO) (two more than on the Orbiter). Together the primary thrusters and verniers formed the Reaction Control System (RSU). The primary thrusters provided both attitude control and three-axis translation, and the verniers only attitude control. They were used for these functions during the launch, separation from the core stage, on-orbit, and re-entry phases of the flight (up to an altitude of 10 km). If needed, some of the UD thrusters could also act as a back-up for the DOM engines.

Orbital maneuvering engines (B. Vis).

The UD thrusters (17D15), built in-house at NPO Energiya, had a thrust of 390 kg and a specific impulse between 275 and 295 s. Unlike the DOM engines, they used gaseous rather than liquid oxygen as an oxidizer. This was obtained with a small turbopump assembly mounted on the ODU LOX tank. First, liquid oxygen from the LOX tank passed through the pump, where its pressure was increased to 78.4 MPa. Then it entered a gas generator where it was ignited with a minute amount of *sintin* fuel (ratio of 100 : 1) to form a mix of gaseous oxygen, carbon dioxide, water vapor, and droplets with a temperature of 60°C. After any residual liquids had been dumped

Aft thrusters (B. Vis).

overboard, the gaseous oxygen was used to drive the turbine and was then stored in separate tanks at pressures ranging from 2.45 to 4.9 MPa. From there it was delivered to the combustion chamber to react with liquid *sintin* through electrical ignition. Each UD thruster could be fired for a duration of anywhere between 0.06 and 1,200 seconds and be ignited up to 2,000 times during a single mission. The thrusters were designed to sustain 26,000 starts and 3 hours of cumulative firing.

The DO verniers (17D16 or RDMT-200K) provided 20 kg of thrust and had a specific impulse of 265 s. They were developed by the Scientific Research Institute of

Machine Building (NII Mashinostroyeniya) in Nizhnyaya Salda, which had been a branch of the Scientific Research Institute of Thermal Processes (NII TP) until 1981 and specialized in small thrusters for spacecraft. The RDMT-200K was probably a cryogenic version of the RDMT-200, a thruster with similar capabilities built for the Almaz space station but burning storable propellants. The verniers were similar in design and operation to the UD thrusters, but used liquid oxygen and a different cooling system. They were intended for short-duration burns with an impulse time between 0.06 and 0.12 seconds and could be ignited up to 5,000 times during a single mission. Thrusters based on the RDMT-200K were supposed to fly on the upper stage of the Yedinstvo/ULV-22 rocket, a launch vehicle studied by the Makeyev bureau in the late 1990s to fly from Australian territory.

Aside from the ODU engines and thrusters, Buran had four small solid-fuel motors (thrust 2.85 tons each) to instantly separate the vehicle from the Energiya core stage in case of a multiple engine or other catastrophic launch vehicle failure. Presumably developed by NPO Iskra, they were situated in the nose section of the vehicle and should have given Buran enough speed to stay clear of the out-of-control rocket after separation. They were *not* supposed to be used in a standard separation from the core stage after main engine cutoff. The solid-fuel motors were apparently not installed on the first flight vehicle that made the one and only Buran mission in 1988.

Layout

Because of the absence of main engines, the layout of the aft part of Buran's on-orbit propulsion system differed from that of the Shuttle Orbiter. The Shuttle's OMS and aft RCS engines are concentrated in separate "OMS pods" on either side of the vertical stabilizer and are each divided into two compartments, one for the OMS and one for the aft RCS. Buran had a single pod ("Base Unit" or BB) for both DOM engines *under* the vertical stabilizer, with two Reaction Control System units ("Control Engine Unit" or BDU) attached to either side of the aft fuselage. The left BDU (BDU-L) and right BDU (BDU-P) each had twelve primary thrusters and four verniers.

The pod housed one big LOX and one big *sintin* tank, two auxiliary *sintin* tanks (exclusively used for the Reaction Control System), and gaseous oxygen tanks (solely used for the primary thrusters). If needed, *sintin* could be transferred from the main tank to the auxiliary ones with a turbopump driven by gaseous oxygen. The helium tanks were immersed in the main LOX tank in order to save space and cool the gas.

The forward reaction control system unit (BDU-N), situated in the nose of the vehicle, had 14 primary thrusters. Unlike the Shuttle Orbiter's forward RCS, it carried no verniers. Also installed in the BDU-N were one gaseous oxygen tank and one auxiliary *sintin* tank. The BDU-N was connected to the aft engine pod via several interconnect lines that allowed gaseous oxygen, *sintin*, and helium gas to be transferred from aft to front. After the orbital phase of the mission was completed, any remaining *sintin* from the front auxiliary tank was transferred back to the aft main tank to satisfy center-of-gravity requirements for landing. Although the Shuttle

Orbiter has always had the capability of cross-feeding propellant between the two OMS pods, it cannot transfer propellant between the OMS pods and the forward RCS. Such an interconnect system was proposed as one of many Shuttle upgrades, but the idea was eventually shelved.

Adaptations for long missions

For long-duration flights or flights requiring extra propellant reserves, it was possible to mount additional tanks in the payload bay. There was room for an additional fuel tank in the front of the bay and for an additional oxidizer tank (big or small) in the aft. These would have been placed such that the vehicle's center of gravity was not disturbed. The additional tanks could have increased Buran's overall propellant load from 7.5 tons to 14 tons, allowing the vehicle to reach an altitude of up to 1,000 km. Plans for a comparable "OMS kit" in the Shuttle Orbiter's payload bay were never implemented.

In order to counter evaporation of the cryogenic oxidizer, Buran's ODU was filled with supercooled LOX at a temperature of −210°C (with LOX having a boiling point of approximately −180°C). This, along with the use of several layers of thermal insulation and LOX-mixing techniques, was enough to prevent any significant boil-off for 15 to 20 days. On longer missions the LOX would have been maintained at proper temperatures by circulating cooled helium through the tank's heat exchanger and also by installing a special cryocooler using the so-called reversed Stirling cycle.

The ODU had an elaborate fault detection and identification system, consisting among other things of about 100 sensors to measure pressures, temperatures, vibrations, etc. The engines could be shut down in a fraction of a second if a dangerous situation developed [21].

AVIONICS

Buran's avionics system performed three main functions:

- Guidance, navigation, and control: input of navigational data into the on-board computers, which in turn sent signals to the engines for attitude control and maneuvering functions in orbit and to the aerodynamic surfaces for control during atmospheric re-entry.
- Sending switch-on/switch-off commands to on-board systems and changing their operating modes in keeping with the flight program.
- Monitoring the operation of on-board systems and in case of anomalies ensure the safety of the crew and the completion of the flight program.

Buran's avionics systems had to meet much higher standards than those of the Soyuz spacecraft because they had to be capable of operating much more independently from ground control stations and needed to ensure a precision landing on a runway rather than a landing in vast stretches of steppe in Kazakhstan. Moreover, they were supposed to enable the vehicle to fly unmanned missions. Buran's avionics systems consisted of 1,256 instruments of 105 different types installed in 59 electronics boxes in the crew module, the mid and aft fuselage.

Computers

The heart of Buran's flight control system were two Soviet-built redundant computer sets known as the Central Computing System and the Peripheral Computing System, each consisting of four identical computers called "Biser-4" ("Beads"). The US Space Shuttle has a single redundant set of four computers and a fifth back-up computer using different software. Weighing 33.6 kg and using 270 watts of power, each Biser was made up of a central processing unit to provide the central computational capability and an input–output processor to transmit commands to vehicle systems and validate response data from those systems. The computers ran in sync with each other, with the computations of each computer being verified by the other. If one of the computers failed, it was voted out by the others. Each redundant set remained fully operational with two computers down. If a third one failed, there was still at least a 75 percent chance of maintaining the same capacity as a full set. Rather than using program synchronization as was the case with the Shuttle's General Purpose Computers, the Biser computers were synchronized by a single quartz clock generator that emitted 4 Mhz clock pulses to all eight computers at intervals of 32.8 milliseconds. The generator had five redundant channels.

Each Biser-4 was equipped with 131,072 32-bit words in random-access memory and 16,384 in read-only memory. The software was divided into system software to operate the computers themselves and applications software to perform the functions required to fly and operate the vehicle. While the operations system software permanently resided in the computer, the applications software was too big to fit in the available computer memory space. Therefore, it was divided into several memory groups corresponding to specific flight phases and stored on a magnetic tape mass memory unit with a capacity of 819,200 32-bit words. In that way, applications software needed for a specific phase of the flight could be loaded into the computers' random-access memory from the mass memory unit when needed. The unit stored two versions of each memory group.

The lead organization for the development of the on-board computers was NPO AP, headed until 1982 by Nikolay Pilyugin, who was subsequently replaced by Vladimir Lapygin. Originally, the software was also to be written at NPO AP, but software development ran into major problems in the early 1980s, which is why several other organizations became involved in 1983. Two new specific software languages known as "PROL-2" (used by the on-board computers) and "DIPOL" (used by ground computers during vehicle testing) as well as a software language enabling those two to interact ("FLOKS") were devised for Buran under the leadership of Mikhail Shura-Bura at the Institute of Applied Mathematics [22].

Navigation systems

The orbiter's primary navigational aids were three so-called "gyro-stabilized platforms" (GSPs). Comparable with the Shuttle's three Inertial Measurement Units (IMUs), they provided inertial attitude and velocity data to the guidance, navigation, and control software. Just like the IMUs, the GSPs were isolated from rotations by four gimbals and used a set of gyros to maintain the platform's inertial orientation. Attitude data was provided by so-called resolvers and velocity data by a set of accelerometers. Whereas the Shuttle's IMUs are mounted on a navigation base forward of the flight deck control and display panels, Buran's GSPs were installed in a small module in the payload bay, attached to the outer wall of the aft flight deck just under the aft-looking window.

Since Energiya's own GSPs were much more accurately aligned prior to launch than those of Buran, the orbiter primarily relied on data from Energiya's navigation sensors for accurate azimuth alignment during launch. Buran's on-board computers continuously compared navigation data originating from Energiya with that obtained by the orbiter's own sensors and then made the necessary corrections.

In-orbit alignment of the GSPs was conducted with star trackers and a radio altimeter, attached to the right and left sides of the GSP module. While the Shuttle Orbiter also uses star trackers for IMU alignment, the radio altimeter was unique to Buran. The star trackers, concentrated in a so-called Stellar–Solar Instrument (ZSP), measured the line-of-sight vector to at least two stars. Using this information, the on-board computers calculated the orientation between these stars and the orbiter to determine the vehicle's attitude. Comparison of this attitude with the attitude measured by the GSP provided the correction factor necessary to null the GSP error. The ZSP had a door which was opened after opening of the payload bay doors.

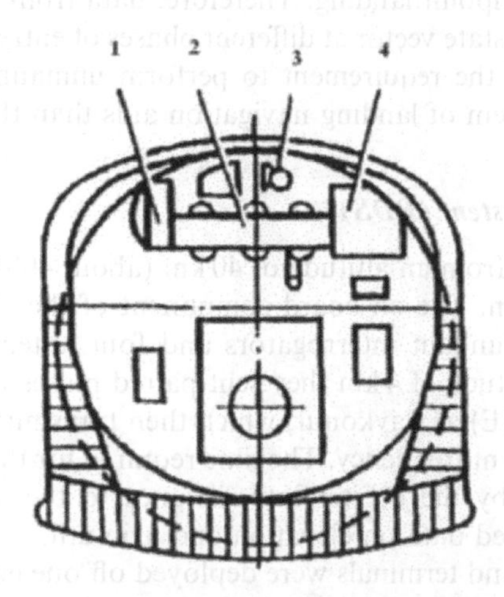

Location of navigation sensors under aft porthole: 1, radio altimeter; 2, GSP module; 3, visual navigation measurement system; 4, stellar–solar instrument (*source*: Yuriy Semyonov/Mashinostroyeniye).

The radio altimeter (Vertical Radio Altimeter or RVV), providing local vertical measurements, acted as a back-up to the star trackers for GSP alignment and thereby increased the reliability of the navigation system. GSP re-alignment required at least two measurements of the local vertical, ideally with an interval of a quarter orbit. However, the RVV only provided reliable information over bodies of water, making it necessary to accurately time the measurements. Shortly after a GSP alignment session, the RVV could also be used for autonomous navigation, updating the vehicle's state vector. State vector updates were also performed with a Sunrise/Sunset Detection Instrument (PRZS), an optical device that compared the expected and actual moments of sunset and sunrise as seen from the orbiter. This was also unique to Buran.

When orientation was lost completely (e.g., due to a computer failure) and could not be restored with the star trackers or radio altimeter, Buran could rely on an infrared horizon sensor system (Local Vertical Sensor or PMV) reacting to the Earth's radiation to re-establish orientation to a point where the radio altimeter could take over. Not available on the Space Shuttle, this system could be used during unmanned missions.

Navigational aids to be used for proximity and docking operations in orbit were a radar system known as the Mutual Measurement System (SBI), the Cosmonaut Visual Rangefinder (VDK), and the Visual Navigation Measurement System (NIVS). The latter was permanently mounted on a special porthole in the aft flight deck and had to be manually aligned with the GSP module by the crew.

Just like the Shuttle Orbiter, Buran had three-axis rate gyro assemblies and body-mounted accelerometers to measure angular rates and accelerations for use in flight control algorithms [23].

For entry the accuracy of the GSP-derived state vector was insufficient to guide the spacecraft to a pinpoint landing. Therefore, data from other navigation sensors were blended into the state vector at different phases of entry to provide the necessary accuracy. Because of the requirement to perform unmanned landings, Buran had a more elaborate system of landing navigation aids than the Space Shuttle Orbiter:

Radio Rangefinder System (RDS)

This measured range from an altitude of 40 km (about 400 km from the runway) all the way to touchdown. The on-board component of the RDS, known as 17M900, consisted of four redundant interrogators and four antennas, weighing a total of 85.5 kg. Up to an altitude of 4 km they sent paired pulses to six distance-measuring equipment units (DME) at Baykonur, which then transmitted paired pulses back to the orbiter on a different frequency. The time required for the round trip of this signal was then translated by the orbiter into distance to the transponder. The system indirectly also provided data on elevation and azimuth.

Three of the ground terminals were deployed off one end of the runway and the other three off the opposite end. One terminal in each set of three was located along the runway centerline, about 20 km away, and the other two were deployed on either side within 60 km from the runway. Each of the six terminals had a unique coded reply, allowing Buran's Biser-4 computers to select and use distance measurements from the three terminals whose positions provided the best accuracy. Each of the three selected terminals was interrogated 60 times per second, nearly four times the rate for standard *en route* distance-measuring equipment used in aviation at the time. Each terminal transmitted via two antennas, one horizontally polarized and the other vertically polarized, enabling Buran to receive a strong signal over its circularly polarized antennas despite extreme pitch or roll maneuvers. When the orbiter reached an altitude of 4 km, the RDS interrogators switched to distance-measuring units on either end of the runway that had the same precision as a microwave landing system. The RDS has no equivalent on the US Space Shuttle Orbiter.

Radiotechnical Short-Range Navigation System (RSBN)

This provided azimuth and range data between altitudes of 40 km and 4 km and consisted of a 40 kg on-board set (17M902) and an E-329 beacon located mid-field off the east side of the runway. The RSBN ground station beacon continuously transmitted pulse pairs on its assigned frequency to Buran's on-board RSBN receiving equipment. RSBN is the equivalent of the Space Shuttle's Tactical Air Navigation (TACAN) system, but unlike TACAN, which is the Shuttle's primary navigation aid during most of the final descent, it only played a back-up role to the RDS. There was only one RSBN set aboard Buran, as compared with three TACAN sets on the Shuttle Orbiter. The RSBN was compatible both with Buran-specific ground station beacons and standard beacons used in aviation. This meant that RSBN was the prime navigation aid for emergency landings on runways not equipped with the RDS system.

Radio Beacon Landing System (RMS)

This was the prime navigation aid for final approach and landing, providing azimuth and elevation data from an altitude of about 7 km. The RMS was a standard all-weather scanning-beam microwave landing system (MLS) similar to those adopted for civil aircraft in the early 1980s by the International Civil Aviation Organization. In microwave landing systems antennas located on the ground transmit a reciprocating beam to an aircraft, while the aircraft measures the interval between a pair of received beams and thereby determines the azimuth and elevation angle. Buran was equipped with three RMS sets (17M901) each containing a transmitter/receiver and a decoder (34.5 kg). The ground-based component incorporated an azimuth and elevation antenna on either side of the runway, providing azimuth coverage of about 30 degrees from the runway centerline and vertical guidance up to 30 degrees. The antennas had a much greater range (at least 25 km) than traditional aviation microwave landing systems.

The RMS was very similar to the Space Shuttle's Microwave Scan Beam Landing System (MSBLS). VNIIRA (the All-Union Scientific Research Institute of Radio Equipment), the Leningrad institute that built the system, was criticized by some for not using an advanced aviation microwave landing system called Platsdarm. This was under development at the institute by the end of the 1970s and featured phased-array

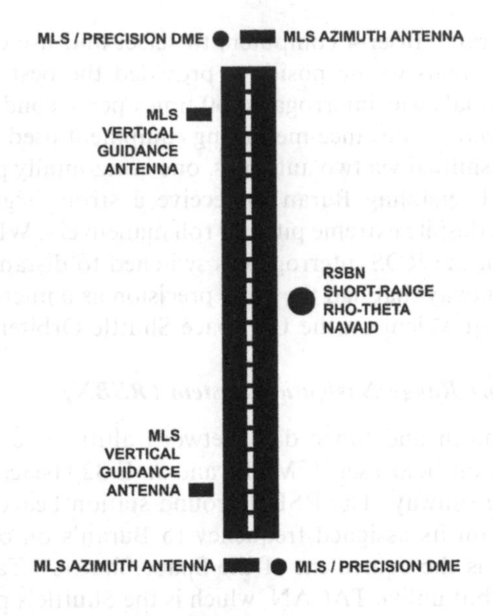

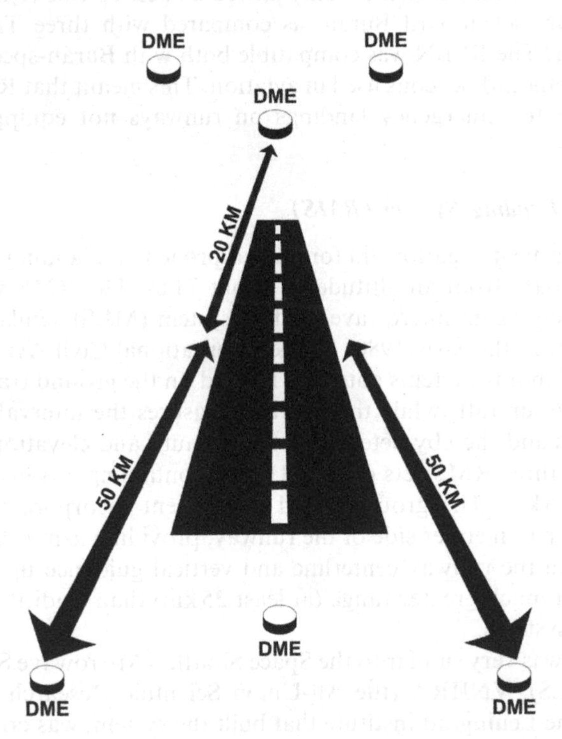

Location of navigation equipment in the landing area (reproduced from *Aviation Week & Space Technology*).

antennas with electronic scanning rather than dish antennas with mechanical scanning as in the RMS. Some felt that the simultaneous work on the Buran system and Platsdarm was a wasteful duplication of effort.

Altitude/Velocity Parameter System (SVSP)

The SVSP consisted of air data probes extended from Buran at an altitude of 20 km to measure barometric altitude, true and indicated airspeed, Mach speed, angle of attack, and dynamic pressure, and display that data for the commander and pilot in the cockpit. The SVSP was only supposed to correct the vertical channel of the inertial navigation system in emergency situations where other navigation aids failed, helping the crew to guide Buran to a manual touchdown. The SVSP was the equivalent of the Shuttle's Air Data System.

High-Altitude Radio Altimeter (RVB) and Low-Altitude Radio Altimeter (RVM)

The RVB was designed for accurate measurements of geometrical altitude using the principle of impulse modulation of an emitted signal. Its information was only supposed to be used for actual flight trajectory changes when the orbiter flew over flat terrain (because the local relief was not necessarily at the same level as the runway) or in emergency situations, where it could have been used to provide elevation data in conjunction with the air data probes.

The RVM accurately measured the altitude above the runway from flare-out at an altitude of 20 m to touchdown as well as absolute flight altitude under 1 km. The RVB has no equivalent on the Shuttle, whereas the RVM performs the same role as the Shuttle's Radar Altimeters.

The on-board and ground-based components of the RDS, RSBN, and RMS were known together as the Vympel ("Pennant") system and were developed by VNIIRA under the leadership of Gennadiy N. Gromov. Vympel also included three ground-based radar complexes that monitored the vehicle's adherence to the calculated flight path during approach and landing. Each of the complexes contained two radars. The first complex (TRLK-10K or Skala-MK) acquired the vehicle at a distance of about 400 km, using both primary (skin-echo) and secondary (transponder) signals, with the transponder reply transmitting such data as altitude, speed, and heading. At a distance of about 200 km an intermediate-range radar complex (E-511 or Ilmen) took over flight path monitoring. Precision approach radars (E-516V or Volkhov-P) monitored the final approach and landing [24].

COMMUNICATIONS

Buran's communication systems performed the following functions:

- two-way voice communications between the orbiter and Mission Control and between the orbiter and other spacecraft;

- intercom between crew members inside the vehicle and between crew members inside and outside the vehicle;
- relay to the ground of television images;
- relay to the ground of telemetry about the crew's health, condition of on-board systems, payload-related activities;
- trajectory measurements to determine the vehicle's exact orbital parameters;
- interaction between ground-based and on-board computers.

There were three independent radio systems, operating in three different wavebands (roughly equivalent to the Space Shuttle's P-band, S-band, and Ku-band communication systems):

- Meter waveband (VHF): for direct line-of-sight communications with ground stations, tracking ships, and the landing facility, and also for intercom. This system used omnidirectional antennas.
- Decimeter waveband (UHF): for communications with ground stations and tracking ships either directly or through geostationary relay satellites. Equipped with three transceivers, this system used two omnidirectional antennas and five active-phased array antennas.
- Centimeter waveband (SHF): solely for communications through geostationary relay satellites using two parabolic narrow-beam antennas. One of these (ONA-I) was mounted on the aft wall of the payload bay, covering the upper hemisphere, and the other (ONA-II) was located in a well on the underside of the aft fuselage, covering the lower hemisphere. ONA-I could be moved off-axis so that its view to the geostationary satellite was not blocked by the vehicle's vertical stabilizer. Depending on the mission objectives and the vehicle's orientation, the antennas could be used either together or individually. Both antennas could only be deployed in orbit and had to be stowed for a safe re-entry. Therefore, they could be pyrotechnically jettisoned if something went wrong during the stowage process. The ONA antennas performed the same role as the Shuttle's Ku-band antenna, the major difference being that the Shuttle has just one such antenna installed on the starboard side of the payload bay that covers both hemispheres. The ONA antennas were not installed on Buran's single mission in November 1988.

The data relay satellites intended for use by Buran were the Luch/Altair satellites, approved by the same February 1976 government decree that had given the go-ahead for the Energiya–Buran program. The equivalent of NASA's Tracking Data and Relay Satellites (TDRS), these were 2.4-ton three-axis stabilized satellites designed to relay communications from and to both Buran and the Mir space station and also to provide mobile fleet communications for the Soviet Navy. They were developed by the Scientific Production Association of Applied Mechanics (NPO PM) near the Siberian city of Krasnoyarsk. Five were launched between October 1985 and October 1995.

Luch/Altair satellite (*source*: *Novosti kosmonavtiki*).

Buran's communication systems were developed by the Moscow-based organization NPO Radiopribor (currently named Russian Scientific Research Institute of Space Equipment Building or RNII KP). Headed throughout the Buran years by Leonid I. Gusev, this organization had a virtual monopoly in developing communication systems for Soviet spacecraft [25].

PAYLOAD DEPLOYMENT AND RETRIEVAL

For satellite deployment missions Buran would have been equipped with an extendable turntable that would first lift the payload out of the confines of the cargo bay. After deploying the payload's appendages and checking all on-board systems, the satellite would then have been spun up and released with the help of springs.

Buran was also supposed to be outfitted with a robotic arm system to deploy and retrieve payloads. One of its primary tasks would have been to lift space station modules out of the vehicle's cargo bay and attach them to available docking ports and also to provide a stable platform for spacewalking cosmonauts. Developed by the Central Scientific Research Institute of Robotic Technology and Technical Cybernetics (TsNII RTK) in Leningrad, the so-called On-Board Manipulator System (SBM) was similar in design to the Shuttle's Canadian-built Remote Manipulator System (RMS). Measuring 15 m and weighing 360 kg, it had six joints and could lift a payload of up to 30 tons. Maximum translation speed was 30 cm per second without a payload and 10 cm per second with a payload. The SBM would be operated manually from a console in the aft flight deck with two joysticks, one to move the arm itself, and

Test model of Buran's mechanical arm (*source*: *www.buran.ru*).

the other to operate the grapple fixture. Three cameras, one on the wrist and two in the cargo bay, would have assisted in these operations. It was also possible to operate the arm automatically using software stored in the on-board computer. During unmanned missions the arm could even have been controlled from Mission Control via the on-board computer system. With the Shuttle never having been designed to fly unmanned, the RMS did not provide that capability, although the technique was later introduced for the International Space Station's Remote Manipulator System (SSRMS), which was first remotely operated from the ground in March 2006.

The major difference with the RMS was that even on standard missions Buran would have carried *two* arms, one on the left, the other on the right longeron to provide more flexibility in loading/unloading operations or provide back-up capability. Although provisions for two arms were incorporated in each Space Shuttle Orbiter, the idea of ever flying two RMS units on a single Orbiter was abandoned in the late 1990s. After the 2003 Columbia accident the remaining Shuttle Orbiters *were* equipped with a second arm known as the Orbiter Boom Sensor System, but this is solely intended to make camera surveys of RCC panels and heat shield tiles.

The SBM was not flown on Buran's only mission in November 1988, but was supposed to be installed on the second flight vehicle for a docking mission with the

Mir space station. A working model of the arm was built and installed at TsNII RTK on a special test stand capable of simulating weightless conditions [26].

DOCKING AND EXTRAVEHICULAR ACTIVITY (EVA)

For space station missions Buran would have carried a Docking Module (SM) in the forward part of the payload bay. It consisted of a spherical section (2.55 m in diameter) topped by a cylindrical tunnel (2.2 m in diameter) with an APAS-89 androgynous docking port, a modified version of the APAS-75 system developed by NPO Energiya for the 1975 Apollo–Soyuz Test Project. The spherical section, bolted to the floor of the cargo bay, had two side hatches, one connecting it to Buran's mid-deck and the other providing access to the payload bay for spacewalking cosmonauts or to a Spacelab-type module. The tunnel provided the actual interface between the Docking Module and the target vehicle and would be extended to its full length after opening of the payload bay doors. With the tunnel fully extended, the adapter was 5.7 m high. If the extendible part of the tunnel became stuck in its

Buran's Docking Module (*source*: *www.buran.ru*).

deployed position, it could be pyrotechnically separated to allow the crew to close the payload bay doors.

At least one flightworthy SM was built for the first mission of flight vehicle nr. 2, which would have featured a docking with Mir and a Soyuz TM spacecraft. The Buran Docking Module served as the basis for a small module that was supposed to be attached to the Mir-2 space station to act as a berthing place for Soyuz, Progress, and Buran vehicles and as an airlock for spacewalks. It would be towed to the station by a detachable Progress-M propulsion compartment. The module was eventually launched as Pirs to the International Space Station in September 2001 (see Chapter 8).

During missions not involving dockings, Buran would have flown with an internal airlock in the mid-deck. The EVA spacesuit used by the cosmonauts would have been a modified version of the semi-rigid Orlan spacesuit, originally developed in the 1960s for the Soviet piloted lunar program. Developed by the Zvezda organization in Tumilino, it would have been worn by the cosmonaut who was supposed to stay behind in lunar orbit aboard the LOK mother ship to assist his colleague in spacewalking to the lunar lander before landing and back to the LOK after ascent from the lunar surface. The Orlan was a simplified, lighter version of the moonwalker's Krechet suit. Unlike the Krechet, it was not completely self-contained (being connected to the spacecraft's power systems with an umbilical) and designed for relatively short spacewalks.

After cancellation of the lunar program a modified version of the suit known as Orlan-D was developed for EVAs from the Salyut-6 space station, launched in 1977. The modifications were mainly related to the fact that the suit had to remain in orbit for a long time, be serviceable, and be worn by different cosmonauts. In October 1980 NPO Energiya and Zvezda reached agreement on using the same Orlan-D for spacewalks from Buran. The suit and airlock could support up to three 5-hour EVAs during a 7-day Buran mission and from six to eight EVAs during a 30-day mission.

In March 1984 Zvezda was ordered by MOM and MAP to start development of a jet-powered backpack, giving cosmonauts more flexibility during spacewalks. Interestingly, the order came just one month after the first use of the analogous Manned Maneuvering Unit (MMU) on Space Shuttle mission 41-B. Called 21KS or SPK ("Cosmonaut Maoeuvering Unit"), the device was intended for spacewalks both from the Mir space station and Buran. One of the main functions that the Russians had in mind for the unit was to allow spacewalking cosmonauts to inspect Buran's heat shield in orbit. Two of the units could be installed aboard Buran, one on the starboard side of the cargo bay, the other on the port side.

The development of the 21KS also required Zvezda engineers to design a compatible, fully self-contained spacesuit called Orlan-DMA. This no longer had an electrical umbilical connecting it to on-board systems and was equipped instead with a special unit containing power supply, radio communications, and telemetry systems. In 1987 the final decision was made to use this suit in the Buran program instead of the Orlan-D.

Although the Orlan-DMA saw extensive use by Mir spacewalkers between 1988 and 1997, the 21KS was flown only twice by cosmonaut Aleksandr Serebrov from

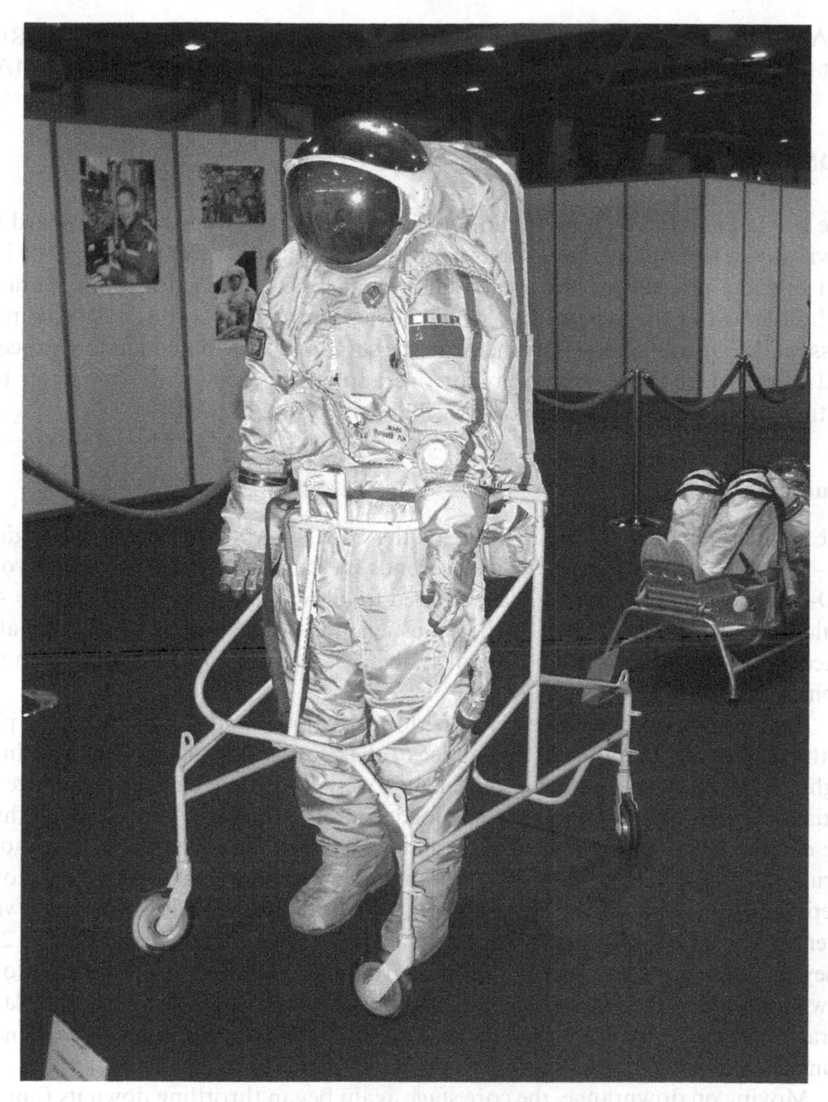

Orlan-DMA spacesuit (B. Hendrickx).

Mir in early 1990. Since the station could not maneuver to retrieve him if he became stranded, Serebrov remained attached to the station by a 60 m long safety tether. Untethered spacewalks with the 21KS would probably have been authorized only for the Buran program, with the cosmonaut being able to venture 100 m from the vehicle.

In 1992 Zvezda and the German Dornier company studied the feasibility of jointly developing a European–Russian spacesuit for the European Hermes spaceplane, Buran, and the then still planned Mir-2 space station. The work on the joint suit ("EVA Suit 2000") continued after cancellation of those programs in 1993, but

ESA backed out the following year because of financial constraints. The Russian suit now used on ISS is the Orlan-M, a further modification of the Orlan-DMA [27].

NOMINAL FLIGHT SCENARIOS

The single mission flown by Buran on 15 November 1988 was not a standard flight. It was flown without a crew on board and with the sole intention of testing the launch and re-entry procedures. No major on-orbit tasks were scheduled and Buran flew without many of the systems that would have been required for a multi-day manned mission. What will be described here are the standard launch and landing procedures and standard on-orbit operations for operational missions with a crew on board. Details of actually planned missions will be given in Chapter 8.

Launch

The launch began with the ignition of the core stage's four RD-0120 engines at $T - 9.9$ seconds, followed at $T - 3.7$ seconds by the ignition of the strap-on rockets' RD-170 engines. The interval was required to allow the core stage engines to slowly build up thrust and thereby ease the acoustic loads on the orbiter. If an anomaly was detected by the rocket's flight control system, all engines could be shut down at any moment prior to $T-$ zero.

As the stack cleared the tower, it performed a pitch and roll maneuver to place it in the proper attitude for the remainder of the ascent. About half a minute into the flight the core stage and strap-on engines were throttled back to minimize aerodynamic pressures and longitudinal loads on the vehicle. After passing through the densest layers of the atmosphere, all engines were throttled back up to nominal thrust, although the Blok-A RD-170 engines were soon again throttled down in preparation for shutdown. The four strap-on boosters shut down in pairs with an interval of 0.15 seconds and were jettisoned about two seconds later at $T + 2$m26s. They continued to fly in pairs, separating from one another somewhat later to come down some 425 km downrange. As mentioned earlier, the strap-ons could land on parachutes for recovery but were not configured as such on the two Energiya launches that were flown.

Moving on downrange, the core stage again began throttling down its four liquid oxygen/liquid hydrogen engines less than a minute before shutdown, which occurred at $T + 7$m47s. The engines were shut down in pairs with an interval of 0.2 seconds. Fifteen seconds later the orbiter separated from the core stage and safely maneuvered itself away with gentle burns of its primary thrusters. The core stage then continued on a ballistic trajectory to burn up over the Pacific Ocean. Not having required orbital velocity yet, Buran then needed two burns of one of its DOM engines about 11 and 40 minutes into the flight to place itself into an initial orbit. The required burn duration was calculated by the on-board computers on the basis of the launch vehicle's performance. The maximum acceleration forces for the crew during launch would not have exceeded $3g$.

Artist's conception of Buran launch (*source*: *www.buran.ru*).

The crew had no active role to play during the launch phase and merely had to monitor the operation of on-board systems on their cockpit displays. The orbiter's computers automatically controlled the operation of the life support, thermal control, power, and monitoring systems as well as that of the hydraulic systems and Auxiliary Power Units, which might be needed in a launch abort to perform an emergency landing. They also opened and closed the vehicle's vent doors at the required moments [28].

Orbital operations

Like the Space Shuttle Orbiter, Buran was a versatile vehicle that could have been used for a wide range of orbital operations. The following possible tasks were later identified by the Russians:

(a) Deployment of satellites or other cargos: the maximum payload was 30 tons into a 50.7° inclination 200 km orbit and 16 tons into a 97° orbit. The payload bay could house a payload with a maximum length of 15 m and a maximum diameter of 4.15 m. Because of the less stringent center-of-gravity requirements resulting from the absence of main engines, Buran's maximum payload capacity was actually higher than that of the Space Shuttle.
(b) Servicing satellites in orbit
(c) Returning satellites back to Earth. The maximum mass that could be returned from a 50.7° inclination 200 km orbit was 20 tons.
(d) Space station missions: resupply, assembly, crew exchange, crew rescue.
(e) Missions to assemble large structures in space.
(f) Autonomous scientific missions.

Three basic types of operational mission durations were envisaged for the vehicle. The first would be short-duration missions (up to 3 days) to place heavy payloads into orbit, deliver emergency supplies to space stations, or rescue space station crews. Such missions would be characterized by multiple operations and maneuvers in a relatively short time span, heavily taxing both the crew and the ground and also requiring many of them to be conducted automatically.

Medium-duration missions (up to 8 days) were expected to be the most frequent ones and would have several objectives or one particularly time-consuming and demanding goal. Typical medium-duration flights would include routine missions to space stations, multiple satellite deployment missions, satellite-servicing missions, assembly flights, etc. Although comparable in the number of operations with the short-duration flights, the longer time in orbit would make it possible to more evenly spread the workload for the crew.

Finally, long-duration missions (9 to 30 days) would primarily be devoted to scientific, materials-processing, and biotechnological experiments, which take a relatively long time to produce the necessary results. For this purpose, the Russians were planning to develop a Spacelab-type module to be placed in the cargo bay. The longest missions would have required the installation of an extra cryo kit for the fuel cells. The number of maneuvers performed during this type of mission would have been very low. In terms of the daily workload for the crew and the ground, such missions would have been comparable with a routine workday on a space station.

Range safety restrictions at the Baykonur cosmodrome, mainly dictated by the impact zones of the strap-on boosters, limited the possible orbital inclinations of the spacecraft to 50.7–83°, 97°, 101–104°, and 110°. The vehicle could have operated at altitudes between 200 and 1,000 km, although the higher of these would have necessitated the installation of extra propellant tanks in the cargo bay [29].

Re-entry and landing

A nominal re-entry could be initiated whenever Buran's ground track carried it over or near one of three runways available in the Soviet Union. The primary landing site was at Baykonur, with back-up sites available in the Soviet Union's Far East and in the Crimea (see Chapter 4). The ship had a maximum cross-range capability of 1,700 km, but that was only required for an emergency return back to Baykonur after a single revolution when the vehicle was launched into a polar orbit. For more common inclinations below 65° a cross-range capability of 1,050 km was sufficient.

Deorbit preparations began with the crew realigning the GSPs, retracting antennas, closing the payload bay doors, and preparing hydraulic systems for re-entry. When descending from an altitude of 250 km, about one hour would elapse between the deorbit burn and touchdown, with the vehicle covering a total distance of about 20,000 km and reducing its speed from Mach 25 to zero. After the deorbit burn, performed with the help of the DOM engines, Buran needed some 25 minutes to reach the official boundary between space and the atmosphere at 100 km, at which point it was still at a range of 8,500 km from the runway. It was only then that the three Auxiliary Power Units were activated.

The return through the atmosphere was divided into three phases: "Descent", "Pre-Landing Maneuvering", and "Approach and Landing". These correspond roughly to the three major phases of a Shuttle Orbiter return ("Entry", "Terminal Area Energy Management", and "Approach and Landing").

"Descent" was the hypersonic phase of the re-entry (Mach 28–Mach 10 at 100 km–20 km altitude) where the vehicle was exposed to the highest temperatures and achieved maximum cross-range. The flight control system guided the orbiter through a tight corridor limited, on the one hand, by altitude and velocity requirements (in order to make the runway) and by thermal constraints, on the other hand. Buran's angle of attack was kept at a high value (39°) during most of this phase to keep the temperatures within acceptable limits, while roll reversals were used to bleed off air speed and thus reduce kinetic energy. When the vehicle reached Mach 12, the angle of attack was gradually lowered from 39° to 10° to increase the lift-to-drag ratio. As the atmosphere thickened, the ship gradually transitioned from 20 aft attitude control thrusters to conventional aerodynamic control surfaces. The thrusters were used up to an altitude of 10 to 20 km. Between altitudes of about 80 and 50 km Buran was enveloped in a sheath of ionized air that blocked all communications with the ground. After coming out of the blackout, the ship's RDS system began beaming pulses to transponders on the ground to furnish the on-board computers with range data. Azimuth and range data from the more traditional RSBN beacon navigational aid system were only used as a back-up to the RDS data.

During the "Pre-Landing Maneuvering" phase (Mach 10–Mach 2 at 20 km–4 km altitude) Buran gradually transitioned from hypersonic to supersonic speeds and lined itself up with the runway for the final approach and landing. At this stage it intercepted one of two so-called "Heading Alignment Cylinders" (TsVK), imaginary cylinders to align the vehicle with the runway. Which of the two was chosen mainly depended on the wind direction. By the end of this phase Buran reached an "entry point" 14.5 km from the runway to begin the final descent. Primary navigational input throughout this phase still came from the RDS rangefinder system, backed up by the RSBN for azimuth and range data and by the RVB high-altitude altimeter and

SVSP air data system for altitude data. The SVSP probes were deployed at an altitude of 20 km.

The Approach and Landing phase saw the orbiter moving from hypersonic to subsonic speeds and finally coming to a stop on the runway. It began with a steep glideslope of −17° to −23° degrees (depending on landing mass), allowing the ship to correct any small trajectory errors it still had at the entry point. At an altitude of 400–500 m a pre-flare maneuver was started to position the vehicle for a shallow glideslope of −2° in preparation for landing. A final flare at an altitude of 20 m led to touchdown some 1,000 m past the runway threshold at a speed between 300 and 330 km/h. Wind speed limits were 5 m/s for tail winds, 20 m/s for head winds, and 15 m/s for crosswinds. After touchdown, speed was brought down to zero by the brake chutes and the main gear brakes, with the speed brakes only used in manual landings. Steering during roll-out was provided by the nose gear steering system and by differential braking. The maximum roll-out distance was 1,800 m. The navigation aids during Approach and Landing were the RMS microwave system for altitude and azimuth, the RDS rangefinder system for range and the RVM low-altitude altimeter for altitude.

The landing could be performed in automatic, flight director, or manual mode. Automatic mode was the preferred mode even for manned missions (see Chapter 7). Flight director systems, also used in aviation, provide visual indications on the pilots' displays of what the autopilot would want to do if it were flying the vehicle under the current settings. In other words, the pilots fly the vehicle manually but are guided by the autopilot. Simulations showed that the use of this mode throughout descent would be monotonous and tiring and should be restricted to the final approach and landing, especially if visibility was poor. Moreover, this mode did not give the crew the necessary psychological comfort because it could not always anticipate unexpected events. In manual control the pilots *themselves* determined the flight path using information on the expected touchdown point and remaining energy and also by relying on navigational aids, outside visual clues, and data uplinked from the ground. If all that information was available to them, they could switch to manual mode at an altitude of about 20 to 30 km. In emergency situations they could land the vehicle using only navigational aids or information provided by Mission Control [30].

EMERGENCY SITUATIONS

Launch and landing emergencies

The Energiya–Buran launch profile offered more rescue options for the crew than that of the Space Shuttle, mainly because of the use of liquid-fuel rather than solid-fuel boosters. Whereas a Solid Rocket Booster (SRB) failure will almost always result in the loss of vehicle and crew (as tragically demonstrated by the Challenger disaster), an engine failure on one of Energiya's four strap-ons would not necessarily have had catastrophic results. Buran cosmonauts had the following escape options.

Pad emergency escape system

In case of an emergency on the launch pad the crew could egress Buran and flee to an underground bunker using the pad emergency escape system. Unlike the slidewire baskets used on the Space Shuttle launch pads, the cosmonauts were to glide down a giant chute and subsequently seek shelter in bunkers under the launch pad (see Chapter 4).

Ejection seats

For its manned test flights Buran was to be equipped with ejection seats, allowing the crew members to escape through two overhead hatches in case of an emergency on the pad, in the early stages of launch and the final phases of landing. Ejection seats were also flown on the four two-man test flights of the US Space Shuttle Columbia, but were disabled for Columbia's first operational mission (STS-5, which carried two mission specialists on the flight deck) and eventually removed altogether. The Russians were planning both two-man and four-man test flights and the intention was to have ejection seats for *all* cosmonauts irrespective of crew size. With a crew of four, the two non-pilots would have been seated in the front part of the mid-deck and could have been ejected via two hatches mounted in between the forward reaction control system and the six forward cockpit windows. In that configuration the front equipment bay in the mid-deck would have been moved to the rear. The ejection seats would have been removed for flights carrying more than four cosmonauts.

Buran's ejection seats (called K-36RB or K-36M11F35) belonged to the family of K-36 seats of the Zvezda organization that are standard equipment on Soviet high-performance combat aircraft. More than 10,000 K-36 seats had been produced by the early 1990s and several hundred real ejections had been made with very high survival rates.

The K-36 seats are based on a modular design to which systems are added or deleted depending on the specific aircraft on which they are installed. The modifications for Buran were needed not so much for the landing phase, but mainly to pull the cosmonauts away to a safe distance from the rocket in case of a pad or launch accident. For that purpose they were equipped with a small solid-fuel rocket that would have been needed only for ejection during launch and on the pad (after retraction of the crew access arm). In the latter scenario the K-36RB would have been able to reach an altitude of 300 m in order to clear the 145 m high rotating service structure which would have been in its path. The pilot was supposed to come down 500 m from the pad in a matter of just 10 seconds.

Another feature unique to the K-36RB was an under-seat stabilization system with drag parachutes that would be used up to an altitude of about 1 km. The system's two booms were separated along with the solid-fuel rocket at a point near the upper portion of the seat's trajectory. This was installed in addition to a standard two-boom upper stabilization system with end-mounted parachutes.

The cut-off altitude and speed for the use of the K-36RB during launch would have been 30–35 km and Mach 3.0–3.5, respectively (compared with 24 km and Mach 2.7 for the Shuttle seats), which roughly equates to $T + 100$ seconds in a normal

K-36RB ejection seat and Strizh pressure suit (B. Vis).

launch profile. That limit was determined by the ability of the Strizh pressure suits to protect the pilot from the thermal stresses experienced in an ejection. The Strizh suits were covered with special heat-resistant material to protect the pilot against heating caused by the high aerodynamic loads during ejection. The K-36RB could also be used during landing, from the moment speed was reduced to Mach 3.0–2.5 all the way to wheels stop. The seats were also installed on the BTS-002 atmospheric test model (as they were on Enterprise for the Approach and Landing Tests). The ejection seats could be activated by the crew, by on-board automatic devices, or by a command from the ground.

Also unique to the K-36RB was a computer linked to Buran's computers that ensured that the seat was configured for one of five ejection modes corresponding to

Ejection seats fired from a mock-up of Buran's cockpit (B. Vis files).

the orbiter's launch or landing phase. Mode 1 was for ejection on the launch pad, mode 2 for ejection during initial ascent, mode 3 for high-Mach/high-altitude ejections, mode 4 for the landing approach, and mode 5 for the final approach, touchdown, and roll-out.

The K-36RB was tested in at least three ways with mannequins clad in Strizh pressure suits. In one method the seat was installed in a ground-based mock-up of Buran's cockpit and in another in the aft cockpit of a converted two-seater MiG-25RU training aircraft (serial number 0101). A rather original way was found of testing the ejection seats and associated escape suits at much higher velocities and altitudes. An experimental version of the seats called K-36M-ESO was carried as a piggyback payload on five unmanned Soyuz rockets carrying Progress resupply ships between September 1988 and May 1990 (Progress 38, 39, 40, 41, and 42). The ejection seats were fitted inside the rocket-powered tower atop the launch vehicle and ejected at altitudes between 35 and 40 km and speeds between Mach 3.2 and 4.1.

Emergency Separation

Emergency Separation (Russian acronym EO) would have been activated if a serious launch vehicle problem had forced shutdown of all engines *above* the altitude where the ejection seats could be used. In that scenario, Buran could have swiftly separated from the core stage using four small solid-fuel motors in its nose section and subsequently attempted to stabilize itself to perform a manual emergency landing on a runway downrange after having dumped excess propellant from the ODU engine system. In case no emergency runway was available, the crew could still have used their ejection seats once Buran was stabilized and had reached a safe altitude.

NASA studied a similar escape option for the Shuttle called "Fast Separation" that would have allowed the Orbiter to separate from the External Tank in approximately three seconds in case of an SRB failure. However, analysis showed that if this is attempted while the SRBs are thrusting, the Orbiter would "hang up" on its aft attach points and pitch violently, probably resulting in the destruction of the vehicle. Similarly, the Russians concluded that Emergency Separation would be hard to achieve without Buran hitting the core stage, especially when the stack experienced high aerodynamic pressures in the early stages of the launch, virtually all the way to separation of the strap-on boosters. Moreover, safely landing the vehicle in such a scenario would have required a very elaborate and costly network of emergency runways all the way downrange.

Return Maneuver

A Return Maneuver (MV) enabled Buran to return to its launch site runway in case of a single-engine failure on the strap-ons or the core stage early in the launch. The "negative return" point would have been reached between $T + 2\text{m}05\text{s}$–2m10s in the event of a strap-on failure and between $T + 3\text{m}00\text{s}$–3m10s if a core stage engine shut down.

If an RD-170 engine in one of the four strap-ons failed, the engine of the diametrically opposed booster would also shut down to make sure that the rocket did not deviate from its trajectory. Subsequently, all the remaining liquid oxygen in both strap-ons would have been dumped overboard to minimize the amount of dead weight carried up by the rocket and also to ensure that conditions at separation were close to the ones originally planned. The LOX could be released via a 600 mm diameter drainage channel, which exited the lower end of the LOX tank, situated some 15.5 m above the engines. Kerosene, which comprised only one-third of the overall propellant mass in each booster, would not have been dumped overboard to prevent the formation of an explosive mix.

The return profile would have been very similar to that of a Return to Launch Site (RTLS) abort during Space Shuttle launches. The vehicle would have continued to fly downrange to expend excess propellant and would have performed a pitch-around maneuver to orient the stack to a heads-up attitude pointing towards the launch site. The core stage would then be separated, allowing Buran to glide to a landing on its cosmodrome runway. In order to improve Buran's weight and center of gravity for the glide phase and landing, excess propellant for the ODU propulsion system was to be expended by simultaneously firing the two DOM engines and dumping liquid and gaseous oxygen overboard.

Single-Orbit Trajectory

If one of the core stage engines failed *after* $T + 3\text{m}10\text{s}$, Buran could still reach orbit, but the exact scenario depended on when the failure occurred and how much propellant Buran's DOM engines needed to achieve that orbit, something that was calculated by the on-board computers. If the failure happened late in the launch, Buran's DOM engines could have boosted the vehicle to its nominal orbit or to a lower but still usable orbit (in NASA parlance the latter scenario is called "Abort to Orbit", performed once by Challenger on STS-51F in July 1985). If it occurred much earlier, the remaining propellant in the core stage would have been burned to depletion, with Buran then firing its DOM engines to reach a very low orbit and somewhat later re-igniting those engines to initiate re-entry. Excess DOM propellant would have been expended prior to entry interface to meet center-of-gravity requirements. This "Single-Orbit Trajectory" (OT) abort is the same as an Abort Once Around (AOA) on Shuttle launches [31].

In-orbit emergencies

In-orbit anomalies posing a serious threat to the crew could have resulted in any of three abort scenarios: "Immediate Return", "Early Return", or rescue by a Soyuz spacecraft. Buran crews would have had checklists instructing them what to do in each of those situations.

Immediate Return

"Immediate Return" was an abort scenario requiring Buran to deorbit between 40 minutes and 3 hours after the occurrence of the anomaly. Forty minutes was considered the minimum time needed to complete all preparations for re-entry such as closing of the payload bay doors, preparing hydraulic systems, donning pressure suits, and loading re-entry software into the on-board computers. One anomaly likely to result in an Immediate Return was considered to be a serious fire in the crew compartment not disabling the vehicle or the crew. After extinguishing the fire with available means, the crew would have donned their Strizh pressure suits to prevent smoke inhalation. If the fire threatened to go out of control after the crew had put on the suits, the crew also had the option of depressurizing the cabin to starve the flames of oxygen. Another such anomaly would be a sudden leak in the ODU propulsion system that within a short period of time would lead to an inability to fire the deorbit engines or position the ship for a Soyuz rescue mission, stranding the crew in orbit.

"Immediate Return" could only end with a safe runway landing if the vehicle's ground track happened to carry it over one of the three available Soviet landing sites within a short time after the occurrence of the anomaly. If this was not the case, the crew would have had to eject from Buran before it crash-landed. Factors to be taken into account here would have been the need to protect public safety and have rescue crews within a reasonable distance of the landing zone (especially if the vehicle had to be ditched in the ocean).

Early Return

"Early Return" was a scenario giving the crew 3 to 24 hours to prepare for deorbit, with a much better chance of reaching one of the three runways. This would have required the same type of action on the part of the crew as "Immediate Return", but only at a more relaxed pace. The most probable event leading to such a situation would have been a dangerous loss of redundancy in critical on-board systems such as the computers, the GSP gyro platforms, the fuel cells, etc. Even depressurization of the crew cabin was ranked as an anomaly that would give the crew several hours or more to prepare for deorbit, assuming that the loss of pressure was slow enough for the crew to have a chance to don their Strizh pressure suits and hook them up to the Personal Life Support System (the minimum time required for this was five minutes). Cabin systems were designed to operate in a vacuum and the suit's life support systems could sustain the crew for 12 hours.

Soyuz rescue

A rescue option unique to the Soviet space program was the ability to send a Soyuz spacecraft to an incapacitated orbiter. That plan could have been set in motion in any scenario where Buran would have been unable to return to Earth, such as a propulsion system failure, major damage to the thermal protection system, etc. In any given situation, it would have taken the Soyuz at least several days to reach Buran, making it necessary for the crew to conserve power and consumables until the rescue craft arrived. After crew evacuation, Buran could then either have been sent on a destructive re-entry over unpopulated regions or—if deemed feasible—safely brought back to Earth unmanned.

Of course, a Soyuz rescue could only have been conducted in certain well-defined circumstances. First, it assumed that Buran was equipped with an APAS docking adapter. Second, the ship needed to have at least some level of control (navigation systems and steering thrusters) enabling it to be positioned for the active Soyuz vehicle to dock with it. Third, the crew should have numbered no more than two cosmonauts, since the three-man Soyuz had to be launched with a "rescue commander" to assist the stranded pilots in boarding the Soyuz. In the late 1980s/early 1990s the Russians had a cadre of "rescue commanders" for emergency flights to Mir who could quite easily have been cross-trained for Buran rescue missions. The Soyuz rescue scenario seems to have been worked out specifically for the early two-man test flights.

Even if Buran carried more than two crew members, a Soyuz rescue was not entirely out of the question, at least if the ship was on a space station mission. Fuel reserves permitting, the Soyuz could have evacuated all crew members by making repeated flights between the stricken Buran and the space station. This was only in the very unlikely event that the vehicle had a problem preventing it from landing *and* could not reach the station or return to it. Otherwise, the Buran crew could simply

Soyuz spacecraft in orbit (*source*: NASA).

have stayed aboard the space station until rescue arrived. Taking into account the fact that the bulk of Buran missions would have been to Mir and Mir-2, this is a luxury that most Buran cosmonauts would have had long before NASA even began thinking about the "safe haven" concept in the wake of the 2003 Columbia accident.

The Russians took the Soyuz rescue option very seriously. They were even planning to simulate it during the second mission, in which the ship would have launched and landed unmanned but would have been temporarily boarded by a Soyuz crew while in orbit (see Chapter 5). If Buran had ever flown its two-man test flights, a Soyuz vehicle would very probably have been on stand-by at the Baykonur cosmodrome to come to the rescue. The early Buran pilots would have needed some limited Soyuz training, even if the Soyuz would be piloted by a rescue commander. This is probably one of the reasons Buran pilots Igor Volk and Anatoliy Levchenko made Soyuz flights in 1984 and 1987, although the primary goal of these flights was to test their ability to fly aircraft after a week in zero gravity (see Chapter 5).

Of course, it should be understood that, while all these abort scenarios were *theoretically* possible, it is far from certain that all of the situations described above would have been survivable. Much would have depended on the exact circumstances. Also, at least several of them were only feasible with a limited number of crew members on board (two to four). On the whole, though, it can be said that Buran crews would have stood a better chance of surviving in-flight emergencies than any Space Shuttle crew to date [32].

REFERENCES

[1] B. Gubanov, *Triumf i tragediya Energii (tom 3: Energiya-Buran)*, Nizhniy Novgorod: Izdatelstvo Nizhegorodskogo instituta ekonomicheskogo razvitiya, 1998, pp. 151–187.

[2] *KB Khimavtomatiki, stranitsy istorii, tom 1*, Voronezh: KBKhA, 1995, pp. 72–73; B. Gubanov, *op. cit.*, pp. 121–136.

[3] B. Gubanov, *op. cit.*, pp. 105–116; S. Konyukhov, *Rakety i kosmicheskiye apparaty konstruktorskogo byuro Yuzhnoye*, Dnepropetrovsk: GKB Yuzhnoye, 2000, pp. 88–89; S. Konyukhov, *Prizvany vremenem: rakety i kosmicheskiye apparaty konstruktorskogo byuro Yuzhnoye*, Dnepropetrovsk: Art Press, 2004.

[4] B. Gubanov, *op. cit.*, pp. 105–106; V. Trofimov, *Osushchestvleniye mechty*, Moscow: Mashinostroyeniye/Polyot, 2001.

[5] B. Gubanov, *op. cit.*, pp. 78–79.

[6] B. Gubanov, *op. cit.*, pp. 230–231.

[7] V. Opanasenko, D. Pyrlik, "Design, preparation, and launch of the reusable rocket space complex Energiya–Buran" (in Russian), article on the website *Aerokosmicheskiy portal Ukrainy*.

[8] B. Gubanov, *op. cit.*, pp. 142–151.

[9] B. Gubanov, *op. cit.*, pp. 248–258.

[10] Y. Semyonov, *Mnogorazovyy orbitalnyy korabl Buran*, Moscow: Mashinostroyeniye, 1995, pp. 116–124; A. Stepanov, "The payload bay doors of the orbital ship Buran" (in Russian)/K. Sergeyev *et. al.*, "Metals used in the airframe of the orbital ship Buran" (in Russian), in: G. Lozino-Lozinskiy, A. Bratukhin, *Aviatsionno-kosmicheskiye sistemy*, Moscow: Izdatelstvo MAI, 1997, pp. 95–102.

[11] Y. Semyonov, *op. cit.*, pp. 140–141.

[12] Y. Semyonov, *op. cit.*, pp. 130–139; V. Sayenko, "The auxiliary power unit of the orbital ship" (in Russian)/N. Fomin, "The hydraulic system and actuators" (in Russian), in: G. Lozino-Lozinskiy, A. Bratukhin, *op. cit.*, pp. 225–234.

[13] Interview with Gleb Lozino-Lozinskiy on the *www.buran.ru* website.

[14] Y. Semyonov, *op. cit.*, pp. 124–130; L. Voynov, "Thermal design of the Buran orbital ship" (in Russian)/V. Timoshenko, "Design and experimental testing of Buran's thermal protection system" (in Russian)/M. Gofin, "The thermal protection system of the reusable orbital ship" (in Russian), in: G. Lozino-Lozinskiy, A. Bratukhin, *op. cit.*, pp. 115–144.

[15] Y. Semyonov, *op. cit.*, pp. 147–150, 186–188; Y. Baturin (ed.), *Mirovaya pilotiruyemaya kosmonavtika*, Moscow: RTSoft, 2005, p. 438.

[16] Y. Semyonov, *op. cit.*, pp. 78–79, 215–220; Y. Mushkaryov, "The information display system and controls of the reusable Buran ship" (in Russian)/V. Fedotov, V. Novikov, "The cockpit" (in Russian), in: G. Lozino-Lozinskiy, A. Bratukhin, *op. cit.*, pp. 201–205, 235–238; Y. Tyapchenko, "The information display system of the orbital ship Buran" (in Russian), *Aviakosmicheskaya tekhnika i tekhnologiya*, 4/1998; Y. Tyapchenko, "Information display systems of piloted spacecraft" (in Russian), online at A. Zheleznyakov's *Entsiklopediya "kosmonavtika"*.

[17] Y. Semyonov, *op. cit.*, pp. 222–230, 253; I. Abramov, I. Skoog, *Russian Spacesuits*, Chichester: Springer/Praxis, 2003, pp. 211–221.

[18] Y. Semyonov, *op. cit.*, pp. 229–231.

[19] Y. Semyonov, *op. cit.*, pp. 247–252.

[20] Y. Semyonov, *op. cit.*, pp. 55, 142–143, 159–160, 183–186; M. Schautz *et al.*, "Testing of a Buran Flight-Model Fuel Cell", *ESA Journal*, 2/1994, pp. 129–137; "The Ural Electro-

chemical Integrated Factory" (in Russian), online at *http://www.midural.ru/ek.ru/dbo.enterprise/78/Default.htm*
[21] Y. Semyonov, *op. cit.*, pp. 190–214; B. Sokolov, A. Sanin, "Frost and flame working together" (in Russian), *Aviatsiya i kosmonavtika*, 1/1991, pp. 44–45.
[22] Y. Semyonov, *op. cit.*, pp. 155–158, 254–269; N. Dubova, "Development of the control system of the rocket carrier Energiya and the space shuttle Buran" (in Russian), *Computerworld*, 23/2000. Some sources claim there were just four computers on Buran.
[23] Y. Semyonov, *op. cit.*, pp. 258–259, 277–284.
[24] P. Klass, "Details of Soviet Shuttle Auto-Land System Disclosed", *Aviation Week & Space Technology*, 12 December 1988, pp. 33–34; I. Lebedev, "Vympel or Platsdarm?" (in Russian)/Y. Nikonov, Y. Belyatskiy, "Land anytime, anywhere" (in Russian), *Tekhnika—molodyozhi*, April 1990, pp. 28–34; Y. Semyonov, *op. cit.*, pp. 305–312, 390–391; Vympel page on the *www.buran.ru* website.
[25] Y. Semyonov, *op. cit.*, pp. 56, 174–177.
[26] Y. Semyonov, *op. cit.*, pp. 32, 55–56; SBM page on *www.buran.ru* website.
[27] Y. Semyonov, *op. cit.*, pp. 53–54 ; I. Abramov, I. Skoog, *op. cit.*; B. Olesyuk, "Why isn't the space motorcycle being used?" (in Russian), *Gudok*, 11 September 1992, p. 3.
[28] Y. Semyonov, *op. cit.*, pp. 66–67; B. Gubanov, *Triumf i tragediya Energii (tom 4 : polyot v nebytiye)*, Nizhniy Novgorod: Izdatelstvo Nizhegorodskogo instituta ekonomicheskogo razvitiya, 1999, p. 8.
[29] Y. Semyonov, *op. cit.*, pp. 34, 60–61; B. Gubanov, (vol. 3), *op. cit.*, p. 47.
[30] Y. Semyonov, *op. cit.*, pp. 37–39, 73–76, 284–291, 294–317; V. Kirpishchikov, "Descent and landing trajectories of the Buran orbital spaceship. Algorithms of the automatic guidance and control" (in Russian)/V. Trufakin, "Guidance and control of the orbital plane" (in Russian), in: G. Lozino-Lozinskiy, A. Bratukhin, *op. cit.*, pp. 46–65.
[31] J. Lenorovitz, "Soviet Ejection Seat for Buran Shuttle Qualified for Deployment at up to Mach 4", *Aviation Week & Space Technology*, 10 June 1991, pp. 44–46; Y. Semyonov, *op. cit.*, pp. 85–90; B. Gubanov, (vol. 3), *op. cit.*, pp. 117–121; I. Abramov, I. Skoog, *op. cit.*, pp. 217–221.
[32] Y. Semyonov, *op. cit.*, pp. 90–100.

4

Organizations and infrastructure

MANAGEMENT

Unlike the American space program, whose military and civilian components were split with the formation of NASA in 1958, the Soviet space program remained firmly rooted in the missile program from which it originated in the 1950s and remained an institutional arm of the defense industry. Therefore, the distinction between military and civilian space projects was much more blurred than in the United States, which was also evident in the management of the Energiya–Buran program.

Communist Party level

At the Communist Party level, space was the responsibility of the Central Committee's Secretary for Defense Matters, a position established by Nikita Khrushchov in 1957 as the focus of power shifted from the USSR Council of Ministers to the Central Committee of the Communist Party. The holder of the post was the most important figure in determining space policy in the USSR between 1957 and 1991, although it should be understood that space was only one of the numerous responsibilities resting on his shoulders.

Dmitriy F. Ustinov, who served as Secretary for Defense Matters from March 1965 until October 1976, was by many accounts the single most important man behind the decision to go ahead with Buran. Following in his footsteps were Yakob R. Ryabov (1976–1979), Andrey P. Kirilenko (1979–1983), Grigoriy V. Romanov (1983–1985), Lev N. Zaykov (1985–1988), and Oleg D. Baklanov (1988–1991).

Government level

The Soviet military industrial complex, consisting of nine ministries, was run by the Military Industrial Commission (VPK), a body residing under the Council of

Ministers. It made key decisions on the development and production of military and space technology, approved timelines, kept close track of R&D work conducted in its subordinate organizations, and ensured smooth cooperation between the various ministries. While the big policy and funding decisions were left to the Central Committee and Council of Ministers, the VPK was the workhorse that made sure those decisions were implemented.

The VPK was headed by the deputy chairman of the Council of Ministers. VPK chairmen during the Buran years were Leonid I. Smirnov (1963–1985), Yuriy D. Maslyukov (1985–1988), Igor S. Belousov (1988–1991), and again Maslyukov (1991).

The "missile and space ministry" was called the Ministry of General Machine Building (MOM) and was set up in 1965. Most of the leading design bureaus and associated manufacturing facilities involved in Energiya–Buran (including NPO Energiya, KBKhA, KB Yuzhnoye) were subordinate to this ministry. The ministry's leading R&D institute was TsNIIMash (Central Scientific Research Institute of Machine Building) in Kaliningrad, which also ran the Mission Control Centre (TsUP) from where Buran was controlled.

The Ministers of General Machine Building were Sergey A. Afanasyev (1965–1983), Oleg D. Baklanov (1983–1988), Vitaliy K. Doguzhiyev (1988–1989), and Oleg N. Shishkin (1989–1991). Within MOM, prime responsibility for Energiya–Buran was initially concentrated under the 3rd Chief Directorate ("Rocket and Space Complexes"), headed by Yuriy N. Koptev (the later head of the Russian Space Agency). In 1982 a Directorate of Experimental Work (UER) was set up under the 3rd Chief Directorate to concentrate specifically on Energiya–Buran. Headed by I.P. Rumyantsev, the UER had its offices at the premises of NPO Energiya and its workforce was actually on the NPO Energiya payroll.

In order to relieve the overloaded 3rd Chief Directorate, a new 11th Chief Directorate was eventually established under the leadership of P.N. Potekhin, with one of its departments (headed by M.V. Sinelshchikov) devoted specifically to Energiya-Buran. UER also became subordinate to this Directorate.

Minister Afanasyev also created a so-called Operational Control Group (GOR) to help NPO Energiya coordinate work on the Energiya–Buran program on a day-to-

Ministers of General Machine Building Sergey Afanasyev (left) and Oleg Baklanov.

day basis. This group was particularly active in the early years of the project in order to solidify the cooperation between the various organizations involved. In 1984 MOM set up a permanent representation at Baykonur to coordinate work there. Staffed by leading MOM officials, it had its offices in the Energiya assembly building.

Given the fact that Buran was a winged vehicle, another key ministry involved in the program was the Ministry of the Aviation Industry (MAP). The leading design bureau in charge of Buran under MAP was NPO Molniya. Major test and research facilities under MAP were the Central Aerohydrodynamics Institute (TsAGI) in Zhukovskiy for wind tunnel tests, and the Gromov Flight Research Institute (LII), also in Zhukovskiy, which was the home base of Buran's cadre of civilian test pilots and provided facilities for simulating Buran landings on aircraft and the BTS-002 atmospheric test model.

MAP ministers in the course of the Energiya–Buran program were Pyotr V. Dementyev (1953–1977), Vasiliy A. Kazakov (1977–1981), Ivan S. Silayev (1981–1985), and Apollon S. Systsov (1985–1991). Prime responsibility for Buran was entrusted to the ministry's 12th Chief Directorate, specifically founded for this purpose in 1977 under the leadership of R.S. Korol.

Military level

The main military organization involved in the Energiya–Buran program was GUKOS (Chief Directorate of Space Assets), which during the early years of the program was subordinate to the Strategic Rocket Forces (RVSN). RVSN had been set up as an independent branch of the armed forces in December 1959 to run the burgeoning strategic missile program, a task earlier performed by the Chief Artillery Directorate (GAU) under the Ministry of Defense. All this was in contrast to the situation in the United States, where missiles were the responsibility of the Air Force. RVSN also inherited the space-related functions of the GAU—namely, pre-launch processing, launch, tracking, and control of both civilian and military satellites. In 1964 RVSN further consolidated its control over space operations with the creation of TsUKOS (Central Directorate of Space Assets), a body that was directly subordinate to the RVSN Commander-in-Chief. In March 1970 TsUKOS was reorganized as GUKOS, which in turn separated from RVSN in November 1981 to become directly subordinate to Defense Minister Ustinov. In November 1986 it was reorganized as a separate branch of the armed forces under the name UNKS (Directorate of the Commander of Space Forces).

In August 1992, after the disintegration of the Soviet Union, it became known as VKS (Military Space Forces). In November 1997 VKS was reabsorbed by the Strategic Rocket Forces, only to be given back its independent status in June 2001 under the name KV (Space Troops).

GUKOS was appointed as the so-called "client" for the Energiya–Buran program on 30 July 1976. This meant that, formally at least, it was responsible for determining the specifications for the system. In this respect, it essentially assumed the same position as NASA in the United States, with the design bureaus and factories playing the role of "contractors". In principle, however, most of the initiatives to develop new spacecraft or work out specifications came from the design bureaus themselves in a bottom-up management style characteristic of the Soviet space program. The relationship between "client" and "contractor" in the Soviet context was also different in that GUKOS did not directly control the purse strings of the Energiya–Buran program. Like most other space projects, the program had to be run with the annual funds allocated to the ministries of the military industrial complex from the state budget, which was a way of covering up actual defense expenditures. Aside from fulfilling its "client" function, GUKOS/UNKS was in charge of operating the launch facilities at the Baykonur cosmodrome and tracking stations across the Soviet Union.

GUKOS/UNKS commanders during the Buran years were Andrey G. Karas (1965–1979), Aleksandr A. Maksimov (1979–1989), and Vladimir L. Ivanov (1989–1996). Several bodies and posts were set up within GUKOS that were specifically related to the Energiya–Buran program. In December 1979 a special coordinating group overseeing work on the program was set up under the leadership of former cosmonaut Gherman Titov, a deputy head of GUKOS at the time. In 1984 Yevgeniy I. Panchenko was named deputy head of GUKOS specifically in charge of Energiya–Buran and "automated control systems". In 1986 a new 4th Directorate in charge of Buran and "special space assets" was established under the leadership of Nikolay E. Dmitriyev. Military R&D work on the Energiya–Buran program was conducted by the Strategic Rocket Forces' TsNII-50 research institute, which became directly subordinate to GUKOS in 1982. It was headed by Gennadiy P. Melnikov (1972–1983), Ivan V. Meshcheryakov (1983–1988), and Eduard V. Alekseyev (1988–1992).

Despite repeated attempts by the Air Force to loosen the Strategic Rocket Forces' stranglehold on the space program, its space-related responsibilities remained largely limited to cosmonaut training. For the Buran project the Air Force trained its own team of test pilots based at the Chkalov State Red Banner Scientific Test Institute (GKNII) in Akhtubinsk.

Scientific level

Although Energiya–Buran was not a scientific program, institutes of the USSR Academy of Sciences did considerable R&D in support of the project. Among them were the Institute of Applied Mathematics (IPM) (software development) and the Institute of Applied Mechanics (IPRIM) (research on heat-resistant materials). Many leading figures in the Soviet space industry were members of the Academy, including Valentin Glushko. Presidents of the Academy of Sciences during the Buran years were Mstislav V. Keldysh (1961–1975), Anatoliy P. Aleksandrov (1975–1986), and Guriy I. Marchuk (1986–1991).

Umbrella organizations

With 73 ministries and departments and about 1,200 organizations and enterprises involved in the Energiya–Buran program, there was also a need for several bodies to manage the program beyond institutional borders. The most important of these was the Interdepartmental Coordinating Council (MVKS). Created in July 1976, this was the leading body overseeing the Energiya–Buran program. It included representatives from all the ministries involved in the program as well as the general and chief designers, the heads of the main manufacturing facilities, and several leading scientists. Originally, it was headed by the first deputy head of the Ministry of General Machine Building B.V. Valmont, but from July 1981 on by the MOM minister himself, with the commander of GUKOS serving as his deputy. During its meetings, usually held at the Baykonur cosmodrome, MVKS discussed key technical and organizational aspects of the Energiya–Buran program. Its decisions had the same authority as those made by the Council of Ministers and needed to be implemented immediately, which is why MVKS was sometimes referred to as a "mini Council of Ministers". The only comparable entity in the Soviet space program until then had been the Council for the Problems of Mastering the Moon (or simply the "Lunar Council"), formed in 1966 to run the N-1/L-3 piloted lunar-landing program, although it is not entirely clear if that had the same powers as MVKS in the Energiya–Buran program.

Also playing an important role in the program was the Interdepartmental Expert Commission (MEK), established in January 1977. Headed by TsNIIMash director Yuriy Mozzhorin, MEK consisted of about 70 leading representatives of the main organizations involved in Energiya–Buran. Its initial task was to review the basic design of the Energiya–Buran system, but it continued to play an important role after the design had been frozen, mainly in safety and quality control. For this purpose working groups were set up by the MEK, which made over 2,000 recommendations to improve the Energiya–Buran system.

Finally, there was the Council of Chief Designers (SGK), a body set up in the late 1940s by Sergey Korolyov that initially consisted of the six chief designers involved in missile programs (Korolyov, Glushko, Barmin, Kuznetsov, Pilyugin, and Ryazanskiy). The Council brought together individuals who were subordinate to different ministries and thereby circumvented the normal chain of command in the industry, facilitating swifter and more efficient work.

While SGK was an influential and rather unorthodox management institution under Korolyov, it gradually evolved into a bureaucratic organization under Mishin and Glushko. The agenda was often set weeks or months in advance without taking into account new developments. Meetings were now also attended by representatives of the Central Committee, the ministries, and the Military Industrial Commission, with the chief designers often electing to send their deputies rather than go themselves. Decisions by the Council were officially unanimous, but were in actual fact made solely by the chairman, even if there were objections from other members. This changed in 1983, when decisions of the Council had to be endorsed by the signatures of all members. Although this move complicated the conduct of the meetings, it did make the decisions more authoritative. Chaired by Valentin Glushko and later by Yuriy Semyonov, the Council met weekly at NPO Energiya to discuss ongoing technical issues, including those regarding the Energiya–Buran program. Glushko and later Semyonov bore personal responsibility for the implementation of those decisions [1].

MAIN DESIGN AND PRODUCTION FACILITIES

NPO Energiya–ZEM

NPO Energiya, the former "Korolyov design bureau", was the organization in charge of the Energiya–Buran project as a whole, performing a role comparable with that of a "prime contractor" in the West. NPO Energiya was responsible for making all key technical decisions and coordinating work between the numerous organizations. Situated in the Moscow suburb of Kaliningrad (renamed Korolyov in 1996), it was initially part of the NII-88 rocket research institute founded in 1946, but split off from that organization along with Factory 88 to form the independent OKB-1 (Experimental Design Bureau 1) in 1956. It was renamed Central Design Bureau of Experimental Machine Building (TsKBEM) in 1965, NPO Energiya (NPO standing for "Scientific Production Association") in 1976, and RKK Energiya (RKK standing for "Rocket and Space Corporation") in 1994. Factory 88 was renamed Factory of Experimental Building (ZEM) in 1967.

Placed in charge of NPO Energiya in May 1974 was Valentin P. Glushko, who thereby relinquished his duties as chief designer of KB Energomash, the rocket engine design bureau that had merged with TsKBEM to form NPO Energiya. Being a member of the Academy of Sciences (since 1953) and a member of the Central Committee of the Communist Party (since 1976), Glushko had considerable political clout and enjoyed almost unconditional support from Dmitriy Ustinov. Initially, Glushko was both "general designer" and "director" of NPO Energiya, but in June 1977 Vakhtang D. Vachnadze was assigned to the newly created post of "general director" to handle the organization's day-to-day administrative affairs. Glushko died in January 1989 and was replaced in August 1989 by Yuriy P. Semyonov, who was initially only general designer, but also took over the post of general director from Vachnadze in March 1991.

By late 1977 work on the Energiya–Buran project at NPO Energiya was concentrated in Department 16. Igor N. Sadovskiy was the chief designer of Energiya–Buran as a whole, with Yakob P. Kolyako, the former head of the heavy-lift launch vehicle section, serving as deputy for the rocket, and Pavel V. Tsybin as deputy for the orbiter. There were changes in the wake of a December 1981 party and government decree calling for organizational improvements in the Energiya–Buran program. Responsibility for the orbiter was transferred to design Department 17 of Yuriy P. Semyonov (Soyuz–Salyut), while Department 16 remained in charge only of the rocket. In January 1982 Sadovskiy, who had been on bad terms with Glushko, was replaced as chief designer of Energiya–Buran by Boris I. Gubanov, a veteran of KB Yuzhnoye in Dnepropetrovsk, who had played a key role in the development of missiles such as the R-14, R-36, and R-36M. From that moment on Gubanov was chief designer of the Energiya–Buran system as a whole and also chief designer of the rocket, while Semyonov was chief designer of the orbiter. Sadovskiy became Gubanov's first deputy, while Vladimir A. Timchenko served as Semyonov's deputy for the orbiter.

On the production side, NPO Energiya's ZEM manufacturing facility was in

Energiya–Buran chief designers Igor Sadovskiy (left) and Boris Gubanov.

charge of building many key systems needed for orbital flight—in particular, the orbital maneuvering engines and primary thrusters of the ODU propulsion system as well as the power supply system. These parts were then shipped either to the Tushino Machine Building Factory or to Baykonur for installation in the vehicle. ZEM also housed a full-scale "electrical analog" of Buran (the so-called "Integrated Stand" or OK-KS).

ZEM also manufactured several parts of Energiya's strap-on boosters. In the mid-1970s an agreement had been reached that KB Yuzhnoye in Dnepropetrovsk would only build the so-called "modular part" of the strap-ons—in other words, the part that was common to the strap-ons and the Zenit first stage. Most of what was unique to the strap-ons would have to be built at ZEM—in particular, the nose and tail sections of the boosters, the parachute containers, drain valves, and actuators. According to original plans, final assembly of the strap-on boosters was to take place at ZEM, but later it was decided to move this work to the Baykonur cosmodrome. The parts manufactured at ZEM were delivered to Baykonur by rail and integrated with the modular part *in situ* at the cosmodrome. Finally, ZEM also manufactured the pneumatic and hydraulic systems for the Energiya core stage. Directors of ZEM during the Buran years were Viktor M. Klyucharyov (1966–1978) and Aleksey A. Borisenko (1978–1999) [2].

NPO Molniya/TMZ

NPO Energiya's major subcontractor for the Buran project was NPO Molniya, situated in Tushino in the northwest outskirts of Moscow. The organization was established on 24 February 1976 under the Ministry of the Aviation Industry by the merger of three existing organizations:

- MKB Burevestnik: set up in 1954 in Tushino as the design bureau aligned with Factory nr. 82, renamed the Tushino Machine Building Factory (TMZ) in 1963. The factory serially manufactured surface-to-air missiles and target drones. In 1966 the design bureau and TMZ became involved in the improvement and production of the Sukhoy design bureau's T-4 supersonic bomber, the airframe of which was built with new titanium alloys using new automated welding techniques. Burevestnik was headed from 1965 to 1986 by Aleksandr V. Potopalov.

- MKB Molniya: established in 1948 in Tushino as OKB-4 to design various types of helicopters, but in the early 1950s began specializing in air-to-air and air-to-surface missiles. Placed in charge of the organization in 1955 was Matus R. Bisnovat, who after his death in 1977 was replaced by G.I. Khokhlov.
- Experimental Machine Building Factory (EMZ): set up in 1966 by the merger of a branch of the Khrunichev factory and KB-90, which until then had been part of Branch nr. 1 of Chelomey's OKB-52 design bureau. Branch nr. 1 was the former OKB-23 design bureau of Vladimir M. Myasishchev, which among other things had worked on the Buran cruise missile and various spaceplane projects, and was absorbed by OKB-52 in 1960. The Khrunichev branch and KB-90 were situated next to one another in the Moscow suburb of Zhukovskiy and had been responsible for modernizing aircraft built at Khrunichev and also performing test flights from a nearby airfield. Myasishchev, who had been placed in charge of TsAGI after OKB-23's merger with Chelomey's bureau, was appointed head of EMZ in June 1967. The bureau was involved in the design of several heavy and high-altitude aircraft. Following Myasishchev's death in 1978, the organization was headed by Valentin A. Fedotov (1979–1986) and Valeriy K. Novikov (1986–2006).

Although several people from the former OKB-23 were apparently transferred to EMZ, none of the three organizations had been involved in any of the numerous spaceplane projects studied in the Soviet Union in the 1960s. Therefore, over 100 Spiral veterans were transferred to NPO Molniya from MMZ Zenit (the Mikoyan design bureau) and its former space branch in Dubna, which in 1972 had merged with MKB Raduga (another former branch of Mikoyan's bureau) to form DPKO Raduga. These people occupied leading positions within NPO Molniya, first and foremost Gleb Lozino-Lozinskiy, who became general designer and director of the organization. He was replaced as general director by Aleksandr V. Bashilov in 1994, but remained in the post of general designer until his death in 2001 at age 91. Also invited to work for NPO Molniya were specialists from Branch 1 of Chelomey's bureau, NPO Energiya, TsNIIMash, and several other organizations.

NPO Molniya was responsible for the "aircraft-related" elements of Buran: the fuselage, crew compartment, aerodynamic surfaces, landing gear, and hydraulic systems. In addition to that, it oversaw the development of the payload bay doors, the thermal protection system, the power distribution system, and the pressurization and ventilation system. The bulk of the work on Buran within NPO Molniya seems to have been assigned to EMZ, which was involved in the development of the crew module shell, manual flight controls, environmental and thermal control systems, the emergency escape system, and the turbojet engines needed for the approach and

Gleb Lozino-Lozinskiy (*source*: *www.buran.ru*).

landing tests with the BTS-002 Buran model. EMZ was also in charge of modifying the VM-T aircraft for ferrying Buran and elements of the Energiya rocket to Baykonur. Potopalov's Burevestnik team was responsible for developing the vehicle's primary load-bearing structure.

Even after being absorbed by NPO Molniya, the individual design bureaus that constituted the organization did not all abandon their former lines of work. EMZ continued to develop a variety of aircraft, and MKB Molniya continued to work on air-to-air missiles, although most specialists involved in this work (including Khokhlov) were transferred to another design bureau (MKB Vympel) in the early 1980s.

Production of the airframe took place at the Tushino Machine Building Factory (TMZ), which had built a wide variety of aircraft, surface-to-air missiles, and target

Lozino-Lozinskiy's grave at the Donskoye cemetery in Moscow (B. Vis).

Buran orbiter under construction at TMZ (*source*: *www.buran.ru*).

missiles since its establishment in 1932, briefly branching out into trams and trolleybuses after the war. Key aircraft manufactured at TMZ were Sukhoy's T-4 from 1966 to 1974 and Mikoyan's MiG-23 from 1975 to 1982. Aside from assembling the airframe and all airplane-related elements of Buran, TMZ was also responsible for installing heat-resistant tiles on Buran's aluminum skin. TMZ received components for the airframe from more than 450 aviation enterprises across the Soviet Union.

Although TMZ was an existing facility, most buildings needed for the construction of Buran seem to have been built from scratch. The most important ones were building nr. 110 (general assembly), nr. 111 (final assembly + production and installation of heat-resistant tiles), nr. 112 (assembly of the crew cabin), and nr. 112a (pressure and strength tests of the crew cabin).

TMZ never delivered flight-ready orbiters, mainly because the VM-T carrier aircraft were not powerful enough to transport fully-equipped orbiters to the Baykonur cosmodrome. Final outfitting was carried out at the cosmodrome's Buran assembly building by engineers of both ZEM and TMZ. Directors of TMZ during the Buran years were I.K. Zverev (1974–1982) and Suren G. Arutyunov (1982–1999). NPO Molniya also had a so-called "Experimental Factory" that among other things built various test stands for Buran (such as the PRSO and PDST landing simulators) and manufactured the Auxiliary Power Units.

One problem in transporting Buran to the launch site was that there was no suitable airfield in the vicinity of TMZ. Therefore, the orbiter had to be transported from Tushino (in the northwest outskirts of Moscow) to Zhukovskiy (southeast of Moscow, some 20 km from the outer ring road around the city). First, a special transportation device moved the orbiter through the streets of Tushino to the banks of the Moscow River. Several streets in the Moscow suburb had to be widened to give the vehicle with its 24 m wingspan enough clearance. The vertical stabilizer was removed for the entire trip from Tushino to Baykonur. Subsequently, the spacecraft was placed on a special barge equipped with ballast tanks, increasing its draught

Buran sails through the heart of Moscow under a giant cover (*source*: *www.buran.ru*).

sufficiently for it to pass under the bridges of the Moscow River. The barge then transported Buran to Zhukovskiy, floating right through the heart of the nation's capital. Most of these transports took place when the Energiya–Buran program was still a state secret, which is why the orbiter was hidden from view by a huge cover that didn't betray its true shape [3].

KB Energomash/OZEM

KB Energomash, situated in the Moscow suburb of Khimki, was responsible for the design of the RD-170 engines of the Blok-A strap-on boosters. The bureau originated as OKB-456 in 1946 and was headed from the beginning by Valentin P. Glushko, who had begun his career as a rocket engine designer at the Gas Dynamics Laboratory in Leningrad in 1929. OKB-456 developed all the engines for the Soviet Union's early ballistic missiles and derived space launch vehicles. In January 1967 OKB-456 was renamed KB Energomash (KBEM). In May 1974 it was united with the old Korolyov bureau (then named TsKBEM) to form the giant conglomerate NPO Energiya. While Glushko became the new head of Korolyov's former empire, he appointed Viktor P. Radovskiy as chief designer of the Energomash subdivision. On 19 January 1990, one year after Glushko's death, Energomash again separated from NPO Energiya and became known as NPO Energomash. The following year Radovskiy was replaced by Boris I. Katorgin, who led the organization until 2005.

KB Energomash had a so-called "Experimental Factory", originally known as "Experimental Factory 456" and renamed OZEM in 1967. It produced test models and the first flightworthy versions of new rocket engines. However, since the factory's production capabilities were relatively limited, serial production of engines was usually farmed out to other organizations. Being the most complex engines designed yet, the RD-170 and its Zenit cousin (the RD-171) were no exception. Even for the experimental engines the manufacture of the combustion chamber was entrusted to the "Metallist" factory in Kuybyshev, which had already built combustion chambers for the N-1 rocket. The production of the chambers was overseen by the "Volga Branch" of Energomash in Kuybyshev.

In 1978 it was decided that serial production of the RD-170 and RD-171 would eventually be transferred to PO Polyot in the Siberian city of Omsk, which until then had specialized in building small satellites and the 65S3/Kosmos-3M launch vehicle. Polyot's task was dual. It delivered components to the KB Energomash factory for the engines manufactured there and at the same time produced complete engines itself. PO Polyot's first RD-170 rolled off the assembly line in 1983 and during that same year KB Energomash set up a branch in Omsk, mainly to produce the blueprints necessary for serial production of the engines. Between 1983 and 1992 PO Polyot manufactured eleven RD-170 and about forty RD-171 engines. While many of those were used in test firings, none of them was ever completely installed on an Energiya or Zenit rocket. However, virtually all individual components of these engines were later used in the assembly of RD-171 engines for the Sea Launch version of Zenit. Energomash's Experimental Factory produced its last RD-170 in 1990. Its director during the Buran years was Stanislav P. Bogdanovskiy (1968–1992) [4].

KB Khimavtomatiki/VMZ

The Energiya core stage's RD-0120 engines were designed by the Design Bureau of Chemical Automatics (KB Khimavtomatiki or KBKhA) in Voronezh. This bureau was founded in 1941 by Semyon A. Kosberg in the city of Berdsk and was transferred to Voronezh as OKB-154 in 1946. Kosberg headed the bureau until he was killed in an automobile accident in 1965 and replaced by Aleksandr D. Konopatov, who remained in charge of the bureau until 1993. The bureau developed engines for surface-to-air missiles, submarine-launched and intercontinental ballistic missiles, and entered the space business in the late 1950s with the development of upper-stage engines for R-7 derived launch vehicles. It also designed the second and third-stage engines for the Proton rocket. KBKhA was a newcomer to the development of cryogenic engines when it was assigned to develop the RD-0120. Chief designer of the RD-0120 was Vladimir S. Rachuk, who would go on to become the general designer of KBKhA in 1993.

Actual manufacturing of the RD-0120 engines took place at the Voronezh Machine Building Factory (VMZ), located on the same premises as KBKhA.

Founded in 1928, VMZ switched to the production of rocket engines in 1957, building all the engines designed at KBKhA [5].

KB Yuzhnoye/YuMZ

The modular part of the Energiya strap-on boosters was designed by KB Yuzhnoye in the Ukrainian city of Dnepropetrovsk. This originated in 1954 as OKB-586 and under the leadership of Mikhail K. Yangel pursued the development of ballistic missiles using storable propellants as well as derived launch vehicles such as the Kosmos and Tsiklon series of boosters. Renamed KB Yuzhnoye in 1966, it was also in charge of developing a wide array of military and scientific satellites. Yangel died in 1971 and was succeeded by Vladimir F. Utkin, who headed the organization until 1990.

Yuzhnoye's production facility was the Yuzhnyy Machine Building Factory (YuMZ or "Yuzhmash"), originally founded in 1944 as the Dnepropetrovsk Automobile Factory. In 1951 it was renamed Factory nr. 586 and ordered to switch to the serial production of OKB-1 missiles (the R-1 and R-2). Several years later it began producing missiles and eventually launch vehicles and satellites for KB Yuzhnoye. During the Buran years Yuzhmash was headed by Aleksandr M. Makarov (1961–1986) and Leonid D. Kuchma (1986–1992), the later President of the Ukraine. Eventually, serial production of the modular part of the Energiya strap-ons was also to be transferred to PO Polyot in Omsk, but this apparently never happened.

NPO Energiya Volga Branch/Progress

The bulk of the design work on the core stage was performed by a branch of NPO Energiya situated in the city of Kuybyshev (renamed Samara in 1991). Situated on the banks of the Volga river almost 1,000 km east of Moscow, this city is home to some of the most important Russian space-related enterprises. The most famous of these is a design bureau that originated as Branch nr. 3 of Sergey Korolyov's OKB-1 in 1959. It oversaw the further development of R-7 based rockets and later also designed the lower stages of the N-1 rocket as well as the nation's photoreconnaissance, biological, and materials-processing satellites. The bureau's chief Dmitriy Kozlov was asked to design the core stage of NPO Energiya's new heavy-lift rockets, but turned down the offer, electing instead to focus on ongoing projects.

As a result, Branch nr. 3 became independent as the Central Specialized Design Bureau (TsSKB) on 30 July 1974, with NPO Energiya setting up a new "Volga Branch" in Kuybyshev to work on the core stage. Actually, the branch was set up at a time when the Energiya rocket as such did not yet exist and NPO Energiya was still working on its early RLA concepts. Receiving basic parameters from NPO Energiya's central design bureau in Moscow, a 1,000-man strong team under the leadership of Boris G. Penzin put out all the necessary blueprints for the construction of the core stage. According to one veteran this was so much that "not even two Energiya rockets could lift it off the ground". The Volga Branch was also responsible for designing the Blok-Ya launch table adapter. Penzin retired in 1987 and was replaced by Stanislav A. Petrenko.

Assembly of core stage elements took place at the Progress factory, also situated in Kuybyshev, although that was primarily aligned with Kozlov's design bureau. The construction of the giant core stage required the construction of several new halls as well as several other facilities, such as chambers for cryogenic tests of the propellant tanks using liquid nitrogen. The Energiya core stage was flown to Baykonur in two separate sections. A branch of the Progress factory was responsible for final core stage assembly and Energiya integration at the Energiya assembly building at Baykonur. The director of Progress for most of the Buran years was Anatoliy A. Chizhov (1980–1997) [6].

TRANSPORTING ENERGIYA AND BURAN TO BAYKONUR

Since the main production facilities for Energiya–Buran were located at great distances from Baykonur, a practical way had to be found of transporting the various elements to the cosmodrome. This was not so much of a problem for the 3.9 m wide strap-on boosters. The first stage of the 11K77/Zenit rocket, which served as the basis for the strap-ons, had already been tailored for rail transport to the launch site and the strap-ons could therefore use the same infrastructure. However, this was not the case for Buran itself and the core stage, which were too big to be transported by conventional means. There was also the problem of returning Buran to Baykonur in case it was forced to make an emergency landing on its back-up landing strips in the Crimea and the Soviet Far East.

One way of avoiding this problem would have been to concentrate the bulk of the assembly work at the cosmodrome itself. In other words, elements of Buran and the core stage would have been transported to the cosmodrome in many small pieces using conventional means of transport and then assembled together at the launch site. This had been done with the massive first stage of the N-1 rocket, the major parts of which were welded together at the cosmodrome. However, for Energiya–Buran this was not considered a viable solution. It would have required the construction of costly new facilities at the launch site and thousands of skilled engineers and workers would have had to be sent away from their home base on lengthy assignments to the cosmodrome.

Transportation by road and/or water was considered, but all proposals were deemed too costly because of the need to perform major construction work and make changes to existing infrastructure. One option studied for the core stage was to transport it by barge over the Volga river from Kuybyshev to the Volgograd/Astrakhan region and from there to Baykonur over a specially constructed railway.

The only solution left was to transport the elements by air, either by helicopter or airplane. Serious consideration was given to using the Mi-26 helicopter of the Mil design bureau, which had become operational just as the Energiya–Buran program got underway in the second half of the 1970s. The Mi-26 is still the heaviest and most powerful helicopter in the world, capable of lifting about 20 tons. In this scenario, the

Testing transportation techniques by helicopter (*source*: *www.buran.ru*).

orbiter airframe or elements of the core stage would have been mounted on an external platform and then lifted by a combination of two to four Mi-26 helicopters (depending on the mass of the payload). Test flights with the mid fuselage of a defunct Il-18 aircraft were staged from the Flight Research Institute in Zhukovskiy, but showed that this transportation technique was cumbersome and even dangerous. During one test flight the pilots were forced to drop the payload after it had begun dangerously swaying from one side to the other due to air turbulence. Another problem was the helicopter's limited range, which would have made it necessary to make several refueling stops on the way to Baykonur.

As foreseen by the government decree of 17 February 1976, the Ministry of the Aviation Industry began looking at a number of aircraft to solve the transportation problem. Two airplanes considered were the Tupolev Tu-95 and Ilyushin Il-76, but it soon turned out they would not be up to the task at hand. The most advanced Soviet cargo plane available at the time was the An-22 Antey of the Antonov design bureau in Kiev. In service since 1965, it was capable of lifting 60–80 tons. Engineers studied the possibility of mounting Energiya–Buran hardware on the back of the aircraft or inside by increasing the diameter of its aft fuselage to 8.3 m (the latter version was known as An-22Sh), but both configurations presented insurmountable aerodynamic and stability problems. A more capable Antonov cargo plane, the An-125 Ruslan, was under development in the late 1970s. However, its single vertical fin made it impossible to install long payloads and its relatively small undercarriage could not handle high-crosswind landings with a big payload installed on the back of the fuselage.

VM-T/Atlant

Yet another proposal came from Vladimir Myasishchev, whose Experimental Machine Building Factory was heavily involved in the Buran program after its incorporation into NPO Molniya in 1976. Myasishchev's idea was to convert his old 3M long-range strategic bomber into a transport plane. Also known as the 201M, 103M, or M6 (with NATO designation Bison-B), the 3M had made its debut back in 1956 and was a modification of the original 2M or M4 ("Bison-A"). With a cargo capacity of just about 50 tons, the converted strategic bomber would not be able to transport a fully outfitted Buran or carry a complete Energiya core stage, leaving much of the final assembly work to be done at the cosmodrome itself. It was therefore only seen as an interim solution until a more capable aircraft came along. Many felt that Myasishchev's plan was outrageous, especially given the fact that the 3.5 m wide 3M would be dwarfed by Energiya's 8 m diameter core stage. Among the skeptics was none less than Minister of the Aviation Industry Pyotr Dementyev himself, but with no better solutions available in the short run, he eventually agreed, reportedly under pressure from Ustinov. Approval for the use of the 3M came in the party and government decree on Buran of 21 November 1977, which was followed by an official order from the Ministry of the Aviation Industry on 30 December 1977.

Even before this, several ways had been studied of adapting the 3M for its new role. The most radical was to widen the fuselage from 3.5 m to 10 m and only retain the 3M cockpit, wings, and engines, giving the plane an appearance reminiscent of a C-5 Galaxy. It would have to be outfitted with a twin-fin tail and unlike the basic 3M would have required a tricycle rather than a bicycle landing gear. Another idea was to transform the 3M into something that more or less resembled a Boeing 377 "Guppy". In this configuration the nose section, mid fuselage, wings, engines, and landing gear of the 3M remained unchanged. Bolted on top of the mid fuselage would have been a cylindrical container that would be an integral part of the aircraft, with the aft section serving as the plane's tail (with two fins). Cargo would have been loaded via the nose section of the container. Both versions were turned down because they essentially turned the 3M into a new airplane, taking many years to develop. Moreover, the orbiter could only be transported by these planes if its wings and vertical stabilizer were removed.

The next idea was to mount a 37.5 m long and 9 m wide removable container on top of the 3M. The big advantages of this approach were that the 3M itself required only minimal modifications (only the double fin) and that the cargo could be loaded into the container on the factory floor and off-loaded inside the assembly buildings at Baykonur. This is the configuration that was approved by the government decree on Energiya–Buran of 21 November 1977. A drawback of the design was that the container alone weighed 17.8 tons. Although by this time the core stage had been changed to a dual-section design with four propellant tanks, it would still take three ferry flights to transport all elements to the cosmodrome.

The following year, as the core stage returned to its original single-element configuration, engineers of the Myasishchev bureau decided to do away with the container and transport both Buran and elements of the core stage exposed to the open air on top of the 3M. Buran would have to be flown to the launch site in a stripped-down state and it would still take two flights to transport all the elements of the core stage, but it was the best solution at hand until a more capable aircraft came along. On 16 November 1978, MOM and MAP put forward a plan calling for the 3M to carry four types of payloads:

- OGT (mass 45.3 tons, later increased to 50.5 tons, length 38.45 m): a stripped-down Buran (among other things without the vertical stabilizer and ODU propulsion system).
- 1GT (mass 31.5 tons, length 44.46 m): the liquid hydrogen tank with forward and aft protective covers.
- 2GT (mass 30.0 tons, length 26.41 m): the liquid oxygen tank, the RD-0120 engine section, the instrument section, and a forward protective cover, with the tip of the LOX tank acting as the aft protective cover.
- 3GT (mass 15.0 tons, length 15.67 m): the protective covers for 1GT and 2GT. After the liquid hydrogen tank had been delivered to Baykonur, the forward and aft covers were taken off, joined together, and flown back to the manufacturer. Installed inside the two covers was the disassembled forward cover used to transport the liquid oxygen tank. The 3GT configuration could also have been used to transport Buran's crew compartment.

After extensive wind tunnel tests of each configuration at TsAGI, the final go-ahead for modifying the 3M for its new role came in mid-1979. Selected for the job were three 3M aircraft that had earlier been modified as tanker aircraft (3MN-2). One (tail number 01504) was to be used only for static tests at TsAGI, while the other two (tail numbers 01402 and 01502) would enter service. Among the modifications were the replacement of the four VD-7B engines by the more powerful VD-7MD (with an afterburner for higher take-off thrust), the installation of a new, longer aft section with two horizontal and two vertical tails, and the use of improved flight control, navigation, and radio systems. Maximum crew size was reduced from eight to six. The refueling hardware was removed from the aircraft, but re-installed on 01402 in 1984 with the aim of canceling the refueling stop required during the long flights with the OGT payload from Zhukovskiy to Baykonur. Although some in-flight "dry" hook-up tests were performed in conjunction with a 3MN-2 tanker aircraft, it appears that the refueling system was never actually used. With a total length of 51.2 m and a wingspan of 53.14 m, the aircraft weighed 139 tons at take-off (minus the payload). Maximum take-off weight (with the OGT payload) was 187 tons. The payloads were hoisted onto the aircraft and off-loaded with a mate–demate device known as PKU-50, which was available in Zhukovskiy, Kuybyshev, and at the Baykonur cosmodrome.

VM-T/Atlant with Buran mock-up (*source*: *www.buran.ru*).

VM-T/Atlant with 1GT payload (*source*: Boris Gubanov).

VM-T/Atlant with 2GT payload (*source*: Boris Gubanov).

VM-T/Atlant with 3GT payload (*source*: Sergey Grachov).

The name originally painted on the planes was 3M-T ("T" standing for transport), but since 3M was a secret designator, the name was changed shortly before one of the planes was demonstrated at a Moscow air show. The most straightforward change was to repaint the 3 as the Cyrillic equivalent of the letter V ("B"), resulting in the name VM-T. These also happened to be the initials of Myasishchev's first name and patronymic (Vladimir Mikhaylovich). The planes were also called Atlant (Russian for "Atlas", the Greek mythological figure who held the burden of Earth on his shoulders).

Before transporting actual flight hardware, the Atlant planes undertook numerous test flights from the Flight Research Institute in Zhukovskiy. These began on 29 April 1981 with the first in a series of 19 flights of Atlant 01402 without a payload. In October 1981 the same plane was loaded with a mock-up 1GT payload, which had been delivered from the Progress factory in Kuybyshev to Zhukovskiy by barge via the Volga, Oka, and Moscow rivers. After several taxi tests and take-off runs, the combination took to the skies on 6 January 1982 with a six-man crew commanded by Anatoliy Kucherenko, climbing to an altitude of 2 km before returning to its home base. To onlookers it seemed as if a giant cylinder was flying in the sky, with the "small" Atlant barely visible under it. After four more flights with the 1GT payload Atlant 01402 flew seven test flights with a mock 2GT payload between 15 March and 20 April 1982, revealing the need to fly this payload at a somewhat slower speed to prevent vibrations. This cleared the way for the first ferry flights of Energiya hardware (2GT and 1GT configurations) from Kuybyshev to Baykonur on 8 April and 11 June 1982.

Meanwhile, Atlant 01502 had begun its own autonomous test flights in March 1982 and flew the mock 1GT payload on 19 April. Both planes then underwent several months of modifications before 01502 flew to Baykonur in December 1982 for the first test flights with a 3GT payload. On 28 December the plane for the first time returned a 3GT from Baykonur to Kuybyshev.

In early 1983 Atlant 01502 was ready for the first test flights with a Buran test model. A total of eight test flights were flown between 1 March and 25 March. The final one ended with the VM-T skidding off the runway in an incident blamed on pilot error. Due to a mistake in the landing gear deployment sequence, the nose gear failed to lock and lost steering capability during the landing roll-out. High crosswinds and the aircraft's own drag chute then pushed the combination off the runway, where it got stuck in the sand. Attempts to tow it back onto the runway caused serious damage to the aircraft's fuselage, which took several months to fix. Eventually, the Buran model had to be removed from the aircraft with two big cranes before the aircraft could be pulled loose. The incident was apparently photographed by American reconnaissance satellites and reported by the American magazine *Aviation Week & Space Technology* less than a month after it happened.

In December 1983 and August 1984 the VM-T ferried the first orbiters from Zhukovskiy to Baykonur (the OK-ML1 and OK-MT full-scale models). The planes were declared operational by a government and party decree in November 1985, and the following month (on 11 December) one of them delivered the first flight vehicle to the cosmodrome. During delivery of the second flight vehicle on 23 March 1988, the aircraft had a close call during the final approach to the runway, when it lost both of its left engines because of a fuel leak and suffered a power blackout in the cabin. Increasing airspeed to compensate for the loss of the engines, pilot Anatoliy Kucherenko managed to safely land the VM-T on the runway, with the airplane coming to a stop after an unusually long landing roll-out.

In all, the VM-T Atlant planes flew more than 150 missions in support of the Energiya–Buran program. In the late 1980s and early 1990s they were also considered for other tasks, such as serving as a launch platform for experimental spaceplanes and rockets and performing ferry flights and drop tests of the European Hermes spaceplane, but these plans never came to fruition [7].

An-225/Mriya

Even as the VM-T Atlant began its test flight program, the Russians continued to study more capable carrier aircraft that could transport elements of the Energiya–Buran system in one piece. Since no existing aircraft was capable of doing that, it was clear that the only way out of this problem was to develop a dedicated airplane. Not only would such an aircraft transport elements of Energiya–Buran, it could also serve as a launch platform for small air-launched reusable spacecraft that NPO Molniya

Mriya carrying Buran piggyback (*source*: *www.buran.ru*).

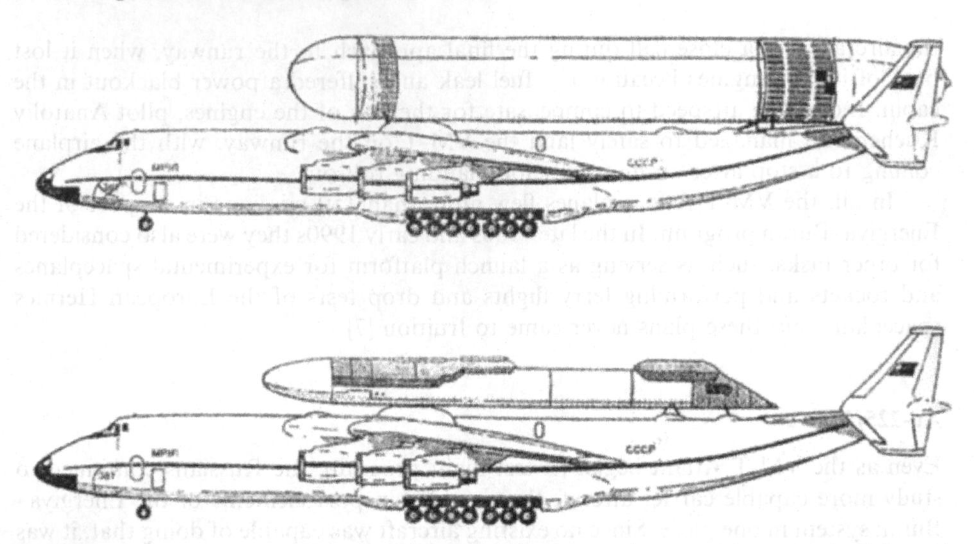

Mriya carrying a complete Energiya core stage and strap-on booster (*source*: Boris Gubanov).

had begun studying in the late 1970s/early 1980s. These studies ("System 49" and "Bizan") initially focused on the use of the An-124 Ruslan, but it eventually turned out that a more capable aircraft would be required (see Chapter 9).

All this led to the idea to build a heavier version of the Ruslan that eventually became known as the An-225 or Mriya (Ukrainian for "dream"). By the summer of 1984, after just about one year of work, engineers at the Antonov bureau had nailed down the basic design details. The plane would have forward and aft fuselage plugs to increase length as well as wing inserts to extend span and allow the installation of two additional Lotaryov D-18T turbofans beyond the four usually flown on Ruslan. The number of main landing gear assemblies was increased from five per side to seven to handle the increased take-off weight. This resulted in a 32-wheel landing gear system (two nose and fourteen main wheel bogies, seven per side, each with two wheels). The conventional tail assembly of the An-124 was changed to a twin-fin assembly to ensure controllability with a large cargo mounted on the back. This also obviated the need for covering the aft section of Buran with a tail cone (as was the case on the VM-T). The rear loading ramp was deleted to reduce weight, but the front loading ramp was retained. Payloads could be installed inside its 47 m long and 6.4 m wide cargo hold or on the back of the plane, in which case they could be 7–10 m in diameter and 70 m long. With a maximum take-off weight of 600 tons and a maximum payload capacity of 250 tons, the An-225 would become the biggest cargo plane in the world. It could easily transport a fully outfitted Buran vehicle, a complete Energiya core stage, or a complete Energiya strap-on booster. Judging by drawings published at the time, there were also plans to transport space station modules atop Mriya in giant cargo canisters.

The An-225 project received strong support from Pyotr V. Balabuyev, who became the new head of the Antonov design bureau in 1984 and played a vital role

Mriya/Buran at the Paris Air Show in 1989 (*source*: Luc van den Abeelen).

in getting it approved. The final go-ahead came in a government and party decree issued on 20 May 1987 (nr. 587-132). Constructed from a production An-125, the first Mriya (tail number CCCP-82060) was first rolled out just 1.5 years later, on 30 November 1988. After several taxi tests and take-off runs, the aircraft made its maiden test flight from the Antonov bureau's airfield at Svyatoshino on 21 December 1988. Piloted by a seven-man crew, it smoothly touched down after a 1 hour 14 minute flight that accomplished all test objectives. Coming just about a month after the inaugural flight of Buran, Mriya's successful debut was announced by the Soviet media the very same day. In early February 1989 it was first shown to Soviet and foreign journalists at the Kiev airport "Borispol", where it was even briefly inspected by Mikhail Gorbachov. On 22 March 1989 the An-225 made an historic test flight that broke more than 100 aviation records, the most important being the highest take-off mass ever achieved. Carrying a payload of 155 tons, the aircraft weighed 508 tons, exceeding the previous record (set by a Boeing 747-400) by more than 100 tons.

Several weeks later the Mriya flew to Baykonur for a series of brief test flights with the flown Buran vehicle in the first half of May 1989. Then, on 21 May, the 560-ton combination took off for a 4 hour 25 minute flight from the cosmodrome to Kiev, covering a total distance of 2,700 km. Two days later the combination flew to the Moscow area for a short stay in Zhukovskiy before returning back to Kiev. On 7 June Mriya and Buran made a 3.5 hour non-stop flight to Le Bourget to become the star attraction of the 38th Paris Air Show. Observers were surprised to see Buran being flown into Le Bourget through light rainfall. NASA's Space Shuttle Orbiter is never flown through rainfall or even through clouds while being ferried by the Boeing 747 Shuttle Carrier Aircraft, with a weather reconnaissance aircraft flying about 150 km ahead to give adequate warning to the Boeing crew to avoid clouds and rain. No such weather reconnaissance aircraft accompanied Mriya/Buran to Paris, although French Mirage fighters met the combination as it entered French airspace and escorted it to Le Bourget. After a week at Le Bourget, Mriya returned Buran to Baykonur and then made a transatlantic flight to Canada in August to an air show in Vancouver.

Although Mriya made several more appearances at Western air shows (without Buran) in the early 1990s, it gradually lost its *raison d'être* as the Soviet Union collapsed and the Energiya–Buran program was canceled. Construction of a second Mriya was discontinued and the only flown Mriya was grounded in April 1994 after having logged 339 flights lasting a total of 671 hours. Fourteen of those flights (28 hours 27 minutes) were with Buran. Mriya never flew any elements of the Energiya rocket. Plans to use Mriya as a launch platform for the British HOTOL spaceplane and NPO Molniya's MAKS spaceplane never materialized either. Instead, Mriya was placed in storage and many of its parts were "cannibalized" for use on the An-124 Ruslan. Around the turn of the century the Antonov bureau spent $20 million to upgrade the aircraft with new avionics and other modern equipment. The updated An-225, operated jointly by Antonov Airlines and the British firm Air Foyle, entered service in May 2001 as a commercial transport for heavy and oversized freight. On 11 September 2001 the An-225 once again made history by carrying a record cargo of 253 tons [8].

BAYKONUR FACILITIES

The party/government decree of 17 February 1976 that approved the Energiya–Buran program stipulated that in order to save costs the program should use as much of the N-1 infrastructure at Baykonur as possible. Exactly the same recommendation was made by a special commission of the Strategic Rocket Forces that visited the cosmodrome in October 1977. On 1 December 1978 the Central Committee of the Communist Party and the Council of Ministers approved funding for this gargantuan undertaking. However, any cost savings by reusing or adapting N-1 infrastructure must have been relatively small. Both the giant N-1 assembly building and the two N-1 launch pads had to be almost completely rebuilt and several other facilities (most notably the Buran processing building and the runway) had to be built from scratch.

Construction work got underway in 1978 and—as had been the case with the N-1 program in the 1960s—was soon spotted by US photoreconnaissance satellites, providing a clear indication that the Russians were embarking on a major new space initiative. Especially, the construction of the runway was a telltale sign that the Soviet Union was working on a response to the US Space Shuttle program. By the

N-1 pads being rebuilt for Energiya–Buran (B. Hendrickx files).

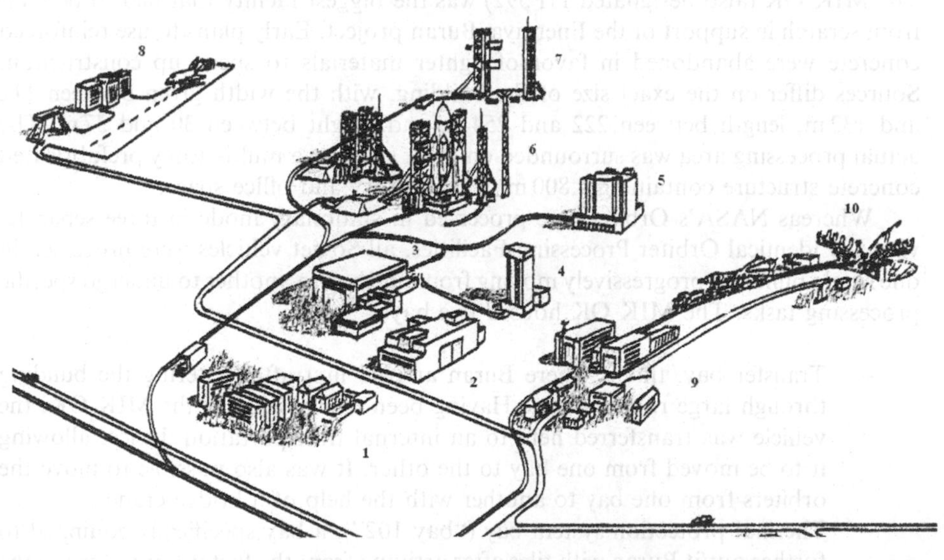

General location of Energiya–Buran and Soyuz facilities at Baykonur: 1, housing area; 2, MIK OK; 3, MIK RN; 4, SDI; 5, MZK; 6, Energiya–Buran launch pads (nr. 37 and 38); 7, UKSS; 8, landing complex; 9, Soyuz assembly buildings; 10, Soyuz launch pad (*source*: Aviatsiya i kosmonavtika).

mid-1980s photographs of the Energiya–Buran facilities made by the French SPOT remote sensing satellite were openly available in the West (see Chapter 7).

The Energiya–Buran facilities were located in the central part of the cosmodrome, some 40 km north of the city of Leninsk and just to the west of the oldest part of the launch site, namely the "Gagarin" launch pad and associated facilities for the Soyuz rocket.

The cosmodrome is divided into so-called "sites". The most important ones dedicated to Energiya–Buran were:

Site 254	The Buran processing building (MIK OK) and a platform for test firings of Buran's propulsion system and Auxiliary Power Units.
Site 112	The Energiya assembly building (MIK RN).
Site 112A	The Assembly and Fueling Facility (MZK) and the Dynamic Test Stand (SDI).
Site 110	Energiya–Buran launch pads 37 and 38.
Site 250	A combined test firing stand and launch pad for Energiya (UKSS).
Site 251	A landing complex with runway and associated facilities.

Sites 254, 112, and 112a comprised the so-called "Technical Zone" (TK) of the Energiya–Buran facilities at Baykonur.

Orbiter Assembly and Test Facility (MIK OK/MIK 254)

The MIK OK (also designated 11P592) was the biggest facility that had to be built from scratch in support of the Energiya–Buran project. Early plans to use reinforced concrete were abandoned in favor of lighter materials to speed up construction. Sources differ on the exact size of the building, with the width given between 112 and 132 m, length between 222 and 254 m, and height between 30 and 37 m. The actual processing area was surrounded on three sides by a multi-storey prefabricated concrete structure containing 4,800 m of laboratory and office space.

Whereas NASA's Orbiters are processed in stationary mode in three separate, virtually identical Orbiter Processing Facilities, all Soviet vehicles were processed in one single building, progressively moving from one bay to another to undergo specific processing tasks. The MIK OK housed five bays:

- Transfer bay: this is where Buran arrived first after entering the building through large rolling doors. Having been transported to the MIK OK, the vehicle was transferred here to an internal transportation device, allowing it to be moved from one bay to the other. It was also possible to move the orbiters from one bay to another with the help of a bridge crane.
- Thermal protection system bay ("bay 102"): a bay specifically equipped to further outfit Buran with tiles after arriving from the factory or to service the thermal protection system in between flights.
- Assembly bay ("bay 103"): here Buran was fitted with parts that could not be installed in the factory either for technical reasons or because of weight

Orbiter Assembly and Test Facility (B. Vis).

constraints imposed by the VM-T carrier aircraft. Among other things, the engines, the power supply system, additional life support systems, and various cables were installed here. This bay was also to be used for repair work in between flights, leak checks, and autonomous tests of various systems.

- Control and test bay ("bay 104"): this was used for electrical tests of individual systems and integrated electrical tests, some additional assembly work that could not be performed in the assembly bay, and also for final close-out work.
- Anechoic chamber ("bay 105"): this bay (measuring 60 × 40 × 30 m) was used for individual and integrated tests of the orbiter's radio systems to make sure that they would not interfere with one another in flight.

The MIK OK would have been capable of supporting a launch rate of up to six missions per year [9].

Orbiter test-firing stand

This test stand is located outside not far from the MIK OK and was used for test firings of the ODU propulsion system (both maneuvering engines and thrusters) and the Auxiliary Power Units. Although these systems also underwent non-integrated tests at other locations in the Soviet Union, the Russians deemed it necessary (unlike

NASA) to conduct test firings with these systems installed on the vehicle. Such tests would not only have been conducted prior to the maiden mission of a new orbiter, but also after each mission, mainly to clean the internal plumbing. The test stand had its own propellant fueling systems [10].

Rocket Assembly and Test Facility (MIK RN/MIK 112)

The Energiya assembly building (11P591) was originally built for the assembly of the N-1 rocket in the 1960s. Measuring 190 × 240 m, the building accommodates five parallel bays, three high bays (heights given between 47 and 60 m) and two low bays (heights given between 27 and 30 m). Externally, the building hasn't changed much since the N-1 days, but because of the fundamentally different design concepts of the N-1 and Energiya, the inside had to undergo a complete overhaul. Low bay 1 was used for assembly of the strap-on boosters, low bay 2 and high bay 3 for integration of the core stage, and high bays 4 and 5 for final assembly of the entire Energiya vehicle and mating with Buran. Bay 1 was operated by NPO Energiya's ZEM factory, and the other bays by a branch of the Progress factory.

The strap-ons arrived at the cosmodrome in several sections: the modular part from Yuzhnoye in Dnepropetrovsk and the aft skirt, nose section, parachute containers, and other elements from ZEM in Kaliningrad. The core stage similarly required much work in the MIK RN, where engineers had to attach the tail section with its RD-0120 engines to the hydrogen tank and subsequently attach the hydrogen tank to the oxygen tank/intertank structure.

The final assembly process began with one left and one right strap-on being placed on a special stand, after which the core stage was inserted in between them. Subsequently, the remaining two strap-ons were placed on top of the others and in the final step Buran was lowered onto the core stage with a special crane. The rocket was attached to a mating unit (Blok-Ya) that connected pneumatic, hydraulic, and electrical systems on the launch vehicle with the launch complex. Measuring 20.25 × 11.5 m and weighing 150 tons, the Blok-Ya was a massive structure, containing 1,123 pipes with a total length of about 12 km. Finally, a crawler transporter parked outside the building was placed under the structure to begin the roll-out [11].

Assembly and Fueling Facility (MZK)

The Assembly and Fueling Facility (11P593) was a building specifically constructed for the Energiya–Buran program. Height has been given between 70 and 80 m, length between 110 and 150 m, and width between 70 and 80 m. Contrary to what the name suggests, it was not really used for assembly work, but primarily for hazardous operations with the entire Energiya–Buran stack or Buran alone. The facility was built around a metal frame designed to withstand possible explosions during such operations. It has no equivalent at the Kennedy Space Center, being used for tasks that NASA performs either out on the pad or in the Orbiter Processing Facility.

The Energiya–Buran stack passed through the MZK before being rolled out to the pad for final launch preparations. Among the activities carried out with the orbiter in this building prior to flight were:

- loading the propulsion system tanks with kerosene;
- loading the Auxiliary Power Units with hydrazine and nitrogen gas;
- loading ammonia into the thermal control system;
- loading of nitrogen into the tanks of the Fire Protection System;
- filling the Pressurization and Depressurization System with air;
- installation of storage batteries and fuel cells;
- installation of cargo into Buran's payload bay.

Another operation conducted at the MZK was the installation of pyrotechnic devices for the separation of the strap-on boosters and Buran from the core stage.

The MZK was also the first facility to receive Buran after landing for removal of any residual fuel and gas, for removal of storage batteries and fuel cells, unloading of cargo from the payload bay, and removal of flight recorders. Residual LOX for the ODU propulsion system and residual LOX and liquid hydrogen for the fuel cells were removed on a special platform near the runway, but residual kerosene in the ODU system and hydrazine for the Auxiliary Power Units were removed in the MZK.

Buran was also delivered to the MZK prior to and after test firings of the propulsion system engines and Auxiliary Power Units on the test-firing platform on Site 254. In the MZK the vehicle was equipped with a special support unit to enable ODU test firings and loaded with hydrazine for test firings of the Auxiliary Power System.

The MZK was also the facility where the Polyus spacecraft, the payload for the first Energiya mission in 1987, was mated with the launch vehicle. It is not clear, however, if future payloads other than Buran would also have been integrated with the launch vehicle in this facility [12].

Dynamic Test Stand (SDI)

Operated and owned by NPO Energiya, the more than 100 m high Dynamic Test Stand (SDI or "Object 858-142D") was built to create and monitor vibrations and resonances similar to those that would be encountered by the Energiya–Buran stack during powered ascent. For this purpose a set of exciters and sensors was placed on the skin of the stacked elements. Data on the behavior of the vehicle was recorded in the facility's computer room and then flown to Kaliningrad for full analysis. Vibration research in 300 channels could be carried out over the range of 0.1 to 20.000 Hz. The exciters could each exert forces from 200 to 5,000 newtons.

The tests performed in the SDI were similar to the "Mated Vertical Ground Vibration Tests" (MVGVT) conducted with the Shuttle Enterprise and a mock-up External Tank and Solid Rocket Boosters at the Marshall Space Flight Center in 1978. Whereas MSFC's Dynamic Test Stand had originally been built for the Saturn V rocket, Baykonur's SDI was constructed specifically for Energiya. There had been some discussion early on in the program to conduct the dynamic tests at the UKSS

Blok-Ya launch adapter (B. Vis).

Assembly of core stage (*source*: Mashinostroyeniye).

Assembly of strap-on boosters (*source*: Mashinostroyeniye).

Assembly of the Energiya rocket (*source*: Mashinostroyeniye).

Attachment of Buran to core stage (*source*: *www.buran.ru*).

The Assembly and Fueling Facility (B. Vis).

Energiya test-firing stand on Site 250 rather than build a dedicated facility. However, the idea was rejected by the designers of the UKSS, who expected they would be too preoccupied with the test-firing program. The military, on the other hand, were against the construction of a facility that would only be used for test purposes.

In the end, NPO Energiya took charge of construction itself, but, since this ran into delays, initial dynamic tests *were* conducted at the UKSS using full-scale Energiya mock-ups known as 4M-D and 4MKS-D in 1983 and 1986 (see Chapter 6). Ultimately, the SDI was not finished until 1989, *after* the two flights of Energiya. The second Buran flight vehicle, attached to a mock-up Energiya, was tested here in June 1991. The SDI was designed to test Energiya in all possible configurations (not just with the orbiter) as well as for tests of Energiya-derived rockets such as Energiya-M and the massive Vulkan rocket. The now abandoned facility still houses a mock-up of Energiya-M built in the late 1980s [13].

Energiya–Buran launch pads

Known as Raskat ("peal of thunder") or 11P825, the Energiya–Buran launch complex consisted of two adjacent pads: pad 37, the "left" pad as seen from the Energiya–Buran Technical Zone, and pad 38, the "right" pad. The pads were situated some 5 km from the Technical Zone. Separated by only a few dozen meters, they were

Energiya being erected inside the Dynamic Test Stand (*source*: *www.buran.ru*).

built on the same site where the two N-1 pads had been built in the 1960s. This decision had not been taken lightly. Many argued the Energiya–Buran pads should be built farther from the Technical Zone, because the cataclysmic on-the-pad explosion of the second N-1 rocket in July 1969 had actually caused damage to the N-1 assembly building. Furthermore, the Energiya–Buran pads required totally new systems such as hydrogen storage tanks and crew emergency escape systems. Although building the pads on the old N-1 complex was not necessarily cheaper or simpler than other options, Vladimir P. Barmin, the head of launch pad design bureau KBOM (Design Bureau of General Machine Building), insisted on maintaining at least some of the colossal work invested in the lunar program.

The N-1 pads consisted of three 23 m deep flame trenches, five-story underground support facilities and a 145 m high rotating service structure. While the underground support facilities had to be almost completely rebuilt, the flame trenches remained largely unchanged, although a new flame deflection system had to be built to make their three-directional design compatible with Energiya's asymmetrically configured propulsion system. Engineers designed a new 1,200-ton heavy launch table compatible with the Blok-Ya mating unit on which Energiya–Buran was mounted.

The N-1's 145 m high rotating service structure remained in place on both pads. Newly installed on the structure were several sets of support arms that embraced the stack for final launch preparations. A lower set interfaced with Buran's mid fuselage and was therefore probably mainly used for fuel cell servicing. Two higher sets of support arms provided electric and other interfaces with the rocket and were used to inspect the rocket's thermal insulation layers.

At some point the rotating service towers on both pads were shortened to about 60 m. This was reportedly done to minimize the chance of the rocket's flames impinging on the tower after lift-off, even though it was at a relatively safe distance in parked position. Since Buran faced the rotating service tower while it sat on the pad, the move may also have been related to the possible use of Buran's ejection seats in the event of an on-the-pad emergency. Although the ejection seats would have lifted the pilots well over the tower (to an altitude of about 500 m), the shortening may have provided an extra margin of safety. At the time of Buran's launch in November 1988, the rotating service structure of pad 37 still had its original height, while that of the (unused) pad 38 had already been shortened.

Flanking the Energiya–Buran stack on either side were two newly erected 64 m high fixed service structures. One of these contained the propellant lines for tanking or detanking of the core stage and strap-on boosters. There were at least three arms connecting it to the rocket. One of the arms was retracted from the rocket only at lift-off to ensure that no hydrogen escaped into the surrounding air, forming a potentially explosive mixture.

The other fixed tower had two arms. One was an arm connected to Energiya's intertank section that contained instruments necessary to correct the rocket's azimuthal orientation gyroscopes. It was retracted with less than a minute left in the countdown. The other was an access arm linking the tower with Buran's crew compartment. Running from the access arm were two pipes leading to two separate

Pad 37 with the 145 m high rotating service structure in place (*source*: *www.buran.ru*).

Pad 37 with shortened rotating service structure (B. Vis).

Trolleys used to ferry crews and personnel to and from the orbiter access arm (*source*: Edwin Neal Cameron).

Underground room where the emergency chute ends (*source*: Edwin Neal Cameron).

underground rooms. To board the orbiter, the crew or launch pad personnel rode on special trolleys inside the top pipe. The trolleys could accommodate about a dozen people. The lower pipe was a giant escape chute to be used by the crew or personnel in the event of an emergency, with a special mattress in the underground room softening their landing. Once there, they would have hermetically sealed themselves in an adjacent blast room where they should have been safe from explosions, leaks, and the like. A special test stand imitating the chute was built in 1986 at the Scientific Research Institute of Chemical and Building Machines (NIIKhSM) in Zagorsk, north of Moscow. The chute at the pad was tested numerous times by engineers. It was reportedly also a favorite playground for off-duty soldiers in the evening hours, as evidenced by the numerous boot imprints on the mattresses.

Soviet engineers may have been inspired by an escape system used at Kennedy Space Center's Launch Complex 39 during the Apollo years. This would have seen crews riding high-speed elevators to Level A of the mobile launch platform, where they would have jumped into a slide tube that would carry them under the launch pad. The slide terminated in a padded "rubber room" which was connected by a massive steel door to a blast room, which could withstand an on-the-pad explosion of the Saturn V launch vehicle. After the Apollo program the slide tube was capped off, although the rooms remain deep down under the pad, serving as "time capsules" of the Apollo program. For the Shuttle program, slide-wire baskets were installed on the fixed service structure at the level of the Orbiter access arm, taking crews down to seek shelter in a nearby bunker or escape from the pad in a small armored vehicle.

Surrounding the pads were four large floodlight towers and two 225 m high lightning protection towers that also supported floodlights. Rust-colored reservoirs containing sound suppression water were located on either side of the fixed towers. During launch huge pipes channeled the water to spray nozzles that sent thousands of liters of water onto the pad during launch.

Just as for their other launch vehicles, the Soviets had a policy of limited pad time for Energiya–Buran, dictated at least partially by the harsh climatic conditions at the cosmodrome, especially during winter. With most hazardous operations and close-out activities performed inside the MZK, the vehicle was in a high state of readiness when it arrived on the pad. Buran was rolled out from the MZK to the pad just 19 days before its first launch attempt on 29 October 1988.

The main hazardous operations remaining to be completed on the pad were the loading of cryogenic propellants into Buran's fuel cells and the ODU propulsion system, and tanking of the core stage and strap-on boosters. A so-called "cryogenic center" serving both pads was built to the north of the launch complex and had huge spherical storage tanks containing liquid oxygen and hydrogen as well as gaseous nitrogen and helium. Fueling of the Energiya rocket was completely automated, with nobody allowed within a 5 km radius of the pad. Unlike the Shuttle pads at the Kennedy Space Center, the Buran pads had no provisions for loading payloads into the cargo bay.

The pad used for the one and only Buran launch on 15 November 1988 was nr. 37. Pad 38, although apparently finished, never hosted an Energiya–Buran stack. It can be distinguished from the used pad by white markings on the fixed and rotating service structures [14].

The Universal Test Stand and Launch Pad (UKSS)

One of the lessons learned from the ill-fated N-1 program was the need to build a test stand for full-scale test firings of the Energiya's rocket stages. The site selected for the test stand was situated several kilometers to the northwest of the Raskat complex, where any explosions would not damage buildings in the Technical Zone. The test stand was designed to support both individual as well as joint test firings of the core stage and strap-on boosters. The opportunity to test the rocket's engines in actual flight configuration minimized the risk of catastrophic launch failures and was one of the reasons the Energiya–Buran pads remained relatively close to the Technical Zone.

Designed to withstand the pressure of an Energiya rocket bolted to the pad and producing 3,600 tons of thrust for dozens of seconds, the test stand could also be easily converted into a launch pad for Energiya rockets with payloads other than Buran and also for the massive Vulkan rocket, an Energiya with eight strap-on

The Universal Test Stand and Launch Pad (*source*: *www.buran.ru*).

boosters. It therefore became known as the Universal Test Stand and Launch Pad (UKSS or 17P31).

The UKSS had one fixed service tower, almost identical to the "fueling tower" of the Energiya–Buran pads. It also featured an equally high mobile tower that provided access to virtually every part of the rocket and was equipped with a crane for hoisting operations. The UKSS had one enormous flame trench with a depth of 40 m that could easily be seen on satellite photographs of the cosmodrome. Being exposed to much higher temperatures and acoustic pressures than the Raskat pads, the UKSS had a much more elaborate sound suppression water system, consisting of three reservoirs containing a total of 18,000 m^3 of water which was sprayed onto the pad at a maximum rate of 18 m^3 per second. The UKSS was surrounded by two lightning protection towers and several floodlight towers. A special propellant storage complex was built several kilometers from the pad. Several support facilities for the UKSS were located at the neighboring Site 250A. The most important of these was a control center some 3 km from the test stand which was designed to withstand an on-the-pad explosion.

The groundbreaking ceremony for the UKSS was held on 20 August 1978. The stand was supposed to have been ready for the first test firings in 1982, but construction ran into serious delays. Initially, the prime contractor for the construction of the facility was NIIKhimmash, which operated several rocket engine test stands at a site north of Moscow not far from Zagorsk. However, because it was of the utmost importance to have the test stand ready *before* the Energiya–Buran pads, it was decided early on to assign the task to the leading launch pad design bureau KBOM, while NIIkhimmash remained in charge of the actual test-firing program.

The first roll-out of an Energiya mock-up to the UKSS took place in early 1983. The pad was used for a series of core stage fueling tests with the Energiya 4M vehicle in 1985. Original plans for individual and joint test firings of the core stage and the strap-on boosters were severely curtailed. In 1986 the UKSS witnessed two test firings of the core stage of the Energiya 5S vehicle. These test firings were to continue with Energiya 6S, but in 1985 a decision was made to turn that vehicle into a flightworthy rocket (redesignated 6SL) and fly it with a payload called Polyus. As a result, the UKSS was converted into a launch pad much earlier than planned. The launch of Energiya 6SL took place on 15 May 1987 and, although the rocket operated flawlessly, the payload was not inserted into orbit due to a navigation error. The pad itself was seriously damaged because the sound suppression water system failed to operate (see Chapter 6) [15].

The landing complex (PK OK)

Very early on in the program a decision was made to build a runway at the Baykonur cosmodrome not only to receive Buran at the end of its missions, but also to deliver Buran and elements of the Energiya rocket to the cosmodrome by the VM-T Atlant and eventually Mriya. NPO Molniya was assigned as prime contractor for the construction of the runway by a party/government decree on 21 November 1977.

Baykonur has had an aerodrome ("Krayniy") since the early days of its existence, but this is situated close to the city of Leninsk, many dozens of kilometers to the south of the launch facilities, and was therefore not suited for this role. Requirements for the location of the new runway were that it had to be outside the "blast zone" of the Energiya pads and be capable of receiving Buran from either side, both during nominal missions and in launch emergencies. The new facility (called PK OK or 11P72) was eventually built some 6.5 km to the northwest of the UKSS complex and 11 km to the northwest of the Raskat complex.

The central part of the landing complex was a 4.5 km long and 84 m wide runway called Yubileynyy ("Jubilee"), capable not only of receiving Buran, but also planes with a take-off mass of up to 650 tons. The surface layer was made of reinforced concrete with a thickness varying between 26 and 32 cm above an 18 to 22 cm sand/cement ground layer. This concrete, which was about 1.5 to 2 times stronger than the type used on ordinary runways, was produced in six factories located at a considerable distance from the runway. This created serious transportation problems since the concrete could remain in liquid state for only one and a half hours before being poured onto the runway. The surface had to be extremely flat, with deviations of no more than 3 mm over a 3 m stretch (compared with 10 mm on ordinary runways). To achieve this, the complete 378,000 m surface of the runway had to be ground like parquet floor with special milling machines.

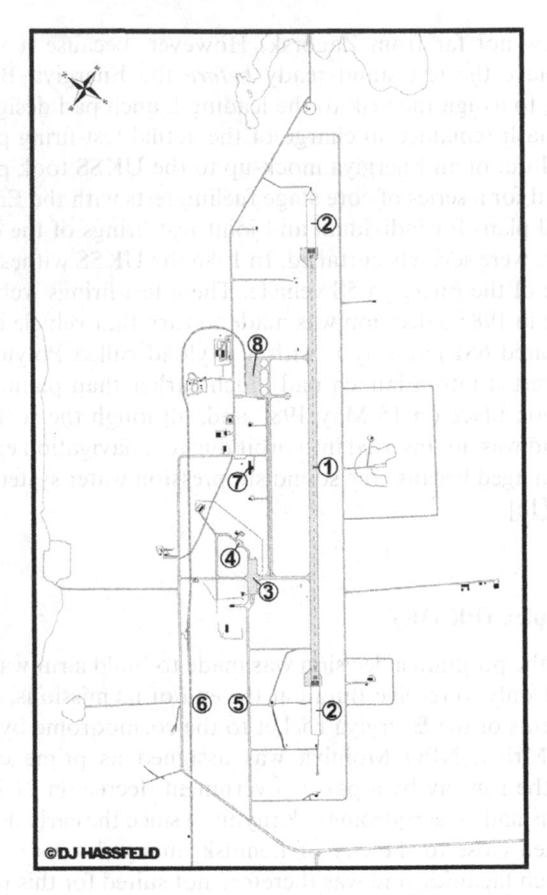

The Buran landing complex: 1, Yubileynyy runway; 2, asphalt stretches; 3, off-loading area; 4, Buran detanking area; 5, main road linking landing complex with other facilities; 6, railway; 7, command and control building (OKPD); 8, airplane parking platform (*source*: Dennis Hassfeld).

At either end of the runway was a 500 m long and 90 m wide stretch of asphalt to give Buran more leeway during emergency landings. Running parallel to the main runway at a distance of some 50 m was a 4.5 km long and 100 m wide dirt runway apparently intended for emergency landings by planes, with no role in the Buran program.

Adjacent to the runway were several facilities:

– A platform to drain liquid oxygen, gaseous oxygen, and liquid hydrogen from Buran's fuel cells and the ODU propulsion system.
– A platform to off-load Buran and elements of the Energiya rocket from their carrier aircraft. This has two mate–demate devices called PKU-50 and PUA-100 capable of handling payloads of 50 and 100 tons, respectively.

Buran being installed atop Mriya using the PUA-100 mate–demate device (*source*: Sergey Grachov).

- A "waiting platform" for vehicles needed to service Buran after landing.
- A parking platform for airplanes.
- An airplane-servicing area.

Also located in the vicinity of the runway was the ground segment of the Vympel navigational aid system (Vympel-N). This included six transponders for the RDS system (only three of which were required for landing), one beacon for the RSBN system, four beacons for the RMS microwave landing system, and a set of radars.

The nerve center of the landing complex was a six-story high command and control building (OKPD) that acted as a control center for the landing phase, working in conjunction with the TsUP Mission Control Center near Moscow. The building had one big control room for Buran and another for ordinary air traffic control tasks [16].

Moving between facilities

Linking the various facilities was an impressive network of roads and railways, some left over from the N-1 days, others built specifically for Energiya–Buran. Twelve meter wide roads connected the MIK OK with the landing facility, the MIK RN, the test-firing stand, and the MZK. Buran was transported with its landing gear retracted on a special 126-ton, 58.8 m long platform with 32 wheels that was pulled by a truck. Maximum speed with the vehicle mounted on top was 10 km/h.

Energiya–Buran on the crawler transporter (*source*: Luc van den Abeelen).

In keeping with Soviet tradition, the Energiya–Buran stack was assembled and rolled out horizontally and then erected after arriving at the launch pad. The transportation device used for this was a giant crawler transporter (TUA) left over from the N-1 days and built by the Novokramatorskiy Machine Building Factory in the Donetsk region (Ukraine). The transporter weighed 2,756 tons (without the stack), measured 56.3 × 90.3 m and was 21.2 m high. It was towed by four 100 horsepower diesel locomotives, moving at a maximum speed of 5 km/h over rail tracks separated 18 m apart. There were two TUA transporters, parked outside high bays 4 and 5 of the MIK RN. The MIK RN was linked by railway with the MZK, the two Energiya–Buran launch pads, and the UKSS [17].

After the cancellation of the Energiya–Buran program in 1993, some of the facilities were mothballed and left to rust, but others have since been modified for new programs (see Chapter 8).

BACK-UP LANDING FACILITIES

Buran had two back-up landing sites, an "eastern" site not far from the Soviet Pacific coast and a "western" site in the Crimea. The eastern site was situated south of Lake Khanka very close to the small town of Khorol, a little over 100 km north of Vladivostok. The aerodrome was originally used in the 1960s as a temporary home base for the Tu-95MR bomber and later hosted Tu-95RTs Navy reconnaissance planes, Tu-16 planes belonging to the Pacific fleet, and fighter jets of the Air Defense Forces. In the 1980s it was modified for its role in the Buran program by lengthening the existing runway and installing equipment of the Vympel navigation system. The runway was 3.7 km long and 70 m wide. The western site was located near the town of Simferopol in the Crimea and featured a 3.6 km long and 60 m wide runway. There is conflicting information on whether the two sites were ready in time for Buran's maiden mission in November 1988, but they should have been available for the first manned missions in the early 1990s [18].

Also considered was the possibility of landing Buran on ordinary runways, not specially adapted for the orbiter and not equipped with the navigation facilities needed to assist in a hands-off landing. A requirement formulated for Buran's test pilots was to land Buran on such runways at nighttime without any illumination [19].

REFERENCES

[1] M. Tarasenko, *Voyennye aspekty sovetskoy kosmonavtiki*, Moscow: Nikol, 1991; B. Gubanov, *Triumf i tragediya Energii (tom 3: Energiya-Buran)*, Nizhniy Novgorod: Izdatelstvo Nizhegorodskogo instituta ekonomicheskogo razvitiya, 1998; A. Siddiqi, *Challenge to Apollo*, Washington, D.C.: NASA, 2000; V. Favorskiy, I. Meshcheryakov, *Voyenno-kosmicheskiye sily (kniga 2)*, Moscow: Izdatelstvo Sankt-Peterburgskoy tipografii, 1998; V. Favorskiy, I. Meshcheryakov, *Kosmonavtika i raketno-kosmicheskaya promyshlennost (tom 2: razvitiye otrasli 1976–1992)*, Moscow: Mashinostroyeniye, 2003; A. Kuznetsov, *Mnogorazovaya kosmicheskaya sistema Energiya-Buran*, Moscow: OmV-Luch, 2004.

[2] Y. Semyonov, *Raketno-kosmicheskaya korporatsiya im. S.P. Korolyova 1946–1996*, Moscow: RKK Energiya, 1996; B. Gubanov, *op. cit.*; V. Filin, *Put k Energii*, Moscow: Logos, 2001.

[3] A. Bratukhin, "Development of the airframe of the orbital ship Buran"/A. Bashilov, "Experimental Factory of NPO Molniya: Main activities"/I. Zverev, S. Arutyunov, "The Tushino Machine Building Factory" (in Russian), in: G. Lozino-Lozinskiy, A. Bratukhin, *Aviatsionno-kosmicheskiye sistemy*, Moscow: Izdatelstvo MAI, 1997, pp. 15–28; A. Bruk, *Illyustrirovannaya entsiklopediya samolyotov EMZ im. V.M. Myasishcheva (tom 3, chast 1)*, Moscow: Aviko Press, 1999; *Tushinskiy mashinostroitelnyy zavod. Ot Stali do Burana: istoriya, tekhnologiya, lyudi*, Moscow: AviaRus XXI, 2001.

[4] V. Trofimov, *Osushchestvleniye mechty*, Moscow: Mashinostroyeniye/Polyot, 2001; B. Katorgin, *NPO Energomash imeni akademika V.P. Glushko. Put v raketnoy tekhnike*, Moscow: Mashinostroyeniye/Polyot, 2004.

[5] *KB Khimavtomatiki, stranitsy istorii, tom 1*, Voronezh: KBKhA, 1995.

[6] Y. Semyonov, *op. cit.*; B. Gubanov, *op. cit.*; V. Filin, *op. cit.*

[7] V. Burdakov, "On the back of a plane" (in Russian), *Krasnaya zvezda*, 12 April 1989; V. Kazmin, "Saving Buran" (in Russian), *Krylya rodiny*, April 1990; V. Fedotov, "Air transport" (in Russian), in: G. Lozino-Lozinskiy, A. Bratukhin, *op. cit.*, pp. 252–257; B. Gubanov, *op. cit.*, pp. 192–196; A. Bruk, *op. cit.*, pp. 209–238; V. Filin, *op. cit.*, pp. 72–73.

[8] Mriya page on the *www.airwar.ru* website; I. Sergeychuk, "The record renaissance of Mriya" (in Russian), article on the website *Aerokosmicheskiy portal Ukrainy*.
[9] Y. Semyonov, *Mnogorazovyy orbitalnyy korabl Buran*, Moscow: Mashinostroyeniye, 1995, pp. 347–350, 374–375.
[10] Y. Semyonov, *op. cit.*, pp. 195, 352–351, 367–368.
[11] B. Gubanov, *op. cit.*, pp. 70, 187–192.
[12] Y. Semyonov, *op. cit.*, pp. 353, 378–379.
[13] *The Russian Space Directory 1994*, Munich: European Space Report, 1993, p. 88; B. Gubanov, *op. cit.*, pp. 69, 229; V. Filin, *op. cit.*, p. 157; A. Kuznetsov, *op. cit.*, pp. 270–271.
[14] P. Mills, "Energia and Buran at Baykonur", *Spaceflight*, November 1989, pp. 380–385; *Startuyet Energiya–The Energia blasts off*, Moscow: Mashinostroyenie, 1990, pp. 5–6; V. Menshikov, *Baykonur: moya bol i lyubov*, Moscow, 1994, pp. 200–202; B. Gubanov, *op. cit.*, p. 52; V. Favorskiy, I. Meshcheryakov, *op. cit.*, pp. 121–122; V. Yermolayev, "Slide" (in Russian), *Rossiyskiy kosmos*, 10/2006, pp. 82–84.
[15] V. Favorskiy, I. Meshcheryakov, *op. cit.*, pp. 120–121; B. Gubanov, *op. cit.*, pp. 208–211, 243, 265, 281; A. Kuznetsov, *op. cit.*, pp. 243, 274.
[16] Y. Semyonov, *op. cit.*, pp. 386–398; V. Favorskiy, I. Meshcheryakov, *op. cit.*, p. 123; A. Kuznetsov, *op. cit.*, p. 246; "Landing complex of the Baykonur cosmodrome" (in Russian), article on the website *www.buran.ru*
[17] Y. Semyonov, *op. cit.*, p. 351.
[18] "Back-up aerodromes for Buran", article on the *www.buran.ru* website.
[19] E. Vaskevich archives.

5

The Buran cosmonaut team

Describing the history of the Buran cosmonaut team is not as straightforward as it may seem at first glance. Unlike the situation in the US, where NASA has always been in charge of selecting and training the (career) astronauts that make up Space Shuttle crews, the Soviet Union's space program lacked a central coordinating NASA-type organization. Several organizations involved in test pilot and cosmonaut training felt they should all independently select their own cosmonaut teams. From these groups, Buran crew members representing those organizations would be assigned.

Three organizations selected cosmonauts specifically for Buran:

- The Cosmonaut Training Center named after Yu. A. Gagarin (TsPK for *Tsentr Podgotovki Kosmonavtov*) based in Star City (*Zvyozdnyy Gorodok*) near Moscow.

  This Soviet Air Force unit, set up in 1960, had been in charge of selecting and training cosmonauts for flights on Vostok, Voskhod, Soyuz, and Salyut.

- The Flight Research Institute named after M.M. Gromov (LII for *Lyotno-Issledovatelskiy Institut*) in Zhukovskiy, some 35 km southeast of Moscow.

  The Flight Research Institute, a civilian research and development entity subordinate to the Ministry of the Aviation Industry (MAP), was founded in 1941 as the leading test center for experimental and production aircraft. LII had a Test Pilot School (ShLI).

- The State Red Banner Scientific Test Institute named after V.P. Chkalov (GKNII for *Gosudarstvennyy Krasnoznamennyy Nauchno-Ispytatelnyy Institut*) in Akhtubinsk, some 130 km west of Volgograd and about 50 km south of the Kapustin Yar cosmodrome in the Volga delta.

  This Air Force unit was set up in 1960 at the same site that had served since 1947 for testing various unmanned flying apparatuses, such as surface-to-surface missiles, air-to-surface missiles, and the Burya intercontinental cruise missile.

The site was sometimes referred to as Vladimirovka, after a nearby railway station. With the establishment of GKNII its role was expanded to testing various aircraft for the Air Force and it also included an Air Force test pilot school known as the Test Pilot Training Center (TsPLI).

In March 1979 MOM, MAP, and the Ministry of Defense jointly decided that a pool of 17 pilots would be required for the Buran test flight program: six from LII, six from GKNII, and five from TsPK [1]. Part of the reason for assigning pilots from three different organizations was no doubt the departmentalism typical of the Soviet space program. However, there appear to have been more rational considerations as well. Since Buran was far more complex than any Soviet spacecraft flown before, the unmatched flying skills of the LII pilots were probably considered necessary to safely guide the vehicle through its initial atmospheric and orbital flight tests, with GKNII becoming involved in the test flights at a somewhat later stage. It was not uncommon in the former Soviet Union for the Air Force team in Akhtubinsk to further test new aircraft once they had been declared airworthy by the LII pilots, and in this respect Buran was no exception [2]. Presumably, the LII pilots were to fly test flights with civilian payloads, and the GKNII pilots test flights with military payloads.

Once their job was completed, these career test pilots would then return to their usual line of work, passing the torch to the "regular" TsPK pilots to finish the test program, and ultimately fly Buran's operational missions. This, at least, appears to have been the original intention when the first pilot teams were selected in the 1970s and Buran was expected to begin flying in the first half of the 1980s. As the orbital test flights slipped into the late 1980s and were spread out over many years, the operational phase became a distant and vague goal. Therefore, in the end the only pilots *seriously* considered to fly on Buran were from LII and GKNII, and further TsPK selections were solely aimed at the mainstream Salyut/Mir space station program.

In addition to the pilots, engineers from both NPO Energiya and the Air Force were assigned to Buran as well, although none of them ever appear to have been specifically selected for the program. Because of all this, it is not possible to really give one single founding date for the Buran cosmonaut team.

SELECTIONS BY TsPK

The Cosmonaut Training Center was the first to select a dedicated group for the Buran program. On 23 August 1976, just six months after the official approval of the Energiya–Buran program, nine pilots were chosen, the first selection by TsPK in six years. They were:

- Leonid Georgyevich Ivanov;
- Leonid Konstantinovich Kadenyuk;
- Nikolay Tikhonovich Moskalenko;
- Sergey Filippovich Protchenko;
- Yevgeniy Vladimirovich Saley;

- Anatoliy Yakovlevich Solovyov;
- Vladimir Georgyevich Titov;
- Vladimir Vladimirovich Vasyutin;
- Aleksandr Aleksandrovich Volkov;

All of them were young, relatively inexperienced Air Force pilots in their mid to late twenties. The rationale behind their selection at this early stage may have been that they would need several years to advance their flying skills while the more experienced LII and GKNII test pilots conducted the early Buran test flight program.

Not surprisingly, it was decided that the cosmonauts would first have to undergo test pilot training before beginning the standard cosmonaut training course. They began studying and training at TsPLI in Akhtubinsk in September 1976, becoming Test Pilots 3rd Class (the lowest test pilot rank) in June 1977. In addition, in August they conducted parachute training. From October 1977 until September 1978 they then underwent the standard basic cosmonaut training course ("General Space Training" or OKP) at TsPK.

After graduation most members of the group (Ivanov, Kadenyuk, Moskalenko, Protchenko, Saley, Solovyov, and Volkov) returned to Akhtubinsk to resume test pilot training with the goal of becoming Test Pilots 2nd Class. It was during this follow-up course that Sergey Protchenko was medically disqualified and dismissed from the cosmonaut team in April 1979 [3]. More than a year later, on 24 October 1980, the group suffered another loss when Leonid Ivanov was killed in the crash of a MiG-27 in Akhtubinsk.

The Air Force's 1976 selection group. From left: Vasyutin, Ivanov, Saley, Kadenyuk, Protchenko, Volkov, Solovyov, Moskalenko, and Titov (B. Vis files).

A remarkable group photo of the 1976 selection group. Although the names of the cosmonauts were still state secrets at the time, it appears to have been made for publicity purposes, as the obelisk and the wall on the right actually are over 150 meters apart (B. Vis files).

On 22 June 1981, Kadenyuk, Moskalenko, Saley, Volkov, and Solovyov were awarded the title Test Pilot 2nd Class. After that, the first four went on to conduct Buran-related training, but Solovyov was transferred to the Salyut space station program, together with Titov and Vasyutin.

Leonid Kadenyuk was the next to be dismissed. He left the cosmonaut team in March 1983 after he had divorced his wife. In the Soviet Union of the 1970s and 1980s, getting a divorce usually resulted in the end of a cosmonaut career for those who were still awaiting their first mission.

The 1976 selection had been limited to the Air Force and it was therefore decided that another screening would take place in the Soviet Navy and Air Defense Forces

Aleksandr Viktorenko (left) and Nikolay Grekov (B. Vis files).

[4]. As a result, on 23 May 1978 one additional candidate from each of these two branches of the military was added to the detachment:

- Nikolay Sergeyevich Grekov (Air Defense Forces);
- Aleksandr Stepanovich Viktorenko (Navy).

In October 1978 the two began training in Akhtubinsk, graduating as Test Pilot 3rd Class on 2 July 1979. They then returned to Star City, where they underwent OKP, finishing that in February 1982. Viktorenko almost died in a bizarre accident during a medical check-up in 1979. He was wearing a band with electrodes around his body to have an ECG made when a 220-volt current was accidentally sent through it. Apparently his heart stopped and he was brought back to life using CPR. According to Viktorenko the incident cost him quite some time in training as the doctors wanted to be 100% certain that he had not suffered any ill effects from the incident [5].

As the training went on, it was becoming increasingly apparent that Buran's first flights would be significantly delayed. Since TsPK was becoming confronted with a shortage of commanders for Soyuz and Salyut, it was decided in late 1983 to transfer all members of the 1976 and 1978 selections to the space station program. In the following years, they would become the core of the cosmonaut detachment, with several of them flying record-breaking missions (for details on further careers see the cosmonaut biographies in Appendix B).

SELECTIONS BY LII

An initial group of civilian test pilots from the Flight Research Institute were selected as candidates for the Buran program on 12 July 1977 by an order of the head of the institute. They were:

- Igor Petrovich Volk;
- Rimantas Antanas-Antano Stankyavichus [6];
- Anatoliy Semyonovich Levchenko;
- Aleksandr Vladimirovich Shchukin;
- Oleg Grigorevich Kononenko;
- Nikolay Fyodorovich Sadovnikov.

A seventh pilot, Aleksandr Ivanovich Lysenko, had been pre-selected to become a member of the group, but while the necessary documents were still being prepared, Lysenko was killed when his MiG-23UB fighter crashed on 3 June 1977.

Selecting the group had been more problematic than anticipated. There was little willingness among LII's pilots to enter the program because of the stringent medical examinations. In fact, initially only Volk and Levchenko had volunteered. Many feared that if anything was found that would disqualify them for Buran, they would also lose flight status for their test pilot work. Because this proved a serious problem in selecting candidates, Igor Volk managed to strike a deal with Oleg Gazenko, who was in charge of the medical screenings. The deal implied that as soon as it would become clear that a pilot was deemed medically unfit for the cosmonaut program, the selection process would be halted, both the pilot and Volk would be informed, and the pilot could return to test flying in LII, no questions asked [7].

Igor Volk can probably be called a logical choice as member of the group. In May 1976 he had already test-flown the 105.11 (nicknamed "Lapot"), an atmospheric test bed of the Spiral spaceplane. However, Volk himself has stressed that he made only one flight on "Lapot" at the invitation of the Mikoyan design bureau and was not involved in the program as such [8].

Late in 1977 the group lost a member when Sadovnikov decided to move from LII to the Sukhoy design bureau and become a test pilot there. Between April and June 1980 he would fly 15 combat missions in Afghanistan on Su-25 fighter jets. After returning to Sukhoy, he went on to become their lead test pilot and a Hero of the Soviet Union before passing away on 22 July 1994, only 47 years old.

Nikolay Sadovnikov (left) and Aleksandr Lysenko (B. Vis files).

Sometime in 1978 Igor Volk became the commander of the group [9]. Eventually, the group would become known as the "Wolf Pack", a tongue-in-cheek reference to Volk, whose name is Russian for "wolf". The cosmonauts from later LII selections would become known as the "Wolf Cubs".

On 3 August 1978 Volk, Stankyavichus, Levchenko, Shchukin, and Kononenko passed the so-called Chief Medical Commission (GMK), which cleared them for preliminary space-related training such as centrifuge tests and parabolic flights on aircraft to simulate zero-*g* [10]. Almost five months later, on 30 December 1978, the five went through the next phase in the selection process, appearing before the State Interdepartmental Commission (GMVK), the top government commission for cosmonaut selection. Its main task was to select cosmonauts on the basis of their political reliability and both moral and human qualities [11]. Headed by Leonid Smirnov, the chairman of the Military Industrial Commission, it consisted of representatives of the appropriate ministries and the KGB and also included Kerim Kerimov, the chairman of the State Commission for manned spaceflight [12]. All five were given the go-ahead to begin their OKP at the Gagarin Cosmonaut Training Center the following April, thereby receiving the official status of "cosmonaut candidates". Unlike Air Force cosmonaut candidates that were selected for flights on Soyuz, they went through their OKP in periodic sessions, continuing their regular test pilot work at LII at the same time.

In December 1980 the Wolf Pack members finished OKP training and passed their final exams. Sadly, Oleg Kononenko was not among them. He had died on 8 September 1980 while testing a naval version of the Yakovlev Yak-38 vertical take-off and landing plane in the South China Sea. Shortly after take-off from the aircraft carrier *Minsk*, the aircraft developed engine trouble and crashed into the ocean before Kononenko had a chance to eject. He was posthumously awarded a second Order of Lenin. One LII pilot has stated that before his death, Oleg Kononenko had been slated to become commander of the LII cosmonaut group [13]. This has not been confirmed by other sources and it should be noted that throughout their careers Volk had been senior to Kononenko.

At this point, the LII test pilots were not yet on an equal footing with the "career cosmonauts" of TsPK and NPO Energiya. The Wolf Pack was considered a specialized group, temporarily assigned to one particular task—namely, to fly Buran's atmospheric and orbital test flights. Their status was somewhat comparable with that of "payload specialists" in the US Space Shuttle program, individuals not employed by NASA who are assigned to particular missions to operate one or more payloads and then return to their usual line of work.

All that changed on 23 June 1981, when MAP issued an order to set up its own cosmonaut team, something which became official on 10 August 1981 with a corresponding order from the chief of LII. This is considered the official date of the formation of the LII Buran cosmonaut team, which now consisted of Volk, Levchenko, Stankyavichus, and Shchukin. The move may have been linked to the success of the first US Space Shuttle mission in April 1981, which in many ways was an eye-opening event for officials in charge of the Energiya–Buran program. It should be noted though that the members of the LII team continued their usual work as test

Oleg Kononenko's remains arrive in Vladivostok on the aircraft carrier *Minsk* on 26 November 1980 (B. Vis files).

pilots throughout their Buran careers [14]. At this point they also received the official title of "cosmonaut-testers", the same title given to the career cosmonauts of TsPK and NPO Energiya after their OKP. Before that they had been named "cosmonaut-researchers", just like scientists of the Academy of Sciences and doctors of the Institute of Medical and Biological Problems. The official formation of the LII cosmonaut team was apparently also related to a new "Statute for Cosmonauts" approved by the Central Committee of the Communist Party and the Council of Ministers on 30 April 1981.

Meanwhile, LII was looking for more pilots to expand its ranks. In November 1979 deputy MAP minister Ivan Silayev had already called for expanding the LII group, but it was not until February 1982 that four more candidatures were presented to MAP:

- Pyotr Vasilyevich Gladkov;
- Ural Nazibovich Sultanov;
- Vladimir Yevgenyevich Turovets;
- Viktor Vasilyevich Zabolotskiy.

Reportedly, the four had already undergone initial screening in 1979. In June 1982 two more names were added to the list of candidates:

- Magomed Omarovich Tolboyev;
- Vladimir Viktorovich Biryukov.

Sadly, Vladimir Turovets had to be dropped very early on in the process, when he was killed in the crash of a Mil Mi-8 helicopter on 8 February 1982. Turovets, born on 20 February 1949 in the Ukraine, had been an LII pilot since June 1977 and a Test Pilot 3rd Class since 1980. His assignment to the Buran team would probably have been problematic anyway because of an incident in which, as a salute to his fellow LII test pilots, he had made a low-altitude pass over the site where Aleksandr Lysenko and fellow pilot Gennadiy Mamontov had crashed in June 1977 [15]. The fact that Turovets was not liked by many people could also have had a negative impact on his selection, even though colleagues have described him as a gifted man and an excellent pilot.

Pyotr Gladkov and Vladimir Biryukov were not selected and returned to test flying. Gladkov, born on 22 July 1949 in Krasnodar, mainly flew state-of-the-art fighter aircraft like the MiG-29, MiG-31, and Su-27, but also heavy transport planes such as the Ilyushin Il-76 and Tupolev Tu-154. For this work he would be awarded the title of Merited Test Pilot of the Russian Federation in December 1997. Vladimir Biryukov, born on 9 June 1950 in the Sverdlovsk region, had become a test pilot at the Flight Research Institute in 1981, upon graduation from test pilot school. In October 1996, he too was awarded the title of Merited Test Pilot of the Russian Federation.

Left over were Tolboyev, Zabolotskiy, and Sultanov. When still in the military, Tolboyev had already tried to become a cosmonaut in the TsPK Buran selection group of 1976, but he hadn't managed to pass the medical commission at the time [16]. Both Zabolotskiy and Sultanov had already been involved in Buran-related research since 1978. This included a series of experiments called "Immersion" (1978–1980), in which they spent some time in simulated zero-*g* in a water tank at

Pyotr Gladkov (left), Vladimir Turovets (center), and Vladimir Biryukov (B. Vis files).

the Institute of Medical and Biological Problems and then performed landings with a Buran-type steep glideslope on Il-18 and Su-7 aircraft [17].

In September 1982 MAP decided that Zabolotskiy, Sultanov, and Tolboyev would undergo medical screening for possible inclusion in the LII cosmonaut team. While Zabolotskiy failed the initial medical, Sultanov and Tolboyev were accepted by the GMK on 25 January 1983, passed the GMVK on 9 March 1983, and were officially included in the LII team as cosmonaut candidates by a MAP order on 25 April 1983. Zabolotskiy was finally declared fit by the GMK on 4 April 1983, but it took almost another year for him to be accepted by the GMVK (15 February 1984) and be included in the team by MAP (12 April 1984).

In spite of the selection of these three new candidates, LII still felt that it needed more cosmonauts. In September 1983 two more pilots were recommended to MAP:

- Sergey Nikolayevich Tresvyatskiy;
- Yuriy Petrovich Sheffer.

Sheffer and Tresvyatskiy passed the GMK on 8 July 1984 and 17 April 1985, respectively, and were both accepted as cosmonaut candidates by the GMVK on 2 September 1985. They were officially included in the team by a MAP order on 21 November 1985. That same month they began their OKP training at Star City together with Tolboyev, Zabolotskiy, and Sultanov, finishing it in May 1987. On 5 June 1987 they were awarded their certificates of cosmonaut-testers.

In the meantime, there had been another organizational change at LII in May 1987, when MAP decided to create the so-called Departmental Training Complex for Cosmonaut-Testers (OKPKI for *Otraslevoy Kompleks Podgotovki Kosmonavtov-Ispytateley*). In a sense this was LII's own cosmonaut training center. Volk was named head of OKPKI, with Levchenko being assigned as his deputy. Stankyavichus

Ural Sultanov climbs out of a Soyuz descent module during a survival training session (B. Vis files).

in turn became the commander of the LII cosmonaut team, with Shchukin acting as his deputy.

OKPKI numbered around 60 people. Apart from the cosmonaut team itself, it consisted of medical, engineering, administrative, and various support departments. One of these support departments had three cameramen, whose job was to film and photograph every test flight that was conducted with the BTS-002 atmospheric test bed and the Tu-154LL and MiG-25 training aircraft.

The LII cosmonaut team spent the next years performing Buran-related test flights. Volk, Levchenko, Stankyavichus, and Shchukin performed take-off runs and approach and landing tests on the BTS-002 between December 1984 and April 1988. Stankyavichus and Zabolotskiy would conduct one additional ground run in December 1989. The others flew flight profiles on the Tupolev Tu-154LL flying laboratory, MiG-25, Su-27, and other types of aircraft.

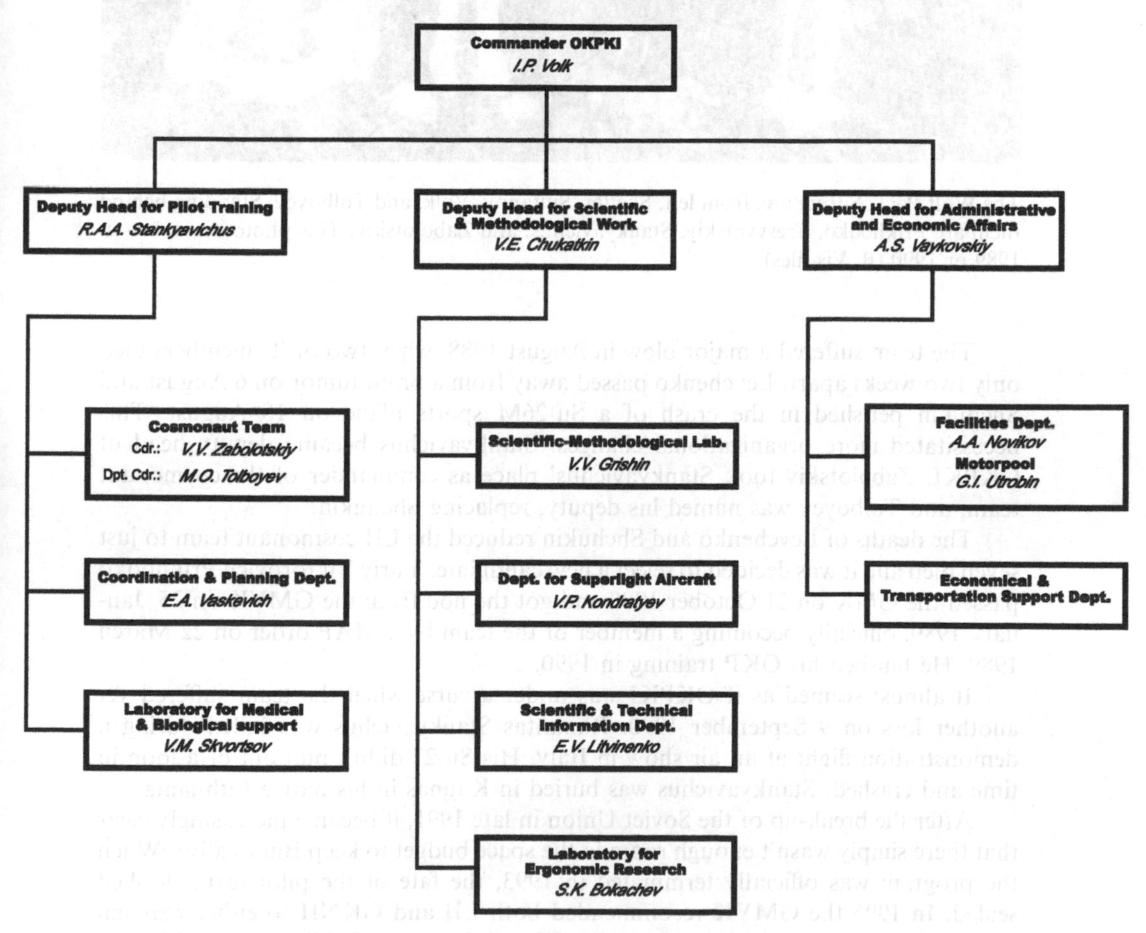

OKPKI organization chart, showing the situation in 1989 (E. Vaskevich archives).

The Wolf Pack. Sitting are, from left, Sheffer, Sultanov, Volk, and Tolboyev. Standing behind them are Prikhodko, Tresvyatskiy, Stankyavichus, and Zabolotskiy. This photo was taken in 1989 or 1990 (B. Vis files).

The team suffered a major blow in August 1988, when two of its members died only two weeks apart. Levchenko passed away from a brain tumor on 6 August and Shchukin perished in the crash of a Su-26M sports plane on 18 August. This necessitated more organizational changes. Stankyavichus became deputy head of OKPKI, Zabolotskiy took Stankyavichus' place as commander of the cosmonaut team, and Tolboyev was named his deputy, replacing Shchukin.

The deaths of Levchenko and Shchukin reduced the LII cosmonaut team to just seven men and it was decided to select a new candidate. Yuriy Viktorovich Prikhodko passed the GMK on 21 October 1988 and got the nod from the GMVK on 25 January 1989, officially becoming a member of the team by a MAP order on 22 March 1989. He finished his OKP training in 1990.

It almost seemed as if OKPKI was under a curse when the team suffered yet another loss on 9 September 1990. Rimantas Stankyavichus was killed during a demonstration flight at an air show in Italy. His Su-27 didn't pull out of a loop in time and crashed. Stankyavichus was buried in Kaunas in his native Lithuania.

After the break-up of the Soviet Union in late 1991, it became increasingly clear that there simply wasn't enough room in the space budget to keep Buran alive. When the program was officially terminated in 1993, the fate of the pilot teams looked sealed. In 1995 the GMVK recommended both LII and GKNII to either reassign their Buran pilots to other programs or disband the teams. While the GKNII team was officially disbanded in late 1996, the LII team officially continued to exist, reportedly at the insistence of Igor Volk, who held out hope that the Buran program would one day be resurrected. Although the LII team was never officially disbanded, it eventually simply dissolved itself as cosmonauts began to leave OKPKI and move on in their careers. The last one to depart LII was Vladimir Tresvyatskiy in late 2004 (for further careers of the LII pilots see Appendix B).

LII's Buran cosmonaut team can probably be regarded as the most diverse group ever selected. No fewer than six nationalities from the former Soviet Union were represented among the eleven men that made up the group. Kononenko, Prikhodko, Sheffer, Tresvyatskiy, and Zabolotskiy were Russians, while Levchenko and Volk were Ukrainians. Shchukin was a Belorussian, Stankyavichus was Lithuanian, Sultanov was a Bashkir, and Tolboyev was an Avar (a people of about 300,000 in Daghestan).

SELECTIONS BY GKNII

GKNII in Akhtubinsk began the selection process in 1978. The job was offered to all pilots working for GKNII and eight pilots displayed interest in the project [18]. They were:

- Ivan Ivanovich Bachurin;
- Aleksey Sergeyevich Boroday;
- Viktor Martynovich Chirkin;
- Vladimir Mikhaylovich Gorbunov;
- Vadim Oleynikov;
- Vladimir Yemelyanovich Mosolov;
- Nail Sharipovich Sattarov;
- Anatoliy Mikhaylovich Sokovykh.

On 1 December 1978 six of them (Bachurin, Boroday, Chirkin, Mosolov, Sattarov, and Sokovykh) were cleared by the GMVK to begin their OKP training in January

Vladimir Gorbunov (B. Vis).

The TsPK, LII, and GKNII cosmonaut teams pose in front of a Soyuz launch vehicle during a visit to Baykonur in 1981. From left are Boroday, Vasyutin, Shchukin, Volk, Grekov, Levchenko, Bachurin, Volkov, Stankyavichus, Moskalenko, Viktorenko, Kadenyuk, Sokovykh, and Mosolov (B. Vis files).

1979. Oleynikov had apparently been medically disqualified [19]. Gorbunov later claimed that he also began OKP, but left the program at his own initiative, having come to the conclusion that there was no future for him in Buran. In fact, he had told his commander in Akhtubinsk that he was signing up for Buran as long as he could combine that with his job in flight testing the MiG-29, a program he felt offered him much more perspective [20].

The six remaining pilots commuted between Akhtubinsk and TsPK in training cycles of between one and two months. While in Akhtubinsk, they continued their test flying for the Air Force. Not all of them would finish OKP. In April 1980, Nail Sattarov was flying a Tupolev Tu-134, a medium-size passenger plane, when he decided to have a little fun and perform a roll maneuver with the aircraft. Although colleagues of his have stressed that this was something everyone had done at least once, Sattarov had the bad luck of being caught. Besides a reprimand and temporary grounding, he was also removed from the Buran training group. Apparently, his commanders felt that such undisciplined behavior disqualified him from being a cosmonaut. Interestingly, although this is the reason that is usually given as the one that ended his cosmonaut career, Sattarov himself insists that the incident had nothing to do with him discontinuing OKP and leaving the group. Instead, he claims that in late March or early April 1980 he had indicated to his commanders at GKNII that he felt his flying career was going nowhere if he would continue to train as a cosmonaut, and that his request to return to test flying full time was granted [21].

Aleksey Boroday, Vladimir Mosolov, Ivan Bachurin, and Anatoliy Sokovykh (from left) in Baykonur, early 1981. After Sattarov and Chirkin had left the GKNII team, these four were the only cosmonauts left in that team until new members were selected in 1985 (B. Vis files).

The remaining five candidates completed OKP in November 1980 and were awarded their certificates of cosmonaut-testers on 12 February 1982. All five resumed their test pilot duties at GKNII, with their Buran-related work as a secondary task.

Shortly afterwards, Viktor Chirkin came to the conclusion that the Buran program was not going the way it should and he seriously doubted the vehicle would ever mature to the point that manned flights by Air Force test pilots would materialize. At his own request, he was relieved of his duties in the cosmonaut group in 1981. Chirkin went on to become a Major General, and in 1995 a Hero of the Russian Federation.

Anatoliy Sokovykh was less lucky. He too returned to test flying but in 1985 became involved in an incident that cost several people their lives. In one version, one of his crew members accidentally shot down another plane, but another version has it that his crew accidentally destroyed the wrong ground target, killing several soldiers. Sokovykh had not made the mistake himself, and it has even been said that the crew itself had only followed instructions from the ground. However, as commander of the aircraft he was held responsible and reportedly was demoted. In addition, with such a blot on his reputation, he could not maintain his position as cosmonaut [22].

By then, it had already been decided that the group needed fresh blood, and in August 1985 the GMVK added three new test pilots (from eight candidates that had originally been considered) to the group. They were:

- Viktor Mikhaylovich Afanasyev;
- Anatoliy Pavlovich Artsebarskiy;
- Gennadiy Mikhaylovich Manakov.

The three joined another group of cosmonaut candidates to undergo OKP in Star City. Most of the others had also been selected for the Buran program. They were, from LII, Magomed Tolboyev, Yuriy Sheffer, Ural Sultanov, Sergey Tresvyatskiy, and Viktor Zabolotskiy and, from NPO Energiya, Aleksandr Kaleri, Sergey Krikalyov, and Sergey Yemelyanov. The final candidate was Yuriy Stepanov, a physician who came from the Institute of Medical and Biological Problems (IMBP).

However, soon after Afanasyev, Artsebarskiy, and Manakov finished OKP in May 1987, they were approached by TsPK cosmonaut training chief Vladimir Shatalov with the request to transfer to the TsPK cosmonaut team. That same offer was extended to Ivan Bachurin, Aleksey Boroday, and Vladimir Mosolov [23]. Shatalov, who was in dire need of new crew commanders, considered the six pilots

The 1985 OKP class. From left (standing) Artsebarskiy, Kaleri, Tolboyev, Sheffer, Sultanov, Tresvyatskiy, Zabolotskiy, Krikalyov, and Manakov. Sitting are Afanasyev, Stepanov, and Yemelyanov (B. Vis files).

readily available candidates who had all finished OKP and could easily move over to the TsPK ranks. In addition, the Buran program was suffering delays, which might make the choice easier for them.

Afanasyev, Artsebarskiy, and Manakov accepted the offer and were officially included in the TsPK team in January 1988. However, Bachurin, Boroday, and Mosolov declined. They were preparing for the approach and landing test program on BTS-002 and felt that was more of a challenge for test pilots. As Mosolov put it: "We didn't want to be like dogs instead of crews. We wanted to fly with techniques that we could control and manage" [24]. In addition, Aleksey Boroday has said that the GKNII commander, wary of losing all his Buran cosmonauts to TsPK, had refused to consider a transfer of the three senior pilots [25].

Meanwhile, Mosolov was dismissed from the GKNII group in 1987 because of his divorce [26]. This left the GKNII team with only two members and it was decided that the group needed to be expanded yet again.

Ivan Bachurin was given the task to pre-select young test pilots and invite them to become members of the group. Seven pilots decided they wanted to join and took the initial medicals, which were held in TsPK's medical department. Four of them didn't pass, and this was a reason for two more to decline. They decided they didn't want to go to TsNIAG, the central military hospital, and risk being grounded from test flying if they would be declared medically unfit for the Buran program. The only one left was Anatoliy Polonskiy [27].

In the end, Polonskiy passed the medical board in February 1988 along with two other candidates and on 25 January 1989 all three were confirmed by the GMVK:

- Anatoliy Borisovich Polonskiy;
- Valeriy Ivanovich Tokarev;
- Aleksandr Nikolayevich Yablontsev.

Two months later, three more pilots passed the medical commissions. They were:

- Valeriy Yevgenyevich Maksimenko;
- Aleksandr Sergeyevich Puchkov;
- Nikolay Alekseyevich Pushenko.

In May 1989, the six began the basic cosmonaut training course in Star City, even though the GMVK would only officially confirm the selection of Maksimenko, Puchkov, and Pushenko on 11 May 1990. The six graduated on 5 April 1991, all of them getting the qualification of cosmonaut-tester with the accompanying certificate.

It should be noted that after all the selections mentioned above the GKNII pilots initially had a status similar to that of the LII pilots until 1981—that is, something comparable with payload specialists in the United States. It was not until 7 August 1987 that the USSR Ministry of Defense officially set up a team of what could be considered GKNII "career cosmonauts". They were based at a branch of GKNII in the Moscow suburb of Chkalovskiy, right next to Star City. The first to be included in the team by that same order were Bachurin and Boroday, with Bachurin named as commander.

Added to the team on 25 October 1988 was Leonid Kadenyuk, who had been a member of TsPK's 1976 Buran selection, but had been dismissed after divorcing his wife. Since he had already undergone OKP, he could begin working with Bachurin and Boroday right away. In fact, all three were assigned two years later to a mission in which a Soyuz vehicle would link up with an unmanned Buran to test some of its systems in orbit.

On 8 April 1992 the Chkalovskiy team was expanded with Puchkov and Yablontsev and now consisted of five members, which was the originally planned number. As members of the team began leaving, the vacant slots were filled by pilots of the 1989–1990 pools, just to maintain the total number at five. With the Buran program in its death throes, this was apparently done more for bureaucratic reasons than anything else.

After Bachurin left in November 1992, his place was taken by Tokarev on 30 January 1993. Following Boroday's departure in December 1993 Pushenko was added to the team on 6 February 1995. All this resulted in the rather bizarre situation that in the late 1980s/early 1990s the GKNII pilots were essentially split into two groups, the "career cosmonauts" based in Chkalovskiy and the "temporary cosmonauts" based in Akhtubinsk. Two of the earlier selected pilots, Polonskiy and Maksimenko, were never included in the Chkalovskiy team.

Just like the LII team, the GKNII team was faced with the choice in 1995 of either ceasing its existence or reassigning its pilots to other space projects. Unlike the LII team, the GKNII team was officially disbanded by an Air Force order on 30 September 1996. Two of the GKNII pilots, Kadenyuk and Tokarev, would eventually go on to fly in space, albeit in other capacities. Kadenyuk flew as a Ukrainian payload specialist on Space Shuttle mission STS-87 in 1997. Tokarev became a TsPK cosmonaut, performing a short-duration Space Shuttle mission (STS-96) to the International Space Station in 1997, and a long-duration mission aboard the ISS in 2005–2006 (for more details on the further careers of the GKNII pilots see Appendix B).

OTHER COSMONAUTS INVOLVED IN BURAN

All cosmonauts mentioned so far were selected specifically to fly on Buran, even though a fair number were transferred to the Soyuz, Salyut, and Mir programs later. In addition to these, several other cosmonauts from both TsPK and NPO Energiya at one time or another conducted training either for flights aboard Buran itself or for Soyuz missions to Buran.

As the "prime contractor" for Buran, NPO Energiya assigned a number of engineers to the program. This was particularly the case for the 1978 class, for which possible flights on Buran were taken into consideration during the selection process, although this was not the sole purpose of their selection. Several of its members spent part of their initial time in the cosmonaut team studying and training for Buran.

Aleksandr Nikolayevich Balandin worked on and off on ergonomics and the design of Buran's cockpit control panels between 1979 and March 1987. Aleksandr Ivanovich Laveykin was involved in Buran training from 1979 to 1984, accumulating 25 hours of flight time on L-29 aircraft and performing 46 parachute jumps. Musa Khiramanovich Manarov prepared for Buran flights from 1979 to 1982, clocking up more than 43 hours of flight time on L-39 aircraft.

There were also several NPO Energiya engineers from earlier and later selections who became involved in the Buran program. They were Valentin Vitalyevich Lebedev (1972 class, assigned to Buran from 1983 to 1986), Aleksandr Sergeyevich Ivanchenkov (1973 class, assigned to Buran from 1983 to 1992), and Sergey Konstantinovich Krikalyov (1985 class, assigned to Buran from 1986 to 1988). Many of these engineers (plus Gennadiy Mikhaylovich Strekalov of the 1973 class) were even put forward by NPO Energiya to fly in the co-pilot seat on the very first piloted missions of Buran.

Also involved in the Buran program were several military engineers originally selected by TsPK in the 1960s and early 1970s. These were Yevgeniy Nikolayevich Khludeyev and Eduard Nikolayevich Stepanov of the 1965 TsPK intake and Nikolay Nikolayevich Fefelov and Valeriy Vasilyevich Illarionov of the 1970 class. All but Illarionov had spent most of their careers training for missions on Chelomey's Almaz military space station and the TKS transport ships, but none of the four had ever flown in space or even received a back-up assignment.

Illarionov was active in the Buran program from 1984 until 1992, performing a multitude of engineering tests. These included tests of Buran equipment in simulated zero-*g*, pre-launch and post-landing evacuation exercises, vacuum tests of the airlock and the Docking Module, and tests of the Strizh pressure suit. The other three engineers were transferred to the Buran program in 1985/1986 after having been part of a training group to operate military instruments on the Kosmos-1686 TKS spacecraft. Khludeyev left the program in 1988, but Fefelov and Stepanov stayed until 1992 [28]. In 1990–1992 Illarionov, Fefelov, and Stepanov were in a training group for the aforementioned Soyuz mission to link up with an unmanned Buran in orbit.

Eduard Stepanov (left), Valeriy Illarionov (center), and Nikolay Fefelov (B. Vis files).

Missing in the Buran cosmonaut team were people with scientific backgrounds. Although the Academy of Sciences had set up its own cadre of cosmonauts in 1967, their hopes of flying in space were soon dashed by the cancellation of the manned lunar program and also by the elimination of the third seat in the Soyuz spacecraft following the Soyuz-11 accident in 1971, limiting space station crews to a military commander and a military or civilian flight engineer. Then, when Soyuz regained a three-man capability with the introduction of Soyuz-T in the early 1980s, the third seat was usually reserved for brief visiting flights by foreign spacemen or other "guest cosmonauts". All that could have changed if Buran had ever reached operational status. Especially, the long-duration Spacelab-type missions that were planned for Buran could have become a long-awaited blessing for Soviet scientists aspiring to fly in space. However, with the cancellation of the Soviet shuttle program, Russian scientist cosmonauts saw yet another opportunity to fly in space go up in smoke. Having said that, there were no significant additions to the Academy of Sciences team in the 1980s indicating that big numbers of scientists were going to fly on Buran anytime soon.

Finally, in the early 1990s French "spationauts" Jean-Loup Chrétien, Michel Tognini, and Leopold Eyharts flew both the Tupolev Tu-154LL and MiG-25 Buran training aircraft in preparation for the European Hermes spaceplane program. There are no indications that they were considered to fly aboard Buran itself [29].

TRAINING FOR BURAN

Simulators

The departmentalism of the Soviet space program was not only evident in the selection of cosmonauts for Buran, but also in the construction of simulators needed for cosmonaut training. About a dozen of these were scattered over various organizations involved in the Buran program, some of them apparently performing similar roles. For the Soyuz and space station programs, all simulator training was and still is concentrated at Star City, but this was hardly the case for Buran. Although there were ambitious plans for Buran simulator buildings at Star City and some simulators *were* eventually placed here, little if any Buran-related simulator training appears to have taken place at TsPK. Presumably, this was due to the fact that the Star City facilities were intended in the first place to prepare for manned orbital flights of Buran, which always remained a distant goal, without any really concrete flight plans ever being drawn up.

The bulk of the simulator training took place at NPO Molniya in Tushino and was aimed at preparing for the atmospheric landing tests with the BTS-002 Buran test vehicle. There where three simulators at NPO Molniya. Two of them were called PRSO ("Full-Scale Equipment Test Stand") and installed on top of each other. PRSO-1 was mainly used for testing the software used during the BTS-002 landing tests. First activated in June 1984, it consisted of a simplified Buran cockpit and a "skeleton" containing the main parts involved in landing. PRSO-2 was supposed to

The PRSO simulators (B. Vis files).

become the principal training device for Buran orbital missions, but was never completed [30].

The third simulator at NPO Molniya was PDST ("Piloting Dynamic Test Stand/ Simulator"), which was also geared to simulating the BTS-002 approach and landing tests. This was a Buran cockpit mounted on a motion platform, and housed all the displays and controls found in BTS-002. Installed behind the cockpit windows was a visual display system showing the surroundings of Zhukovskiy, where the test flights were conducted. A three-degrees-of-freedom motion platform was later replaced by a six-degrees-of-freedom motion platform, capable of imitating the movements made by the vehicle. PDST was used to familiarize crews both with nominal and off-nominal flight situations. Before the first approach and landing test in November 1985 the first four pilots involved in the tests (Volk, Stankyavichus, Levchenko, and Shchukin) each spent about 230–240 hours training on PDST, simulating about 160 off-nominal flight scenarios. In between training sessions PDST was also used to test new manual and flight director landing modes.

Another test stand at NPO Molniya was PSS ("Piloting Static Test Stand"). Completed in March 1984, it consisted of a Buran flight deck in a dome-shaped structure where images of the landing area were displayed on the walls with the help of a wide-angle projection system. It was solely used for research purposes, more particularly to test algorithms for manual flight control. It is not entirely clear if the LII test pilots were involved in this work [31].

The PDST simulator (B. Vis files).

Buran training at Star City was to take place in two buildings. One of these was called KTOK ("Orbiter Simulator Building"), which was planned to house a full-scale mock-up of Buran. Although that mock-up never appeared, three other simulators did end up in the facility.

The first was a motion base simulator designed and built by TsAGI that could be used to practice the controlled flight portion of the landing. In spite of the fact that this was probably one of the highest-performance Buran simulators built, it is remarkable that members of the various cosmonaut groups have stressed that they never trained on it [32]. The motion-base simulator was still intact in 1999, but by 2003 it had been largely dismantled and it has now been removed from the training hall.

The second simulator was a full-scale Buran crew cabin, consisting of both the flight and mid-decks. In addition, a Docking Module was added to the simulator that was to be used for docking to the Mir space station. This Docking Module was equipped with an APAS-89 docking system.

The third simulator was a fixed-base flight deck that was prominently positioned in the training hall. By 2003 it was still standing in the hall, although all equipment had been disconnected and the entrance door was sealed off.

After cancellation of the Buran program, other simulators were placed in the KTOK building in support of the Mir and ISS programs. These were full-scale

The KTOK was constructed particularly for the Buran program (B. Vis).

training models of Mir's Spektr and Priroda modules and of the Russian ISS modules Zarya and Zvezda.

Construction of a second large Buran simulator building was begun at TsPK, but abandoned as the future of the program became uncertain. Over 20 m high, it nevertheless is still one of TsPK's most conspicuous buildings. If completed, it should have been able to house a complete Buran orbiter and would have been used among other things to train cosmonauts in operating the remote manipulator arm.

Another Buran cockpit simulator at TsPK was called Pilot-35 ("35" referring to the 11F35 designator of Buran), adapted from a Spiral simulator called Pilot-105. This was used mainly to test the placement of control and display systems in the cockpit and to compare automatic and flight director landing modes. It was also used in conjunction with the TsF-7 centrifuge to test manual landing techniques under simulated flight conditions. However, Pilot-35 appears to have been intended primarily for engineering purposes and it is not clear if it was ever used by cosmonauts [33].

The Buran pilots also conducted extensive training at TsAGI in Zhukovskiy, using a simulator known as PSPK-102. Constructed in 1983, it was a dynamic simulator mounted on a six-degrees-of-freedom motion platform and was later modified as a simulator for various aircraft [34].

TsPK's motion-base Buran simulator (B. Vis).

Buran's Docking Module, which was part of the full-scale Buran crew cabin in TsPK (B. Vis).

The fixed-base flight deck simulator (B. Vis).

Due to the cancellation of the program, the largest Buran training facility in Star City was never completed and left to the elements (B. Vis).

Buran pilots also simulated manual approach and docking techniques on a simulator called Pilot at IMBP. In addition, there were test stands at several organizations that were primarily built for engineering purposes, but were at least partially intended for cosmonaut training as well, although it is unclear whether they were ever actually used for that purpose. These included the full-scale Buran mock-up OK-KS and the crew cabin mock-up MK-KMS at NPO Energiya as well as the crew cabin mock-up MK-M at Myasishchev's EMZ (see Chapter 6). At least three tests stands intended partially for cosmonaut training (KS-SU, ATsK, and Anomaliya) were situated at NPO AP in Moscow, the bureau that was responsible for Buran's computers.

The dispersion of simulators over so many organizations was obviously not convenient for the LII pilots themselves, who were based in Zhukovskiy. Especially after the formation in 1987 of OKPKI, which was supposed to become LII's equivalent of TsPK, there were calls to concentrate simulator training there, but to no avail [35].

Training aircraft

Besides simulator training, a lot of training was conducted by both the LII and GKNII pilots on many types of aircraft. This was mainly in preparation for the atmospheric landing tests on the BTS-002 and also to test the automatic landing systems in preparation for the first flight of Buran in 1988. The training took place both at LII in Zhukovskiy and at the Baykonur cosmodrome.

The most extensively used type of aircraft were Tupolev Tu-154 passenger planes converted as "flying laboratories" (*Letayushchiye Laboratorii* or LL) and therefore also known as Tu-154LL. These were the equivalents of the Shuttle Training Aircraft (STA) in the Space Shuttle program: Gulfstream II business jets which had their cockpit layouts modified to resemble that of the Shuttle. On the STA the left-side instrument panel was modified with a set of Orbiter displays and controls, while the right side contained the normal Gulfstream instruments as a safety measure. The Tu-154LL similarly had a "split-personality" cockpit, but here the Buran displays and controls were in the right side of the cockpit, with the windows draped to simulate the view out of Buran's cockpit. As on the STA, an on-board computer system translated the pilot's inputs into control movements largely mimicking those of Buran. In order to match the descent rate and drag profile of Buran, the thrust of the two side-mounted engines was reversed. Opening of the speed brake was simulated by controlling the thrust of the center engine. The Tu-154 flying labs were used to simulate both manned and automatic landings [36].

Although several Tu-154 aircraft were flown in support of the Buran program, only two had the modified cockpits (serial numbers 083 and 119, also known as LL-083 and LL-119, tail numbers CCCP-85083 and CCCP-85119). Other Tu-154 aircraft used by the Buran pilots had serial numbers 024 and 108 [37]. At least one of the aircraft reportedly also had a Buran-type cockpit installed in the front part of the passenger cabin [38].

Rimantas Stankyavichus at the helm of a Tu-154LL flying laboratory with Buran cockpit lay-out (B. Vis files).

Also actively used were several MiG-25 jets that simulated landings from much higher altitudes than the Tu-154 (over 20 km compared with about 10 km). One type was a modified version of the MiG-25RBK reconnaissance bomber, which had its standard equipment replaced by communication systems, telemetric equipment, and the like. Special containers with equipment could be mounted on pylons under the wings. Painted under the cockpit of these aircraft was the number 02.

The other was a modified version of the two-seater MiG-25PU training aircraft. It was known as MiG-25-SOTN (SOTN standing for optical/TV surveillance) and served the purpose of escorting other Buran-related training aircraft as well as Buran itself to the runway, with a cameraman seated in the front cockpit shooting video. The MiG-25-SOTN, piloted by Magomed Tolboyev, was in the air both for the launch and landing of Buran on 15 November 1988. Apart from serving as a chase aircraft, the MiG-25-SOTN was also used as a Buran training aircraft in its own right. It had the number 22 painted under the cockpit [39].

LII pilots conducted Buran approach and landing flight profiles on numerous other types of aircraft as well. As a training exercise, unpowered landings were not only performed on the Sukhoy Su-7 and Su-27 fighters, but also on heavy bombers such as the Tupolev Tu-16 and Tu-22M, and the Ilyushin Il-62 passenger plane

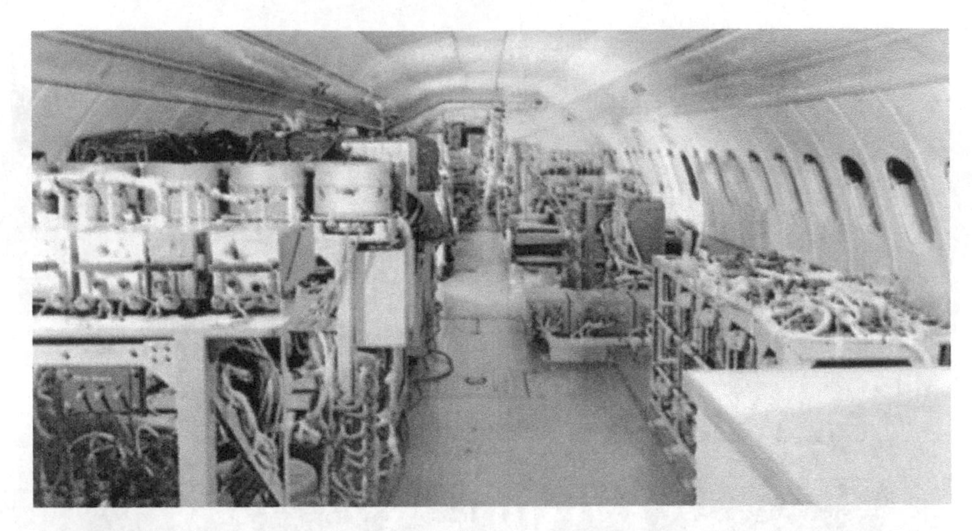

A view of the cabin of a Tu-154LL with the instrumentation to collect data on Buran-type landing profiles (B. Vis files).

(reportedly the most difficult to fly under such conditions). Igor Volk and Anatoliy Levchenko even made unpowered landings from an altitude of 22 km on the supersonic Tupolev Tu-144 (the twin of Concorde), although it is not entirely clear if this was in support of Buran [40].

CREWING FOR BURAN'S FIRST MANNED MISSIONS

Soviet planners envisaged an extensive orbital test flight program for Buran, which at one point included as many as 10 missions, both unmanned and manned. By comparison, the US Space Shuttle flew just four (manned) test flights in 1981–1982 before being declared operational.

For safety reasons, crews for the initial Buran test flights would have been restricted to just two cosmonauts. First, it was not practical to install ejection seats for more than two crew members and, second, if a life-threatening emergency arose in orbit, a Soyuz would have to be able to come to the rescue. Since that Soyuz needed a "rescue commander", only two seats would be left in the vehicle for the stranded Buran crew.

Throughout the 1980s, there was disagreement on the composition of the initial Buran crews. LII in Zhukovskiy, backed by the Air Force, argued that both seats should be occupied by its experienced test pilots. However, NPO Energiya, intent on not being sidelined, pushed to fly one of its engineers in the co-pilot seat rather than an LII test pilot. Therefore, two types of crews were considered for most of the duration of the program: crews consisting of two LII pilots, on the one hand, and crews composed of one LII pilot and one NPO Energiya flight engineer, on the other hand.

It should be stressed that as Buran never came anywhere close to flying a manned mission, none of the crews mentioned below was ever *officially* assigned. The flight plan for the first manned mission remained vague until the end of the program and none of these crews performed any dedicated mission training.

LII crews

From the very beginning LII had the following crews in mind for the first flight:

Prime crew	**Back-up crew**
Igor Volk	Anatoliy Levchenko
Rimantas Stankyavichus	Aleksandr Shchukin

The preference for Volk–Stankyavichus and Levchenko–Shchukin was reflected in the fact that the two crews flew the bulk of the atmospheric Horizontal Flight Tests with the BTS-002 Buran analog. A total of 24 such flights were performed between November 1985 and April 1988. Volk and Stankyavichus were paired for 11 of the missions and Levchenko and Shchukin jointly flew 4 missions (see Chapter 6).

Volk (right) and Stankyavichus during a survival training session (B. Vis files).

The original crewing plan was completely disrupted in August 1988, when in a bizarre twist of fate Buran lost its entire back-up crew with the deaths of both Levchenko and Shchukin. As a result, Volk and Stankyavichus were split up and both got new co-pilots from the LII ranks [41]. Volk has claimed that GKNII pilots Ivan Bachurin and Aleksey Boroday, veterans of six BTS-002 flights, were also considered as the back-up crew [42]. The new crews were:

Prime crew	**Back-up crew**
Igor Volk	Rimantas Stankyavichus
Magomed Tolboyev	Viktor Zabolotskiy

Internal LII documents show that another option considered was to retain the Volk–Stankyavichus team, with Zabolotskiy and Tolboyev acting as back-ups. This plan assumed that Zabolotskiy would first fly a Soyuz mission to give him the necessary spaceflight experience to command Buran if the need arose [43].

When Stankyavichus was killed in a plane crash in September 1990, LII was forced once again to change the composition of the back-up crew [44]. The new crews were:

Prime crew	**Back-up crew**
Igor Volk	Viktor Zabolotskiy
Magomed Tolboyev	Ural Sultanov

These are the last crews known to have been considered by LII for the first piloted Buran mission.

LII/NPO Energiya crews

Internal documents obtained by the authors show that there was fierce debate between LII/MAP and NPO Energiya/MOM in the 1980s over crewing for the first manned missions. It was all very reminiscent of similar disagreements between the Korolyov design bureau and the Air Force over crewing for Voskhod and Soyuz missions in the 1960s. The documents show that the Council of Chief Designers decided on 26 January 1983 to assign only LII test pilots to the first two manned Buran missions, but that NPO Energiya disagreed with the plan in September 1983, putting forward its own flight engineers to occupy the second seat. By the autumn of 1985 NPO Energiya had mustered enough support to secure a joint decision from MOM, MAP, and the Ministry of Defense on the formation of four preliminary crews for the first two manned flights:

Volk	Levchenko	Stankyavichus	Shchukin
Ivanchenkov	Lebedev	Strekalov	Balandin

Somewhat later the pairings were changed as follows:

Volk	Levchenko	Stankyavichus	Shchukin
Ivanchenkov	Strekalov	Balandin	Krikalyov

Another source claims the crews initially were Volk–Ivanchenkov, Levchenko–Strekalov, Stankyavichus–Balandin, and Shchukin–Lebedev, with Lebedev being replaced by Krikalyov in 1986 [45].

On 6 December 1985 the Military Industrial Commission (VPK) went along with the plan, ordering formation of final crews by December 1986 based on the training results obtained by then. The first phase of training for the NPO Energiya engineers would see theoretical, simulator, and aircraft training. LII demanded that the flight engineers fly a total of 398 hours on five different aircraft, but in the end five engineers (Ivanchenkov, Strekalov, Balandin, Krikalyov, and Lebedev) accumulated just 11 hours of flying time during 26 flights on four aircraft in November 1986. In June 1987 Volk and Ivanchenkov flew 10 different landing profiles on the PDST simulator at NPO Molniya, which according to Volk's official protocol showed that the engineers would not be able to safely land Buran in case of an emergency.

Based on the preliminary results of the training program, both MAP and the Air Force recommended in 1987 only to fly experienced LII test pilots on the first Buran missions. With their limited aircraft training, the engineers were not even considered capable of flying in the co-pilot seat of the Tu-154LL training aircraft or the BTS-002. Although the prime landing mode even for manned missions was automatic, MAP and the Air Force argued that the crew would have to take manual control if they were diverted to an emergency landing site not equipped with the necessary navigation equipment to support hands-off landings. Moreover, it was felt that the second crew member needed flying skills equal to those of the commander in order to deal with various off-nominal scenarios. Among those were malfunctions in the commander's flight displays and control panels and a situation where the commander was partially disabled by space motion sickness. A joint LII/TsPK research program called "Dilemma" had shown that the engineers would not be able to render the necessary assistance to the commander in case of these and other emergencies.

Predictably, NPO Energiya and MOM, citing the December 1985 VPK decision, ignored the conclusions of MAP and the Air Force and insisted on a continued training program for the engineers, including simulated flights on the PDST simulator and real flights on the Tu-154LL and BTS-002. One of the arguments in favor of including a flight engineer on the first manned flight (then scheduled to be mission 1K2) was that it would be a conservative 3-day flight, with most systems operating in the automatic mode. On the other hand, LII used the same argument to claim that the limited engineering tasks planned for the flight might just as well be performed by a test pilot. At any rate, the Council of Chief Designers ordered on 23 March 1988 to draw up a new training schedule for the NPO Energiya engineers, but it looks as if NPO Energiya pursued its plans with less vigor as the months went on. The launch date for the first manned mission kept slipping and the exact flight plan remained vague, complicating the formation of a training program. Moreover, by the end of the 1980s virtually all of the NPO Energiya flight engineers involved in Buran had either been reassigned to the Mir program or left, with only Ivanchenkov remaining until 1992 [46].

In later interviews the LII pilots did not hide their opposition to Energiya's push to include engineers in the first crews. Volk said that at one point he went to Minister of General Machine Building Oleg Baklanov, asking him what the use of flying engineers was. According to Volk, Baklanov quoted Glushko as saying that "they would keep an eye on the devices." Losing his temper in a subsequent argument with Glushko over the crew assignments, Volk told the chief designer: "Then let Strekalov and Ivanchenkov fly! And if there is a crash or whatever, then of course the news will be all over the world" [47].

BURAN-RELATED SOYUZ MISSIONS

Although seasoned test pilots, none of the LII Buran pilots had any spaceflight experience. Assuming Buran's first manned mission would be flown by two LII pilots, both would be space rookies. This became problematic after events in October 1977, only a couple of months after the first LII pilots had been selected.

On 9 October 1977 cosmonauts Vladimir Kovalyonok and Valeriy Ryumin blasted off aboard the Soyuz-25 spacecraft to become the first crew to board the Salyut-6 space station. However, one day later, the cosmonauts, both first-time flyers, failed to dock their transport craft with the station and in the end were forced to abandon their attempts and return to Earth. Judging by what happened after the mission, the crew was at least partially blamed for the failure. Upon their return they were not awarded the title of Hero of the Soviet Union, as was customary with cosmonauts who had completed their first mission. Even the crews of Soyuz-15 and Soyuz-23, who had also been unsuccessful in docking their transport ships to the stations they were supposed to occupy, had been awarded the prestigious title.

Besides denying Kovalyonok and Ryumin their Hero of the Soviet Union Gold Star medals, it was decided that from that moment on every Soviet space crew had to include at least one crew member with at least one space mission under his belt. This decision must have been made almost immediately after the landing of Soyuz-25, since Soyuz-26 was launched only three months later, carrying a new crew in accordance with the new rule (the original crew consisted of rookies Yuriy Romanenko and Aleksandr Ivanchenkov, but Ivanchenkov was replaced by veteran cosmonaut Georgiy Grechko).

The 1977 decision also had implications for the all-LII crews assigned to the first manned Buran mission. In preparation for that flight, it was necessary to give at least one of the crew members in both the prime and the back-up crews spaceflight experience. Therefore, the Council of Chief Designers decided on 10 March 1982 that both the prime and back-up crew commanders would occupy the third seat of a Soyuz that was scheduled to fly in the Soyuz–Salyut program [48].

Not only would that give the pilots a taste of the zero-*g* environment, they would also fly several types of aircraft immediately after landing in order to determine to what extent their flying abilities would be affected by their stay in weightlessness (a research program known as *Nevesomost* or "Weightlessness"). Similar experiments (under the name *Tonkost* or "Precision") had already been conducted by Vladimir Dzhanibekov after Soyuz-39 (March 1981) and Soyuz T-6 (June 1982), and by Leonid Popov after Soyuz-40 (May 1981) and Soyuz T-7 (August 1982). Both flew non-Buran landing profiles on a Tu-134 aircraft [49]. However, the LII pilots would be faced with a much more grueling flight schedule after landing. An additional reason for including Buran pilots in Soyuz crews was probably to acquaint them with the spacecraft in preparation for a possible Soyuz rescue mission during the early Buran test flights.

The Soyuz mission of Igor Volk

As the leader of the LII "Wolf Pack", Igor Volk was eyed from the start as the commander for the first manned Buran mission and was therefore the first candidate eligible for a Soyuz "warm-up mission".

In September 1982 Volk was teamed up with cosmonauts Leonid Kizim and Vladimir Solovyov for a brief visiting mission to the Salyut-7 space station in late 1983. Their hosts were supposed to be Salyut-7's third Main Expedition crew (EO-3) of Vladimir Lyakhov, Aleksandr Aleksandrov, and Aleksandr Serebrov, who were scheduled to fly a six-month mission from June until December 1983 after having replaced the EO-2 crew (Titov–Strekalov–Pronina) in orbit.

Crewing for the visiting flight looked as follows:

Prime crew	**Back-up crew**	**Second back-up crew**
Leonid Kizim	Vladimir Vasyutin	Aleksandr Viktorenko
Vladimir Solovyov	Viktor Savinykh	Vitaliy Sevastyanov
Igor Volk	Anatoliy Levchenko	Rimantas Stankyavichus

The 1983 flight schedule was thrown into disarray when the EO-2 crew (Pronina having been replaced by Serebrov) failed to dock their Soyuz T-8 spacecraft with Salyut-7 in April 1983. The new plan was for Lyakhov and Aleksandrov to fly to the station aboard Soyuz T-9 in June 1983 and be relieved by Titov and Strekalov in August for a 100-day mission to complete some of the original EO-2 mission objectives. Volk's mission was scrapped for 1983 since no Soyuz vehicle would be available in time to fly a visiting mission to Salyut-7. The crews for the visiting flight were disbanded in May 1983 and Kizim and Solovyov moved to the training group for long-duration missions [50].

On 26 September 1983, their mission delayed several weeks, Titov and Strekalov were poised for launch again when a fire broke out at the base of their launch vehicle with less than a minute to go in the countdown. Only seconds before the launch vehicle exploded, the Soyuz was pulled away to safety by the emergency escape system. Rather than return to Earth, Lyakhov and Aleksandrov remained aboard the station until late November to complete some of the tasks originally planned for their replacements. Salyut was left behind unmanned, waiting for the next long-duration crew to arrive aboard Soyuz T-10 in February 1984.

Volk, Kizim, and Solovyov relax after a training session in the Soyuz simulator. This is the only known photo of Volk's original crew (B. Vis files).

This time it was the turn of Volk's former crewmates Kizim and Solovyov, who were joined by doctor Oleg Atkov for a record 8-month mission. Two visiting missions were planned, one (Soyuz T-11) carrying a Soviet–Indian crew and the second (Soyuz T-12) with Volk in the passenger seat.

Volk was still without a crew, but all that changed on 17 November 1983, when NASA announced that Kathryn Sullivan would become the first woman to conduct a spacewalk late the following year on Space Shuttle mission STS-41G [51]. It was too tempting for the Soviets not to try and steal this space first, one of the last to be clinched. Under pressure from NPO Energiya chief Valentin Glushko it was quickly decided to include a woman in the second visiting crew to conduct an EVA just weeks before Sullivan's [52].

The crew of Soyuz T-12 (B. Vis files).

This decision may have been a blessing for the LII team, because there had been opposition to a dedicated visiting mission with an LII pilot, amongst others from Glushko himself. In contrast to LII, Glushko was apparently in favor of automatic Buran landings and was not keen on organizing a flight just for the LII pilots to gain flight experience [53].

Within a month of the NASA announcement, crews had been formed. Commander would be Vladimir Dzhanibekov, one of the most experienced active commanders around. His flight engineer would be Svetlana Savitskaya, who had already flown an 8-day mission in August 1982. That too was believed to have been a rush assignment for her, as she flew less than a year before Sally Ride became the first American woman to fly in space in June 1983. The third seat, which basically was up for grabs, was given to Volk, who had been in line to fly the mission anyway.

The crewing for Soyuz T-12 was:

Prime crew	**Back-up crew**
Vladimir Dzhanibekov	Vladimir Vasyutin
Svetlana Savitskaya	Viktor Savinykh
Igor Volk	Yekaterina Ivanova

Volk, Dzhanibekov, and Savitskaya shortly after landing (B. Vis files).

Minutes after the traditional post-landing crew photo was taken, Volk left to begin the most important part of his mission—flying aircraft along the flight path of a Buran shuttle returning from space—leaving Dzhanibekov and Savitskaya behind (B. Hendrickx files).

The back-up assignments raised some eyebrows when they finally became known to Western space analysts in 1988 [54]. No second back-up crew was named and that, together with the composition of the first back-up crew, was a clear indication that this was a "crew of opportunity" and not one that was part of the overall, long-term mission planning for Salyut-7 expeditions. Judging by the absence of an LII pilot in the back-up crew, it looked as if the importance of flying a woman cosmonaut to perform an EVA far outweighed the need to give one of the LII pilots his mandatory spaceflight experience. Vladimir Dzhanibekov claims that Rimantas Stankyavichus was "in the reserve" for the mission, but denied that he had been a back-up for Volk and there aren't any official sources that say he was [55].

It isn't even certain that the back-up crew actually would have flown in case either Dzhanibekov or Savitskaya had become disqualified for some reason. It has been assumed that the flight would only proceed if the woman conducting the EVA was Savitskaya, not Ivanova. Rumors have it that her father, Soviet Air Force Marshal Yevgeniy Savitskiy, had been one of the driving forces behind the whole flight, and although no confirmation has ever been given, several cosmonauts have not ruled out that possibility [56]. As the crew was not supposed to exchange Soyuz vehicles, there was no real *operational* need to fly the Soyuz T-12 mission.

Training for the mission began on 26 December 1983 and was completed on 4 July 1984. Having taken their final exams, the crew was declared ready for the flight. Soyuz T-12 was launched on 17 July 1984 and reached orbit to begin a rather uneventful flight to Salyut-7. The TASS news agency reported that "the spaceship's flight program envisaged a link-up with the Salyut-7/Soyuz T-11 orbital complex", after which its crew "were to carry out scientific and technical research and experiments together with [the station's resident crew]" [57].

With the Buran program still a state secret, nowhere was it reported or even hinted that Volk's presence on board had anything to do with a Soviet shuttle program, nor was any indication given that his main task would come only after landing. Almost six months after the mission, the British Interplanetary Society's *Spaceflight* magazine raised the question: "One puzzle: why did Volk, an experienced test pilot, occupy the passenger seat of a Soyuz T which is normally occupied by non-pilot researchers or foreign cosmonauts?" [58]. The answer, however, could not be given.

On 18 July Soyuz T-12 successfully docked with the Salyut station and Dzhanibekov, Savitskaya, and Volk were welcomed on board by Kizim, Solovyov, and Atkov. In its reports, TASS said that the program of joint operations "included technical and technological experiments, medical, biological, astrophysical and other studies, and Earth photography and observations in the fields of meteorology, geology and environmental protection" [59].

During the joint operations, news services did give details about the experiments that were conducted, but only very limited information was given about Volk's activities. One interesting bit of information came from flight director Viktor Blagov, who told reporters that Volk was not taking part in any physical exercises to counter the effects of weightlessness. Instead, he was taking special tablets for that purpose, while being continuously monitored by Atkov and by doctors on Earth, both during the mission and after the flight. According to TASS, the results of this experiment would help understand how the human body reacted to spaceflight [60]. Besides this, Volk conducted two experiments that studied his eyesight. One focused on in-depth vision and the eye's resolving power, while the other analysed his eye's color perception, its ability to discriminate between various shades of color.

Although not reported at the time, Volk also carried out an experiment called "Pilot" intended to see if his adaptation to zero-*g* would affect his ability to operate flight controls. For this purpose, several flight controls and display panels similar to those used on Buran were installed in the Soyuz T-12 orbital module [61].

Volk's presence on board was almost ignored by the media, especially when Svetlana Savitskaya and Vladimir Dzhanibekov performed an EVA on 25 July that lasted a little over three and a half hours. This was the first EVA by a woman and it took place less than three months before Kathy Sullivan's spacewalk on STS-41G.

On 29 July Dzhanibekov, Savitskaya, and Volk landed safely on Earth, 140 km southeast of Dzhezkazgan in Kazakhstan, after a flight lasting 11 days, 19 hours, and 14 minutes. Volk spent about 20 minutes suspended upside down inside the descent capsule as recovery crews struggled to remove him from the capsule. Afterwards, the cosmonauts were put in chairs to relax a little, as was tradition. But right after their initial medical check-ups, Volk was about to begin his principal experiment (not reported by the media at the time). He was taken to a helicopter that would fly him to Dzhezkazgan. Although not planned, Volk was granted permission by the pilot to occupy the co-pilot seat and take control of the helicopter. Only at that point was it realized that no one had thought of bringing Volk's flying boots. As a result he was forced to fly the helicopter in his socks.

Immediately after arriving in Dzhezkazgan, Dzhanibekov and Savitskaya underwent the traditional welcome by Kazakh government representatives, while Volk, still without his boots, boarded a Tu-154LL Buran training aircraft and flew it to his LII home base in Zhukovskiy near Moscow. The approach and landing were performed following the flight path of a Buran shuttle returning from space. To achieve that, the engine thrust was reversed, the landing gear was lowered, and all flaps were put in such a position that they would give maximum braking effect. Under these conditions, the Tupolev almost fell from the sky, just like Buran would do when returning from space. As soon as he had parked the Tupolev on the tarmac, Volk donned a pressure suit, climbed aboard a MiG-25 fighter, and together with an instructor flew all the way back to Baykonur. It turned out that the space mission had not in any way adversely influenced his flying abilities, so there were no objections for cosmonauts to fly Buran back from orbit [62].

The Soyuz mission of Anatoliy Levchenko

Having flown Soyuz T-12, Igor Volk now had the spaceflight experience necessary to command the first manned mission of Buran. Next in line for a Soyuz mission was Anatoliy Levchenko, scheduled to be Volk's back-up for the Buran flight. Levchenko began intensive training for his mission in March 1987. Two months later he was joined by his crewmates Vladimir Titov and Musa Manarov, who were slated to fly a record-breaking one-year mission as the third Main Expedition (EO-3) aboard the new Mir space station. Levchenko would go up with Titov and Manarov on Soyuz TM-4 in December 1987 and after a short period of handover activities would land with the EO-2 crew aboard Soyuz TM-3.

Crewing for the mission was:

Prime crew	**Back-up crew**	**Second back-up crew**
Vladimir Titov	Aleksandr Volkov	Vladimir Lyakhov
Musa Manarov	Aleksandr Kaleri	Andrey Zaytsev
Anatoliy Levchenko	Aleksandr Shchukin	None

(Kaleri had replaced the original back-up flight engineer Sergey Yemelyanov, who was medically disqualified in May 1987 and would die of a heart attack in 1992 at the age of just 41.)

In keeping with the new policy of *glasnost* that was developing in the Soviet Union under General Secretary Mikhail Gorbachov, the names of the crew members were announced to the public prior to the mission, on 9 December. The press was also told that both Levchenko and Shchukin were test pilots, although no explanation was initially given for their assignment. In an interview two days before launch, the chairman of the State Commission Lieutenant-General Kerim Kerimov was asked to describe the role of the test pilots in the Soyuz passenger seat. Although he revealed that the inclusion of both Volk and Levchenko was at the request of MAP, he didn't elaborate on the reasons to send them into space. All he said was that Levchenko's flight would last a week and that he thought "that period presumably would be

The Soyuz TM-4 crew during a training session. Standing are Vladimir Titov and Anatoliy Levchenko, sitting in front of them is Musa Manarov (B. Vis files).

Anatoliy Levchenko relaxes during his stay on the Mir station (B. Vis files).

sufficient for the comrades who were sending him there to establish his qualities as a test pilot working in space" [63].

However, in this new era of openness journalists no longer would accept such a non-answer. The next day Vladimir Shatalov, asked directly what the reason for Levchenko's presence in the crew was, said that new systems were being developed, including reusable ones that would land like an airplane. Landing such spacecraft called for completely new techniques and it was necessary to investigate the technology of piloting these spacecraft to a safe landing [64]. The Soviet Union had already officially acknowledged that it was working on a reusable spacecraft on the eve of the maiden launch of the Energiya rocket in May 1987.

The following day, 21 December 1987, Soyuz TM-4 lifted off from Baykonur and reached orbit without problems. After a two-day flight, the Soyuz docked with the Mir space station, where Titov, Manarov, and Levchenko were greeted by the resident crew of Yuriy Romanenko and Aleksandr Aleksandrov. That same day Radio Moscow finally confirmed what had been obvious all along, saying "Levchenko has been sent on this mission to try out in zero gravity his skills for the future piloting of a shuttle spacecraft" [65]. For the next seven days Romanenko and Aleksandrov handed over the complex to the new expedition crew. Besides this, a joint program was conducted, although few details were given other than that it consisted of "scientific, technical, medical and biological experiments" [66].

After the handover operations had been completed, Romanenko, Aleksandrov, and Levchenko said their goodbyes to Titov and Manarov on 29 December and landed aboard Soyuz TM-3 some 80 kilometers from Arkalyk in Kazakhstan. Levchenko's flight had lasted 7 days 21 hours 58 minutes. It was reported that winds at the landing site were so strong that it was difficult to set up the tent for the first medical check-ups, forcing the crew to be directly evacuated to the nearby helicopter of the medical staff [67]. Anatoliy Levchenko, supported by two men, was brought to a separate helicopter and within half an hour after landing was on his way to the airport of Arkalyk. In a repeat of Volk's experiment, he boarded a Tupolev Tu-154LL, flew it to LII in Zhukovskiy near Moscow, and then returned to the Baykonur cosmodrome on a MiG-25, performing Buran-landing profiles to test his flying abilities after more than a week in zero gravity.

The planned Soyuz mission of Rimantas Stankyavichus

When Levchenko died of a brain tumor just eight months after his space mission, the new Buran back-up crew (Stankyavichus–Zabolotskiy) was again without spaceflight experience. Therefore, LII started pursuing another Soyuz mission, with Stankyavichus as the prime candidate.

The final decision to go ahead with the mission was jointly made in February 1989 by the Ministers of General Machine Building, the Aviation Industry, Public Health, and Defense as well as the Commander-in-Chief of the Air Force, the head of UNKS (the "military space forces"), and the President of the Academy of Sciences. The plan was for Stankyavichus to go up with Mir's EO-6 resident crew (Solovyov–Balandin) aboard Soyuz TM-9 in September 1989 and return a week later with the EO-5 crew (Viktorenko–Serebrov) aboard Soyuz TM-8. Stankyavichus and Zabolotskiy got down to training at Star City in March 1989. At the time, the EO-6 crew was busy performing back-up duties for the EO-5 mission (then scheduled for launch in April 1989), which is why the two LII pilots were temporarily teamed up with the EO-6 back-up crews:

Viktor Afanasyev	Gennadiy Manakov
Vitaliy Sevastyanov	Gennadiy Strekalov
Rimantas Stankyavichus	Viktor Zabolotskiy

As Stankyavichus got down to training, the Mir flight schedule underwent changes that would jeopardize his Soyuz mission. In February, due to delays in the launch of Mir add-on modules, the EO-5 prime and back-up flight engineers had swapped places, with Viktorenko and Balandin now scheduled to go up in April to be replaced by Solovyov and Serebrov in September. That in itself was no problem for Stankyavichus, but in March the Soyuz spacecraft he was supposed to fly in September was seriously damaged during testing in a vacuum chamber at the Baykonur cosmodrome and had to be sent back to NPO Energiya for repairs. This meant that it would not be available as a back-up vehicle for the launch of Soyuz TM-8 in April. As a result, a decision was made that the EO-4 crew (Volkov, Krikalyov, Polyakov) would return to Earth in April and leave Mir behind unmanned until September, when the originally planned EO-5 crew

Р Е Ш Е Н И Е

Минобщемаша СССР, Минавиапрома СССР, Минобороны СССР (ВВС, УНКС)
Минздрава СССР и Академии наук СССР

О включении в состав экипажей космонавтов-испытателей Минавиапрома СССР для подготовки к полету на орбитальную станцию "Мир".

В связи с необходимостью дальнейшей подготовки экипажей к проведению ЛКИ системы "Буран" и в развитие совместного решения Минобщемаша СССР, Минобороны СССР (ВВС, УНКС), Минздрава СССР и Академии наук СССР от 5 января 1989 г. N 3-2932с принимается предложение Минавиапрома СССР о включении в состав экипажей для подготовки к полету на орбитальную станцию "Мир" в 1989 г. в качестве космонавтов-исследователей:

Станкявичюса Риминтаса Антанаса Антано – космонавта-испытателя,
летчика-испытателя МАП;
(в основной экипаж)

Заболотского Виктора Васильевича – космонавта-испытателя,
летчика-испытателя МАП.
(в дублирующий экипаж)

Министр общего машиностроения СССР — 20.02.89 — **В.Х.ДОГУЖИЕВ**

Министр авиационной промышленности СССР — **А.С.СЫСЦОВ**

Министр здравоохранения СССР — **Е.И.ЧАЗОВ**

Президент Академии наук СССР — **Г.И.МАРЧУК**

Главнокомандующий ВВС Маршал авиации — **А.Н.ЕФИМОВ**

Начальник УНКС Генерал-полковник — **А.А.МАКСИМОВ**

The document assigning Stankyavichus to a 1989 Mir mission (E. Vaskevich archives).

Anatoliy Zhernavkov (a doctor from TsPK and one-time cosmonaut candidate himself), Stankyavichus, Afanasyev, Sevastyanov, and Yevgeniy Khludeyev (former cosmonaut and head of the TsPK department responsible for survival training, and crew recovery) are shown during sea recovery training in preparation for Soyuz TM-9 (B. Vis files).

(Viktorenko–Serebrov) would be launched on Soyuz TM-8. The EO-6/Soyuz TM-9 launch was delayed until February 1990.

At first sight, the only implication for Stankyavichus was that his mission would now take place in February 1990 rather than September 1989. However, because of the postponement, NPO Energiya now refused to fly Stankyavichus, arguing that the Soyuz TM-8 descent capsule would be needed to return 100 kg of additional cargo. In correspondence with MOM and NPO Energiya, LII officials and test pilots as well as the Minister of the Aviation Industry strongly urged to fly Stankyavichus in February 1990 anyway, citing several reasons.

First, the additional cargo could be returned in newly developed ballistic capsules called Raduga that were scheduled to begin flying on Progress spacecraft later in 1990. Second, the February mission possibly was Stankyavichus' last opportunity to fly for several years. A Japanese journalist was scheduled to fly during the EO-7/EO-8 handover in late 1990 and there was public pressure to fly a Soviet journalist *before* that during the EO-6/EO-7 handover in the summer of 1990. After that, all Soyuz passenger seats were reserved for foreign cosmonauts until at least 1993, by which time Buran was expected to make its first manned flight. An additional argument for urgency was that Stankyavichus was an ethnic Lithuanian, all this at a time when the spillover from the 1989 upheavals in Eastern Europe began reverberating throughout the Baltic republics. Indeed, Lithuania would become the first Baltic republic to proclaim its renewed independence in March 1990 [68].

Despite the pleas from MAP and LII, Stankyavichus and Zabolotskiy were forced to suspend their training at Star City in September 1989. Both were assigned to conduct one or more approach and landing tests on the BTS-002. Although the two did a preparatory ground run in December 1989, they would not take the vehicle to the skies. Another opportunity for Stankyavichus to get his space legs did present itself during the EO-6/EO-7 handover in August 1990. The third seat on Soyuz TM-10 became available due to delays in the Soviet journalist-in-space project, but for unknown reasons Stankyavichus was not offered the ride. Tragically, Stankyavichus died in a plane crash in Italy in September 1990. With no further Soyuz slots immediately available and the future of Buran ever more hanging in the balance, no further Soyuz familiarization flights were planned for the Buran back-up crew members that replaced him.

CREWING FOR A SOYUZ MISSION TO BURAN

By mid-1989, several months after Buran's maiden flight on 15 November 1988, plans were finalized for a second mission that would far exceed the first one in complexity. The mission would use the second flight vehicle (2K, sometimes called "Buran-2") and was therefore dubbed 2K1. The plan was for the orbiter to be launched unmanned and fly to the Mir space station, where it would dock with the axial APAS-89 docking port of the Kristall module. Before that, Kristall would be relocated from its lateral port on the Mir multiple docking adapter to the station's front axial port. After docking, the Mir resident crew would board the orbiter to determine the state of its on-board systems, with one of the possible objectives being to use the vehicle's remote manipulator arm to move a payload from the payload bay to Kristall's lateral APAS docking port. One NPO Energiya official said that the payload was a small one-ton module housing a Fosvich X-ray telescope similar to the one on Mir's Kvant module. See [69]. Also installed in the payload bay would have been a pressurized module (37KB) about the size of the Kvant module with instrumentation to record various flight parameters.

Subsequently, the orbiter would undock and continue its flight autonomously. Around the same time, a manned Soyuz equipped with an APAS-89 docking port would be launched to dock with the orbiter. The crew would transfer to the orbiter and perform one day of testing. After the Soyuz undocked, it would fly on to Mir to link up with Kristall, while the unmanned 2K orbiter returned back to Baykonur after a one-week mission [70].

In the late 1980s NPO Energiya was ordered to build three Soyuz spacecraft (serial numbers 101, 102, 103) with APAS-89 docking ports. These vehicles were intended in the first place for possible rescue missions to stranded Buran crews during the test flight program, but it was decided to use the first one in the framework of the

2K1 mission [71]. The flight was partially seen as a dress rehearsal for such a potential rescue mission.

LII demanded that at least one of its Buran pilots be included in the Soyuz crew to give him the necessary experience for the first manned Buran mission [72]. With no or few Soyuz seats available in the mainstream Mir program, this was the ultimate opportunity for a Soyuz familiarization flight, the more so because it involved Buran itself. However, in 1990 a training group was formed for the Soyuz mission consisting of three GKNII pilots and three TsPK military engineers:

Pilots	**Engineers**
Ivan Bachurin	Eduard Stepanov
Aleksey Boroday	Valeriy Illarionov
Leonid Kadenyuk	Nikolay Fefelov

Stepanov and Fefelov were assigned in April 1990 and the others in October/ November 1990. It is not entirely clear if training advanced to the point that actual crews were formed, although Kadenyuk has claimed he was in the second back-up crew with Fefelov [73]. The most active training was performed by the three pilots, who faced the unprecedented task of docking Soyuz with Buran. All three spent many hours in TsPK's Soyuz simulators, practicing dockings both with Buran and Mir. The three engineers reportedly never underwent any dedicated mission training [74].

Bachurin and Boroday undergoing sea training (B. Vis files).

During a break from training, Aleksey Boroday relaxes for a moment with his son besides a small lake in Star City (B. Vis files).

The 2K1 mission was originally scheduled for 1991, but kept slipping as future prospects for the Buran program grew ever dimmer. Officially, the three pilots and Illarionov remained assigned until March 1992, and Fefelov and Stepanov until October 1992 [75]. Kadenyuk has said the mission was officially canceled in August 1992 [76].

Soyuz craft nr. 101 was eventually launched as Soyuz TM-16 on 24 January 1993, carrying another resident crew (Gennadiy Manakov and Aleksandr Poleshchuk) to the Mir space station. Equipped with an APAS-89 docking port, it was the only Soyuz vehicle ever to dock with the Kristall module. Soyuz "rescue" vehicles nr. 102 and 103, which had been only partly assembled, were modified as ordinary Soyuz TM spacecraft with standard "probe" docking mechanisms and were given new serial numbers [77].

REFERENCES

[1] E. Vaskevich archives.
[2] Bert Vis interviews with Nikolay Pushenko, Star City, 9 April 2000, and Igor Volk, Moscow, 13 June 2003.
[3] Letter from Sergey Protchenko to Bert Vis, 31 March 2000; Bert Vis interview with Leonid Kadenyuk, Houston, 28 April 1997.
[4] I. Marinin *et al.*, *Sovetskiye i rossiyskiye kosmonavty 1960–2000*, Moscow: Novosti kosmonavtiki, 2001, pp. 10–11.
[5] Bert Vis interview with Aleksandr Viktorenko, Cologne, 14 July 1994.
[6] Stankyavichus was an ethnic Lithuanian. Since the Lithuanian language uses the Latin alphabet, Soviet/Russian sources needed to transliterate his name from Lithuanian into Cyrillic. For this book, the authors had to transliterate that back into Latin characters. Officially, in Lithuanian, his name is written "Stankevičius", but when transliterating the Cyrillic spelling, it reads "Stankyavichyus". However, in a letter to Bert Vis, dated 19 March 1990, he wrote his own name as "Stankyavichus". The authors have decided to stick by Stankyavichus' own transliteration throughout this book.
[7] Bert Vis interview with Igor Volk, Berlin, 5 October 1991.
[8] Bert Vis interview with Igor Volk, Moscow, 13 June 2003.
[9] I. Marinin, *op. cit.*, p. 322. Another source says it was in 1980. See: V. Vasin, A. Simonov, *Ispytateli LII*, Zhukovskiy, 2001, p. 81.
[10] S. Shamsutdinov, "The Selection of Cosmonauts", *Journal of the British Interplanetary Society*, August 1997, pp. 311–316.
[11] *Ibid.*
[12] Bert Vis interview with Igor Volk, Moscow, 13 June 2003.
[13] Bert Vis meeting with Sergey Tresvyatskiy, Zhukovskiy, 13 June 2003.
[14] S. Shamsutdinov, "The LII cosmonaut team no longer exists" (in Russian), *Novosti kosmonavtiki*, 11/2002, pp. 18–21.
[15] Bert Vis e-mail correspondence with Lida Shkorkina, 29 November 2006.
[16] Bert Vis interview with Magomed Tolboyev, Zhukovskiy, 2 April 2001.
[17] I. Marinin, *op. cit.*, pp. 326, 331.
[18] Bert Vis interviews with Vladimir Gorbunov, Zhukovskiy, 3 April 2001, and Vladimir Mosolov, Moscow, 14 June 2003.

[19] Russian on-line cosmonaut encyclopedia *ASTROnote*. This source claims that a ninth candidate, Nikolay Pavlovich Belokopytov, was also medically disqualified.
[20] Bert Vis interview with Vladimir Gorbunov, Zhukovskiy, 3 April 2001.
[21] Bert Vis interview with Nail Sattarov, Zhukovskiy, 18 April 2002.
[22] Bert Vis interview with Vladimir Mosolov, Moscow, 14 June 2003.
[23] Bert Vis interviews with Aleksey Boroday, Star City, 28 June 1992, and Vladimir Mosolov, Moscow, 14 June 2003.
[24] Bert Vis interview with Vladimir Mosolov, Moscow, 14 June 2003.
[25] Bert Vis interview with Aleksey Boroday, Star City, 10 April 2000.
[26] Bert Vis interview with Vladimir Mosolov, Moscow, 14 June 2003.
[27] Bert Vis interview with Anatoliy Polonskiy, Star City, 13 April 2002.
[28] I. Marinin, *op. cit.*
[29] Bert Vis interviews with Jean-Loup Chrétien, Berlin, 30 September 1991; Igor Volk, Berlin, 5 October 1991; Michel Tognini, Cologne, 18 January 2007; French TV documentary "Ushuaïa", 19 December 1992.
[30] Bert Vis interview with Igor Volk, Moscow, 13 June 2003.
[31] Y. Semyonov, *Mnogorazovyy orbitalnyy korabl Buran*, Moscow: Mashinostroyeniye, 1995, pp. 317–319; O. Nekrasov, "The PRSO and PDST test stands" (in Russian), in: G. Lozino-Lozinskiy, A. Bratukhin, *Aviatsionno-kosmicheskiye sistemy*, Moscow: Izdatelstvo MAI, 1997, pp. 66–69; V. Trufakin, *Orbitalnyy korabl Buran i proyektno-nauchnyy otdel "Dinamika polyota" v NPO Molniya*, 2002, pp. 15–21; E. Vaskevich archives.
[32] Bert Vis interviews with Aleksey Boroday, Star City, 10 April 2000; Magomed Tolboyev, Zhukovskiy, 2 April 2001; Igor Volk, Moscow, 13 June 2003; and Vladimir Mosolov, Moscow, 14 June 2003.
[33] V. Gorbatenko *et al.*, "Simulation on the piloting-research complex in the Cosmonaut Training Centre" (in Russian), in: G. Lozino-Lozinskiy, A. Bratukhin, *op. cit.*, pp. 70–77.
[34] Bert Vis interview with Magomed Tolboyev, Zhukovskiy, 2 April 2001; "Engineering simulator PSPK-102" (in Russian), website of the "Dinamika" company at *http://www.airshow.ru/expo/1145/prod_4551.htm*
[35] E. Vaskevich archives.
[36] Y. Semyonov, *op. cit.*, pp. 321–323.
[37] Authors' correspondence with Magomed Tolboyev, February 2007; E. Vaskevich archives.
[38] Bert Vis interviews with Magomed Tolboyev, Zhukovskiy, 4 April 2007, and Viktor Zabolotskiy, Moscow, 5 April 2007.
[39] On-line aviation encyclopedia *Ugolok neba*.
[40] Bert Vis interview with Magomed Tolboyev, Zhukovskiy, 2 April 2001, and Vladimir Mosolov, Moscow, 14 June 2003.
[41] Bert Vis interview with Magomed Tolboyev, Zhukovskiy, 2 April 2001; S. Shamsutdinov, "The LII cosmonaut team no longer exists", *op. cit.*, p. 20.
[42] Bert Vis interview with Igor Volk, Groningen, 2 July 1990.
[43] E. Vaskevich archives.
[44] S. Shamsutdinov, "The LII cosmonaut team no longer exists", *op. cit.*, p. 20.
[45] S. Shamsutdinov, "The LII cosmonaut team no longer exists", *op. cit.*
[46] E. Vaskevich archives.
[47] Bert Vis interview with Igor Volk, Moscow, 13 June 2003.
[48] E. Vaskevich archives.

[49] Bert Vis interview with Vladimir Dzhanibekov, Star City, 17 October 1991; A. Tarasov, "From space to a plane" (Volk interview) (in Russian), *Pravda*, 4 March 1988; E. Vaskevich archives.
[50] Y. Baturin, *Mirovaya pilotiruyemaya kosmonavtika*, Moscow: Izdatelstvo RTSoft, 2005, pp. 304–305.
[51] NASA News Release 83-046.
[52] V. Savinykh, *Zapiski s myortvoy stantsii*, Moscow, 1999, p. 6; Y. Baturin, *op. cit.*, pp. 308–309.
[53] Bert Vis interview with Igor Volk, Moscow, 13 June 2003.
[54] Letter from the TsPK Information Group to Bert Vis, 11 August 1988.
[55] Bert Vis interview with Vladimir Dzhanibekov, Star City, 17 October 1991. The original crewmates considered for Volk may have been Vasyutin and Savinykh—Bert Vis interview with Viktor Savinykh, Washington, D.C., 25 August 1992.
[56] Bert Vis interviews with Vladimir Dzhanibekov, Star City, 17 October 1991; Igor Volk, Berlin, 5 October 1991; and Viktor Savinykh, Washington, D.C., 25 August 1992.
[57] TASS news report, 17 July 1984.
[58] N. Kidger, "Salyut Mission Report", *Spaceflight*, December 1984, p. 464.
[59] TASS news reports of 17 July and 18 July 1984.
[60] TASS news report of 31 July 1984.
[61] Y. Baturin, *op. cit.*, p. 312.
[62] A. Tarasov, *op. cit.*; Bert Vis interview with Igor Volk, Moscow, 13 June 2003. In the interview Volk claimed that he also landed at the Air Force site in Akhtubinsk as part of the post-flight flying exercise.
[63] Soviet television news report, 19 December 1987.
[64] Soviet television news report, 20 December 1987.
[65] Radio Moscow World Service, 23 December 1987.
[66] TASS news report, 23 December 1987.
[67] Soviet television report, 29 December 1987.
[68] E. Vaskevich archives; S. Shamsutdinov, "The legendary Soyuz ship" (in Russian), *Novosti kosmonavtiki*, 6/2002, pp. 66–69.
[69] P. Bond, "Mir : Where Next", *Spaceflight News*, November 1990, pp. 20–21.
[70] S. Shamsutdinov, I. Marinin, "Flights that didn't take place" (in Russian), *Aviatsiya i kosmonavtika*, 7/1993, pp. 48–49; "Soviets' Next Shuttle Flight to Include Docking at Mir", *Aviation Week & Space Technology*, 10 December 1990, p. 24.
[71] S. Shamsutdinov, "The legendary Soyuz ship", *op. cit.*
[72] E. Vaskevich archives.
[73] Bert Vis Interview with Leonid Kadenyuk, Houston, 28 April 1997.
[74] Bert Vis interviews with Aleksey Boroday, Star City, 28 June 1992; Valeriy Illarionov, Star City, 1 July 1992; and Nikolay Fefelov, Star City, 3 July 1992.
[75] I. Marinin, *op. cit.*
[76] Bert Vis interview with Leonid Kadenyuk, Houston, 28 April 1997.
[77] S. Shamsutdinov, "The legendary Soyuz ship", *op. cit.*

6

Testing the hardware

For most rocket, satellite, and spacecraft programs, the Soviet philosophy was to limit ground testing to the bare minimum and "fly the bird and see how it behaves", no matter how many test flights were required before declaring it operational. For Energiya–Buran the Russians could hardly afford to do the same, if only because of the astronomical cost of the system and the serious implications of losing one or several vehicles in a small fleet of precious reusable spacecraft. Moreover, the Energiya–Buran system represented a leap in technology the likes of which had not been seen in the Soviet space program. It featured the country's first reusable spacecraft (and a big one at that), the world's most powerful liquid-fuel rocket engine (the RD-170), the first big domestic-built cryogenic engine (the RD-0120), the use of a vast array of new materials, and so on. Finally, the N-1 debacle, at least partly attributable to a lack of ground testing (particularly of the first stage), was firmly etched in everyone's memories and a sure sign of the need to approach things differently when the next program of comparable proportions came along.

Therefore, for Energiya–Buran, the Russians had no choice but to shift the emphasis from in-flight to ground-based testing, requiring a major investment in infrastructure and hardware. Although the Russians undoubtedly benefited from more than seven years of US experience with Space Shuttle missions before Buran was finally launched, they clearly left no stone unturned when it came to testing their hardware, even for systems that were very similar to those flown on the Shuttle. In fact, in many respects the Energiya–Buran test program was more extensive than the Shuttle's. Still, even that didn't stop the Russians from sticking to their tradition of flying a piloted vehicle unmanned on its first mission, unlike NASA, which for the first time in its history put a crew on a first-flight vehicle with Columbia in 1981.

ENERGIYA COMPONENT TESTING

The lead research institute for aerodynamic, dynamic, and structural testing of launch vehicle components was the Central Scientific Research Institute of Machine Building (TsNIImash) in Kaliningrad. Studies of the aerodynamic behavior of the Energiya–Buran stack during various phases of the launch and also of the strap-ons after separation were conducted in several wind tunnels and saw the use of more than 80 scale models in about 11,000 experiments overall. Dynamic tests involved the use of 1:10 and 1:5 scale models of the Energiya–Buran system. Individual RD-0120 and RD-170 engines were subjected to vibration tests.

For structural testing of core stage components, the Russians were forced into a different strategy than NASA. The US space agency built full-size test articles of the External Tank's LOX tank, LH_2 tank, and intertank, which underwent individual structural load tests at the Marshall Space Flight Center in 1977. However, the Russians did not have the facilities to do the same with their core stage elements

Core stage test component being airlifted to Kaliningrad (*source*: Boris Gubanov).

and therefore elected to do their structural tests (called "2I") using smaller pieces of the tanks. These included top and bottom sections of the LH_2 tank joined together, a half-size LOX tank, an intertank section with parts of the LOX and LH_2 tanks attached, a tail section with a mock-up LH_2 tank bottom section attached, and a shortened LH_2 tank. All these sections were manufactured at the Progress plant in Kuybyshev and tested either there or at TsNIImash. For the TsNIImash tests, which also included thermal testing, the sections were transported from Kuybyshev to the Moscow area by barge, the same one used to transport Buran orbiters from Tushino to Zhukovskiy. Subsequently, they were picked up by heavy MI-10 helicopters and airlifted to Kaliningrad.

Another institute playing a key role in Energiya–Buran testing was the Scientific Research Institute of Chemical Machine Building (NIIkhimmash) at Novostroyka near Zagorsk (now Sergiyev Posad), some 100 km north of Moscow. Having originated in 1948 as a branch of the NII-88 rocket research institute and independent since 1956, this remains the largest test facility for rocket engines on Russian territory.

SOM-1 test stand (B. Vis files).

It has over 50 test stands for rocket engines and their components and also several thermal vacuum chambers and other facilities for spacecraft testing. When it came to simulating conditions during launch, one of NIIkhimmash's main tasks was to study the acoustic environment during lift-off. This was done using a test stand called SOM-1 in which a 1:10 model of Energiya–Buran sat on a simulated launch table and was lifted several meters above the ground by a special hydraulic system. Another facility at NIIkhimmash ("Stand R") simulated the separation of the Blok-A strap-on boosters from the core stage using full-scale mock-ups of the boosters.

Also involved in simulating the lift-off environment was the Scientific Research Institute of Chemical and Building Machines (NIIKhSM), situated in Zagorsk itself. This had a test stand called SVOD, which featured another 1:10 model of Energiya–Buran equipped with small solid-fuel rockets to mimic conditions during lift-off. NIIKhSM was also responsible for testing various launch pad systems such as the sound suppression water system and the fueling systems.

The Russians never carried out full-fledged fueling tests of core stage propellant tanks until full-scale models of the Energiya rocket were placed on the launch pad at Baykonur in the mid-1980s. This also differed from the situation in the United States, where NASA did fueling tests of complete External Tanks at the Marshall Space Flight Center in 1977. In order to save costs, the Progress factory only built a test stand designed to fill the tanks with liquid nitrogen at temperatures of −180°/−190°C, which was significantly lower than the −255°C required for liquid hydrogen [1].

TESTING THE RD-0120

Before the beginning of the Energiya–Buran program the only liquid oxygen/liquid hydrogen engines developed in the Soviet Union had been the 7.5-ton thrust 11D56 of KB Khimmash and the 40-ton thrust 11D54 and 11D57 of KB Saturn, intended to be used on upper stages of the N-1 rocket. Therefore, the development of the 190-ton thrust RD-0120 for Energiya's core stage, assigned to KBKhA in Voronezh, was a major challenge, partly because much of the infrastructure needed for testing the engine was not yet in place. This required a step-by-step approach to certifying the engine for flight. The first step was to test individual components of the engine, followed by test firings of experimental engines at increasing rates of thrust, and ultimately test firings of flightworthy engines both individually and mounted in clusters of four on Energiya's core stage at Baykonur's UKSS pad.

Since the RD-0120 was the first powerful Soviet cryogenic engine, many components that would usually be installed straightaway in experimental engines now had to be tested individually first. Particular attention was paid to the ignition system in the combustion chamber and the gas generator as well as to the operation of the low-pressure turbopumps. The components for these tests were built both by KBKhA's own Experimental Factory and by the Voronezh Machine Building

Factory and tested at various locations in the Soviet Union, including KBKhA's own experimental base in Voronezh.

While KBKhA had its own rocket engine test stands near Voronezh, none of these could be converted for test firings of this radically new engine, nor was it deemed expedient to build the new facilities there. Instead, test firing stands for the RD-0120 were built by two other organizations. One was the Scientific Research Institute of Machine Building (NII Mashinostroyeniya or NIIMash) in Nizhnyaya Salda north of Yekaterinburg. This organization had become independent in 1981 after having been a branch of the Scientific Research Institute of Thermal Processes (NII TP), which was how the former NII-1 was renamed in 1965 (in 1995 it was again renamed as the Keldysh Research Centre). NIIMash mainly specialized in reaction control thrusters for satellites and manned spacecraft (including Buran).

NIIMash had two test stands (nrs. 201 and 301) for firing the RD-0120 in vertical position. Construction got underway in 1977 and 1981, respectively. The test stands allowed the engines to be gimbaled and had diffusers to simulate operating conditions at higher altitudes. Stand nr. 301 was capable of handling longer duration test firings (over 1,000 seconds) and maximum gimbal angles. RD-0120 engines were delivered to Nizhnyaya Salda from the manufacturing plant in Voronezh by Antonov-8 cargo planes.

The other organization was NIIkhimmash near Zagorsk. It had a cryogenic complex called KVKS-106, already used before the Energiya–Buran program for testing the 11D56 and 11D57 cryogenic engines. KVKS-106 comprised five test facilities, one of which had two test stands (V-2A and V-2B) for full-scale test firings of the RD-0120 in horizontal position. Others were used to test individual components of the engine and the core stage. The engines were transported from Voronezh to NIIkhimmash by road.

The bulk of the testing was to be carried out at Nizhnyaya Salda, but, as the construction of the facilities there ran into delays, it was decided to test the first engines at the existing cryogenic complex of NIIkhimmash. The first test engine (with

RD-0120 test stand at NIIkhimmash (*source*: Russian Space Agency).

a shorter-than-nominal nozzle) was delivered to NIIkhimmash in the autumn of 1978 and underwent a first brief test firing (4.58 seconds) at the V-2B stand in March 1979. The initial tests at NIIkhimmash were conducted at just 25 percent of rated thrust, but allowed to test such things as the ignition and shutdown sequence. The adjacent V-2A, capable of supporting tests at 100 percent thrust, was ready in 1984.

The first test firing at NIIMash's test stand nr. 201 took place on 19 January 1980, also at low thrust. By early 1981 the engine was being tested at 70 percent thrust, and it wasn't until May 1984 that the RD-0120 worked at 100 percent thrust for 600 seconds (with the nominal operating time during launch being 467 seconds). Considerable delays in reaching nominal thrust were caused by problems with the impeller of the liquid hydrogen pump, a rotating disk with a set of vanes that produces centrifugal force within the pump casing. The problem was eventually solved by using a different type of titanium to manufacture the component.

Although the RD-0120 test program took longer than expected, it does not appear to have been a major factor in the delays of Energiya's first launch. By the time of that maiden launch in May 1987 the Russians had accomplished 523 test firings of 103 different RD-0120 engines with a total duration of 73,891 seconds. NIIMash's test stand nr. 301 saw its first test firing on 30 July 1987. Records set during that same period were a 100-second test at 123 percent rated thrust in September 1987 and a maximum-duration burn of 1,202 seconds in January 1988. By early October 1988, about a month before the maiden flight of Buran, a total of 126 engines had undergone 635 test firings lasting a total of 120,454 seconds. By comparison, NASA accumulated 110,253 seconds of burn time on the Space Shuttle Main Engines in 726 test firings before STS-1.

The RD-0120 test program continued at NIIkhimmash even after the cancellation of the Energiya–Buran program under a deal between KBKhA and the US Aerojet company to study the feasibility of using the engine on future American launch vehicles. Several test firings were also performed in the mid-1990s in the framework of the joint European–Russian RECORD program (Russia–Europe Cooperation on Rocket Engine Demonstration), led by the French SEP company. In those tests, the engine was equipped with European instrumentation to allow European engineers to create a detailed software model of the engine and gain experience with staged-combustion cycle cryogenic engines for possible use in future reusable systems. The final test firing of the RD-0120 was conducted at NIIkhimmash in 1997. The maximum accumulated test time for a single engine was about 5,000 seconds and the maximum number of ignitions for a single engine was thirty [2].

TESTING THE RD-170

Like the RD-0120, Energomash's RD-170 engines for the strap-on boosters underwent a step-by-step test program, moving from autonomous tests of individual components to full-scale test firings at increasing rates of thrust. Bench tests were not only carried out with the engines alone (at Energomash's own test facility), but also with the engines mounted on the modular part of the strap-ons (at

NIIkhimmash). Test firings of complete strap-ons at Baykonur's UKSS pad were ultimately canceled, at least partially because the RD-171, the near twin of the RD-170, had already undergone flight testing on the first stage of the Zenit rocket. The road to success for the RD-170 proved much more arduous than for the RD-0120, with several setbacks plunging the program into deep crisis in the early 1980s.

Component tests

One of the main motives for the choice of a four-chamber rather than a single-chamber LOX/kerosene engine in 1973 was the possibility to test major components of the engine (primarily the combustion chamber) individually and only later to assemble them for test firings of the complete engine. This followed from the negative experience with the single-chamber 640-ton thrust hypergolic RD-270 engine for Chelomey's UR-700 rocket, where engineers had moved to all-up tests straightaway. All the 27 test firings carried out in 1967–1969 had ended in some kind of failure before work on the engine was discontinued.

The component tests were conducted between 1974 and 1980 using test models known as "oxygen installations" (UK). Most of these were built on the basis of blueprints and components developed in the early 1970s for the RD-268, a 100-ton thrust engine burning unsymmetrical dimethyl hydrazine (UDMH) and nitrogen tetroxide (N_2O_4). This was possible because UDMH/N_2O_4 engines use virtually the same ratio of propellants as LOX/kerosene engines. It did require the use of new materials compatible with LOX/kerosene and modifications to two test firing stands of Energomash on the banks of the Khimka river in the northwest outskirts of Moscow. These were completed in the first eight months of 1974.

The first two of these test models (1UK and 2UK) were essentially 100-ton thrust experimental model engines to test various aspects of the RD-170, such as the ignition sequence, mixing of the propellants in the combustion chamber and gas generator, cooling of the combustion chamber, and the use of reusable materials. A modified version known as 1UKS burned recycled oxidizer gas produced in a gas generator, as was the case for the RD-170. Between August 1974 and November 1977 as many as 346 test firings of these three types of engines were conducted lasting a total of 19,658 seconds.

The next series of tests involved an installation called 3UK, designed to test the RD-170's gas generator. This consisted of a full-size gas generator, two turbopumps, and a mock-up combustion chamber, making it possible to simulate the pressure, propellant expenditure, and temperature in the gas generator at levels between 30 and 80 percent of nominal values. The tests were conducted between June 1976 and September 1978. A total of 77 3UK installations underwent 132 test firings lasting a total of 5,193 seconds. About 60 mixing heads were tested, with two being chosen for test firings of complete RD-170 engines.

Also built were experimental engines called 2UKS that closely imitated the operating conditions of the RD-170's combustion chamber, but inherited their turbopumps from earlier designs. Therefore, the chamber developed only 80 percent of the nominal thrust at a pressure of 200 rather than 250 atmospheres. Also tested was the gimbaling system and several of the engine's automatic systems. A total of 42 2UKS engines accumulated about 6,000 seconds of burn time in 68 tests from May 1977 until June 1978. Interestingly, the 2UKS served as the basis for the development of the 85-ton thrust RD-120, which would later power the second stage of the Zenit rocket.

Finally, Energomash engineers built the 6UK, which essentially was a real RD-170 without a combustion chamber, the main purpose being to test the turbopump assembly. The installation underwent 31 tests between June 1978 and December 1980. The tests revealed that the turbopump was susceptible to burn-throughs and vibrations. Although as many as 23 6UK installations were used, they accumulated just 280 seconds of testing time. Since the 6UK was nearly as expensive as a complete RD-170/171, the test program was limited and the problems with the turbopump assembly were not debugged by the time the full-scale RD-170 test firings got underway. Therefore, the 6UK was much less effective in paving the way to those test firings than the other UK installations, setting the stage for a major crisis in the Energiya program in the early 1980s.

Full-scale test firings

By the middle of 1980 preparations had been completed for the long-awaited inaugural test firing of a complete RD-170. Mounted on Energomash's test-firing stand nr. 2, the engine was ignited on 25 August 1980, but shut down just 4.4 seconds later. It was only the first in a long string of setbacks for the RD-170/171. The next 15 test firings were also less than satisfactory, leading to a decision to perform the 17th test firing at a lower thrust of 600 tons. This resulted in a first successful, full-duration 150-second test firing of the RD-170 on 9 June 1981.

Subsequent test firings at the same thrust rate also produced satisfactory results, giving Energomash engineers enough confidence to move on to ground tests of the nearly identical RD-171 integrated with a Zenit first stage. These tests were carried out at the IS-102 test stand of NIIkhimmash, originally used in the 1950s for testing the first stage of the R-7 missile and later the scene of test firings of the Proton first stage and the second, third, and fourth stages of the N-1. The engine earmarked for the test (serial nr. 18) had already undergone a successful test firing at Energomash's facilities in September 1981. Later analysis did show that a turbopump rotorblade had been damaged by particles that had somehow entered the turbopump assembly, but this was considered benign enough to press on with the test firing of the Zenit first stage on 26 June 1982. To the amazement of onlookers, the test ended in disaster near the end of its scheduled 6-second duration, when the turbopump assembly burnt through and caused a massive explosion that completely destroyed the stage and the entire test stand.

The disaster raised serious questions about the fundamental design of the RD-170/171, the more so because the test had been performed at only 600 tons of thrust rather than the nominal 740 tons. It led to the creation of an interdepartmental commission to look into the status of the RD-170 development program and consider

Energomash engine test-firing stand (*source*: NPO Energomash).

possible alternatives for powering the Zenit first stage and Energiya's strap-on boosters. Headed by Valentin Likhushin, the head of NII TP, the commission included such luminaries of the Soviet rocket industry as Arkhip Lyulka, Nikolay Kuznetsov, and also Valentin Glushko himself.

One idea, proposed by I.A. Klepikov at Energomash, was to equip each combustion chamber with its own, smaller turbopump assembly, transforming the RD-170 into four engines with 185 tons thrust each (hence their designation MD-185, with the "M" standing for "modular", because the idea was to use the engine on a variety of rockets). Actually, an order to study such an engine had already come from the Minister of General Machine Building Sergey Afanasyev as early as 11 October 1980. Wary of witnessing a repeat of the N-1 fiasco, Afanasyev had ordered to set up a complete department within Energomash to design such an engine in order to safeguard against any major development problems with the RD-170/171. It was felt that the 2UKS experimental engine, successfully tested in 1977–1978, could serve as a prototype for the MD-185.

Another option was to use the NK-33 engines developed by the Kuznetsov design bureau (under the Ministry of the Aviation Industry) for a modified version of the N-1 rocket. Although the N-1 had been canceled before the NK-33 engines ever had a chance to fly, forty of these reusable engines had undergone an extensive series of test firings up to 1977, proving their reliability. By making small modifications to the turbopumps, Kuznetsov's engineers had managed to uprate the NK-33's thrust from about 170 tons to just over 200 tons, meaning that four would be sufficient to replace the RD-170. Energiya's chief designer Boris Gubanov flew to Kuznetsov's plant in Kuybyshev, where he was shown more than 90 such engines lying in storage.

The most radical alternative studied was to replace the Blok-A strap-ons with solid-fuel boosters. That task was assigned to NPO Iskra in Perm (chief designer Lev N. Lavrov), an organization specialized in solid-fuel motors that had already built several small solid-fuel systems for Energiya–Buran. NPO Iskra devised a plan for a 44.92 m high booster consisting of seven segments. Weighing 520 tons (460 tons of which was propellant), the booster would produce an average thrust of 1,050 tons (specific impulse 263 s) and operate for 138 seconds before separating from the core stage.

In the end, none of the three proposals was accepted. Although the MD-185 was probably the least radical alternative, research showed that it would not solve the turbopump burn-through problems as the temperature of the generator gas would be virtually the same as in the RD-170/171. A major problem with both the MD-185 and NK-33 was that they increased the total number of engines on Energiya from eight to twenty, leaving more room for failure.

One can also safely assume that Glushko had second thoughts about using the NK-33 engines. After all his efforts to erase the N-1 from history, it is hard to imagine he would have accepted using engines that had originally been built for this rocket. What's more, in 1977 Glushko had secured a decision from the Council of Ministers to ban all work on powerful liquid-fuel rocket engines not only at Kuznetsov's design bureau, but at *any* organization under the Ministry of the Aviation Industry. Understandably, Kuznetsov was not about to come to Glushko's rescue just like that.

RD-171 in test stand (*source*: NPO Energomash).

One of the conditions he laid down for participating in the Energiya program was that his team be officially rehabilitated after the abrupt and humiliating cancellation of its efforts several years earlier.

It was even easier to find arguments against NPO Iskra's solid rocket motors. Aside from the safety and ecological concerns inherent in solid-fuel rockets, the Soviet Union had no experience in building solid rocket boosters of this size. Moreover, they would not have been reusable and it would have been difficult to operate them in the temperature extremes of Baykonur. It would have taken an estimated 8 years to get them ready for flight.

In fact, any of the three alternative proposals would probably have delayed the first flight of Energiya by many years and would only have added to the already soaring costs of the program. In September 1982 the interdepartmental commission decided to continue test firings of improved versions of the RD-170/171 and at the same time continue research work on the MD-185. The official investigation into the June 1982 accident had concluded that it was probably the direct result of the engine being tested in a vertical position (as opposed to the near-horizontal position for the

Energomash tests). However, Energomash engineers disagreed and believed it had been caused either by aluminum particles entering the turbopump assembly from the propellant tanks or by high vibrations of the turbopump assembly.

Among the measures taken to prevent a repeat of the accident were the installation of filters to prevent particles from entering the turbopump assembly and the strengthening of certain components of the turbopump. Those efforts paid off with the first successful full-duration 142-second test firing of the RD-170 at nominal thrust (740 tons) on 31 May 1983, which by many was considered a make-or-break test for the engine. In the following months, the engine performed better and better, clearing the path for another test of the RD-171 as part of a Zenit first stage. Bearing in mind the disastrous outcome of the first such test, a commission was set up to decide if it could proceed. In October 1984 the commission gave a negative recommendation (even KB Yuzhnoye chief Vladimir Utkin), but that was overruled by the new Minister of General Machine Building Oleg Baklanov, who had replaced Afanasyev in the spring of 1983 and proved to be a more avid supporter of the RD-170 than his predecessor. In the end, the Zenit first stage operated flawlessly in a test firing at the refurbished IS-102 test stand of NIIkhimmash on 1 December 1984, repeating that performance at the end of the same month.

The Zenit flies

A further series of bench tests at the Energomash facilities in early 1985 finally paved the way for the first test flight of the Zenit rocket, which took place from the Baykonur cosmodrome on 13 April 1985 after a scrub the previous day. Although the second stage failed to place the mock-up payload into orbit, the RD-171 and the Zenit first stage performed brilliantly. Another launch with similar outcome was carried out on 21 June 1985. The first completely successful Zenit launch occurred on 22 October 1985, with the rocket placing into orbit a Tselina-2 electronic intelligence satellite, which would be its primary payload for many years to come. After another partial failure in December 1985 (not related to the first stage), the Zenit chalked up another five successful flights before Energiya's maiden launch on 15 May 1987 (for the further history of the Zenit see Chapter 8).

Further test firings

Even as the Zenit was slowly overcoming its teething problems, tests continued of the RD-170 in preparation for the first flights of the Energiya rocket. In November 1985 the engine was test-fired for the first time as part of an Energiya strap-on booster "modular section" at NIIkhimmash's IS-102 test stand. In all, the RD-170/171 underwent fifteen test firings as part of a Blok-A or Zenit first stage at NIIkhimmash. By the time of Energiya's maiden launch in May 1987, a total of 148 RD-170 engines had undergone 473 test firings totaling 51,845 seconds. By early October 1988, only weeks before the first attempted launch of Energiya–Buran, these numbers had increased to 186 engines, 618 test firings, and 69,579 seconds of accumulated burn time.

Zenit launch (B. Hendrickx files).

Test firings of the RD-170 continued after the two flights of Energiya and were mainly aimed at further improving the engine so that it could be reused on as many as ten Energiya missions, a capability that was demonstrated by 1992. Meanwhile, PO Polyot's serial production plant in Omsk opened its own test-firing stand at Krutaya Gorka (55 km north of Omsk), carrying out six tests of RD-170 engines beginning on 29 December 1990. The test stand was reported to be the scene of a major explosion on or around 20 November 1991, which probably rendered it useless for further test firings [3].

When the Energiya program was canceled in 1993, a total of 14 flightworthy RD-170 engines had already been installed on Blok-A strap-on boosters awaiting their missions at Baykonur's Energiya assembly building. In 1996–1997 the engines were removed and shipped back to Energomash to be modified as RD-171 engines for use in the Zenit rocket as part of the Sea Launch program [4].

ENERGIYA PAD TESTS

When it came to testing the whole stack on the launch pad, NASA and the Russians had different strategies because of the presence vs. absence of main engines on the spaceplane. With the US Orbiter being an integral part of the Space Shuttle stack, all pad-related tests were carried out with the Orbiter in place. In 1979 NASA performed fit checks on the pad of a stack consisting of OV-101 Enterprise and a mock-up External Tank and Solid Rocket Boosters. Pad tests of the main engines were conducted during 20-second "Flight Readiness Firings" several weeks prior to the maiden flight of a new Orbiter (except Endeavour).

Since Buran lacked main engines and was only one of several possible payloads for Energiya, most of the early pad testing at Baykonur focused only on the rocket. Various specially adapted experimental versions of the rocket were rolled out without any payloads attached to undergo dynamic tests, fueling tests, and engine test firings at the Universal Test Stand and Launch Pad (UKSS). Only at a later stage were full-scale mock-ups of Buran used for pad tests of the complete stack.

First roll-outs

As early as 1979 crude full-scale mock-ups of the core stage and one of the strap-ons ("EUK-13") were built *in situ* at Baykonur just to get a feel of things to come [5]. During that same year the Progress factory in Kuybyshev manufactured a core stage called 4M that was to be used for a variety of pad tests at Baykonur. However, the beginning of those pad tests hinged on the completion of the UKSS test stand as well as the availability of the VM-T Atlant carrier aircraft to fly elements of the core stage from Kuybyshev to the cosmodrome. After completing its test flights at Zhukovskiy, the VM-T delivered the 4M core stage to Baykonur in two ferry flights on 8 April and 11 June 1982. At about the same time KB Yuzhnoye shipped four mock-up modular sections of strap-on boosters to the launch site.

By the autumn of 1982 the Interdepartmental Coordinating Council (MVKS) set the goal of assembling the first Energiya rocket before the end of the year, which was expected to be a major morale-booster for cosmodrome personnel. This was easier said than done, because much of the equipment needed for this at the Energiya assembly building was not yet in place and NPO Energiya's ZEM factory was running late in supplying the nose and tail sections for the mock-up boosters. In order to meet the deadline, tail sections for the mock-up boosters were quickly manufactured by the Atommash factory in Volgodonsk, which produced large-scale components for the Soviet Union's nuclear power program. Mustering all their improvisation skills, workers managed to complete the assembly of the first so-called "packet" in the final days of 1982, using a specially ordered crane to mount the final two boosters on the stack.

A first demonstration roll-out took place in the late winter, but the exact date is unknown and it is not clear if the vehicle was actually placed on the UKSS. US reconnaissance satellites observed the rocket outside the assembly building in March 1983 [6]. The next step would have been to conduct fueling tests of the 4M's LOX and LH_2 tanks, but much of the 4M core stage's internal plumbing had not yet been supplied by the Progress plant and the UKSS had not been completely finished either. However, engineers came up with an alternative plan to use the 4M stack for dynamic tests that would normally be done much later at the Dynamic Test Stand, the construction of which was running many years behind schedule. The purpose was to learn more about the effects of longitudinal and transverse vibrations on the core stage, the boosters, and the mechanical systems joining them. Another objective was to study the effects of an emergency shutdown of two RD-0120 engines. For the tests

Energiya 4M-D roll-out in May 1983 (*source*: *www.buran.ru*).

the stack was equipped with a wide array of sensors capable of monitoring 85 different parameters.

For the longitudinal vibration tests, a cable would be suspended between the top and bottom sections of the core stage, where one of the engine nozzles was removed. Pyrotechnic bolts would then be fired to release the cable either at the bottom end or top end of the core stage, thereby creating longitudinal vibrations. Dubbed 4M-D (D for "dynamic"), the stack was rolled out to the UKSS for these tests on 7 May 1983. Dynamic tests were later continued with the same stack in horizontal position in the Energiya assembly building. Several months later the UKSS was finally ready for fit checks, and launch pad chief designer Vladimir Barmin insisted on doing another roll-out. In October 1983 the 4M stack once again slowly made its way to the UKSS and all its systems were hooked up to the pad, which had not been the case in the earlier tests [7].

Fueling tests

It was not until 14 March 1985 that the 4M stack was once again erected on the UKSS for the long-awaited fueling tests of the core stage's oxygen and hydrogen tanks. While the Russians had plenty of experience with loading liquid oxygen tanks, they were newcomers to fueling big hydrogen tanks, the more so because they used a special type of subcooled liquid hydrogen. After an initial series of tests in which the hydrogen tank was conditioned with nitrogen and hydrogen gas, the core stage was declared ready for the fueling tests. Between mid-April and late September 1985 the 4M core stage underwent nine fueling cycles. These included both partial and full loads of the hydrogen and oxygen tanks separately as well as one complete load of the entire core stage. After each test, engineers carefully checked the condition of the core stage's outer insulation layer. While the insulation remained intact during and after fueling, some debonding was observed during draining of the tanks.

The 1985 pad tests were rounded out in the first days of October with several other tests of the core stage, including a nitrogen purge of the stage's tail section and two test firings of the hydrogen igniters, needed to burn off any excess hydrogen gas accumulating on the launch pad prior to engine ignition. A derailing incident during the roll-back of the 4M stack on 5 October did not result in any damage to the vehicle [8].

Engine test firings

While the facilities of NIIkhimmash near Zagorsk allowed full-scale test firings of entire Blok-A modular sections to be carried out, there was no infrastructure at the site to do the same with the core stage. NASA's method of testing the Space Shuttle Main Engines in a realistic structural environment was by mounting an Orbiter aft fuselage and a truss simulating the mid fuselage on an External Tank at the National Space Technology Laboratory (now the Stennis Space Center) in Mississippi and, ultimately, by firing the engines with the entire stack on the pad during the Flight Readiness Firings.

The Soviet approach to integrated testing of the RD-0120 was to fire all four engines with the core stage and strap-on boosters bolted to the UKSS test stand/launch pad at Baykonur. For this purpose, two core stages were built, designated 5S and 6S. These were to undergo 17 test firings lasting a total of 3,700 seconds. All the engines involved in these tests had already been test-fired at the NIIMash test stands in Nizhnyaya Salda before being mounted on the rockets.

The 5S core stage, flanked by inert strap-on boosters, was rolled out to the UKSS on 23 January 1986. As the core stage was being prepared for the first test firing, other events were shaping the future of world space programs. America was in mourning following the loss of Challenger and its seven-person crew on 28 January 1986 and the Soviet Union successfully launched its Mir space station on 20 February. Two days after that landmark event, the 5S core stage was ready for an initial 17.8-second test firing of its four RD-0120 engines. However, just 2.58 seconds after ignition, as the engines were building up thrust, on-board automatic devices shut down all four engines due to high temperature readings in one of the gas generators.

Barely had ground controllers realized what had happened, when they found themselves faced with a problem of catastrophic potential. As the engines shut down, a leak occurred in the pneumatic lines that supplied helium to operate the rocket's fill and drain valves, making it impossible to drain the core stage. Loaded with 600 tons of liquid oxygen and 100 tons of liquid hydrogen and with pressure in the cryogenic tanks gradually building, Energiya 5S was slowly turning into a bomb with an explosive potential of 450 tons of TNT. Ground controllers had no choice but to send a crew of volunteers out to the pad to hook up a back-up helium supply system to the rocket. Working in hazardous conditions under the launch table, they suc-

Energiya test model on the UKSS (*source*: Mashinostroyeniye).

ceeded in finishing the job in just an hour's time, allowing detanking operations to begin.

Subsequent analysis traced the cause of the helium leak to a damaged pipe measuring just 20 mm in diameter. As a result of the incident, back-up helium lines were introduced for future vehicles as well as additional means of controlling the valves electrically. The engine shutdown itself was blamed on a faulty low-pressure hydrogen pump in engine nr. 1, which had apparently been inadvertently damaged during repair work in the Energiya assembly building needed after the test firings in Nizhnyaya Salda. In the following weeks the pump was successfully swapped with

Energiya 5S shortly after delivery to the UKSS (*source*: *www.buran.ru*).

another one, an operation that had never been done on the pad before. On 25 April Energiya 5S was ready for another test firing, scheduled to last 390 seconds. This time all four engines operated flawlessly, throttling up and down as scheduled and going through a full gimbaling program. With Energiya virtually ready to fly, the focus shifted to pad tests of the entire Energiya–Buran system (see Chapter 7) [9].

THE FIRST FLIGHT OF ENERGIYA

6S becomes 6SL

According to original flight plans drawn up in the 1970s, Energiya was to begin its test flights in 1983 with two suborbital missions carrying full-scale Buran mock-ups, followed in 1984 by the first launch of an unmanned flightworthy orbiter [10]. By the early 1980s those timelines had changed significantly, as had the flight plans themselves. The idea was now to launch an unmanned Buran into orbit on the first mission of Energiya (rocket 1L) following the completion of fueling tests with the 4M core stage and pad test firings of the 5S and 6S vehicles.

The original plan for 17 pad test firings lasting a total of 3,700 seconds was quickly laid to rest. Satisfied with the results of the 5S test firing on 25 April 1986, the MVKS decided on 5 May to significantly curtail the test-firing program and conduct just one more test firing with the 6S vehicle to reach an accumulated pad burn time of

423 seconds prior to the maiden launch of Energiya. The plan was to turn the remaining 30-second test into a combined test firing of the four RD-0120 core stage engines and the four RD-170 strap-on booster engines, something which would also have been the culmination of the original test-firing program. Consideration was also given to strapping the OK-ML1 Buran mock-up to the core stage for that test.

However, as these events unfolded, an alternative proposal from Energiya chief designer Boris Gubanov had been steadily gaining ground. That was to skip test firings of the 6S vehicle altogether and turn it into a flightworthy rocket for a test mission that would precede the flight of Energiya 1L with the Buran orbiter. In a way, it was a return to the "test-as-you-fly" philosophy so common in other Soviet space projects. Gubanov's main argument was that if one of the test firings ended in a cataclysmic explosion, it would take two to three years to rebuild the unique test stand. Not only was the UKSS later supposed to become a launch pad, it would also continue to serve as a test stand for core stages and strap-on boosters to be flown on operational Energiya missions.

The risk of an accident would be even higher if the strap-on boosters were going to be involved in the tests as well. Rather than test-fire the rocket on the ground, Gubanov argued, it would simply be test-fired in flight. The minimum mission objective would be to fly safely for at least 30 seconds, allowing the rocket to reach a safe distance from the test stand. This would achieve the same goal as a combined 30-second ground-based test firing of the core stage and strap-ons, without running the risk of wiping out the UKSS.

The idea originated in early 1984, but it would take Gubanov almost two years to get it accepted. Gubanov made his first overture to the highest authorities in early 1985, putting forward the idea to Grigoriy V. Romanov, who as the Central Committee Secretary for Defense Matters was the highest political figure in charge of the space program. However, Romanov was not convinced, electing instead to divert more resources and personnel to the space station program with the goal of launching the Mir core module by the next Party Congress in February 1986. The proposal initially also met with stiff opposition from NPO Energiya general designer Valentin Glushko, who at one point even said that "one wouldn't come up with such an idea even when drunk." Also opposed to the plan was launch pad chief designer Vladimir Barmin, whose organization (KBOM) would now have to turn the UKSS into a launch pad much earlier than expected. Also favoring a full-scale ground-based test firing of both the core stage and the strap-on boosters was the military community.

However, as the months progressed, events gradually turned to Gubanov's favor. In July 1985 Romanov, once considered a leading candidate to become the next General Secretary of the Communist Party, was removed from the Politburo and from his post as Secretary for Defense Matters as part of a Party management shake-up following the election of Mikhail Gorbachov as General Secretary in March of that year. By the end of the year Gubanov had garnered support from Minister of General Machine Building Oleg Baklanov (also the head of the MVKS), who in turn convinced his ally Glushko. On 2 January 1986 Baklanov flew to the cosmodrome with a large number of leading space officials, giving them the order not to return home until an Energiya had been launched.

Long before getting the needed political support, Gubanov had secretly been making arrangements to convert the 6S core stage into a flight vehicle called 6SL ("L" standing for "flight"). He had already asked the people of the NPO Energiya Volga Branch to study this possibility during a visit to Kuybyshev in November 1984. An official industry order followed on 16 August 1985 and allowed engineers to "cannibalize" parts of the first flight-rated rocket (1L) to speed up launch preparations. As a result, all elements of the core stage were in place at Baykonur by the beginning of 1986.

A key argument in getting approval for the 6SL launch was that Buran was suffering more and more delays, further pushing back the launch of Energiya 1L. An early demonstration launch of the 6SL vehicle would not only be a boost to the team, but could also help convince the country's political leadership of the program's feasibility. With a new wind beginning to blow through Soviet politics in the mid-1980s, the Energiya–Buran program was finding itself on increasingly shaky ground and was in dire need of a major success. Somehow, Gubanov's original argument for the launch—namely, to reduce the risk of a catastrophic explosion on the UKSS—had moved to the background and a 30-second combined static test firing of the 6SL core stage and strap-ons remained on the agenda even after the successful test firing of 5S in April 1986. The Military Industrial Commission set up an independent commission headed by Konstantin V. Frolov, the Vice-President of the Academy of Sciences, to look into the need for additional test firings, but this failed to give a clear-cut recommendation. However, a continuing string of successful Zenit launches and test firings of Blok-A and Zenit first stages at NIIkhimmash gradually made the test firing redundant. Energiya was ready to fly [11].

Skif-DM/Polyus

When the idea to launch 6S on a shakedown flight emerged in 1985, the question also arose as to what payload to strap to the side of the rocket. An early suggestion was to fly an empty steel canister (4 m in diameter and 25 m long) which would remain attached to Energiya's core stage and re-enter together with it. This would have required no modifications to the UKSS to service the payload. However, the Ministry of General Machine Building insisted on flying some kind of operational payload on 6SL. In the summer of 1985 the choice fell on an existing design for a nearly 100-ton spacecraft to test laser weapons in space.

The history of this project can be traced back to 1976, when NPO Energiya was tasked to start research on various types of "Star Wars" technology. Not coincidentally, this was around the same time that the Energiya–Buran program was initiated. The research, which essentially was a violation of the 1972 Soviet–American Anti-Ballistic Missile Treaty, covered three broad areas: anti-satellite missions, space-based anti-ballistic missile defense, and destruction of high-priority air, sea, and land-based targets using space-based assets. Early anti-satellite efforts at NPO Energiya focused on two types of Salyut-derived "battle stations", initially to be launched by Proton and later in the cargo bay of Buran. One of these (Skif or "Scythian") (17F19) was to use laser weapons to destroy low-orbiting satellites, the other (Kaskad or "Cascade") (17F111) missiles to destroy satellites in medium and geostationary orbits.

Because of the heavy workload at NPO Energiya, the Skif and Kaskad projects were transferred in 1981 to the Salyut Design Bureau (KB Salyut) in the Moscow suburb of Fili, which during that year had become a branch of NPO Energiya after having split off from the rival Chelomey design bureau. For the laser project KB Salyut came up with an entirely new design—namely, a 40 m long 95-ton object that was to be built at the Khrunichev factory and launched by Energiya. The engine section would be a modified Functional Cargo Block (FGB), the main part of the Transport Supply Ships (TKS) originally designed by KB Salyut to transport crews and cargo to Chelomey's Almaz military space stations, but eventually flown as heavy cargo ships to the Salyut-6 and Salyut-7 space stations.

The ultimate goal was to develop a whole series of Skif battle stations with various types of laser installations. One of these was an infrared laser system called Stilet ("Stiletto"). Developed by NPO Astrofizika, this was intended to knock out the optical systems of enemy satellites, thereby rendering them useless. However, development of these laser systems ran into delays because it was difficult to keep them within the required mass limits. Spurred on by President Ronald Reagan's announcement of the Strategic Defense Initiative (SDI) in March 1983, the Soviet military decided to develop an interim demonstration version (Skif-D) equipped with a much lighter, 1 MWt carbon dioxide gas laser built at the Kurchatov Institute of Atomic Energy that was already undergoing tests on a modified Ilyushin-76MD aircraft. The plan was to fly Skif-D1 with various auxiliary systems but *without* the laser itself and fly Skif-D2 *with* the laser system and use that to knock out small targets deployed from the vehicle itself. The main focus of the Soviet "Star Wars" program by now was not anti-missile defense, but to counter SDI by developing a capability to disable the planned American SDI battle stations. This would deprive the US of its missile shield and enable the Soviet Union to launch a pre-emptive nuclear strike.

When the opportunity arose to mount a payload on Energiya 6SL, Baklanov ordered KB Salyut in July 1985 to build a mock-up version of Skif-D called Skif-DM ("M" standing for *maket* or "dummy"). The initial idea was just to build a mock-up of Skif-D filled with sand or water and keep that attached to Energiya or separate it from the rocket for subsequent re-entry. The next suggestion was to place it into orbit for a week-long mission, requiring the inclusion of an FGB assist module and a set of batteries. Finally, Baklanov insisted on a month-long mission to demonstrate some of the capabilities of Skif-D, with the final order coming on 19 August 1985. The hope was to fly Skif-DM (serial nr. 18201) on Energiya 6SL in September 1986, followed by Skif-D1 (18101) in June 1987 and Skif-D2 (18301) in 1988.

In its final design, Skif-DM was actually quite similar to the Skif-D1 demonstration vehicle, equipped with various auxiliary systems needed to operate a space-based laser, but not carrying the laser itself (contrary to many rumors in the West). The spacecraft was 36.9 m long, had a maximum diameter of 4.1 m and a mass of 77 tons. It consisted of a modified FGB section called FSB (Functional Service Block) and a Payload Module (TsM). The FSB was an available FGB section that had originally been planned to act as a space tug for a now canceled Mir module. It contained all the housekeeping equipment that could not be exposed to the vacuum of space and the engines needed for orbit insertion and attitude control. Mounted on the outside of the FSB were two solar panels. Skif's FSB section was protected during the early stages of launch by a newly developed fiberglass payload shroud, which had to be jettisoned such that it would not hit Skif or the Energiya rocket.

The Payload Module was made up of a Gas Compartment (ORT), an Energy Compartment (OE), and a Special Equipment Compartment (OSA). In the final Skif-D design, the ORT was to house canisters with carbon dioxide to feed the laser, but in order not to arouse suspicion in the West, the canisters on Skif-DM (42 in all) were filled with xenon and krypton instead. The gases would be released into space and their interaction with the ionosphere could then be explained as a geophysical experiment. The OE was intended to carry two 1.2 watt electric turbogenerators, but since these would not be ready in time for the Skif-DM launch, this compartment was virtually empty. It did have a special exhaust system using gas vanes that would make it possible to release the xenon and krypton without imparting momentum to the spacecraft. The OSA did not have the carbon dioxide laser system, but did carry the acquisition, tracking, and pointing mechanisms needed to find targets and keep the laser pointed at them. This included a radar system for rough pointing and a small low-energy laser for fine pointing. The pointing mechanism was supposed to be mounted on a rotatable platform, but this was not ready for Skif-DM either. The data were processed by an Argon-16 computer similar to the one flown on the Mir space station.

In order to calibrate the sensors of the acquisition, tracking, and pointing system, Skif-DM carried 34 small targets (both inflatable balloons and angled reflectors) that would be released from two small modules almost resembling strap-on boosters attached to either side of the OSA. Fourteen of the inflatable balloons would release barium to simulate the exhaust trails from ballistic missiles and spacecraft. Officially, the deployment of the targets would be explained as a test of an experimental approach and docking system and the release of the barium as a geophysical experiment to study the interaction of plasma with the ionosphere.

Skif-DM also carried four technological and six geophysical experiments not directly related to Skif-D. The technological experiments (VP-1, VP-2, VP-3, and VP-11) were aimed at studying techniques for launching and operating large-size spacecraft. One set of geophysical experiments (Mirazh-1, Mirazh-2, and Mirazh-3) was designed to study the interaction of rocket combustion products with the upper atmosphere and ionosphere during launch and deorbit. Another set (GF-1/1, GF-1/2, and GF-1/3) studied the interaction of artificial gas and plasma formations with ionospheric plasma during operation of the FSB engines. Observations of the geophysical experiments were to be conducted from the ground, sea, and air.

Since the Payload Module contained relatively few operating instruments, temperatures inside could drop to unacceptably low levels, which is why the outer surface was painted black to ensure maximum absorption of solar heat. Painted on the side was the name Polyus ("Pole"), which is apparently how the vehicle was supposed to be announced to the world after launch. After the Payload Module's

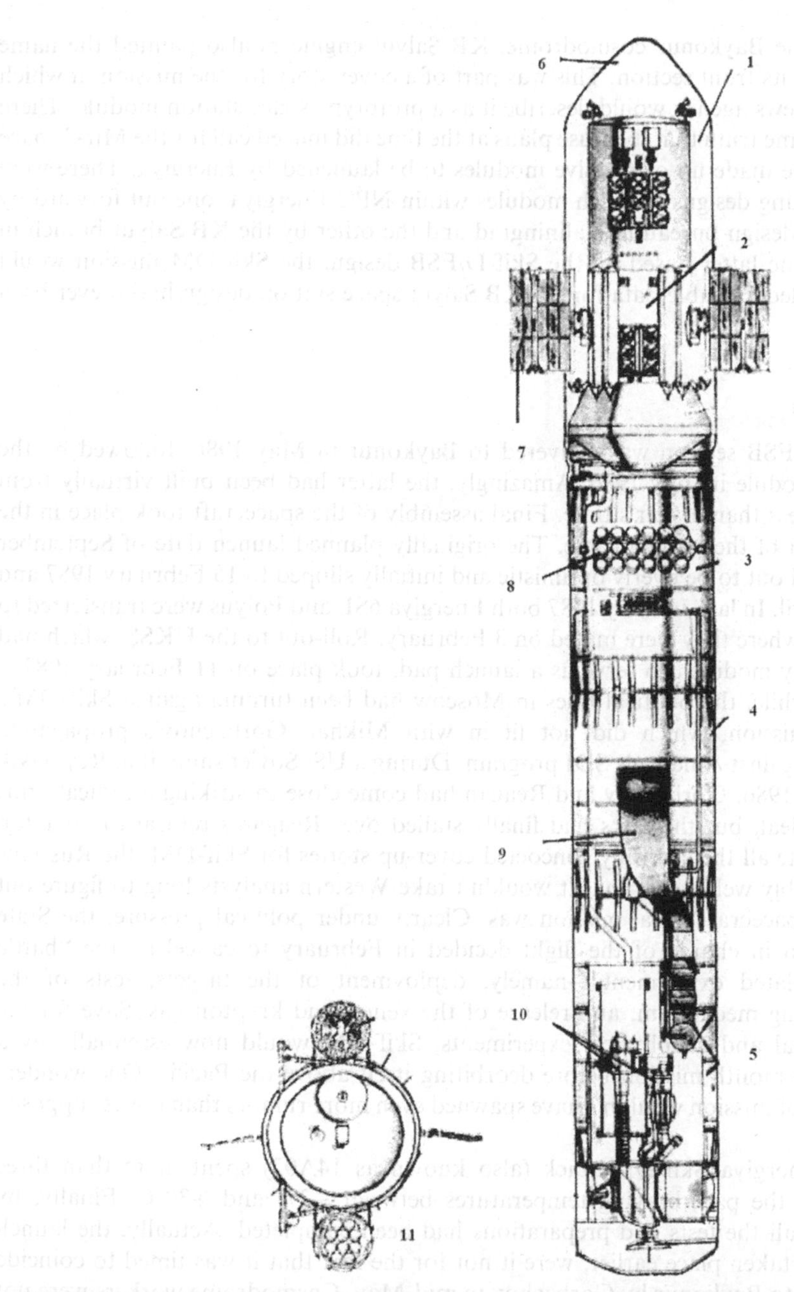

The Skif-DM/Polyus spacecraft: 1, FSB engine section; 2, FSB instrument and payload section; 3, Gas Compartment; 4, Energy Compartment; 5, Special Equipment Compartment; 6, payload shroud; 7, FSB solar panels; 8, gas canisters; 9, momentless exhaust system; 10, acquisition, tracking, and pointing system; 11, bottom view of Skif-DM showing the two modules stowed full with targets (*source*: Zemlya i vselennaya).

arrival at the Baykonur cosmodrome, KB Salyut engineers also painted the name "Mir-2" on its front section. This was part of a cover story for the mission in which the TASS news agency would describe it as a prototype space station module. There was even some truth to it, because plans at the time did indeed call for the Mir-2 space station to be made up of massive modules to be launched by Energiya. There were two competing designs for such modules within NPO Energiya, one put forward by the central design bureau in Kaliningrad and the other by the KB Salyut branch in Fili. With the latter based on the Skif-D/FSB design, the Skif-DM mission would have provided valuable data for the KB Salyut space station design had it ever been selected.

The mission

Skif-DM's FSB section was delivered to Baykonur in May 1986, followed by the Payload Module in July 1986. Amazingly, the latter had been built virtually from scratch in less than a year's time. Final assembly of the spacecraft took place in the Proton area of the cosmodrome. The originally planned launch date of September 1986 turned out to be overly optimistic and initially slipped to 15 February 1987 and later to April. In late January 1987 both Energiya 6SL and Polyus were transferred to the MZK, where they were mated on 3 February. Roll-out to the UKSS, which had been quickly modified to serve as a launch pad, took place on 11 February 1987.

Meanwhile, the political tides in Moscow had been turning against Skif-DM's intended mission, which did not fit in with Mikhail Gorbachov's propaganda campaign against America's SDI program. During a US–Soviet summit in Reykjavik in October 1986, Gorbachov and Reagan had come close to striking a radical arms reduction deal, but the talks had finally stalled over Reagan's refusal to abandon SDI. Despite all the carefully concocted cover-up stories for Skif-DM, the Russians were probably well aware that it wouldn't take Western analysts long to figure out what the spacecraft's real mission was. Clearly under political pressure, the State Commission in charge of the flight decided in February to cancel all the "battle station" related experiments—namely, deployment of the targets, tests of the laser-pointing mechanism, and release of the xenon and krypton gas. Save for the technological and geophysical experiments, Skif-DM would now essentially fly a passive one-month mission before deorbiting itself above the Pacific. One wonders if this type of mission wouldn't have spawned even more rumors than it was supposed to avoid.

The Energiya–Skif-DM stack (also known as 14A02) spent more than three months on the pad, braving temperatures between −27° and +30°C. Finally, by early May all the tests and preparations had been completed. Actually, the launch could have taken place earlier, were it not for the fact that it was timed to coincide with a visit to Baykonur by Gorbachov in mid-May. Cosmodrome workers were not informed of the impending visit, but became suspicious when they were asked to repeat the same checks over and over again. There is conflicting information as to whether Gorbachov was supposed to watch the launch or not. One version has it that he was offered the opportunity to witness the launch, but declined. Other sources say

Skif-DM/Polyus on the UKSS (*source*: *www.buran.ru*).

Gorbachov touring Baykonur facilities in May 1987 (*source*: *www.buran.ru*).

the State Commission in charge of the mission decided not to push its luck and delayed the launch from 12 May to 15 May, by which time Gorbachov would have returned to Moscow. Officially, Gorbachov would be told the launch had been postponed for technical reasons.

Gorbachov arrived at the space center's Yubileynyy runway on 11 May and watched the launch of a Proton rocket with a Gorizont communications satellite later that day. On 12 May he was treated to a tour of the Energiya–Buran facilities. After inspecting the Energiya–Polyus poised for launch on the UKSS, Gorbachov was taken to one of the Energiya–Buran pads, where an Energiya rocket with the OK-MT Buran model strapped to its side had been erected for a series of tests. He also visited the Energiya and Buran assembly buildings.

As Boris Gubanov recalls in his memoirs, Gorbachov made several remarks during the tour that raised serious doubts about his support for the program and left a bitter aftertaste among the space officials accompanying him. He openly questioned whether Buran would have any future applications and on several occasions voiced his opposition to the militarization of space, of which Buran was supposed to be part and parcel. He even expressed skepticism about the readiness of Energiya, although later that day he informed the launch team that the Politburo had given its official approval for the launch. He also approved Glushko's suggestion to officially call the rocket Energiya. Up till then the rocket had received no individual name, with "Buran" being used to refer to the combination of the rocket and the orbiter and "Buran-T" for the combination of the rocket and an unmanned payload canister. On 13 May, Gorbachov was on his way back to Moscow, after having watched the launch of a Zenit rocket with a Tselina-2 electronic intelligence satellite.

In another demonstration of *glasnost*, the TASS news agency issued a statement on Gorbachov's visit later that day, saying:

> "Right now preparations are underway at the cosmodrome to launch a new universal rocket carrier, capable of placing into near-Earth orbits both reusable orbital ships and large-size spacecraft for scientific and economic purposes, including modules for long-term space stations."

This was the first confirmation by TASS of the existence of a new heavy-lift launch system and a Soviet shuttle vehicle. However, there were still limits to openness when it came to the launch of an untried launch vehicle. While Soviet television and radio had begun carrying live coverage of manned Soyuz launches in March 1986, the exact launch date for Energiya was kept secret and TV images of the launch would not be released until after it had taken place.

The launch of Energiya was targeted for 15 May at 15:00 Moscow winter time (12:00 GMT). Fueling of the rocket got underway at 8:30 with the loading of liquid oxygen in the core stage and strap-on boosters. However, problems with the gaseous helium supply to one of the strap-on boosters and also a stuck valve in the core stage's liquid hydrogen tank pushed back the launch 5.5 hours. By that time the sun had set at Baykonur, but, since the payload did not impose any launch window constraints, the launch team decided to press ahead. At $T-10$ minutes the countdown entered the so-called "pre-launch phase", with all operations being controlled automatically. Any hold during this final part of the countdown would automatically lead to a scrub. A major malfunction did occur with less than a minute to go in the countdown, when the sound suppression water system could not be activated. Although this would lead to higher than usual thermal and acoustic loads on the vehicle and the pad, tests conducted at NIIkhimmash had shown that Energiya could safely lift off *without* the sound suppression water. Therefore, computers had been programmed not to stop the countdown in the unlikely event such a failure took place.

With about 9 seconds left in the countdown, Energiya's four RD-0120 engines roared to life. As the huge cryogenic engines built up thrust, the four RD-170 engines of the strap-on rockets were ignited at $T-3$ seconds and, with all engines having reached full thrust, Energiya 6SL leapt off the pad at 21:30 Moscow time (17:30 GMT), lighting up the night sky at Baykonur. Just moments later onlookers saw to their consternation how the rocket significantly leaned over in the direction of Polyus, only to stabilize itself as it cleared the tower. Energiya's automatic stabilization system had been programmed to remain inactive until $T+3$ seconds to ensure that it would not command the engine nozzles to gimbal as they emerged from the Blok-Ya launch table adapter. Therefore, the deviation from the trajectory had been more or less expected, but not everyone watching the launch was aware of this and for a few hair-raising moments it seemed as if Energiya would befall the same fate as its

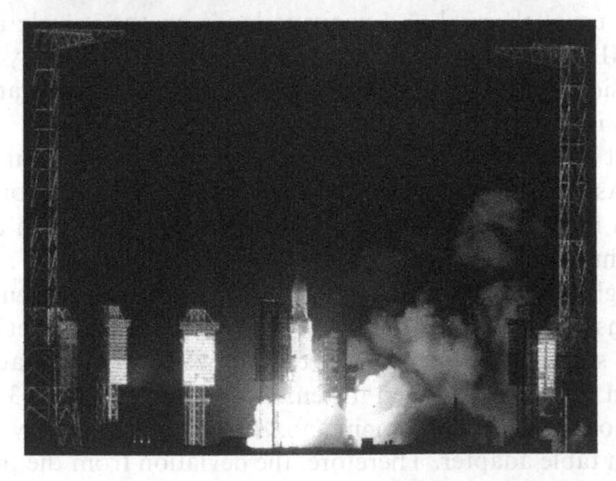

Energiya 6SL lifts off. Note vehicle leaning over to the left in the middle frame (*source*: *www.buran.ru*).

illustrious predecessor, the N-1. For the next launch the stabilization system was programmed to kick in earlier to prevent a repeat of this scenario.

Having cleared the tower, Energiya initiated a roll and pitch maneuver to place itself on the proper azimuth for a 64.6° inclination orbit. There had been some debate in the months prior to launch whether to put Polyus into a 50.7° or 64.6° orbit. A launch resulting in the lower inclination would have allowed Energiya to be about 5 tons heavier, but at the same time would briefly carry the rocket over the territories of Mongolia, China, and Japan. An argument against a 64.6° inclination orbit was that launches were not possible between mid-May and August because the strap-on boosters and Polyus' payload shroud would impact in the nesting grounds of the pink flamingo, a protected species that has its breeding season during that time of the year. In the end, concerns about Energiya debris raining down on foreign territory seem to have outweighed any environmental arguments.

Managers breathed a sigh of relief at $T + 30$ seconds, by which time Energiya 6SL had moved far enough downrange to prevent damage to the UKSS in case of an explosion. Also, the objectives of the originally planned combined static test firing of the core stage and strap-on boosters had now been achieved. Any success beyond that was a bonus as far as managers were concerned. To their delight, Energiya continued to perform outstandingly. The four strap-ons were separated from the core stage at $T + 2$m26s, and at $T + 3$m34s Polyus shed the shroud that had protected its upper FSB section against the aerodynamic pressures experienced during the early stages of launch. At $T + 7$m39s the RD-0120 engines shut down, followed moments later by the separation of Polyus from the core stage, which according to unconfirmed reports came down relatively intact in the Pacific Ocean [12]. The Energiya control room at Baykonur erupted into applause, but while Energiya had completed its job, Polyus still had some critical maneuvers to do.

Not having reached orbital velocity yet, Polyus now was to perform two burns of its FSB main engines to place itself into a circular 280 km orbit. With the FSB section placed on top (as during a TKS launch on Proton), Polyus first had to carry out a 180° flip maneuver around its z-axis so that the engines would face aft for the burns, followed by a 90° roll around its x-axis. Unfortunately, due to a programming error, the FSB's thrusters failed to stop the flip maneuver and by the time the main engines were ignited Polyus was not oriented properly and as a result deorbited itself. Later analysis showed that the thrusters had been deactivated by a command usually issued during a TKS launch that somehow had not been erased for the Polyus launch.

The official TASS launch statement released the following day acknowledged the failure to place the payload into orbit:

> "The second stage of the rocket delivered a satellite mock-up to the required point, but due to a malfunction of its on-board systems the mock-up did not go into the planned orbit and splashed down in the Pacific."

The day after the launch Soviet television viewers were treated to spectacular shots of the super-booster on its launch pad. One of the television shots offered a side view of the payload, revealing it to be a black pencil-shaped object. However, photographs released subsequently only showed the aft part of Energiya, hiding Polyus from view. It was not until many years later that detailed photographs and descriptions of the payload became available.

The loss of Polyus was a bitter pill to swallow, especially for the designers and engineers of KB Salyut and Khrunichev, who had managed to get the improvised payload ready for launch in such a short period of time. Another setback was the significant damage to the UKSS launch pad, caused by the problem with the sound suppression water system and the rocket's deviation from its trajectory shortly after lift-off. The Blok-Ya launch table adapter, designed to support at least 10 launches, was rendered useless because thermal protection covers had been either torn loose or closed too late.

Still, the primary goal of the launch had been to test Energiya and demonstrate its capability to carry a heavy payload and both these objectives had been accomplished. Particularly useful for the subsequent Buran mission was the use of the same systems needed to separate the vehicle from the core stage. Moreover, all the four technological experiments and even some of the geophysical experiments planned for launch and the post-separation phase were actually carried out successfully. Energiya had performed better than anyone could have hoped and was declared ready to carry Buran on its next mission [13].

BURAN PROPULSION SYSTEM AND VSU TESTING

Testing of Buran's ODU propulsion system was the prime responsibility of the so-called Primorskiy Branch of NPO Energiya in the Leningrad region on the shores of the Gulf of Finland. This was set up in 1958 as a branch of Glushko's OKB-456, mainly to test engines with exotic propellants such as the RD-301 fluorine/ammonia engine destined for a Proton upper stage. When OKB-456's successor KB Energomash merged with TsKBEM in 1974 to form NPO Energiya, the Primorskiy Branch became part of the new conglomerate and remained subordinate to it even after Energomash regained its independence in 1990. Its first assignment as part of NPO Energiya was to test the RD-120 engine for the second stage of Zenit. The old RD-301 test stand was refurbished for a series of horizontal test firings of 11D58M engines for the Proton rocket's Blok-D upper stage in 1978–1982, which were probably seen as precursors to similar tests with the Orbital Maneuvering Engines (DOM or 17D12) for Buran. Between May 1985 and September 1988 six 17D12 engines underwent 114 horizontal test firings lasting a total of 22,311 seconds.

Meanwhile, in 1981 construction had begun of a new vertical test stand called V-1 to test complete ODU engine units called EU-597, containing not just the 17D12 engines, but also thrusters and verniers. The first such ODU unit (nr. 10S) began testing in June 1986 but was destroyed in a fire in February 1987, seriously damaging the test stand. V-1 was refurbished for a series of tests with a new unit (nr. 12S) between September 1987 and April 1988 that underwent the complete ODU firing program planned for the first Buran mission. Those tests uncovered a problem that would delay the Buran flight for several months (see Chapter 7). More tests were conducted with unit nr. 31L between June and December 1988 and unit nr. 11S between January 1991 and March 1993. After cancellation of the Energiya–Buran program the unit was mothballed and eventually removed from the test stand. The 17D15 thrusters and 17D16 verniers apparently also underwent individual tests at NIIkhimmash near Zagorsk. Test firings of the ODU integrated in Buran were conducted at Baykonur's test-firing platform [14].

The Auxiliary Power Units (VSU) underwent a test program at the IS-104 and IS-105 test stands of NIIkhimmash, which included simulated hydrazine leaks to test the fire suppression system. The VSU hydrazine tank was put to the test in simulated weightless conditions aboard an Ilyushin-76 aircraft and also at various *g*-levels at Star City's TsF-18 centrifuge. The VSU test program culminated in the units being installed on Buran and activated at the Buran test-firing stand at Baykonur.

FULL-SCALE AND CREW CABIN MODELS

Rather than rely heavily on computer modeling, the Russians built at least seven full-scale test articles of Buran to investigate a variety of manufacturing, assembly, and flying quality characteristics as well as handling procedures. A similar approach was followed for space stations and their modules. By contrast, NASA built just two full-scale Orbiters for test purposes—namely, Enterprise (OV-101) and Structural Test Article 099 (later turned into OV-099 Challenger).

The seven full-scale vehicles were:

- OK-M (serial nr. 001): a full-scale model for structural tests at NPO Molniya. It mainly served as a test bed for the 002 vehicle used in the approach and landing tests. It had the same mass characteristics as the real vehicle, carrying mass models of on-board equipment. Later, it was supposed to be used for underwater EVA training in a hydrotank facility at Star City, but it was rebuilt as a tourist attraction and delivered by barge to Gorkiy Park in Moscow in 1993, where it can still be seen today.
- OK-GLI (serial nr. 002): a full-scale model used for approach and landing tests in 1985–1988 (see pp. 297–309).
- OK-KS (serial nr. 003): a full-scale model for electric and software tests, delivered to NPO Energiya in August 1983. Also used for electromagnetic interference tests. OK-KS served as a test bed to troubleshoot numerous problems that cropped up during the construction of the first flight vehicle. Various software programs for the maiden flight were tested on OK-KS.
- OK-ML1 (serial nr. 004): a full-scale model flown to Baykonur by the VM-T Atlant in December 1983 for preliminary fit checks of ground equipment in the Buran assembly building and on the runway. On one occasion it was mated with an Energiya for dynamic tests both at the UKSS and the left Energiya–Buran pad. For a while it had mock-up turbojet engines installed on either side of the vertical stabilizer.

OK-KS at NPO Energiya (*source*: *www.buran.ru*).

- OK-TVA (serial number 005): a model for thermal, acoustic, and static vibration tests at TsAGI. To facilitate testing, OK-TVA was not assembled as a single vehicle, but split into several real-size sections that could be tested individually: a forward fuselage with crew cabin, mid and aft fuselage, two wings, a vertical stabilizer, elevons, a body flap, a nosecap, and several sections of the leading edges of the wings. The components were covered with standard thermal protection material, among other things to see whether that would be affected by slight deformations in the underlying aluminum skin.

OK-ML1 and OK-MT in tandem at Baykonur (*source*: *www.buran.ru*).

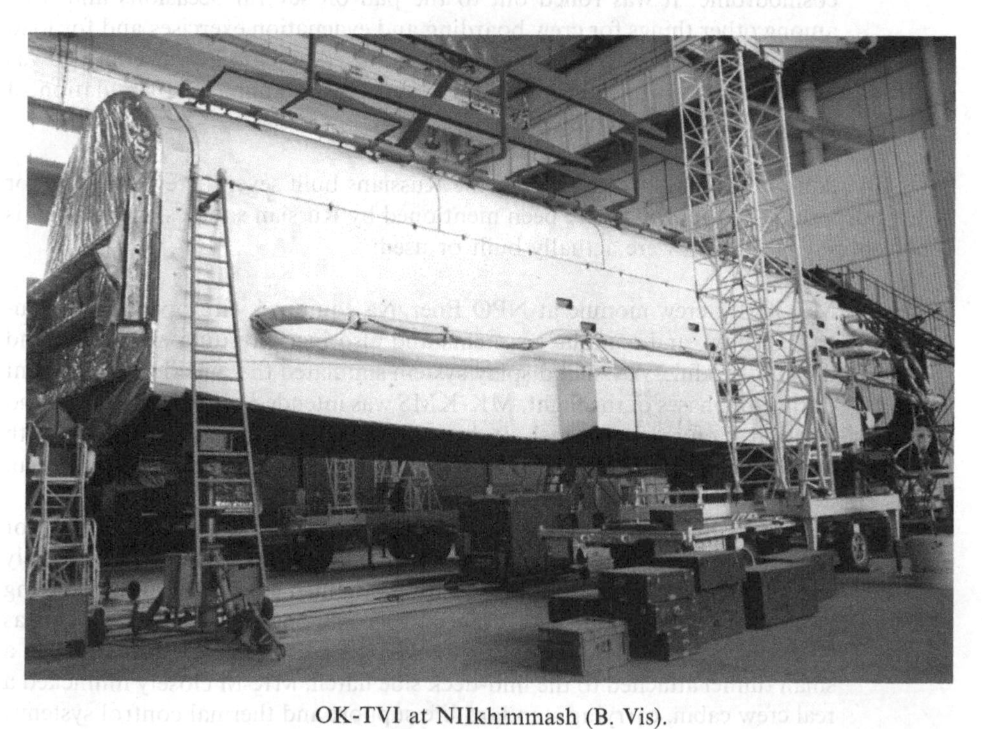

OK-TVI at NIIkhimmash (B. Vis).

Thermal and static vibration tests took place in the TPVK-1 vacuum chamber. It was 13.5 m in diameter and 30 m long and exposed components to temperatures ranging from −150°C (using a liquid nitrogen cooling system) to +1,500°C (using 10,000 quartz lamps with a total capacity of 13,000 kWt). The test rig could apply 8,000 kN of force horizontally and 2,000 kN vertically and took the airframe to 90 percent of design load limits. Acoustic tests were carried out in the RK-1500 acoustic chamber. With a floor space of 1,500 m^2, it was equipped with 16 sound generators that subjected the components to 162 dB sound levels at frequencies of 50 to 2,000 Hz.

- OK-TVI (serial nr. 006): a model for thermal vacuum tests at NIIkhimmash. This consisted of a mid and aft fuselage, a vertical stabilizer, and payload bay doors with radiator panels. Some sources also mention a forward fuselage with crew cabin, although that is not seen in photographs and must therefore have been tested individually. The fuselage sections were equipped with Buran's thermal control system to see whether that could deal with the temperature extremes in space. The components were installed in the KVI thermal vacuum chamber. With a volume of 8,500 m^3, this is the largest such facility in Europe.
- OK-MT/OK-ML2 (serial nr. 015): a full-scale model flown to Baykonur by the VM-T Atlant in August 1984 for fit checks of ground equipment at the cosmodrome. It was rolled out to the pad on several occasions and used among other things for crew boarding and evacuation exercises and for load tests of the ODU engine compartment and the Auxiliary Power Units. It was also transported to the runway for crew egress training and simulation of other post-landing activities.

In addition to the full-scale models, the Russians built several crew modules for test purposes. The following have been mentioned by Russian sources, although it is not entirely clear if all were actually built or used:

- MK-KMS: crew module at NPO Energiya equipped with operational control, display, and computer systems and also incorporating an airlock and docking module. A visual display system simulated the outside environment during all phases of the flight. MK-KMS was intended for training crews and Mission Control personnel. It had the same communication links with Mission Control in Kaliningrad as the ones available to Buran during an actual flight.
- MK-M: a Buran crew module placed in a vacuum chamber (VU-1000) for tests of the life support systems and medical support systems (presumably located at Myasishchev's EMZ). It also contained an airlock and a docking module. The crew module was placed vertically in the chamber, which was 10 m wide and 11 m high. Crews entered and egressed the crew module via a small tunnel attached to the mid-deck side hatch. MK-M closely mimicked a real crew cabin, carrying standard life support and thermal control systems.

Any of those systems located outside the crew cabin were also installed in the chamber in roughly the same position with respect to the crew cabin as in a real orbiter.

The cabin carried mock-ups of equipment not related to the life support system. MK-M allowed crew members to wear Strizh pressure suits that could be immediately pressurized in case of a leak. The water delivered to the crew compartment was produced in actual Buran fuel cells. It is not clear if the test stand was ever used for crew training. The plan was eventually to turn it into a so-called ground-based "analog" of vehicles in orbit, among other things to facilitate troubleshooting activities.

- MK-1KA: a crew module with nose section mounted vertically on a turntable to practice crew evacuation from the vehicle.
- MK-KB: a crew module containing mock-ups of equipment needed for the 002 vehicle.
- MK-GN: a crew module placed in a hydrolab for EVA training.
- MK-KB.E: a crew module for electrical tests, later integrated into the OK-KS vehicle.
- MK-KB.U: a crew module to study the placement of equipment and crew work stations in the cabin [15].

THE BOR-4 TEST VEHICLE

Heat shield testing went much further than the experiments with the OK-TVA and OK-TVI models at TsAGI and NIIkhimmash. Smaller pieces of thermal insulation were tested in plasma generators at TsNIIMash in Kaliningrad, the Institute of Mechanical Problems in Moscow (IPMekh), and other test stands at NPO Molniya in Tushino and the Siberian Scientific Research Institute of Aviation (SibNIA) in Novosibirsk. Besides that, Ilyushin-18D and MiG-25 aircraft were used to test tiles and felt reusable surface insulation at subsonic and supersonic speeds. The thermal insulation was installed on areas of the aircraft that were subjected to the highest dynamic pressure and acoustic loads from the engines. NASA similarly tested Shuttle tiles on F-15 Eagle and F-104 Starfighter jets.

Since the US Space Shuttle's Thermal Protection System was very similar to that of Buran, the Russians probably watched the first Shuttle missions with more than casual interest. In a 1984 National Intelligence Estimate on the potential for the transfer of US space technology to the Soviet Union, the CIA concluded that the Soviets had benefited considerably from surface heating data from the STS-2 and STS-3 missions publicly released by NASA in June 1982. The report quoted NASA officials as estimating that the data could save the Soviets the equivalent of $750 million in R&D cost and considerably reduce development time [16].

Whether that was true or not, it didn't stop the Russians from pursuing their own test program. Unlike NASA, the Russians had the unique opportunity to test Buran's heat-resistant materials during actual re-entries from orbit using scale models of the canceled Spiral spaceplane. In the late 1960s and early 1970s they had already flown scale models known as BOR (*Bespilotnyy orbitalnyy raketoplan* or "Unmanned Orbital Rocket Plane") on suborbital trajectories. These were BOR-1 (a wooden mock-up), BOR-2 (a 1 : 3 scale model), and BOR-3 (a 1 : 2 scale model) with ablative heat shields. In 1975 plans were completed at the Flight Research Institute (LII) for an orbital test bed known as BOR-4, which was a 1 : 2 scale model of Spiral. Even though the future of the Spiral program was very much in limbo by this time, BOR-4 test flights would also have been applicable to the giant lifting body proposed by NPO Molniya in early 1976. When that was dropped in favor of the delta-wing Buran, it looked as if BOR-4 would remain on the drawing boards forever.

However, the Russians soon realized that in order to test the heat shield they did not necessarily need a vehicle that exactly copied Buran's outlines. The most important thing was to ensure that the heat-resistant materials would be exposed to the same type of temperatures for about the same period of time. Moreover, the nose section of the BOR-4 test vehicle *did* more or less match the contours of Buran's nose section. Therefore, it was decided in 1977 to develop two types of scale models in support of the Buran program:

– BOR-4, a 1 : 2 scale model of the Spiral spaceplane, to test Buran's heat shield materials.
– BOR-5, a 1 : 8 scale model of Buran, to test its aerodynamic characteristics.

Design

Just like the Spiral spaceplane, BOR-4 was a flat-bottomed lifting body with a vertical fin and foldable wings. It was 3.859 m long with a launch mass of about 1,450 kg

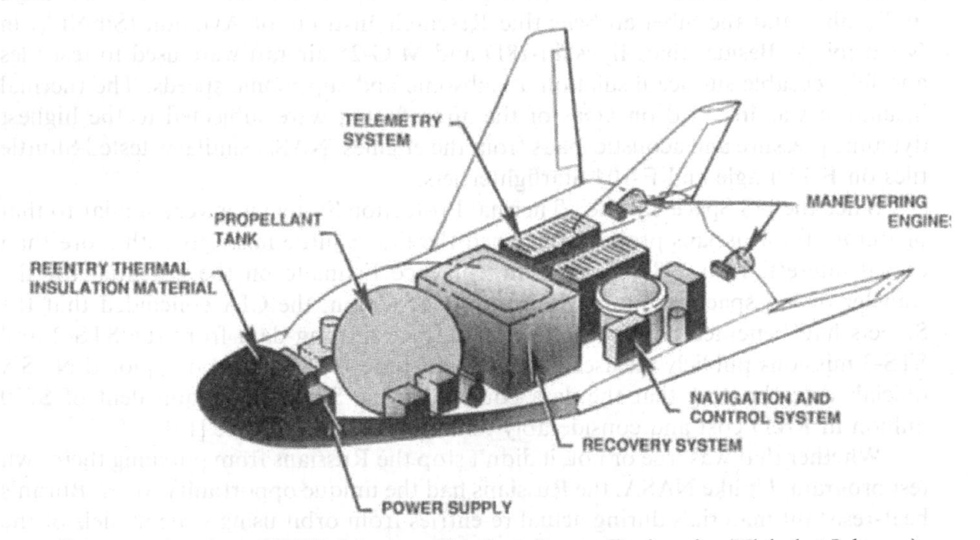

Cut-away drawing of BOR-4 (*source*: Teledyne Brown Engineering/Nicholas Johnson).

and a landing mass of 795 kg. The BOR-4 vehicles made it possible to test the three main types of thermal insulation used on Buran—namely, tiles, felt reusable surface insulation, and reinforced carbon–carbon. Black tiles (using both the TZMK-10 and TZMK-25 substrate) covered the belly, white tiles (with TZMK-10 substrate) were installed on the sides, and ATM-19 felt insulation protected the upper part of the vehicle. Carbon–carbon GRAVIMOL material was used only on the nosecap, since the wing leading edges were too thin for installation of such material.

The tiles were not applied directly to the BOR's airframe, but to a thin layer of aluminum of the same composition as that used in Buran's airframe. In between this aluminum layer and the actual airframe was an ablative heat shield material (PKT-FL) that had been planned for the original BOR-4 vehicle to be flown in the framework of the Spiral program. This provided the necessary redundancy in case any of the Buran heat shield material burnt through during re-entry. The area between the nosecap and the airframe was filled with insulating material made of heat-resistant fibers. Since the wings were much thinner than the rest of the airframe, they were filled with a porous felt material impregnated with a water-based substance. Evaporation of that substance provided enough cooling for the wing during re-entry in case the Buran thermal protection material proved ineffective.

BOR-4 was equipped with 150 thermocouples, installed mainly on the airframe and just under the coating of some of the tiles. In addition to that it had accelerometers, angular velocity sensors, pressure sensors, and sensors that indicated the position of the wings. Information obtained from the sensors was recorded on board and sent back to earth in "packages" to tracking ships and also to a ground station during re-entry.

Missions

A typical BOR-4 mission would begin with a launch from the Kapustin Yar cosmodrome near Volgograd using a modified two-stage Kosmos-3M booster known as K65M-RB5. Baykonur no longer supported that rocket at the time and although Plesetsk did it was situated too far north to place the spaceplane models into the proper inclination. The rockets used for the BOR-4 missions had originally been earmarked for other missions, but had already exceeded their guaranteed "shelf life" and would have been used for test launches anyway.

The vehicle would be launched with its two wings completely folded so that it fitted under the rocket's fairing. After release from the launch vehicle, the wings were unfolded to a position that would keep the vehicle stable during re-entry at an angle of attack of between 52° and 57° between altitudes of 70 and 60 km. Orientation in orbit was carried out with the help of eight microthrusters burning hypergolic propellants. After a single revolution of the Earth, BOR-4 initiated its descent back to Earth, firing what is believed to have been a jettisonable solid-fuel motor mounted on top of the vehicle. At an altitude of 30 km the on-board control system sent BOR-4 on a steep spiralling trajectory to decrease speed and at 7.5 km the spacecraft deployed a parachute that reduced the vertical landing speed to 7–8 m/s.

Since the vehicle was not equipped with landing gear, it needed to land on water to ensure that its heat shield remained intact for post-flight analysis. The only major bodies of water on Soviet territory that would be in the BOR's flight path were the Black Sea and Lake Balkhash. However, the Russians had never returned a winged vehicle or lifting body from orbit and were not confident they could aim the spacecraft for precision splashdowns in the Soviet Union. Therefore they opted to land the first vehicles in the Indian Ocean, where they would still come down in water even if they fell short of or overshot the planned landing area. That did, however, significantly increase the cost of the recovery operations, which, moreover, would be hard to conceal from the prying eyes of Western reconnaissance aircraft.

After splashdown a conically shaped float was inflated on top of the spacecraft to improve its buoyancy. The float also had flashing lights and antennae to make it easier for the recovery forces to locate the BOR. Before being hoisted on board a recovery ship, a crew was sent out to the vehicle to disarm an on-board self-destruct system.

While BOR-4 was designed in Zhukovskiy under the leadership of LII chief Viktor V. Utkin, the vehicle was manufactured and covered with heat-resistant materials at the Tushino Machine Building Factory. The man in charge of the BOR-4 program at NPO Molniya was Stepan A. Mikoyan, a deputy of chief designer Gleb Lozino-Lozinskiy. The BOR-4 test flights were coordinated by a State Commission headed by former cosmonaut Gherman Titov, then serving as a deputy head of GUKOS. Titov, incidentally, had also been part of the Spiral cosmonaut training group at Star City in the 1960s.

The orbital flights were preceded by an experimental suborbital launch on 5 December 1980 in the direction of Lake Balkhash. This mission was intended to test the rocket, the aerodynamic characteristics of the vehicle, and the performance of the aerodynamic surfaces and rocket thrusters. Designated BOR-4S (serial nr. 401), the vehicle only had the original ablative heat shield. The final part of the flight was monitored by two Ilyushin 18RT aircraft flying in the vicinity of Lake Balkhash. These were modified Ilyushin 18D aircraft specially adapted to perform tracking in areas that were not covered by Soviet ground-based or sea-based tracking means.

In the spring of 1982 seven Soviet ships set sail for the Indian Ocean to support the first orbital mission of a BOR-4 vehicle. These included two vessels to ensure communications between the fleet and the home front, namely the Navy's *Chumikan* and the Academy of Sciences' *Kosmonavt Georgiy Dobrovolskiy*. In no time Royal Australian Air Force P-3C Orion reconnaissance aircraft deployed from RAAF Base Williams in Point Cook were circling overhead to monitor the vessels' activities.

Finally, on 3 June 1982 the first BOR-4 covered with Buran's heat shield materials (serial nr. 404) was successfully placed into orbit. After a single orbit the vehicle fired its deorbit engine and re-entered the Earth's atmosphere, performing a cross-range maneuver that took it about 600 km to the south of its orbital path. The craft gently splashed down some 560 km south of the Cocos Islands, which was about 200 km from its intended landing point. The recovery operation was seriously hampered by stormy seas. Battling high waves, recovery forces on board the vessel *Yamal* needed several attempts to hoist the BOR-4 on deck. During one of those

Kosmos-1374 being hoisted aboard the *Yamal* (*source*: Royal Australian Air Force).

attempts, the vehicle accidentally bumped into the *Yamal*, causing significant damage to the spacecraft's nose section. The entire operation was photographed by an Australian Orion aircraft, which according to Russian eyewitnesses flew so low that the slipstream nearly knocked them off their feet.

The TASS news agency issued a routine statement saying a satellite called Kosmos-1374 had been launched for "the further study of outer space", providing no hint of its real mission. It only differed from the standard Kosmos launch announcement by adding that "the scientific research envisaged by the program had been carried out." Within a week US media reports were suggesting the mission had been a test of a small shuttle vehicle, although some argued it had been a test of a prototype spaceborne nuclear weapon targeted on US and British naval forces in the Indian Ocean.

State Commission leader Gherman Titov, who had already pushed for a Black Sea landing on the *first* mission, now turned to the Military Industrial Commission with a request to have the *next* BOR-4 land in the Black Sea, expressing his fear the ship could be captured by the Americans. However, he was overruled by his superiors, possibly because Kosmos-1374 had landed well off target and a splash-down in the much smaller Black Sea could not yet be guaranteed. On 15 March 1983, Kosmos-1445 (serial nr. 403) was launched on a repeat mission, coming down 556 km south of the Cocos Islands. On station in the Indian Ocean apart from Navy vessels of the Black Sea fleet were the tracking ships *Kosmonavt Vladislav Volkov* and *Kosmonavt Pavel Belyayev*. Two Il-18RT tracking aircraft were in the skies over

Kosmos-1445 on board the *Yamal* (*source*: Royal Australian Air Force).

Afghanistan to monitor the final part of the re-entry. Coming in the middle of the Soviet–Afghan war, their missions were not without risk and they were protected by a whole squadron of Soviet fighter jets.

Kosmos-1445 was again retrieved by the *Yamal*. Better prepared than during the Kosmos-1374 mission, the Australian Air Force once again sent out P-3C Orion aircraft to monitor the recovery operation and obtained even better pictures than before, some of which were released to the public by the Australian Ministry of Defense in April 1983. Also keeping a close eye on events were several Australian Navy vessels, which reportedly came so close that the Soviet crew members could use their binoculars to catch a glimpse of the movies shown on giant screens on the upper decks in the evening.

Confident enough now they could bring back the BORs with sufficient precision, the Russians decided to land the next two BORs in the Black Sea just west of Simferopol. The first was launched on 27 December 1983 as Kosmos-1517 (serial nr. 405) and the second was orbited on 19 December 1984 as Kosmos-1614 (serial nr. 406). The TASS launch announcements differed from the earlier ones in acknowledging that the satellites "had performed a controlled entry into the atmosphere and

Kosmos-1517 shows the effects of re-entry (*source*: *www.buran.ru*).

landed in the pre-designated area of the Black Sea." While Kosmos-1517 was successfully retrieved by the *Yamal*, it was later revealed that Kosmos-1614 was lost, having either burned up in the atmosphere or sunk in the Black Sea. The recovery vessels, aircraft, and helicopters searched the 70 × 30 km landing ellipse for about a week, but to no avail. Talking about the cause of the mishap many years later, State Commission leader Gherman Titov said that "while fixing one problem, engineers had created another."

Despite the failure to recover the final vehicle, it was felt that enough data had been gathered during the four orbital flights that a fifth mission was reportedly canceled. The Russians later said the missions had allowed them to test the effects of aerodynamic, temperature, and acoustic loads as well as vibrations on the heat shield between altitudes of 100 and 30 km and speeds of between Mach 25 and Mach 3. Particularly helpful had been the temperature data obtained in critical areas such as the nosecap and the underbelly of the vehicle. The BOR-4 missions had helped to determine the ideal size of gaps between the tiles, measure the "catalytic activity" of the heat shield in real plasma conditions, and also to study the risks associated with losing one or more tiles. The flights had also made it possible to "outline measures to reduce the mass of Buran's heatshield", although there is no evidence those measures were actually implemented [17].

The legacy of BOR-4

For most of the 1980s the BOR-4 vehicles were widely interpreted in the West as subscale models of a military spaceplane to be launched by the Zenit rocket, a program that was believed to run parallel to the Energiya–Buran effort (see Chapter 7). But, even as BOR-4 led a life of its own in the imagination of Western analysts, the Russians were considering using the spacecraft for other missions. One vehicle, possibly the one that had originally been supposed to fly the fifth BOR-4 orbital flight, was modified for an experiment to evaluate radio transmissions during atmospheric re-entry from a suborbital mission. Dubbed BOR-6, it was equipped with two large antennas extending out and downward from the nose. Using a special cooling system, these antennas were designed to see if radio signals could penetrate the plasma sheath that envelops spacecraft during re-entry and causes radio blackouts. Construction of the spacecraft was finished by 1990, but it was never launched due to the collapse of the Soviet Union and the ensuing shutdown of the Buran program. The Russians also looked at the possibility of converting BOR-4 type vehicles into space-to-ground weapons as part of a Soviet "Star Wars" program (see Chapter 8).

With the political climate changing and the Russians scrambling to find new customers for their space technology, BOR-4 was offered on a commercial basis to the international community in the early 1990s. The European Space Agency weighed the possibility of using BOR-4 vehicles to test the heat shield of Europe's Hermes spaceplane, but these plans never materialized.

Interestingly enough, as early as 1983 the BOR-4 recovery images inspired engineers at NASA's Langley Research Center to clandestinely build small models of the subscale spacecraft. Over subsequent years they analysed and improved the design in over 1,200 wind tunnel and computer tests to refine the shape of the outer mold line. This resulted in plans for a 10-ton spaceplane called HL-20, a lifting body closely resembling BOR-4 and considered in the early 1990s as a crew transportation system and crew rescue vehicle for the Freedom space station. Also known as the Personnel Launch System (PLS), it would be launched by an expendable rocket such as the Titan-4 and be capable of carrying a crew of 10. Although a full-scale mock-up of the HL-20 was built, the design was not selected for further development as the Russian Soyuz spacecraft was picked as the lifeboat for Freedom and eventually the International Space Station. Later in the 1990s Langley proposed a 42 percent dimensional scale-up of the HL-20 called the HL-42, but this seems to have been a short-lived effort.

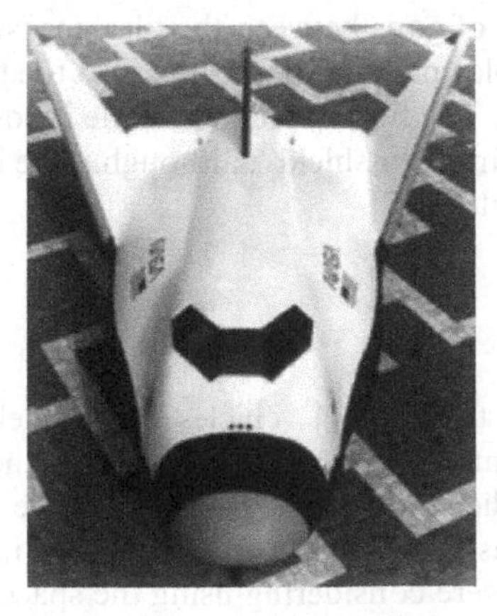

The HL-20 (*source*: NASA).

The BOR-4/HL-20 design was once again picked up by the Orbital Sciences Corporation in the late 1990s for a "Space Taxi" proposed initially under NASA's Space Launch Initiative and later as a candidate for the Orbital Space Plane that would complement the Shuttle by carrying crews to and from orbit, but the project was canceled after the February 2003 Columbia accident.

In January 2006 NASA announced a program called Commercial Orbital Transportation Services (COTS) in which two industry partners would receive a combined total of approximately $500 million to help fund the development of a reliable, cost-effective commercial transportation system to support the International Space Station after the retirement of the Space Shuttle in 2010. One of the vehicles studied was SpaceDev's Dream Chaser. Having the same size and outer mold line as BOR-4 and the HL-20, it would fly six rather than ten passengers in order to save weight. Although Dream Chaser was eventually not selected, SpaceDev founder Jim Benson created a new company called the Benson Space Company that intends to purchase multiple Dream Chaser vehicles from SpaceDev to become the first-to-market with a spaceship designed for both suborbital and eventually orbital flights. Benson hopes it will also be used to transport people and cargo to the International Space Station and to a variety of emerging private-sector orbital destinations [18].

THE BOR-5 TEST VEHICLE

The aerodynamic behavior of Buran was studied using 85 different scale models (ranging from 1:3 to 1:550) in 25 wind tunnels simulating Mach 0.1 to 2.0. These wind tunnels were situated at TsAGI in Zhukovskiy, and at SibNIA and the Institute of Theoretical and Applied Mechanics (ITPM), both in Novosibirsk. A total of 36,630 wind tunnel tests were conducted prior to the maiden flight of Buran.

The Russians also developed 1:8 scale models of Buran called BOR-5 to study the vehicle's behavior at re-entry speeds. Unlike BOR-4, they simulated the shape of Buran itself and were launched on suborbital trajectories. The purpose of these flights was:

- to determine major aerodynamic characteristics in real flight conditions at high velocities;
- to determine aerodynamic coefficients, the lift-to-drag ratio, balancing characteristics, roll and pitch stability and to compare them with calculated characteristics;
- to investigate pressure distribution along the vehicle's surface;
- to determine heat and acoustic loads;
- to check the adequacy of the techniques used to calculate aerodynamic characteristics.

The BOR-5 models weighed 1,450 kg and were 3.856 m long. Because of their small size and the specifics of their trajectory, they were exposed to much higher temperatures than Buran and therefore were covered with an ablative heat shield rather than

Wind tunnel model of Buran (*source*: *www.buran.ru*).

tiles. The nosecap was made of a tungsten–molybdenum alloy. Just like the BOR-4 vehicles, they were equipped with a wide range of sensors to measure temperatures, aerodynamic characteristics, and orientation. The data obtained by these sensors were sent back in real time via telemetry.

The BOR-5 models were launched from Kapustin Yar by the Kosmos-3M-RB5 rocket and launched in the direction of Lake Balkhash, covering a distance of about 2,000 km. Having reached a maximum altitude of 210 km, the second stage of the Kosmos booster pitched down to accelerate the model to Mach 18.5 at 45 degrees before separation. After separation from the second stage, the model used small gas thrusters for orientation, switching to aerodynamic surfaces as it entered the denser layers of the atmosphere. Beginning at an altitude of 50 km, it followed the same changes in bank angle and angle of attack as Buran, albeit at much higher speeds than the full-scale orbiter. At an altitude of 7 km a parachute was deployed to reduce the vertical landing speed to 7–8 m/s.

The BOR-5 vehicles were built at NPO Molniya's EMZ factory with the assistance of specialists from other divisions of NPO Molniya and also from the Flight Research Institute. Like the BOR-4 missions, the test flights were supervised by a State Commission headed by Gherman Titov.

The first BOR-5 (serial nr. 501) was launched on 6 July 1984, but was lost when it failed to separate from the second stage due to an electric fault. The first successful mission took place with vehicle nr. 502 on 17 April 1985. Post-flight analysis did

Final BOR-5 vehicle in Florida (*source*: Rudolf van Beest).

reveal significant damage to the nosecap and leading edges of the wings, which altered the vehicle's aerodynamic characteristics. Therefore, on subsequent missions those areas were protected with a special molybdenum alloy and a special anti-oxidation coating. Three more successful missions (using models nr. 503, 504, 505) were conducted on 27 December 1986, 27 August 1987, and 22 June 1988. Models 501 to 504 were outfitted with small mock-up turbojet engines on either side of the vertical stabilizer, but these were no longer mounted on the final BOR-5 vehicle because by then it had been decided to fly Buran without turbojet engines.

Vehicle 505 was unsuccessfully put up for auction in the United States in 1991 and was stored in the Mojave Desert for about four years before being put on display at the Santa Barbara Museum of Flight. In 1997 it was purchased by a person in Merritt Island, Florida, who still owns the vehicle [19].

HORIZONTAL FLIGHT TESTS

Like NASA in 1977, Buran program managers considered it necessary to conduct an approach and landing test program to investigate the performance of the orbiter during the final atmospheric portion of the mission. Key objectives were to check the ability of the cosmonauts to fly Buran to a controlled landing and to demonstrate the possibility of conducting automatic landings. The Soviets referred to these tests as "Horizontal Flight Tests" (*Gorizontalnye Lyotnye Ispytaniya* or GLI).

Although the objectives of the program closely matched those of NASA's Approach and Landing Tests (ALT) with Space Shuttle Enterprise, the Russians faced one huge obstacle. They lacked an airplane that was big enough to carry an orbiter piggyback for drop tests. NASA had used a Boeing-747 carrier plane to bring the Enterprise to the desired altitude, after which it was released so it could glide to a landing at Edwards Air Force Base in California. However, at the time the Russians didn't have the Antonov An-225 Mriya available yet and were still relying on the VM-T Atlant. Atlant's limited capability posed a serious problem for program managers. It was not able to lift a complete Buran orbiter and, in order to get off the ground, the orbiter had to be stripped of many of its systems, including the tail. It was clear that conducting an approach and landing test program the way NASA had done was virtually out of the question, although the possibility was considered, among other things by using the An-22 Antey aircraft.

Instead, it was decided that an orbiter would have to be built that could take off from a runway by itself. That was an immense challenge for designers since Buran, like the Space Shuttle Orbiter, was never designed to take off like a conventional airplane. Nevertheless, a modified vehicle was constructed that would meet the requirements. Officially named OK-GLI, it was described as an "analog" of Buran and would become commonly known as BTS-002 (or BTS-02), with BTS standing for "Big Transport Airplane" (*Bolshoy Transportnyy Samolyot*). It got the registration number CCCP-3501002.

The BTS-002 atmospheric test model (*source*: Timofey Prygichev).

Fuel tank in the payload bay of BTS-002 (*source*: *www.buran.ru*).

First of all, nacelles were added to the aft fuselage that would house afterburner-equipped Lyulka AL-31F turbojet engines like those that were standard on Sukhoy Su-27 jet fighters [20]. This was in addition to two Lyulka AL-31 engines without afterburners on either side of the tail section which at the time were scheduled to be installed on spaceworthy orbiters as well. But, while BTS-002 had those two engines, in the end plans to install them on the "real" orbiters were dropped (see Chapter 7). The presence of engines on the BTS also afforded longer flight times (more than 30 minutes) and consequently more time to test flying characteristics than was the case with Enterprise's Approach and Landing Tests, which lasted no longer than 5.5 minutes.

Wind tunnel tests had to be conducted to see whether or not the addition of these engines would have any influence on the vehicle's aerodynamics, which was determined as minimal. Since BTS-002 would not be subjected to the high temperatures of re-entry, no thermal protection system was needed. Instead, foam plastic tiles were used to cover the craft. The fuel tank for the turbojet engines was placed in the otherwise empty payload bay. Maximum take-off weight was 92 tons.

Another modification needed on BTS-002 was a system to retract the landing gear shortly after take-off. Also, the nose gear strut was slightly lengthened to increase its ground angle to 4°, which was required to facilitate take-off. As a result, BTS-002's nose was considerably higher from the ground than Buran's.

Just like Enterprise, BTS-002 had an air data system mounted on a boom extending from the nose of the vehicle (on spaceworthy vehicles this was embedded in the heat shield for protection during re-entry). The cockpit contained work stations for a commander (RM-1), co-pilot (RM-2), and flight engineer (RM-3), although the latter was never used. The pilots wore standard flight overalls and helmets and were strapped in ejection seats designated K-36L. BTS-002 had four on-board computers.

The presence of the jet engines and the landing gear retraction system were the main external differences between BTS-002 and the "real" Buran. The airframe configuration was similar to that of the orbiters that were destined for space. Center of gravity and other flight dynamics criteria were deemed within acceptable limits.

The BTS-002 pilots trained extensively for the missions on a wide variety of "flying laboratories" and also in the PRSO-1 and PDST simulators at NPO Molniya. All in all, during training sessions, the crews spent about 3,200 hours in the flight simulators, which given the eventual success of the flights clearly paid off [21].

In 1983 BTS-002 was transported by barge from the Tushino Machine Building Factory to Zhukovskiy, where it underwent further testing in a newly built facility at the premises of EMZ [22]. The approach and landing tests took place at the neighboring Flight Research Institute. Before the flights started, the infrastructure consisting of beacons, radars, and transponders was modified to make it similar in set-up to that of the Yubileynyy field at the Baykonur cosmodrome. In late 1984, all was set for the first tests. Whereas NASA had conducted a relatively short program consisting of only five flights between August and October 1977, the Ministry of the Aviation Industry took one small step at a time [23].

As was common practice when new airplanes were tested in the Soviet Union, the flights were preceded by a number of taxi tests and take-off runs with increasing speeds. During most or all of the flights the crew flew two approach trajectories. First, they would descend to an altitude of some 15 to 20 m and then take the vehicle back to an altitude of 4,000 m for a second approach. On each flight BTS-002 was escorted by one or two airplanes. In all, four different chase planes were used during the tests: the L-39, Tu-134, Su-17, and MiG-25-SOTN.

This is an overview of all the ground runs and landing tests:

Ground run | **Date:** 29 December 1984
Crew: Volk–Stankyavichus | **Duration:** 5 minutes (14:30–14:35) (Moscow time)

During this first short taxi test, a maximum speed of between 40 and 45 km/h was reached, after which BTS-002 was subjected to a series of full-scale equipment tests.

Ground run | **Date:** 2 August 1985
Crew: Volk–Stankyavichus | **Duration:** 14 minutes (18:56–19:10)

During this second ground run the crew conducted two take-off runs down the runway. During the first one they tested the nose gear steering system at speeds of 30–40 km/h and performed braking at a speed of 100 km/h. Then they turned BTS-002 around and took it to a maximum speed of 205 km/h before deploying the drag chutes.

Ground run **Date:** 5 October 1985
Crew: Volk–Stankyavichus **Duration:** 12 minutes (15:31–15:43)

Maximum speed 270 km/h. One of the left main gear tires blew out due to skidding during braking.

Ground run **Date:** 15 October 1985
Crew: Volk–Stankyavichus **Duration:** 31 minutes (14:44–15:15)

With a speed of 300 km/h, Volk and Stankyavichus almost reached the minimum take-off speed and briefly lifted the nose gear into the air.

Ground run **Date:** 5 November 1985
Crew: Volk–Stankyavichus **Duration:** 12 minutes (13:40–13:52)

Maximum speed during this run was 170 km/h.

GLI-1 **Date:** 10 November 1985
Crew: Volk–Stankyavichus **Duration:** 12 minutes (14:06–14:18)

On 10 November 1985, after taking an 1,800 m run and reaching a speed of 320 km/h, BTS-002 took off from Zhukovskiy's runway for its first flight, during which an altitude of 1,500 m and a

BTS-002 takes off (*source*: *www.buran.ru*).

speed of 480 km/h were reached [24]. The flight, primarily intended to determine the craft's stability and handling, was a complete success, and upon their return Volk and Stankyavichus were greeted by their colleagues and ground crews in the traditional Soviet test-pilot way: by being tossed in a blanket. After that it was back to business with a debriefing by a commission of the Ministry of the Aviation Industry, headed by LII chief A.D. Mironov. Such debriefings would take place after each of the subsequent flights.

GLI-2 — **Date:** 3 January 1986
Crew: Volk–Stankyavichus — **Duration:** 36 minutes (14:19–14:55)

Second "general" test flight. A speed of 520 km/h was reached while the analog climbed to an altitude of 3,000 m. As had been done on the first flight, a conventional 3 degree glideslope was used and BTS-002 was manually brought back to the runway.

Ground run — **Date:** 26 April 1986
Crew: Levchenko–Shchukin — **Duration:** 14 minutes (15:17–15:31)

The second projected Buran crew conducted a ground run in preparation for its own flights on the analog. One of the right main gear tires blew out due to skidding during braking.

GLI-3 — **Date:** 27 May 1986
Crew: Volk–Stankyavichus — **Duration:** 23 minutes (13:34–13:57)

Third "general" test flight. Altitude 4,000 m, speed 540 km/h.

GLI-4 — **Date:** 11 June 1986
Crew: Volk–Stankyavichus — **Duration:** 22 minutes (07:42–08:04)

During the fourth and final "general" test flight an altitude of 4,000 m and speed of 530 km/h were reached. It was also the first flight during which the standard landing mode with a steep glideslope of about 20 degrees was worked out. All three channels needed to fly the orbiter towards landing in an automatic mode were tested sequentially. Leveling out began at approximately 500 m, so the final angle of approach was only two to three degrees.

GLI-5 — **Date:** 20 June 1986
Crew: Levchenko–Shchukin — **Duration:** 25 minutes (07:40–08:05)

On the fifth flight, the crew took things one step further by simultaneously switching on all three channels needed for an automatic landing.

GLI-6 — **Date:** 28 June 1986
Crew: Levchenko–Shchukin — **Duration:** 23 minutes (09:30–09:53)

All three channels were used to make BTS-002 glide automatically to an altitude of 100 m. At that altitude, Levchenko took over the controls for final approach and landing.

GLI-7 — **Date:** 10 December 1986
Crew: Volk–Stankyavichus — **Duration:** 24 minutes (13:07–13:31)

The automatic landing system controlled BTS-002 until the final second before touchdown. At that point, Volk switched the system off and performed a manual landing.

GLI-8 **Date:** 23 December 1986
Crew: Volk–Stankyavichus **Duration:** 17 minutes (12:43–13:00)

GLI-8 saw the first landing considered to have been automatic, although the system was switched off once the main gear had touched down. Roll-out was controlled by the pilots.

GLI-9 **Date:** 29 December 1986
Crew: Levchenko–Shchukin **Duration:** 17 minutes (12:57–13:14)

BTS-002's complete approach and landing took place in automatic mode from an altitude of 4,000 m until coming to a complete stop. The only thing that was still done manually was lowering the nose gear to the runway.

GLI-10 **Date:** 16 February 1987
Crew: Volk–Stankyavichus **Duration:** 28 minutes (13:30–13:58)

First fully automatic landing, in which the pilots didn't undertake any action from the initiation of the approach from 4,000 m until coming to a full stop on the runway.

Ground run **Date:** 25 March 1987
Crew: Shchukin—Levchenko **Duration:** 2 minutes (16:22—16:24)

Tests of the braking system.

Ground run **Date:** 30 March 1987
Crew: Stankyavichus–Volk **Duration:** 25 minutes (16.40–17.05)

Tests of the braking system.

GLI-11 **Date:** 21 May 1987
Crew: Levchenko–Shchukin **Duration:** 20 minutes (10:17–10:37)

Approach and landing took place in automatic mode.

GLI-12 **Date:** 25 June 1987
Crew: Stankyavichus–Volk **Duration:** 19 minutes (14:34–14:53)

Approach and landing took place in automatic mode.

GLI-13 **Date:** 5 October 1987
Crew: Shchukin–Volk **Duration:** 21 minutes (13:50–14:11)

Approach and landing took place in automatic mode.

Air Force test pilots Ivan Bachurin and Aleksey Boroday were scheduled to take BTS-002 to the skies for GLI-14. But, after starting up the turbojet engines, warning lights indicated that a problem had been detected. After consultation with the test director, they decided to taxi to the runway and start a take-off run. If the engines indeed weren't functioning the way they should after they had been throttled up, the flight would be aborted. When it turned out that the warning lights were still on, the test director scrubbed the flight and ordered Bachurin and Boroday to return to the platform. This was the only scrub in the program after the vehicle's engines had been started up.

After the problem had been solved, the flight got a new designation and the crew got another opportunity.

GLI-14B **Date:** 15 October 1987
Crew: Bachurin–Boroday **Duration:** 19 minutes (08:12–08:31)
Approach and landing took place in automatic mode.

After the unmanned spaceflight of Buran, Ivan Bachurin wrote the following report on GLI-14B as part of a paper on the GLI program:

> "We were informed about the upcoming flight a week in advance. We prepared for the tasks we were to perform by flying the mission profile on the simulator. After that, we made the plotting charts, divided the various tasks between the two crew members, etc.
>
> On the eve of the flight, we attended a session of the commission that determined the readiness of the aircraft, the ground facilities and infrastructure, and the crew. The reports were all fairly straightforward and only a few questions were asked. The ground facilities and the aircraft were ready, and the crew was fully prepared to perform their duties. The chairman of the commission then asked: 'Is there a need for the commander to rehearse the flight on the Tu-154LL?' 'Yes, there is.' 'Is the plane ready?' 'Yes, it is.' 'Then you will perform that flight after the meeting.'
>
> We performed the rehearsal flight on the flying laboratory without any problems, completed the training in the cockpit of the analog and performed a start-up of the engines for training purposes.
>
> We spent the night at the airfield since the flight was scheduled to take place early in the morning. We didn't talk about the upcoming flight: we had had good discussions about that subject for a week.
>
> In the morning, we looked out the window to check the conditions. They had forecast that the wind would pick up in strength. We washed and shaved, had breakfast and underwent a medical check-up. After that, we sat and waited for the order to go. In my mind, I went through the whole upcoming flight again. Then, after a few minutes, came the signal: 'Everything is ready. The bus is on its way to pick you up.' 'OK, we're on schedule.' We then took our gear and left for the bus, and I felt that usual pleasant feeling of being ready to perform the flight.
>
> On our way over to the operations building, we passed 'Number Two' as we called the analog amongst ourselves. Inside the building, everybody was busy with his tasks. We walked into one of the dressing rooms, and weren't disturbed by anybody for the next 15 minutes. Then the order came: 'The crew is to take its positions.'
>
> We went to the steps leading up to the vehicle, where a single cameraman was recording all our activities. In the small room at the end of the steps experts helped us don our personal parachute harnesses, after which we crawled through the side hatch into the cockpit and took our seats.

By then two planes, one to escort us and the other to shoot video, reported that they were ready. We could begin.

While constantly consulting the mechanic and the Flight Experiment Control Post (FECP) we prepared and started the engines, and turned on all the ship's systems. The engineers at the FECP supervised the commands we gave to the on-board systems and could step in at any time to assist us. In the meantime [the two] planes took off.

We disconnected the external power sources and started to taxi out to the runway. The aircraft handled well, braking was very effective. I tried to remember what our altitude from the ground would be, which was unusually high.

On the runway we warmed up the engines. By then, the [two] planes took their positions in the air so that at take-off they would be flying beside us. Then, at the command of the escort plane's pilot, we put our engines in the take-off mode, did a final check of the parameters for the engines and other systems, and began taking off. The run along the runway was steady and easily controllable. Exactly at the right speed and almost immediately after I deflected the stick, the nose wheel lifted from the ground. Then we were airborne. I reduced the deflection of the stick and the plane maintained its planned climb angle.

The co-pilot in the right seat, Aleksey Boroday, reported: 'I'm pulling up the landing gear. The temperatures of engines two and three are gradually reaching their pre-set limits.' The commander says: 'Do not exceed.' The co-pilot replies: 'The temperature is now constant.' It is good that the pilot has the capability to

BTS-002 in flight, accompanied by a chase plane (*source*: *wwww.buran.ru*).

take part in the control of the ship, and is constantly ready to assist the commander. 'Wheels up.'

I found that as far as stability and controllability were concerned the real ship differed little from the simulator. The aircraft 'was tightly in our hands'.

'You're right on schedule,' we were told by the pilot of the escort plane. I looked around and saw the fighter not far from us, together with a Tu-134 that was shooting video. I warned the pilot of the escort plane that I was about to carry out the standard maneuvers used in these test flights for defining flying characteristics. Also, I checked the air brake.

The altitude was the predetermined one and I turned to get to our entry point. The FECP navigation officer gave us our exact position. We reached the entry point. I switched the engines to idle and activated the automatic control unit. Very eagerly, perhaps too eagerly, 'Number Two' executed the desired maneuver to begin the planned descent trajectory. We kept an eye on the steep descent trajectory, the performance of the on-board systems, and the air brake. The speed was what had been calculated and the plane quickly descended to the ground. Then, the plane began to level off and 'Number Two' smoothly decreased its speed. The landing gear was lowered and my hand was near the plane's control stick. The fact that we were flying in automatic mode didn't mean that we were sitting idle. The Tu-154LL would have been 'scattered all over the ground' had it not been for the intervention of Aleksandr Shchukin when during one of the automatic flights the plane dived right to the ground!

The altitude decreased to 200 meters, then 100, then 50. At that point, the plane was on glideslope. 'Thirty meters ... twenty meters', read the co-pilot. 'Let's go up again'. I turned off the automatic control unit and increased the engine thrust. The co-pilot closed the air brake and turned off the landing mode.

The second pass was carried out in the same sequence: in automatic mode up to full stop on the runway. 'Altitude ten ... five ... three, two, one meter ... contact!', reported the pilot of the escort plane. 'Drag chute deployed', the co-pilot confirmed.

Roll-out was steady and we had no more than a two-meter deviation from the runway's centerline. The fighter that had served as the escort plane finished its activities with a beautiful zoom climb.

Lowering the nose wheel was smooth and the braking on the wheels was effective. Jettisoning of the parachute occurred at the right speed. 'Number Two' rolled to a stop. We taxied in for parking and after turning off the engines we left our seats and disembarked via the steps to the platform. Technicians and engineers came to the plane, and I looked with gratitude to those who had spent many hours the previous night preparing 'Number Two' for flight.

...

Finally, we gave our report and the flight was analysed. At the end of that debriefing the test director announced the date for the next flight" [25].

Subsequent flights simulated situations where a returning orbiter would not be at the ideal point when the final approach maneuvers were initiated. For this, the crew would bring BTS-002 to different altitudes, or fly at speeds or in directions that differed from the calculated flight paths. In all cases, the control system corrected the deviations and brought the vehicle safely back to the runway.

GLI-15 **Date:** 16 January 1988
Crew: Volk–Stankyavichus **Duration:** 22 minutes (13:06–13:28)
Approach and landing took place in automatic mode.

GLI-16 **Date:** 24 January 1988
Crew: Bachurin–Boroday **Duration:** 22 minutes (14:00–14:22)
Approach and landing took place in automatic mode.

GLI-17 **Date:** 23 February 1988
Crew: Bachurin–Boroday **Duration:** 22 minutes (13:34–13:56)
Approach and landing took place in automatic mode.

GLI-18 **Date:** 4 March 1988
Crew: Volk–Stankyavichus **Duration:** 32 minutes (12:40–13:12)
Approach and landing took place in automatic mode.

BTS-002 landing (*source*: *www.buran.ru*).

GLI-19 **Date:** 12 March 1988
Crew: Boroday–Bachurin **Duration:** 21 minutes
Approach and landing took place in automatic mode.

GLI-20 **Date:** 23 March 1988
Crew: Boroday–Bachurin **Duration:** 21 minutes
Approach and landing took place in automatic mode.

GLI-21 **Date:** 28 March 1988
Crew: Boroday–Bachurin **Duration:** 22 minutes
Approach and landing took place in automatic mode.

GLI-22 **Date:** 2 April 1988
Crew: Stankyavichus–Shchukin **Duration:** 20 minutes
Approach and landing took place in automatic mode.

GLI-23 **Date:** 8 April 1988
Crew: Shchukin–Stankyavichus **Duration:** 21 minutes
Approach and landing took place in automatic mode.

GLI-24 **Date:** 15 April 1988
Crew: Volk–Stankyavichus **Duration:** 19 minutes
Final flight in the GLI program. Approach and landing took place in automatic mode.

With this, the first phase of the approach and landing tests was completed, but plans were drawn up for follow-on test flights. In March 1988 the Council of Chief Designers ordered the possibility of including NPO Energiya engineers in future BTS-002 crews to be studied, but it appears this option was not seriously considered. A later plan called for 13 more flights with pilots from both LII and GKNII. After the deaths of Levchenko and Shchukin in August 1988, LII proposed two crews consisting of Volk–Tolboyev and Stankyavichus–Zabolotskiy. Those two teams would fly the bulk of the missions, while the GKNII pilots would get just three or four flights [26].

In late 1989 Volk declared that he was still expecting to participate in the new series of BTS flights [27]. A first ground run to kick off the new phase was conducted by Rimantas Stankyavichus and Viktor Zabolotskiy, who had already been training for a possible Soyuz "warm-up flight". During the test BTS-002 blew both tires of its right main landing gear.

Ground run **Date:** 28 December 1989
Crew: Stankyavichus–Zabolotskiy **Duration:** Unknown

Many years later Igor Volk would explain that this had been "an attempt [by Lozino-Lozinskiy] to renew the program. But, unfortunately, after the program had been stopped for a year and a half, it appeared they needed to correct many things and they stopped again" [28]. One source claims there were two more take-off runs on 23 November and 6 December 1990 just to keep BTS-002 in working order [29]. However, BTS-002 would never fly again. Still, all 24 flights had taken place without encountering any significant problems and played an important role in paving the way for Buran's first orbital missions [30].

REFERENCES

[1] B. Gubanov, *Triumf i tragediya Energii (tom 3: Energiya/Buran)*, Nizhniy Novgorod: Izdatelstvo Nizhegorodskogo instituta ekonomicheskogo razvitiya, 1998, pp. 181–184, 270; V. Filin, *Put k Energii*, Moscow: Logos, 2001, pp. 66–67; A. Kuznetsov, *Mnogorazovaya kosmicheskaya sistema Energiya-Buran*, Moscow: OmV-Luch, 2004, pp. 180–213.

[2] A. Makarov, "NIIkhimmash: Leading center of the Russian Space Agency for integrated testing of rocket and space technology" (in Russian), *Rossiyskiy kosmos*, 1/1995, pp. 27–29; *KB Khimavtomatiki, stranitsy istorii, tom 1*, Voronezh: KBKhA, 1995, pp. 71–80; I. Afanasyev, "Microthrusters from Nizhnyaya Salda" (in Russian), *Novosti kosmonavtiki*, 21–22/1998, pp. 54–55; B. Gubanov, *op. cit.*, pp. 132, 166, 266; A. Kuznetsov, *op. cit.*, pp. 143, 182, 201–203, 212–213.

[3] N. Kidger, "Another Zenit Failure?", *Spaceflight*, January 1992, p. 8.

[4] B. Katorgin, F. Chelkis, "The RD-170 engine: Main stages and results of development" (in Russian), paper presented at the Tsiolkovskiy readings in Kaluga in September 1993; B. Gubanov, *op. cit.*; V. Rakhmanin, *Odnazhdy i navsegda*, Moscow: Mashinostroyeniye, 1998; V. Trofimov, *Osushchestvleniye mechty*, Moscow: Mashinostroyeniye, 2001; B. Katorgin, *NPO Energomash imeni akademika V.P. Glushko. Put v raketnoy tekhnike*, Moscow: Mashinostroyeniye/Polyot, 2004; V. Rakhmanin, "On the history of the development of the first stage of the Energiya rocket carrier" (in Russian), *Dvigateli*, 5/2003, 6/2003, 1/2004.

[5] Y. Semyonov, *Raketno-kosmicheskaya korporatsiya Energiya im. S.P. Korolyova 1946–1996*, Moscow: RKK Energiya, 1996, p. 367.

[6] *The Soviet Space Program. National Intelligence Estimate (NIE 11-1-83) Volume II: The Estimate*, 18 July 1983, p. V-8.

[7] B. Gubanov, *op. cit.*, pp. 227–228, 267, 269; V. Filin, *op. cit.*, 2001, pp. 74–105.

[8] B. Gubanov, *op. cit.*, pp. 269–270; V. Filin, *op. cit.*, pp. 112–117.

[9] *KB Khimavtomatiki, stranitsy istorii*, tom 1, *op. cit.*, p. 80; B. Gubanov, *op. cit.*, pp. 286–287, 291–292; V. Filin, *op. cit.*, pp. 128–140; S. Solovyov *et al.*, "The first test firing" (in Russian), *Novosti kosmonavtiki*, 7/2002, pp. 64–66.

[10] A. Borisov, "Buran: Flight to nowhere?" (in Russian), *Novosti kosmonavtiki*, 23–24/1998, pp. 68–69.

[11] B. Gubanov, *op. cit.*, pp. 196–208, 273–290; V. Filin, *op. cit.*, pp. 144, 153–155, 158–159.

[12] D. Vorontsov, "About Deytron and others: Notes of a rank-and-file engineer" (in Russian), *Novosti kosmonavtiki*, 9/2006, p. 72.

[13] B. Gubanov, *op. cit.*, pp. 293–310; V. Sorokin, "10 years since the first launch of Energiya" (in Russian), *Novosti kosmonavtiki*, 11/1997, pp. 38–44; A. Loktev, "The unknown Buran around Energiya" (in Russian), *Po Yaroslavke*, 31 May 2002; K. Lantratov, "The 'Star Wars' which never happened" (in Russian), detailed history of Skif published on-line at *http://www.buran.ru/htm/str163.htm*, translated into English in *Quest* magazine, 1/2007, 2/2007.

[14] Y. Semyonov, *op. cit.*, pp. 542–547.
[15] Y. Semyonov, *Mnogorazovyy orbitalnyy korabl Buran*, Moscow: Mashinostroyeniye, 1995, pp. 239–245, 326–344; V. Fedotov, V. Novikov, "The cabin" (in Russian); V. Shabanov, "Experimental testing of Buran" (in Russian), in: G. Lozino-Lozinskiy, A. Bratukhin, *Aviatsionno-kosmicheskiye sistemy*, Moscow: Izdatelstvo MAI, 1997, pp. 235–245, 261–267; O. Urusov, "Berth for 'birds'" (in Russian), *Novosti kosmonavtiki*, 3/2004, p. 72; "Frequently asked questions" page on Vadim Lukashevich's Buran website at *http://www.buran.ru/htm/chavo.htm* It should be noted that sources differ on the exact designations and serial numbers of the test vehicles. The data used here are from Lukashevich.
[16] *Potential for the Transfer of US Space Technology to the Soviet Union, National Intelligence Estimate (NIE 11-1/7-84). Key Judgments.* 26 November 1984, p. 4.
[17] S. Mikoyan, "The Spiral spaceplane and the BOR-4 and BOR-5 flying models" (in Russian), in: G. Lozino-Lozinskiy, A. Bratukhin, *op. cit.*, pp. 296–302; V. Dmitriyev, "Space launch" (in Russian), *Morskoy sbornik*, November 2004; BOR page on Vadim Lukashevich's Buran website at *http://www.buran.ru/htm/bors.htm*; *Lyotno-issledovatelskiy institut imeni M.M. Gromova*, Moscow: Mashinostroyeniye/Polyot, 2001, pp. 153–157.
[18] J. Asker, "NASA Design for Manned Spacecraft Draws on Soviet Subscale Spaceplane", *Aviation Week & Space Technology*, 24 September 1990, p. 28. A detailed history of BOR-4 can be found in B. Hendrickx, "The Soviet BOR-4 Spaceplanes and Their Legacy", *Journal of the British Interplanetary Society*, March 2007, pp. 90–108.
[19] S. Mikoyan, *op. cit.*; BOR page on Vadim Lukashevich's Buran website; some sources mention a failed launch attempt on 4 July 1983.
[20] D. Fink, J. Lenorovitz, "Soviets Fly Jet-Powered Space Shuttle Testbed", *Aviation Week & Space Technology*, 12 October 1987.
[21] V. Golovachov, "Those magnificent men ...", *Soviet Weekly*, 8 April 1989.
[22] A. Bruk, *Illyustrirovannaya entsiklopediya samolyotov EMZ im. V.M. Myasishcheva*, tom 3, chast 1, Moscow: Aviko Press, 1999, p. 51.
[23] S. Mikoyan, "Horizontal flight tests of the Buran ship" (in Russian), in: G. Lozino-Lozinskiy, A. Bratukhin, *op. cit.*, pp. 78–83; "The flying analog plane BTS-002 GLI" (in Russian) at *http://www.buran.ru/htm/anabst.htm*; correspondence with Aleksey Boroday (December 1990) and Lida Shkorkina (between 1990 and 1992); Bert Vis interviews with Igor Volk (2 July 1990, 5 October 1991, 23 November 1998, and 13 June 2003); Aleksey Boroday (10 April 2000); and Magomed Tolboyev (2 April 2001).
[24] Remarks by Igor Volk during a visit to the USA, April–May 1989 (James E. Oberg Cosmogram Newsletter #42, 5 May 1989).
[25] I. Bachurin, "Buran with a crew on board" (in Russian), written in November 1988, published in: G. Lozino-Lozinskiy, A. Bratukhin, *op. cit.*, pp. 84–87.
[26] E. Vaskevich archives.
[27] "Landing tests", *Spaceflight News*, December 1989, p. 6.
[28] Bert Vis interview with Igor Volk, Moscow, 13 June 2003. Volk made his remarks after consulting Zabolotskiy by phone and putting the question to him.
[29] BTS-002 chronology at *http://www.buran.ru/htm/hrono.htm*
[30] Information on the BTS-002 test program was compiled from: correspondence with Lida Shkorkina (1990–1991); correspondence with the TsPK Information Group (1991); *Vzlyoty i posadki Burana*, undated booklet on the BTS-002 missions; BTS-002 chronology at *http://www.buran.ru/htm/hrono.htm*

7

Buran in the spotlight

By 1988, twelve years after the approval of the Energiya–Buran program, the stage was finally set for the Soviet space shuttle to make its orbital debut. While earlier test flights of piloted spacecraft had been prepared in utter secrecy and conveniently disguised under the all-embracing "Kosmos" label, the Russians no longer had the luxury of doing the same with Buran. Times had changed after General Secretary Mikhail Gorbachov's rise to power in the spring of 1985. The new policy of *glasnost* was sweeping through all ranks of Soviet society, including the country's space program.

Disclosing the existence of a Soviet equivalent to the US Space Shuttle in some ways must have been an embarrassing move for the Russians. Not only did the maiden flight of Buran come seven years after the first mission of the Space Shuttle, the Soviet media had always been very critical of the Shuttle program, portraying it as just another tool of the Pentagon to realize its ambition of militarizing space. This tradition began with the very first Shuttle launch on 12 April 1981, which entirely by coincidence overshadowed the 20th anniversary of the mission of Yuriy Gagarin. Reporting on the launch, Radio Moscow World Service said:

> "The United States embarked on the Shuttle program some 10 years ago. Its military pin on the program far-reaching hopes for transferring the arms race to space. One of the main missions in the first few flights of the Shuttle will be testing a laser arms guidance system."

Even though the Shuttle eventually flew only a handful of dedicated Defense Department missions, no Shuttle flight went by without the Soviet media reminding the world of the ship's military potential, the more so after President Ronald Reagan's announcement of the Strategic Defense Initiative in March 1983. Even when Challenger exploded in January 1986, Radio Moscow warned its listeners that:

> "a similar failure in the SDI system the American Administration is so anxious to create would cause a global disaster" [1].

Many Soviet space officials and cosmonauts had also denounced the Space Shuttle program as a wasteful effort, emphasizing that a fleet of expendable rockets was a much more economical way of delivering payloads to orbit. At the same time, some also stopped short of flatly denying that reusable space transportation systems were being studied, although no technical details or timelines were given. Until 1987 the Energiya–Buran program was a closely guarded state secret, requiring a cover-up operation comparable in scale with that for the Soviet manned lunar program in the 1960s and early 1970s.

However, as had been the case with the N-1 Moon rocket, there was no way the Russians could conceal Buran-related construction work and tests from the all-seeing eyes of US reconnaissance satellites. Long before the Russians opened the information floodgates, US intelligence had a very good understanding of the system's configuration and capabilities, although some serious misjudgments were made as well, at least based on what has been declassified so far. Significantly, the information was publicly released on a much wider scale than it had been during the Moon race in the 1960s.

WESTERN SPECULATION ON SOVIET SHUTTLE PROGRAMS

Western knowledge of Spiral

As the Space Shuttle moved to the foreground as the next big step in NASA's manned spaceflight program in the late 1960s/early 1970s, Western experts began speculating on the existence of a Soviet equivalent. Few realized that the shuttle effort the Soviet Union was involved in at that time was Spiral, the 9-ton military spaceplane to be launched from the back of a hypersonic aircraft. One person who did get the basic concept right was Peter James, a Pratt & Whitney engineer and intelligence informant, who talked to leading Soviet aviation and space experts during the annual congresses of the International Astronautical Federation and summarized his findings in a controversial 1974 book called *Soviet Conquest from Space*.

During his private discussions with Soviet officials, James clearly picked up some shreds of information on Spiral. One of the specialists he talked to at the 1969 IAF congress in Argentina was none less than Gennadiy Dementyev, identified in the book as "one of the heads of the Soviet space effort and affiliated with the Moscow Aviation Institute". As is now known, Dementyev (the son of the Minister of the Aviation Industry) had indeed worked at the institute, but in 1967 was named Lozino-Lozinskiy's deputy for the Spiral project at Mikoyan's space branch in Dubna. In his book, James correctly pointed out that the Soviet Union was working on an air-launched spaceplane which in an initial stage would be launched by an expendable rocket while the design of the carrier aircraft was finalized. However, he grossly overestimated the size of the shuttle vehicle, claiming it would have a payload capacity of 35–45 tons, more than the Space Shuttle. He also got the organizational background completely wrong [2].

Meanwhile, some die-hard armchair analysts of the Soviet space program were doing their own research on Soviet shuttle activities in the best traditions of "space sleuthing". As early as the mid-1970s a group of Dutch space enthusiasts, having carefully analysed obscure Soviet technical publications and isolated statements from Soviet officials, also came to the conclusion that the Soviet Union was developing an air-launched shuttle system. Unaware of James' publication, they described the system as follows in one of their articles:

> "The Soviet shuttle system, ... baptized recently 'ALBATROS', seems to be much more advanced than the US type currently under construction. It will possibly be in active duty well ahead [of] its American counterpart. The system consists of ... twin recoverable craft, both ... delta wing vehicles, and uses the horizontal liftoff principle from specially adapted SST [supersonic transport] runways. The booster-plane is an improved SST type, in size about the Tupolev 144 SST. It is however a much more advanced type, a HST (Hypersonic Transport) using air breathing engines to ride the piggyback orbiter towards 30 km altitude. There the ALBATROS is separated and uses its own system of chemical and electrical engines to propel itself into Earth orbit [3].

Whether by coincidence or not, this was a fairly accurate description of the Spiral system, although few believed them at the time. Like James, the Dutch space sleuths also overestimated the size and capabilities of the shuttle vehicle, but had got the basic concept right. "Albatros" later turned out to be the name of an unrelated two-stage-to-orbit shuttle system to be launched from a hydrofoil that was studied by students at the Bauman technical university in Moscow [4].

Given the dearth of declassified CIA reports on the Soviet space program, it is difficult to say at this stage exactly what was known about Spiral inside the US intelligence community. A 1983 CIA report said a spaceplane effort had begun at the Mikoyan design bureau in 1969 (four years after it actually got underway) [5]. In the few documents that have been released so far, there is no mention of the fact that the spaceplane was eventually supposed to be air-launched, but that doesn't necessarily mean the CIA wasn't aware of those plans. Since the hypersonic carrier aircraft never got further than the planning stage, any information on it must have been gathered via private conversations or human intelligence.

What the Russians definitely could not conceal from US reconnaissance assets were the test flights flown in support of Spiral, beginning with the BOR-1/2/3 suborbital missions in 1969–1974. What is known for sure is that US intelligence picked up signs in the second half of the 1970s of the test flights of the 105.11 Spiral atmospheric subsonic test bed in Akhtubinsk (the "Vladimirovka Advanced Weapons and Research Complex" or VAWARC as the CIA called it). Those tests were reported in the trade press in early 1978 [6]. However, in a classified assessment of Soviet space capabilities released to authorized persons in August 1980, US intelligence experts wrongly concluded that these had been tests of a delta-wing spaceplane to be orbited by the three-stage version of the Proton rocket, capable of putting 20 tons into low Earth orbit. Although Spiral had been canceled by this time, the small spaceplane was believed to be under development for future military missions such as reconnaissance, satellite inspection and neutralization, although it could also be developed into a crew ferry vehicle to support space station operations.

At the same time, the report linked the construction of a runway and new launch pads at Baykonur's former N-1 launch complex to a *separate* effort to build a new family of heavy-lift launch vehicles, capable among other things of orbiting a reusable spacecraft the size of the Space Shuttle Orbiter. This is the first reference in the declassified US intelligence literature to the Energiya–Buran program. The motives for building such a vehicle were believed to include a desire to economize on space launches, particularly in the area of large space station construction, manning, and supply, as well as the general desire to compete with the United States for prestige. The spaceplane was expected to be fielded in the early to mid-1980s, followed by the larger vehicle in the early 1990s [7].

The phantom spaceplane

The 1980 CIA report marked the beginning of a rumor that persisted in the West throughout the 1980s—namely, that the Soviet Union was simultaneously developing *two* shuttle systems, a small spaceplane orbited by a conventional rocket and a large shuttle similar to its American counterpart. The speculation entered the public domain in the early 1980s via annual Defense Department publications known as *Soviet Military Power* and America's leading aerospace magazine *Aviation Week & Space Technology* (sometimes jokingly called *Aviation* Leak).

Speculation about the spaceplane was fueled by a series of mysterious test flights in 1977–1979 in which the Proton rocket deployed two heavy objects that re-entered after a single orbit (Kosmos-881/882, 997/998, 1100/1101). Many observers interpreted these "Double Kosmos" missions as re-entry tests of a spaceplane. It wasn't until the early 1990s that the Russians revealed that these had been test flights of the return capsules of the TKS spacecraft, which were transport ships for the Almaz military space station of the Chelomey design bureau. There are no indications, however, that the missions were linked to the spaceplane program by US intelligence analysts. In fact, the classified 1980 CIA report had correctly identified the missions as re-entry tests of the TKS return vehicles, although it wrongly interpreted the TKS vehicles as successors to the military Almaz space stations rather than transport vehicles serving those stations.

The first irrefutable evidence for the existence of a Soviet shuttle program came in April 1983, when the Australian Air Force publicly released images of the Indian Ocean recovery of the BOR-4 vehicle Kosmos-1445, one of the Spiral scale models that had been modified to test heat shield materials for Buran. Unaware of BOR-4's roots in the canceled Spiral program, analysts quite logically concluded that the vehicle, which was aerodynamically completely different from the big shuttle, must be a subscale model of the rumored spaceplane.

By the early 1980s US intelligence was aware of the development of the Zenit medium-lift launch vehicle, which it called SL-X-16. The spaceplane was now linked to *that* booster rather than the Proton, putting it in a somewhat lighter class (roughly 15 tons). Once again a series of mysterious test flights lent credence to this idea. In 1986–1987 the Zenit flew four missions in which it deployed heavy, inert payloads into low Earth orbits (Kosmos-1767, 1820, 1871, 1873), interpreted by some outside the intelligence community as being mass models of the spaceplane. Not until the turn of the century did the Russians reveal that the heavy Zenit payloads had been mass models of the Tselina-2 electronic intelligence satellite with an additional mock payload attached to see how the Zenit would perform when placing heavy payloads into orbit.

This is not to say that there was unanimity among Western observers about the existence of the spaceplane. A report in May 1986 said it was now thought the BOR-4 test flights could have been merely tests of the thermal protection system for the large shuttle [8]. Others interpreted the BOR-4 flights as pure technology development tests analogous to the American PRIME and ASSET programs in the 1960s, not connected with any specific follow-on project. It was also noted that the Soviets had never before employed orbital flight tests of subscale models [9].

When Soviet officials finally began disclosing details about the Energiya–Buran system in 1987–1988, there still was no mention of the spaceplane. As preparations for the first flight of Buran were nearing completion and the maiden mission of the

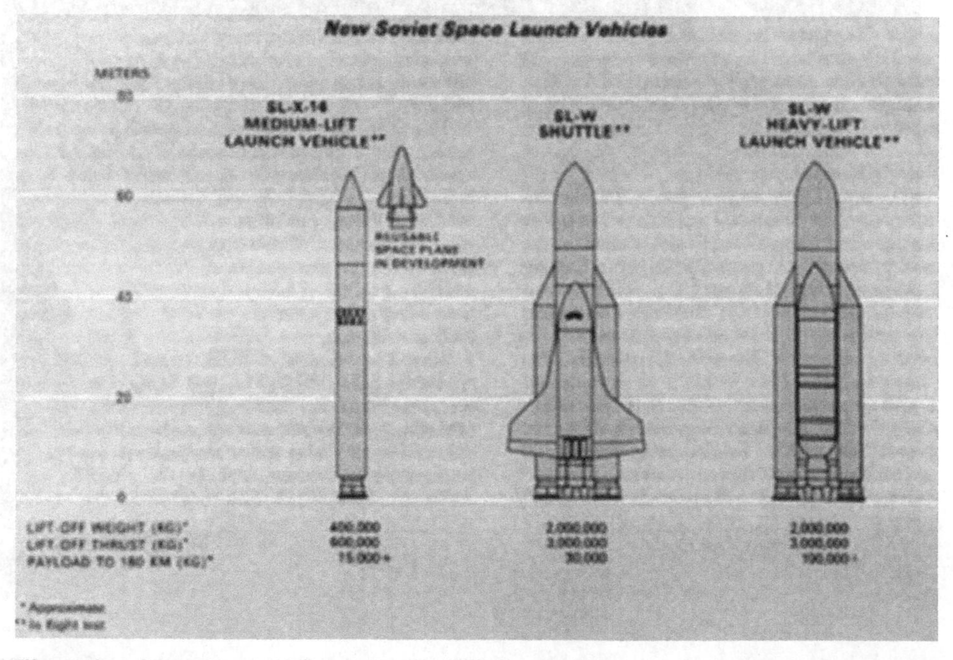

Illustration from *Soviet Military Power 1987* shows purported Zenit-launched spaceplane (*source*: US Department of Defense).

spaceplane had still not materialized, the US intelligence community was beginning to have some doubts as well about the program's existence. In a classified assessment of the Soviet shuttle program in September 1988, just two months before the flight of Buran, the CIA did not exclude the possibility that BOR-4:

> "is only a test vehicle used to gather aerodynamic, aerothermal, and materials data for the larger shuttle orbiter."

However, the overall consensus among CIA experts still was that a separate spaceplane program *was* underway. Unlike the large shuttle, the spaceplane was believed to have significant military potential. It was expected to be able to change its orbital inclination by as much as 15° and change its orbital altitude by about 4,200 km, making it ideal for reconnaissance, inspection, and combat missions. Its expected cross-range capability of up to 2,400 km would provide many additional opportunities each day to return to selected military airfields. It was also expected to have limited space station support capability, being used for rapid return of high-priority cargo or crew rescue missions.

The report did acknowledge that the spaceplane had apparently taken a backseat to the large shuttle for several reasons. Two of its primary missions—real-time

Purported spaceplane attacking an enemy satellite. Illustration from *Soviet Military Power 1985* (*source*: US Department of Defense).

reconnaissance of critical targets and post-strike reconnaissance—were by now being fulfilled by newly developed near real-time imaging satellites. Furthermore, resource constraints had possibly forced the Russians to complete the two costly programs sequentially rather than simultaneously. Finally, Soviet attempts to inhibit American anti-satellite and SDI efforts, including a self-imposed moratorium against ASAT testing, were expected to keep the program at a low level at least into the early 1990s. [10].

The Russians elected not to disclose the purpose of the BOR-4 missions until after Buran had flown. One week after the mission, an article in *Pravda* officially described them as test flights of Buran's heat shield [11]. However, in February 1989 *Scientific American* magazine published an article on the Soviet Union's space program, which again identified the BOR-4 vehicles as scale models of a small spaceplane. With nothing to hide anymore, the Russians were quick to react. Soviet deputy Defense Minister Vitaliy M. Shabanov called the story about the spaceplane a "canard", not ruling out the possibility that it was just a ploy to obtain funding for a new Dyna-Soar type program. Asked what kind of vehicle was shown in the BOR-4 picture published in the magazine, Shabanov said:

> "Well, this is obvious. In order to test the Buran reusable spacecraft four scale models were launched. They were placed into orbit with the designations Kosmos-1374, 1445, 1517, and 1614. The models were used to test elements of the heat shield, control systems, and so on. One of them was photographed by the Australians" [12].

What Shabanov failed to mention, however, was that the vehicles had not been scale models of Buran, but of a spaceplane canceled back in the 1970s.

Even in subsequent years the rumored spaceplane, which some claimed was called Uragan ("Hurricane"), occasionally resurfaced in Western publications. One article in 1995 said that Richard Ward, a noted international technology analyst based with Lockheed, had been told the story of the 1980s space fighter in private discussions with Soviet engineers in May 1990. Ward had been part of an American delegation visiting aerospace centers in Moscow and Kiev, where he talked to several representatives of the aviation industry. He was told that the BOR-4 missions had indeed served as a test series for a full-scale interceptor. Launched by Zenit, the operational vehicle would have had a crew of two and would have been armed with a recoilless gun for on-orbit attacks. The project had reportedly been given impetus after the US announcement that military Shuttle launches from Vandenberg were slated to begin in the autumn of 1986 [13].

Despite the persistent rumors, twenty years on not a single shred of convincing evidence has appeared to counter the notion that the Zenit-launched BOR-4 derived spaceplane was no more than a figment of the imagination of Western analysts. All indications are that BOR-4 was indeed flown for the official reason given by the Russians—namely, to test Buran's heat shield. It is also known now that there was a parallel effort to convert BOR-4 vehicles into space-to-ground weapons as part of a

Soviet "Star Wars" program (see Chapter 8), but, again, here its role was not that of a subscale model for a piloted spaceplane.

After the cancellation of Spiral in the late 1970s, the Soviet Union did continue conceptual studies of various other small spaceplanes (notably LKS, MAKS, and OK-M), but all of these were aerodynamically different from BOR-4 and its alleged full-scale version. NPO Energiya's OK-M *was* intended for launch by Zenit, but primarily seen as a space station support system. However, new evidence shows that NPO Molniya's air-launched MAKS was supposed to carry out many of the same military tasks that had been eyed for Spiral (see Chapter 9). If there was a need for a military spaceplane in the 1980s, MAKS perfectly fitted the bill. It inherited the military advantages of Spiral, being more flexible and less vulnerable than a Zenit-launched spaceplane. The most plausible conclusion at this stage is that the Russians *did* consider a military spaceplane in the 1980s, but it was not the one that many Western analysts believed was under development and it was never given the same priority as Buran. Although the BOR-4 missions indirectly provided data applicable to MAKS, they were not seen as precursors to MAKS.

Snooping on Buran

Although the Energiya–Buran program remained shrouded in secrecy for much of the 1980s, US intelligence specialists had a fairly good idea of the system's characteristics and capabilities, mainly thanks to detailed American reconnaissance satellite images of both Baykonur and the Flight Research Institute (LII) in Zhukovskiy. The latter was identified in the intelligence literature as Ramenskoye, which is the name of LII's airfield in Zhukovskiy and (confusingly) also of a neighboring town and railway station.

The first clear evidence for the existence of a large shuttle came in the late 1970s, when construction of the runway and launch pads at Baykonur got underway. The Energiya–Buran pads were identified as "Complex J" (the same code name given to the N-1 pads) and the UKSS pad as "Complex W". By 1982 spy satellites had even spotted the construction of the back-up runway in the Soviet Far East. [14].

The first public assessment of the system's capabilities was given in *Soviet Military Power 1983*, published in early 1983. Since the first test models of Energiya were yet to be rolled out to the pad, analysts still had a poor understanding of the system's configuration and capabilities. Drawings showed the Soviet shuttle mounted on an external tank with two rocket boosters, with the main engines apparently on the orbiter itself. The lift-off weight of the system was estimated to be just 1,500 tons compared with the NASA Shuttle lift-off weight of 2,220 tons. Combined with an estimated lift-off thrust of between 1,800 and 2,700 tons (compared with roughly 3,000 tons for the Shuttle), this translated into a staggering payload capacity of 60 tons, twice that of the Space Shuttle Orbiter. The Soviet shuttle was believed to have a substantially different wing design with an 80-degree sweep. The heavy-lift launch vehicle (HLLV) was depicted as a 95 m high core vehicle with three 35 m high liquid propellant boosters and a top-mounted payload with a maximum mass of between 130 and 150 tons [15].

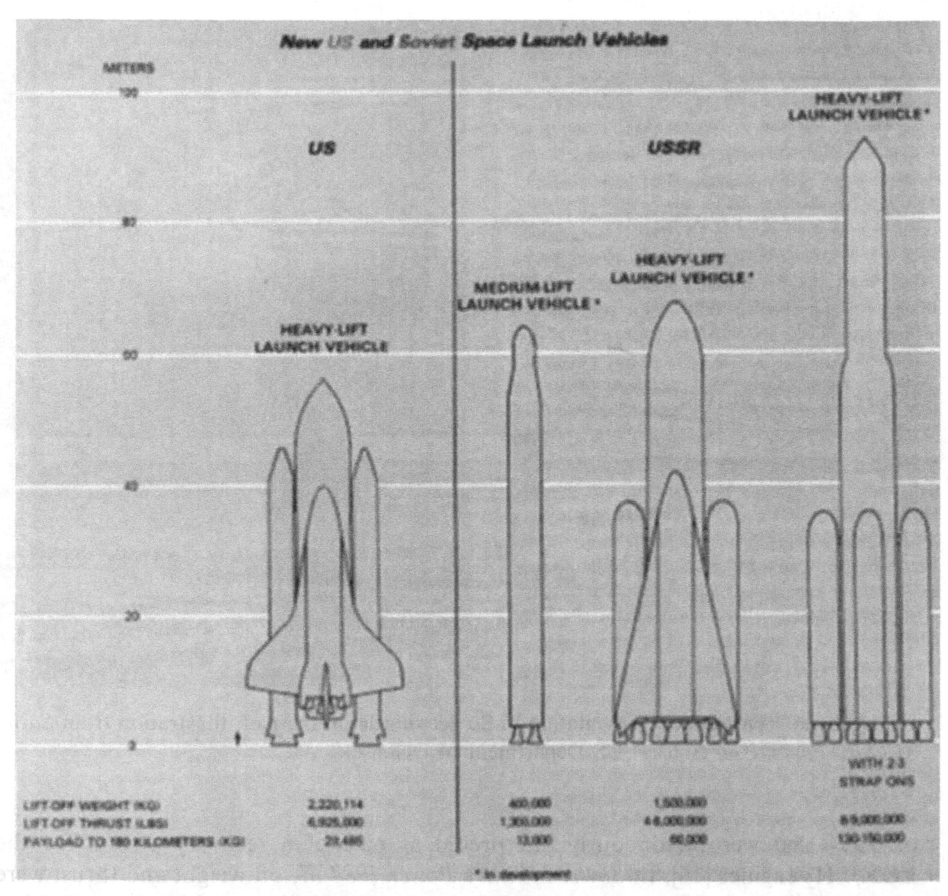

Assessment of Soviet heavy-lift launch vehicle capabilities in *Soviet Military Power 1983* (*source*: US Department of Defense).

By the middle of 1983 several events led to a much better understanding of the Soviet shuttle system. Spy satellites had acquired detailed images of a test orbiter sitting atop a VM-T carrier aircraft during tests earlier in the year and had also spotted an incident in which the pair accidentally skidded off the runway in March 1983. Moreover, the first test versions of Energiya had been rolled out of the assembly building in the first half of the year. A CIA National Intelligence Estimate in July 1983 now correctly concluded that the orbiter had a configuration very similar to that of the US Space Shuttle Orbiter and that the main engines were on the core rather than on the orbiter. The report was wrong in stating that the rocket had only two strap-on boosters and that the core was outfitted with "at least two and probably three engines". This may have been related to the fact that the Energiya rolled out in May 1983 had only three nozzles installed on the core stage (see Chapter 6). The report referred to the spherical sections above the core stage nozzles as "pod-like

US Defense Department representation of Soviet shuttle on the pad. Illustration from *Soviet Military Power 1986* (*source*: US Department of Defense).

objects" that were erroneously interpreted as part of a recovery system for the LOX/LH_2 engines [16]. In *Soviet Military Power 1984* lift-off weight and thrust were now estimated at 2,000 tons and 3,000 tons, respectively, resulting in an orbiter payload capacity of 30 tons. The HLLV was still expected to be a nearly 100 m high rocket with six or more strap-on boosters and a payload capacity of 150 tons, a configuration that was actually more reminiscent of the Vulkan rocket.

It was not until the 1986 edition that *Soviet Military Power* published a drawing of a 100-ton capacity HLLV where the orbiter was replaced by a side-mounted cargo pod. This was also the first edition that got the dimensions of the rocket/orbiter stack more or less right, although the first Energiya–Buran combination did not make its appearance on the Baykonur launch pads until summer/autumn of 1986, *after* the report had been published. The following year potential payloads for the cargo version of the rocket were said to be modules for large space stations, components for a manned or unmanned interplanetary mission, and even directed-energy ASAT and ballistic missile defense weapons.

Reconnaissance satellite images of Baykonur also gave some idea of how the testing proceeded. In 1984 and early 1985 the SL-X-16 ("Zenit") medium-lift booster had been observed being alternately removed from and erected on the pad, suggesting Soviet dissatisfaction with the ground test results. This in turn had implications for

US Defense Department representation of Soviet shuttle atop the VM-T aircraft. Illustration from *Soviet Military Power 1985* (*source*: US Department of Defense).

the HLLV/shuttle program, which used common engines. Apparently, the belief at this time was that the SL-X-16 was powered by liquid hydrogen/liquid oxygen engines and that these same engines were used both in the strap-ons and the core stage of the HLLV. The fact that no Energiya had yet been seen with an orbiter strapped to the side was also seen as an indication that the program was suffering delays [17]. When a test orbiter did finally undergo the first pad tests in August–October 1986, the news was reported only weeks later [18]. US reconnaissance assets apparently also picked up signs of the Energiya core stage test firings in the first half of 1986, but these were not openly reported until a year later [19].

Buran-related test activities were not always correctly interpreted, especially when the Soviet approach to testing was different than NASA's. Just weeks before the first ground run of the BTS-002 atmospheric test bed at LII in late December 1984, *Aviation Week* correctly reported that approach and landing tests of the shuttle were imminent, but wrongly concluded that the vehicle would be dropped from the VM-T carrier aircraft in similar fashion to the test flights of Enterprise in 1977 [20]. In April 1986, by which time BTS-002 had performed numerous ground runs and two

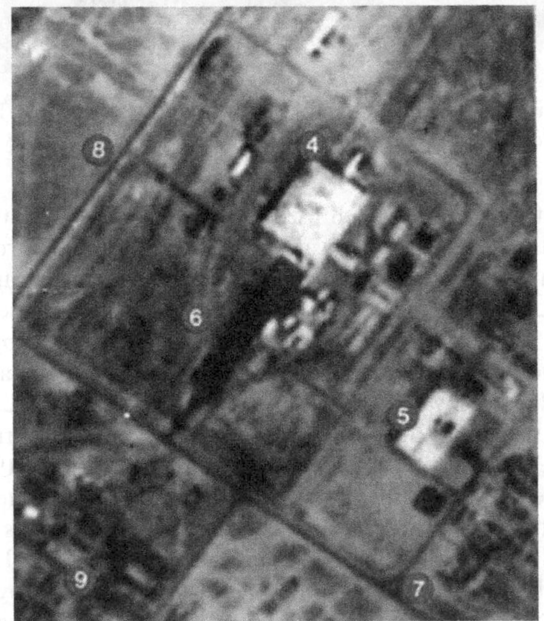

SPOT images of Baykonur runway (top) and assembly buildings. MIK RN and MIK OK are in positions 4 and 5 (*source*: SPOT Image).

landing tests, *Aviation Week* referred to reconnaissance photography showing jet engines mounted on either side of the tail, but still believed the vehicle was being dropped from the VM-T. The jet engines were thought to be on board only to test their ability to correct an orbiter's flight path when returning from space and might or might not be lit prior to separation from the aircraft depending on test objectives [21].

Of course, the information that leaked out via *Aviation Week* did not necessarily reflect what the US intelligence community *really* knew. Other observers, taking into account the known capabilities of the VM-T, correctly concluded that it could hardly carry a full-size, full-weight orbiter to sufficient altitude for a safe free flight [22]. *Aviation Week* did not report the correct flight profile of the BTS-002 until late 1987 after having been informed by Soviet space officials at an international space congress in Moscow. The only mistake remaining was that the tests were said to take place at Baykonur [23].

Western observers had more to go on than just the intelligence community's interpretation of reconnaissance satellite imagery. Pictures taken of Baykonur by civilian remote-sensing satellites such as the US Landsat and the French SPOT had sufficient resolution to show the construction work going on in support of the Energiya–Buran program. Unlike the spy satellite pictures, these were openly available to the public.

It is not clear how much information on the program leaked to the West through breaches in the Soviet censorship and security apparatus or via human intelligence. One piece of information that did slip through was that the name of the Soviet shuttle was Buran. The name first appeared in a 1983 CIA National Intelligence Estimate and also surfaced in several open Western publications the following years, well before the Russians officially announced it [24]. This information could not possibly have been gleaned from spy satellite photography, because the name was not on any of the test models and was not painted on the first flight vehicle until 1988. Actually, before 1988 Buran was not the name of a specific orbiter, but a generic name used by the Russians to refer to the combination of rocket and orbiter (see Chapter 2).

Missions

The only known detailed intelligence analysis of Buran's potential missions came in a classified CIA report about two months before the maiden flight took place. Analysts felt that the only mission that would justify the resource commitment required for its development was to provide logistic support to existing and future manned space stations, supplementing or replacing Soyuz-TM and Progress spacecraft. The orbiter could return to Earth entire modules or large payloads, including products obtained in materials-processing experiments. It could also provide efficient crew rotation and cargo transportation for the large 10 to 20-man space stations that were expected to succeed Mir. While the individual modules were likely to be launched by Proton or Energiya rockets, large solar panels, girders, and other structural components necessary to transform the modules into an operational complex could be carried in the orbiter's cargo bay.

Satellite deployment was not expected to become one of Buran's primary missions, because the extensive assortment of Soviet expendable launch vehicles could provide more cost-effective launch services than the shuttle. The fact that the engineless Buran itself was the payload of a heavy launch vehicle was seen as an indication that it was not primarily designed as a satellite launch vehicle. The addition of the orbiter dramatically reduced the payload capacity of the rocket and increased the total launch cost as well. The diversity of operational orbits employed by the Soviets would also require additional kick stages, tugs, or propulsion modules, increasing mission complexity. CIA analysts also believed Buran would be stymied in any effort to break into the Western commercial launch market because many payloads originally configured for launch by the US Shuttle had already been modified for launch aboard expendable vehicles after the Challenger accident.

Other missions were likely to include retrieval and in-orbit repair of malfunctioning satellites, which could reduce the cost of Soviet space operations, at least if combined with other orbiter missions. Soviet satellites historically had limited operational lifetimes, often because of premature failure of electronic components. The ability to replace major components aboard the orbiter or at a manned space base, without having to return the satellite to Earth, could reduce the annual launch rate requirement (about 100 satellites at the time) and provide considerable savings in associated launch costs.

CIA analysts also saw limited military potential for Buran. The long launch preparation times and the vehicle's limited orbital maneuverability restricted its usefulness for intelligence collection and combat missions. Military use of the orbiter was expected to be restricted to covert military research similar to that performed aboard Soviet space stations. Unlike expendable rockets, Buran could also be used to deploy certain military satellites in order to make initial detection and identification more difficult [25].

PLANNING THE FIRST FLIGHT

The US Defense Department estimated in the early 1980s that the HLLV would fly first in 1986–1987, followed by the Soviet orbiter in 1987–1988. Ironically, this prediction was more realistic than what was being planned by the Russians, who had a history of setting optimistic timelines for their space projects. When the Energiya–Buran program was approved in February 1976, the goal had been to fly the maiden mission in 1983, but this date started slipping soon. A government decree in December 1981 moved the target date to 1985 and another one on 2 August 1985 set the mission for the fourth quarter of 1986 [26]. Even that must have been a completely unrealistic goal given the progress made by that time in rocket and orbiter testing. A major factor in the delays probably were the serious problems with test firings of the RD-170/171 engines in the early 1980s, although other technical as well as budgetary issues must also have come into play.

As mentioned in the previous chapter, original plans apparently called for launching the first two missions of Energiya with mock-up orbiters that would remain attached to the rocket and re-enter together with it. Later those plans were dropped in favor of launching a real orbiter on the first Energiya mission. Then, as Buran ran into delays, it was decided to turn a test model of Energiya (6S) into a flightworthy version (6SL) and launch that with the Polyus/Skif-DM payload, moving the Buran mission to the second flight of Energiya.

Manned or unmanned?

Irrespective of whether the first Energiya should carry a flight-rated orbiter or not, the question also arose whether to fly the real orbiter unmanned or manned, *whenever* it was ready to go. All earlier Soviet piloted spacecraft (Vostok, Voskhod, and Soyuz) had flown unmanned test missions before being cleared to fly cosmonauts. NASA had done the same in the Mercury, Gemini, and Apollo programs, but departed from the practice for the first Space Shuttle flight, flown by astronauts John Young and Robert Crippen in April 1981.

The Russians, on the other hand, decided to play it safe and fly their first mission unmanned. In fact, at least *two* unmanned flights were envisaged in the test flight schedules known to have been drawn up in the late 1980s. While this came as a surprise to many in the West (simply because NASA had done it the other way), it was a completely logical decision in light of what had happened in earlier Soviet manned space projects.

Not only had it always been a tradition to put new piloted spacecraft through their paces in the unmanned mode, even when cosmonauts were on board, their role was often secondary to that of automatic systems. Vostok, the Soviet Union's first piloted spacecraft, was a highly automated vehicle, partly because of early concerns over the effects of zero-*g* and other factors on a cosmonaut's ability to control a spacecraft, but also because its design was unified with that of an unmanned spy satellite. While the same can be said of America's Mercury spacecraft, mission success increasingly depended on human involvement as NASA moved on to the Gemini and Apollo programs, in no small part due to pressure from the astronauts themselves. In the Soviet Union, on the other hand, the high degree of spacecraft automation continued in the Voskhod and Soyuz programs, even though experience had shown by that time that people could perfectly operate in weightlessness.

All this regularly resulted in fierce man-vs.-machine debates between Air Force officials, on the one hand (in favor of manual control), and the design and industry teams, on the other hand (in favor of automatic control), especially when it came to deciding whether Soyuz dockings should be performed in manual or automatic mode. However, the Air Force and its cosmonaut team had little say in the design of piloted spacecraft. This was considered the exclusive domain of the design bureaus themselves and of the space branch of the Strategic Rocket Forces, which officially placed orders for both manned and unmanned space vehicles (including Buran) and lacked the Air Force tradition of emphasizing the need to have a man in the loop. The results of this policy are still evident today. Dockings of Soyuz spacecraft with the International Space Station continue to take place in automatic mode, with the commander allowed to take over manual control only in emergency situations.

When the Russians were faced with the decision whether to fly Buran manned or unmanned, launch and orbital operations were probably not a major issue in the discussion. Launch is a highly automated process anyway with little or no crew involvement (as it is in the Space Shuttle program) and orbital operations could be limited to an absolute minimum if a very conservative test flight were planned.

The biggest difference with earlier ballistic spacecraft was that Buran would land like an airplane, an operation usually entrusted to pilots. Indeed, Space Shuttle pilots have so far always taken control of the Orbiter for the final approach and landing, although autoland capability has been present from the beginning of the program. However, for Buran, automatic landings would have been the preferred technique even for *manned* missions, meaning there was little point in risking the lives of cosmonauts on a first-flight vehicle.

Among the official reasons given was that with the limited amount of landing opportunities the ship would have to be able to land in adverse weather conditions such as snow and fog. Also, there was a need to safely bring the vehicle down if the pilots suffered from the effects of zero-*g* or became incapacitated for some reason [27]. Another factor may have been that, by the time Buran flew, there was more confidence in the ability of microwave landing systems to ensure safe hands-off landings than there had been almost a decade earlier when the Space Shuttle was gearing up for its test flights [28]. However, the *real* underlying reason appears to have been a preference in the space industry for highly automated spacecraft that was firmly rooted in the history of the Soviet space program.

Cosmonaut Oleg Makarov, one of the more outspoken members of the cosmonaut corps, voiced his view on the matter in an interview shortly after Buran's flight. He said the country's space program had been built primarily on expendable launch vehicle technology, whereas the US program had evolved from both launcher and winged vehicle experience, and had included significant participation of personnel with aviation backgrounds. He said:

> "The aviation industry is strong in the US, but it is just the opposite in my country. While the US places more confidence in the crew, the Soviet space program places full reliance on totally automatic missions for initial tests, such as the first launch of Buran. The [most important role] in the Soviet space program is [played] by launch vehicle engineers" [29].

Despite the industry's preference for dead stick landings, there can be little doubt that the man-vs.-machine debate would have flared up again if Buran had ever got to the point where it was ready to carry a crew. Actually, there was skepticism in the cosmonaut corps not only about automatic landings, but also about the wisdom of flying Buran unmanned at all. According to the official history of NPO Energiya a number of cosmonauts, including Igor Volk and Aleksey Leonov, sent a collective letter to the government several months before the launch, saying that Buran could not reliably fly in unmanned mode and should be manned on its first mission.

However, a special commission set up to investigate this matter concluded that Buran should fly unmanned. One concession that was apparently made as a result of the discussion was to fly a conservative two-orbit mission rather than a more ambitious three-day flight considered earlier [30].

The formation of the commission may have been a symbolic move more than anything else. Speaking after the flight, Igor Volk stated that there had never really been an option to fly the first mission of Buran with a crew. He called the timing of the flight political, saying it had been demanded by management to demonstrate the vehicle in competition with the return to flight of the Space Shuttle after the Challenger accident in 1986. Volk rated the system's maturity for the first flight as "near-zero", saying the two-orbit mission was flown because that was all the ship's computers could handle at the time [31].

Vehicle configuration

The decision to fly a two-orbit rather than a three-day mission allowed the Russians to significantly reduce the number of on-board systems and thereby move up the launch date. Apart from requiring less sophisticated software, the shorter flight obviated the need for installing such systems as fuel cells, a payload bay door opening mechanism, payload bay door radiators, etc. The only objective of the flight was to see if Buran could safely reach orbit and return back to Earth. With no crew on board, few of the life support systems needed to support humans were carried. For instance, Buran had a 90 percent nitrogen/10 percent oxygen atmosphere to minimize the risk of fire.

Original plans called for the space-rated orbiters to be equipped with two Lyulka AL-31 turbojet engines to provide flight path modification capability during the return to Earth. For this purpose Buran had two niches on either side of the vertical stabilizer to house the engine pods. However, in late 1987/early 1988 a decision was made *not* to install the engines, fill the niches with panels, and cover them with ATM-19PKP flexible thermal insulation.

There is conflicting information on the reasons for this decision. One source claims the atmospheric landing tests performed with the full-scale BTS-002 vehicle had shown that control was sufficient without these engines [32]. Another says the engines were not ready for the first flight. Although they had been flown on the BTS-002, they had never been ignited in flight, nor had the thermal protection covers for the engine inlet and outlet been tested. On top of that, there were mass-related issues that needed to be addressed before the engines were flown. Not only did the engines weigh about 400 kg each, they also required support systems such as a kerosene tank (probably to be placed in the mid fuselage under the payload bay), fire suppression systems, etc. However, once those issues had been resolved, the engines might well have flown on future missions [33]. One may also speculate that the presence of the engines would have unnecessarily complicated the automatic flight program for the maiden mission. Interestingly, the throttle lever for the AL-31 engines was *not* removed from Buran's cockpit for the first flight. The removal of the AL-31 engines slightly changed the vehicle's center of gravity and placed higher aerodynamic loads

Post-flight picture of Buran shows one of the engine niches covered with flexible thermal insulation panels (B. Vis).

on the vertical stabilizer. Therefore, additional wind tunnel tests were run to make sure that the absence of the engines posed no unexpected problems.

Buran's cargo bay was not empty during the first flight. Sitting in the middle of the cargo bay was a pressurized module known as Unit for Additional Instruments (BDP for *Blok Dopolnitelnykh Priborov*) or 37KB. BDP performed a role similar to

The 37KB/BDP payload (B. Vis).

the Development Flight Instrumentation (DFI) on the orbital test flights of the US Space Shuttle. It was stowed full with instrumentation to record about 6,000 parameters during the flight and also carried support equipment such as batteries to compensate for the absence of fuel cells on Buran's maiden mission.

Its design was based on a series of modules (37K) originally planned for the Mir space station, only one of which (Kvant or 37KE) was eventually flown. On 19 April 1982 the KB Salyut design bureau (then a branch of NPO Energiya) received an order to develop a series of such modules for Buran that would carry out a variety of functions. Built at the Khrunichev factory, the first such module (serial nr. 37070) was shipped to Baykonur in February 1986 to be flown on the maiden mission of Buran. After having been tested in the Proton area of the cosmodrome, it was transported to the MIK OK orbiter-processing facility for installation into Buran. Weighing 7,150 kg, it was 4.1 m wide, 5.1 m long, and had an internal volume of 37 m^3. The ultimate plan was to turn these modules into small scientific laboratories that could either remain in the cargo bay of Buran (like Spacelab) or be temporarily attached to space stations (see Chapter 8) [34].

PREPARING FOR THE MISSION

First orbiter roll-outs

After the Energiya pad fueling tests and core stage test firings in 1985–1986, the focus shifted to pad tests of the entire Energiya–Buran system. For this purpose the Russians used two full-scale mock-ups of Buran called OK-ML1 and OK-MT, delivered to the cosmodrome by VM-T carrier aircraft in December 1983 and August 1984. The work began in January–February 1986 when OK-ML1 was mated with the 4M core stage (a stack known as 4MP1) for a series of tests at the Assembly and Fueling Facility (MZK). Several weeks later 4M was united with OK-MT (a stack called 4MP or 11F36P) for more tests at the MZK from 13 to 16 May.

After roll-back to the Energiya assembly building, 4M was reconfigured for a series of fueling and dynamic tests with the OK-ML1 vehicle on both the UKSS and Energiya–Buran pad 37. Both the core stage and strap-on boosters were loaded with simulated propellants. The dynamic tests were necessary because the huge Dynamic Test Stand was still under construction and saw the use of small solid-fuel rockets on the core stage to create vibrations. Dubbed 4MKS-D, the stack spent two weeks on the UKSS (13–28 August 1986) and over a month on pad 37 (29 August–4 October 1986) [35]. This was the first time ever that an Energiya–Buran combination had spent time on the pad.

Next up was the 4MP/11F36P stack with the OK-MT vehicle for various loading tests of the orbiter both in the MZK and on launch pad 37. The combination was

Roll-out of OK-MT full-scale model (*source*: *www.buran.ru*).

rolled out to the pad on 5 May 1987 and shown to General Secretary Gorbachov during his visit to the cosmodrome in mid-May. 4MP returned to the MZK on 14 May, one day prior to the launch of Energiya 6SL, to make sure that it wouldn't be damaged in case the Energiya blew up during lift-off. Afterwards, it spent one more month on the pad (28 May–29 June 1987) to complete the tests. Similar tests with the 4MP stack were conducted on pad 37 in October–November 1987. After that, the flight hardware for the first Energiya–Buran mission was ready to make its appearance on the pad.

Preparing Buran

The orbiter for the first space mission (vehicle 1K, mission 1K1) arrived at the Baykonur cosmodrome on a VM-T carrier aircraft on 11 December 1985. Although eventually called Buran, it had the name Baykal painted on its side until at least April 1988 (see Chapter 2). The MIK OK orbiter-processing building on Site 254 was only 50/60 percent ready, with only one of the five bays (bay 104) available. The decision to send Buran to Baykonur so early came in an apparent response to the August 1985 government/party decree calling for a maiden mission in late 1986. The move was also designed to stimulate and speed up work at Baykonur. For the same purpose MOM minister Oleg Baklanov flew to Baykonur in January 1986 and set up three teams to prepare for the flight: the first team (headed by Yuriy Semyonov) had to ensure that Buran was ready for flight in the third quarter of 1987, the second team (led by Boris Gubanov) had to concentrate on readiness of Energiya–Buran as a whole, and the third team (led by Baklanov's deputy S.S. Vanin) focused on readying

1K orbiter with the name Baykal painted on the side (*source*: *www.buran.ru*).

launch facilities and other ground equipment. Between January and March 1986 the number of people working at the Buran processing facility rose from a mere 60 to 1,800.

Not only the Buran facilities but also the orbiter itself was far from ready when it arrived at the cosmodrome. This was related not only to the earlier than planned shipping date, but also to the limited payload capacity of the VM-T aircraft. Many systems needed for orbital flight as well as major components such as the vertical stabilizer and landing gear had not yet been installed. Furthermore, only 70 percent of the thermal protection tiles had been installed, making it necessary to set up a special tile-manufacturing facility in the Buran processing facility.

In the summer of 1987 leading program officials were invited to attend a meeting of the Defense Council, the supreme decision-making body on national security issues. Chaired by General Secretary Mikhail Gorbachov, it included the highest party and military officials in the Soviet Union such as the Minister of Defense and the Minister of Foreign Affairs. At the meeting, the officials pledged to launch Buran in the first quarter of 1988, although some (including Boris Gubanov) privately had strong doubts the orbiter would make that target date [36]. Presumably, Soviet space officials were under pressure to launch Buran before the post-Challenger return to flight of the US Space Shuttle, then planned for June 1988.

Final assembly of Buran was not officially completed until 15 October 1987, after which the vehicle was transferred to MIK OK's control and test bay for electric tests and to the anechoic chamber for tests of the radio systems. It had already spent some time in those bays *before* completion of assembly to uncover any potential problems at an early stage. On 15 February 1988 preparations began for test firings of the ODU propulsion system and the Auxiliary Power Units at the orbiter test-firing stand,

Orbiter undergoing test firings (*source*: *www.buran.ru*).

located in open air right next to the MIK OK. On the whole, the tests, conducted between 25 April and 9 May, produced satisfactory results. During the two weeks that Buran spent outdoors, several communication tests were carried out between the orbiter and Mission Control in Kaliningrad near Moscow using relay satellites.

Preparing the stack

The next step in the launch preparation process was for Buran to be mated with its launch vehicle (Energiya rocket 1L) for an experimental roll-out to pad 37. The 1L rocket had always been well ahead of Buran in its launch preparations. Assembly of the core stage in the Energiya assembly building had begun back in October 1986, shortly after work with the core stage for vehicle 6SL had been completed. In early 1988 (14 January–2 February) the 1L rocket had already spent about three weeks on pad 37 for a variety of tests, including firing tests of the hydrogen igniters and retraction tests of the various platforms connecting the launch towers with the rocket.

The Energiya 1L–Buran stack arrived on the pad in the third week of May (the roll-out date has been given both as 19 May and 23 May). Once again a multitude of tests were performed, although none of them involved actual fueling of the rocket or the orbiter. One goal of the pad tests was to see if various sources of electromagnetic radiation at Baykonur did not interfere with the operation of on-board systems. The main problems uncovered during the pad tests were with the interaction between the orbiter and rocket computers and with the ground software needed to analyse telemetry at the cosmodrome and in Mission Control.

Actually, the pad tests in May–June were only part of a broader series of exercises at the cosmodrome intended to simulate pre-launch and post-landing operations, including numerous off-nominal situations. Involved in the exercises were not only the launch and recovery teams, but also the LII pilots, who simulated automatic landings on board Tu-154LL aircraft, with the MiG-25-SOTN performing the role of escort aircraft as it would during Buran's final descent. The exercises also offered the opportunity to test virtually the entire communication network for the mission, including tracking stations, Mission Control in Kaliningrad, and orbiting communications satellites. Ground crews rehearsed post-landing operations and were trained how to deal with a return-to-launch-site abort during ascent. For this purpose, the OK-MT Buran mock-up was transported to the Yubileynyy runway.

The Energiya–Buran stack returned to the assembly building after about 3–4 weeks of tests (the roll-back date has been given both as 10 June and 19 June). Apparently, the original plan was for the orbiter and rocket to undergo some additional tests and then return to the pad for launch in the summer of 1988. Internal planning documents show that in early 1988 the launch was scheduled for July [37]. However, program managers felt that several problems that had surfaced during testing over the preceding weeks needed to be dealt with and decided to remove Buran from the rocket and return it to its MIK OK processing facility.

The most serious problem had cropped up in April during test firings of an ODU propulsion module at the Primorskiy Branch of NPO Energiya near Leningrad. A valve used in the liquid-oxygen gasification system of the primary thrusters failed

Energiya 1L during pad tests in January 1988 (*source*: Mashinostroyeniye).

to close when commanded to do so, a problem that could jeopardize the operation of the thrusters in flight. Because of this and other issues with the ODU, it was deemed necessary to remove Buran's ODU module and partially disassemble it to carry out modification work. This also required changes to the flight software, which had already been adapted numerous times in the preceding months, a penalty the

Energiya–Buran inside the MZK building (*source*: *www.buran.ru*).

Russians had to pay for flying Buran unmanned. In the end, Buran went into orbit with the 21st version of the flight software.

After repairs to the ODU and integrated electrical tests with the final version of the flight software, Buran was moved back to bay 4 of the Energiya assembly building on 29 August for reintegration with Energiya 1L. With that work complete, the stack was rolled over to the nearby MZK building on 13 September for a series of hazardous and other operations. These included various loading operations (kerosene for the Buran propulsion system, hydrazine fuel and nitrogen gas for the Auxiliary Power Units, ammonia for the thermal control system, air for the cabin repressurization system), installation of batteries aboard Buran, solid-fuel separation motors on the strap-on boosters, and pyrotechnics for the Buran/core stage separation system.

Finally, the large doors of the MZK were opened in the early hours of 10 October and four diesel locomotives began pulling the impressive 3,500-ton combination of Energiya, Buran, and transporter to launch pad 37. In an old tradition, coins imprinted with the roll-out date were placed on the rails before the assembly passed

Energiya–Buran being erected on the pad (*source*: *www.buran.ru*).

by and collected afterwards as souvenirs. It took the assembly some 3.5 hours to inch its way to the launch pad. Then another three hours were required to place the stack into vertical position and another hour to connect the Blok-Ya launch adapter to the launch table. All was now ready for final launch preparations to begin [38].

MEDIA BUILD-UP TO THE FIRST FLIGHT

None of these preparatory activities were reported by the Soviet media as they happened. Although Buran was no longer a state secret, Soviet space officials adopted a carefully limited posture concerning their plans. As the new policy of openness came into effect, some space officials and cosmonauts had begun confirming the existence of a shuttle in interviews and informal conversations in the mid-1980s. The *official* disclosure of the Soviet shuttle was left to Glavkosmos, an organization often described in the West as a Soviet equivalent to NASA, although it was actually the international relations arm of the Ministry of General Machine Building. Speaking at a Moscow press conference on 8 April 1987, Glavkosmos official Stepan Bogodyazh finally acknowledged that the Soviet Union was developing a reusable spacecraft and would announce the launch in advance. Bogodyazh's statement was confirmed on 13 May 1987, when the TASS news agency reported the imminent launch of the first Energiya rocket and added it would be used in the future to orbit reusable spacecraft. In January 1988 Glavkosmos chief Aleksandr Dunayev told another Moscow news conference the first Soviet shuttle would be launched soon. Two months later he said the mission was still expected shortly, although engineers were encountering problems daily. He also promised that (unlike the maiden Energiya launch) the shuttle launch would be broadcast live on Soviet television.

Despite these occasional statements, Soviet officials provided little if any technical details on their shuttle system. All this changed with the release of a government decree in early July 1988 that officially declassified the Energiya–Buran program [39]. By the end of that month the newspaper *Pravda* published an in-depth article on the Energiya–Buran system by chief designer Boris Gubanov, who confirmed earlier statements that Buran's first flight would be unmanned:

> "The role of manned flights on such carriers is not yet fully clear, such is the opinion of many specialists. A blind imitation of air travel is not relevant here. Space technology has gone its own way. Automatic ships were the first to enter space and humans followed only later. In the future, space will mainly be the working field of automatic spacecraft and transportation systems. The role of humans will probably be linked to research and specific maintenance and repair work ... Today's task is to accomplish the landing of an orbital ship in automatic mode without the involvement of pilots, and later that of separate units and stages. Nowadays, automatic flights from take-off to touchdown are also performed with aircraft, such as the Tu-204."

First picture of Energiya–Buran officially released to the public. The name "Buran" has been made unreadable (*source*: TASS).

Almost certainly, the flight had originally been timed to upstage the return to flight of the US Space Shuttle (mission STS-26), but the ODU problems and several other issues had thwarted those plans. However, unaware of those developments, Western media speculated the Soviet shuttle might still blast off before Discovery. On 20 September the *Washington Times* reported US spy satellites had photographed the Soviet shuttle on the pad earlier that month and that some US officials expected a launch within a week. This clearly was a mistake, because the stack had been inside the Energiya assembly building since late August. However, the newspaper also quoted Soviet space expert Saunders Kramer as saying the odds of that happening were all but zero. Kramer correctly stated the vehicle had been wheeled out to the pad in the spring and "then inexplicably removed", attributing the delay to problems with the computer software [40].

On 29 September 1988 all eyes were turned to Cape Canaveral in Florida, where the Space Shuttle Discovery was poised to return America to space 2.5 years after the January 1986 Challenger disaster. Lift-off occurred at 15:37 GMT, and eight minutes later Discovery safely entered orbit. However, the Russians were intent on stealing at least some of the thunder from NASA's success, taking advantage of the occasion to finally unveil their counterpart of the Space Shuttle to the world. Television viewers around the Soviet Union were surprised when the evening news program *Vremya* opened by showing a shot of the Energiya–Buran stack on the launch pad (taken during the May–June pad tests). The name Buran, painted on the side of the vehicle, had been carefully retouched so as not to be visible to the television audience, although it had leaked to the West several years earlier. The picture was accompanied by a terse TASS statement saying preparations for the launch were underway and that the mission would be unmanned. The Buran lead story was followed by footage of a conversation between the orbiting Mir crew and East German leader Erich Honecker. *Vremya* completely downplayed the news from Cape Canaveral by showing a brief clip of the Discovery launch just before going off the air. Interestingly, the following day some Soviet newspapers published exactly the same photograph with the name Buran erased altogether.

The next comment on launch preparations came from LII lead test pilot Igor Volk. Speaking at a meeting of the Association of Space Explorers in Bulgaria in early October, he revealed that 23 October was the target date for the launch when he left the Soviet Union. The Soviet shuttle moved into the background again until 23 October, when the TASS news agency once again repeated that final launch preparations were underway and revealed that the orbiter was called Buran. That same day *Vremya* showed the first ever footage of Energiya–Buran, including spectacular shots inside the MZK, during the roll-out and on the launch pad.

THE FIRST LAUNCH ATTEMPT

On 26 October the State Commission in charge of test flights of the Energiya–Buran system met at Baykonur to set a launch date for the mission. Such commissions were set up routinely in the Soviet Union to oversee launch preparations for specific projects. The composition of the Buran State Commission reflected the importance attached to the flight. Established in December 1985, it was headed by none less than the Minister of General Machine Building himself, initially Oleg Baklanov, replaced in April 1988 by Vitaliy Doguzhiyev. In all, the commission numbered 44 people, including 9 ministers, 10 deputy ministers, the President of the Academy of Sciences, leading Ministry of Defense officials, and several general and chief designers (Glushko, Gubanov, Semyonov, Lozino-Lozinskiy, Konopatov, Radovskiy, Barmin, Andryushenko, and Lapygin). The "technical leader" of the State Commission, somewhat comparable with a "launch director" in the US, was NPO Energiya head Valentin Glushko. However, the 80-year old Glushko, who had suffered a stroke only months earlier, was recovering in a Moscow hospital and had to be replaced by his deputy Boris Gubanov. Final preparations at the pad were the responsibility of the military teams of the so-called 6th Test Directorate under the leadership of Major-General Vladimir E. Gudilin.

Despite last-minute concerns over problems with another ODU test firing near Leningrad on 19 October, the commission declared Energiya and Buran ready to go. With meteorologists predicting excellent weather conditions, the commission set the launch for 29 October at 6:23.46 Moscow time (8:23.46 local time at Baykonur, 3:23.46 GMT), a decision announced by TASS the same day. Lift-off was timed such that the launch could be observed by the orbiting crew of the Mir space station (Vladimir Titov, Musa Manarov, and Valeriy Polyakov). Mir would be a minute or two short of a Baykonur flyover at lift-off time, allowing the crew to watch virtually the entire launch. Mir's orbit had been adjusted on 16 October to permit close coordination during the brief Buran flight, including launch and retrofire. However, the observations from Mir were not a strict requirement and the launch was not likely to be scrubbed if that objective could not be met. The main constraint for the launch window was to ensure a landing at Baykonur well before local sunset, ideally around noon. Speaking to reporters later that same day, Doguzhiyev did not hide the tension felt around the cosmodrome:

> "No one is indifferent or passive at the cosmodrome ... Behind outward calm there is much nervous pressure. Even we [State] Commission members find it difficult to answer questions" [41].

Among the final tasks to be accomplished at the pad in the last three days prior to launch was the retraction of the 145 m high rotating service structure, a relic of the N-1 days, which was now moved back to its parking position, fully exposing Buran to the elements. Strict safety measures were in place to protect personnel against any potential accidents on the pad. The region around the launch complex was divided into four safety zones. Zone 1 (2 km radius around the pad) was completely evacuated 12 hours before launch. By that time any personnel involved in final countdown operations were required to go to hermetically sealed and heavily armored bunkers, from where all final launch preparations (including fueling) were controlled. The bunkers were said to be capable of surviving impacts of rocket debris. Zone 2 (5 km radius) was cleared of personnel at $T-8$ hours as final preparations got underway for loading of liquid hydrogen. Zones 3 and 4 (8.5 km and 15 km radius) were evacuated at $T - 4$ hours and $T - 3$ hours to ensure safety of people in case of an explosion during engine ignition and during the early stages of ascent. The rules were much stricter than at the Kennedy Space Center, where people are allowed to watch Shuttle launches in open air from a distance of just about 5 km.

Two days before the planned launch, concern arose over some equipment in Buran's automatic landing system. VNIIRA, the design bureau in charge of the system, requested installing back-up equipment aboard the orbiter and first test that aboard a Tu-134B aircraft. Although this required activation of all the navigation and landing support systems at the Yubileynyy runway, the landing tests were authorized given the potentially catastrophic consequences of a failure in the automatic landing system. The back-up equipment was successfully tested during several approaches to the Yubileynyy runway on 28 October.

On the eve of the launch, Soviet officials backed down from their earlier promises to provide live television coverage of the launch and were now planning to show the recorded launch 35 minutes after the event. The landing would not be carried live either. As expected, the planned launch time of 3:23 GMT went by without any comment from the Soviet media. Anxiety grew as nothing was heard in the following 45 minutes or so. Finally, shortly after 4:00 GMT, TASS broke the silence by issuing a brief two-line statement:

> "As has been reported earlier, the launch of ... Energiya with the orbital ship Buran had been planned for 6.23 Moscow time on 29 October. During pre-launch preparations a four-hour delay of the launch has been announced."

This now theoretically put the launch time at 7:23 GMT, raising serious doubts among observers that the launch would take place that day at all. According to the original flight plan, landing would have taken place at 6:49 GMT, but the launch delay would now move this to around 11:00 GMT (16:00 local time at Baykonur), close to sunset. At around 7:30 GMT, shortly after the rescheduled launch time, TASS reported what had been obvious all along:

> "During final launch preparations for the rocket carrier Energiya with the orbital ship Buran there was a deviation in one of the launch support systems. As a result of this an automatic command was issued to stop further work. At the present time work is underway to eliminate the problems. A new launch date and time will be announced later."

It wasn't until later in the day and in press interviews the following days that officials began providing details about the exact cause of the scrub. It turned out the countdown had been halted at $T - 51$ seconds because a platform had failed to properly retract from the rocket. When it came to describing the exact nature of that platform, Soviet space officials, not used to communicating problems to the media, did a poor job. Talking to a *Pravda* reporter, Vladimir Gudilin, the head of launch pad operations, said:

> "51 seconds before the launch one of the servicing platforms did not move away from the rocket. To be more precise, it visually moved away, but the signal confirming this did not reach the computer checking the launch readiness of all systems. Until the last seconds this platform holds an aiming platform, controlling the gyroscopes. The computer [did not receive the retraction signal] and instantaneously stopped the launch program" [42].

A TASS reporter quoted Gudilin as saying that:

> "the platform of the cosmonaut emergency evacuation unit—this is where the system which ensures that the rocket's gyroscopes are installed precisely is situated—did not move away to a safe distance" [43].

Reporting on the scrub several days later, *Aviation Week* interpreted these statements as follows:

> "The . . . launch attempt was scrubbed when the orbiter access arm on the launch pad's left service structure failed to retract as commanded . . . The arm extends to the orbiter's side hatch and allows cosmonauts and technicians to enter the vehicle on the pad. The arm also carries an umbilical connection which provides

Image of Energiya–Buran showing the arm connected to the azimuthal alignment plate (*source*: *www.buran.ru*).

ground support to the orbiter's guidance and navigation system—specifically, its gyroscopes'' [44].

The platform in question actually was a 300 kg black plate mounted outside the intertank area of the core stage. It held three instruments needed for pre-launch azimuthal alignment of the core stage's inertial guidance platforms. The plate, about the size of a small automobile, was installed on the core stage in the Energiya assembly building. After the rocket arrived on the pad, the plate was connected to a swing arm extending from the launch tower. This arm was situated several levels *above* the orbiter access arm. The retraction process took place in two steps: first, the plate had to disconnect itself from the intertank in three seconds' time and only after that would the swing arm come into action to safely retract it from the rocket. What happened on 29 October 1988 was that the plate needed forty rather than three seconds to disengage. Sensing the sluggish movement of the plate, the rocket's on-board computers stopped the countdown and no retraction signal was sent to the swing arm, which obediently remained in place.

Gudilin's rather confusing statements, which created the impression that an arm had failed to retract from the rocket, led to some discussion at the State Commission the day after the scrub. Particularly unhappy with the confusion was Vladimir Barmin, the chief designer of the launch pad design bureau KBOM, which would have been held responsible if a swing arm problem had *really* been the culprit. However, the problem was with the plate disconnect mechanism, which was considered the responsibility of the rocket team.

Despite the TASS statement on the 4-hour launch delay, another launch attempt later in the day was never seriously discussed. According to information released much later, the countdown could not be recycled for another attempt the same day if it was halted after the orbiter had switched to internal battery power at $T-80$ seconds. Furthermore, by the time the countdown was halted, the umbilicals for thermostatic control of the hydrogen tank had been disconnected and temperatures inside the tank were slowly rising. Therefore, some 10 minutes after the hold was called and the problem had been identified, the team decided to delay the launch for several days and begin the lengthy process of draining the core stage and strap-ons.

Even that did not go entirely by the book. Draining of one strap-on booster's liquid-oxygen tank went agonizingly slowly, a problem later attributed to a filter that had become clogged by contaminants in the oxidizer tank. Access to the filter at the launch pad was very difficult, causing fears the rocket might have to be rolled back to the assembly building, but in the end the problem was fixed on the pad. One eyewitness later said the clogged filter would probably have caused a catastrophe during launch. If true, the scrub had been a blessing. Also uncovered during post-scrub operations was that an accelerometer in Energiya's tail section had been inadvertently mounted upside down [45].

The problem with the plate disconnect mechanism had been completely unexpected. There had been no azimuth orientation system on the Energiya 6SL vehicle launched in May 1987 because there were less stringent requirements for precise orbital insertion of the Polyus payload. Neither had the problem surfaced during the simulated countdowns at the pad in January–February and May–June. The causes of the mishap were investigated by a team headed by Vyacheslav Filin, a deputy of Gubanov. Part of the troubleshooting was to simulate the forces needed to detach the plate. This was done at the Energiya assembly building using the already assembled Energiya 2L vehicle and also a mock-up core stage intertank structure. In the end, the problem was traced to rubber dust covers on the plate that had somehow become sticky, possibly because of exposure to variable temperatures in the preceding months. In no time, engineers at the Progress plant in Kuybyshev redesigned the dust covers as well as the plate disconnect mechanism to ensure that the problem would not recur [46].

BURAN FLIES

Defying the weather

The troubleshooting was not finished until after the October Revolution holiday, celebrated on 7 November. However, as Gudilin said in one interview:

> "the time has gone by when launches were hurried along to fit in with holiday dates."

On 12 November TASS announced that the launch had been rescheduled for 15 November at 6:00 Moscow time (3:00 GMT, 8:00 local time at Baykonur). Visual observations from Mir were no longer a factor in determining the launch time, probably because orbital precession had shifted the station's flight path such that it now passed over the cosmodrome much earlier than was acceptable for the Buran launch.

The biggest concern as launch time drew closer was the weather. While skies had been crystal clear for the launch attempt on 29 October, a low-pressure front bringing rain and strong winds was now approaching Baykonur from the Aral Sea. At 17:00 local time (12:00 GMT) on 14 November meteorologists reported they were seeing a tendency for the front to bypass Baykonur, although nothing could be guaranteed. Four hours later the forecast had remained unchanged and the State Commission decided to press ahead with fueling of the rocket. First to be loaded were the liquid-oxygen tanks of both the strap-on boosters and core stage, followed about two hours later by the kerosene tanks of the strap-ons and the core stage liquid-hydrogen tank. Soviet media made no secret of the iffy weather conditions. On the eve of launch, a correspondent of the *Vremya* evening television news program reported:

> "Everything that depends on people has been done. But the weather is worsening with each passing hour. If the wind rises into a squall and the orbital vehicle ... becomes covered with a crust of ice, then the launch time will be changed again."

В/ч 11284 АВ-4

ШТОРМ-ПРЕДУПРЕЖДЕНИЕ № 1

15. ноября 1988 г.

1. В период с «07⁰⁰» часов до «12⁰⁰» часов.
2. По району (маршруту) Юбилейный площадка 110.
3. Ожидается:
Туман при видимости
600–1000 м.
Усиление юго-западного ветра
9–12 м/сек порывы временами
до 20 м/с
4. Время составления «06» ч. «15» м. Подпись составившего
5. Время вручения «08» ч. «17» м. Подпись получившего

Gale warning issued at 6:15 AM local time (*source*: wwww.buran.ru).

At midnight local time, with fueling of the liquid-oxygen tanks underway, the forecast took a turn for the worse. The low-pressure front had broken up in two parts, one of which was now headed straight for the launch site. Less than 2 hours before launch the chief weather officer handed over a gale warning to Gudilin. Conditions expected between 7:00 and 12:00 local time (2:00–7:00 GMT) both at the launch pad and the runway were strengthening southwest winds with speeds of 9 to 12 m/s, gusting to 20 m/s. Meanwhile, weather balloon data also showed unstable conditions up to altitudes of 25 km, with highly variable wind speeds (maximum 70 m/s) and wind directions at different levels. With just 30 minutes left in the countdown, observed conditions were overcast skies with a cloud ceiling at 550 meters, drizzle, winds of 15 m/s gusting to 19 m/s, a temperature of +2.8°C, and visibility of 10 km. The agreed wind speed limit for launch was 15 m/s, while for landing the maximum allowed wind speeds were 5 m/s for tail winds, 10 m/s for crosswinds, and 20 m/s for head winds.

The marginal weather conditions were a matter of concern for several reasons. The combination of drizzle and low temperatures posed the threat of significant ice build-up on the rocket, orbiter, and launch pad. Chunks of ice falling off the rocket during launch could cause significant damage to Buran's fragile thermal protection system. This risk has always been well understood in the Space Shuttle program, where specialized ice inspection teams are routinely sent out to the pad in the final hours before launch. The available information suggests that the Russians considered a 2 mm ice layer on the rocket acceptable and decided to go ahead based on the prediction *prior to fueling* that the thickness of the layer would not exceed 1.7 mm. All indications are that no ice inspection teams were sent to the pad and that any later estimates were based solely on close-up television shots of the launch vehicle. Apparently, those images were not always reassuring. As Gudilin later recalled:

> "we could see relatively big chunks of ice falling from the rocket and the orbiter."

Aside from ice build-up on the vehicle, there were worries about ice formation on the runway and the general effects of cold weather on vehicle performance. Although Energiya had no solid rocket boosters, the Challenger accident, where cold temperatures had contributed to the failure of an O-ring seal in one of the solids, was "in the back of our minds", as Gudilin puts it in his memoirs.

Another issue were the strong winds both at ground level and in the upper atmosphere. There were fears that ground-level winds could cause the vehicle to hit one of the launch pad structures during lift-off and that unstable upper-level winds could knock the stack off course. Weather officers at Baykonur continuously sent the latest wind data to a team of specialists at NPO Elektropribor in Kharkov, the design bureau that was responsible for Energiya's guidance, navigation, and control systems. Computer simulations there convinced the team there was enough margin to go ahead, although the observed conditions were clearly outside the experience base for this type of launch vehicle. Winds were also near or above prescribed limits for a return-to-launch-site abort or a nominal landing. That problem was addressed by having Buran approach the runway from the northeast rather than the southwest, turning an out-of-limits tail wind into an acceptable head wind, although even that was on the limit.

The weather on 15 November 1988 violated just about every imaginable meteorological launch commit criterion for a Space Shuttle launch. Leaving aside the temperatures and the wind, two other showstoppers for a Shuttle launch that day would have been the precipitation and the low cloud cover. No NASA launch or flight director in his right mind would even consider launching or landing a Shuttle Orbiter if there is only the slightest chance of precipitation in the vicinity of the launch pad or runway. With the Orbiter moving at high speeds, precipitation has the potential of causing significant damage to the vehicle's thermal protection system. However, for reasons that are not entirely clear, precipitation was no safety issue for the Russians, even though Buran's thermal protection system was very similar to that of the Orbiter. Even hail was said to be an acceptable condition, although this may have been bluff more than anything else. In fact, Buran's tiles suffered serious damage when the vehicle ran into a hail storm during a trip atop the Mriya carrier aircraft in 1989.

Cloud cover was not an issue for the Buran launch because there were no pilots on board who needed a clear view of the runway for a return to launch site or manual landing. The only clouds that meteorologists kept a close eye on were those with lightning potential. The overcast skies did prevent good ground-based optical tracking during launch and landing, which can be a critical factor in post-flight analysis of anomalies.

Even though conditions were close to violating launch commit criteria, the team decided to fly anyway, despite another gale warning issued just 13 minutes before launch. True, the Russians' launch weather rules in general were more relaxed than those adopted by NASA or the US Air Force, with some launches known to have taken place in near-blizzard conditions. However, the Buran mission was different from a conventional rocket launch in that the spacecraft was supposed to land like an aircraft.

All this begs the question why officials didn't wait one or more days for the weather to clear, especially because this was the maiden flight of a vehicle vastly different from anything the Russians had flown before. Speaking shortly after the mission, former cosmonaut Gherman Titov said:

> "We deliberately refused to postpone the launch and wait for ideal conditions. The value of the flight is that its program included the maximum sum of real and rather difficult tasks" [47].

Still, one can only wonder if the team didn't suffer from what is sometimes referred to in the US as "launch fever". Testifying to this is an eyewitness report of one member of the meteorological support team, who claims that some of the observations that morning showed wind gusts of up to 25 m/s. However, the chief weather officer, under pressure to report good news, only presented the launch team with the weather updates that showed the lower wind speed values. The same person notes that Gudilin's main argument in favor of launching that day was that another scrub could delay the flight until spring. It would require more testing and take them further into late autumn and possibly winter, when weather conditions can get far worse than the ones observed that morning [48].

Another concern with a lengthy delay may have been that the already frail support for the Buran program from the Soviet leadership might dwindle even further and could put the flight on indefinite hold, particularly now that the US Space Shuttle had returned to flight. Still, whatever the real motives were for launching that day, it was a decision fraught with risk [49].

Flight control and communications

Buran's mission was controlled from the Mission Control Centre (TsUP) in Kaliningrad near Moscow, the same facility from where Soviet manned space missions had been monitored ever since the joint US–Soviet Apollo–Soyuz mission in 1975. For the Buran mission a new big control room with modernized computer systems was inaugurated. It had the same layout as the neighboring space station control room, with several rows of consoles and a "balcony" where invited guests and media representatives could follow events. Later the Buran control room was modified for controlling the Russian segment of the International Space Station, while the Mir control room was closed down after the space station's re-entry in 2001. Flight

Four Soviet tracking ships (*Belyayev*, *Volkov*, *Patsayev*, *Dobrovolskiy*) moored side by side in Leningrad (*source*: Simon Vaughan).

director for the Buran mission was V.G. Kravets, although overall supervision was in the hands of former cosmonaut Valeriy Ryumin, who also served as flight director for Mir at the time. Working in conjunction with TsUP during the approach and landing phase was the command and control building (OKPD), located right next to the Yubileynyy runway at Baykonur.

TsUP received and relayed information via an elaborate communication network consisting of six ground stations on Soviet territory, four vessels of the Soviet space communications fleet, and several communications satellites in geostationary and highly elliptical orbits. Combined, these facilities provided about 40 minutes of coverage during a single 90-minute orbit.

The ground stations, part of the so-called Command and Measurement Complex (KIK), were situated in Yevpatoriya (Crimea), Shcholkovo (near Moscow), Dzhusaly (near Baykonur), Ulan-Ude, Ussuriysk, and Yelizovo (near Petrapavlovsk-Kamchatskiy). All received broadband information (television and telemetry) from Buran and relayed that real-time to TsUP via Molniya-1 satellites and/or ground lines.

The communication vessels were the *Kosmonavt Georgiy Dobrovolskiy* and the *Marshal Nedelin* in the South Pacific and the *Kosmonavt Vladislav Volkov* and *Kosmonavt Pavel Belyayev* in the South Atlantic.

The *Dobrovolskiy* had moved to the South Pacific (45° southern latitude, 133° western longitude) from its usual location in the South Atlantic. Just like the KIK ground stations, it relayed broadband information from Buran real-time to TsUP. The signal traveled more than 120,000 km to reach Mission Control. First, the received data were relayed from the *Dobrovolskiy* to the geostationary Gorizont-6 satellite, which had been relocated from 140°E to 190°E between July and September in support of the mission. From Gorizont the data went to a ground station of the Orbita network in Petropavlovsk-Kamchatskiy, from there to the neighboring KIK station in Yelizovo, subsequently to an orbiting Molniya satellite, and from there to a station near Moscow, which finally transmitted the data to TsUP.

The *Nedelin* had left the port of Petropavlovsk-Kamchatskiy on 5 October, reaching its final location (same coordinates as the *Dobrovolskiy*) on 25 October. It served in a back-up role to the *Dobrovolskiy*, being capable of receiving only telemetry. The telemetry was processed on board and then relayed to the Raduga-16 communications satellite, stationed at 190°E right next to Gorizont-6. From there it went to the ground station in Petropavlovsk-Kamchatskiy, which relayed it to TsUP via ground lines.

Just like the *Nedelin*, the *Volkov* (5° northern latitude, 30° western longitude) and the *Belyayev* (16° northern latitude, 21° western longtitude) received only telemetry from Buran, relaying that to TsUP via Raduga satellites.

A crucial link in the network was Kosmos-1897, the second satellite in the Luch/Altair series, the Soviet equivalent of the US Tracking Data and Relay Satellites. After its launch in November 1986 the satellite had been stationed at 95°E to support Mir operations, but on 26 July 1988 it began moving westward in preparation for the Buran launch, reaching its ultimate destination of 12°E on 26 August. Its footprint stretched from the middle of the Atlantic Ocean to the central Soviet Union. Unlike the Molniya, Raduga, and Gorizont satellites, it was used for direct two-way communications between TsUP and Buran via a station near Moscow. The satellite had three antennas, one for the link with the ground and two for direct line-of-sight communications with Buran (one in the centimeter waveband, the other in the decimeter waveband). However, the centimeter waveband system, mainly needed for television, was not activated for the mission, because Buran was not equipped with parabolic narrow-beam ONA antennas. Television images from a camera installed in the cockpit were relayed directly to ground stations when the vehicle passed over Soviet territory. Although only one Luch/Altair was available during Buran's mission, plans were to deploy two more for 100 percent coverage of future Buran flights [50].

The launch

Fueling of Energiya was completed three hours before launch and that of the LOX tanks of Buran's ODU propulsion system at $T - 2h45m$. A critical point came at $T - 10$ minutes, when the countdown switched to automatic control. Controllers breathed a sigh of relief when the balky azimuthal alignment plate responsible for the 29 October scrub retracted as planned at $T - 51$ seconds. With sound suppression water gushing onto the pad, the four RD-0120 engines of the Energiya 1L rocket roared to life at $T - 9.9$ seconds and smoothly built up thrust, clearing the way for ignition of the four strap-on boosters at $T - 3.7$ seconds. With all engines at full thrust and no problems detected, Energiya–Buran slowly lifted off the pad exactly as planned at 6:00.00 Moscow time. It was a highly emotional moment for the thousands of people who had dedicated many years of their lives to this program ever since its approval in February 1976, although the atmosphere in the nearby control bunker was said to be business-like as all eyes were focused on the performance of the rocket and orbiter.

Buran clears the tower (*source*: *www.buran.ru*).

As it cleared the tower, the stack performed a 28.7° roll maneuver to place it in the proper position for ascent. For onlookers the launch proved to be rather anticlimactic. Just seconds after clearing the launch tower, Energiya–Buran disappeared into the low cloud deck. "What a pity for the photographers," *Pravda* wrote the following day. "Standing out there freezing in the steppes all night and then everything is over in the blink of an eye" [51].

The only persons to maintain visual contact with the vehicle after that were the crews of an An-26 weather reconnaissance plane and the MiG-25 SOTN chase plane. The task of the chase plane during launch was not only to shoot video of the stack, but also to accompany Buran to the runway in the event of a return-to-launch-site abort. Behind the controls of the chase plane, which had taken off ten minutes before launch, was LII pilot Magomed Tolboyev, accompanied by cameraman Sergey Zhadovskiy. "It's on its way! It's going!" Tolboyev enthusiastically radioed to the ground as the stack broke through the clouds. Somewhat later he called out: "Engine operating mode changing." He was referring to a reduction in thrust of both the core stage's RD-0120 engines (between $T + 30$s and $T + 1$m11s) and the strap-on boosters' RD-170 engines (between $T + 39$s and $T + 1$m15s) as the rocket and orbiter passed through the phase of maximum aerodynamic pressure.

At $T + 2$m23.95s the four strap-ons shut down their RD-170 engines and at $T + 2$m25.85s separated in pairs from the core stage. The separation was clearly visible from the MiG-25, with Tolboyev reporting:

"The strap-ons have separated! They're on their way back to the ground ... Great. We can see them falling together, in parallel."

Not long after separation from the rocket each pair of boosters split in two, with all four now headed back to Earth individually. The boosters were not equipped with parachute systems for this mission and crashed into the steppes some 420 km from the launch pad about 7 minutes after separation.

Not long after booster separation the MiG-25 lost sight of the stack as it moved further downrange and eventually disappeared behind the horizon. All eyes were now focused on the telemetry being received from Soviet tracking stations and relayed to TsUP near Moscow. About 3 minutes into the launch, Buran reached the point where it could no longer return to the Baykonur cosmodrome for an emergency landing. As the stack sped further towards orbit, a camera installed behind one of Buran's cockpit windows began sending back images of the Earth. At $T + 6$m53s the core stage's RD-0120 engines began slowly throttling down and eventually shut down at $T + 7$m47.8s. Energiya's job done, members of the rocket team quietly shook hands beneath the table, celebrating the second successful flight of the launch vehicle in as many attempts. Now it was up to the orbiter team to finish the job. Buran was now in a theoretical -11.2×154.2 km orbit and, if nothing were done, would re-enter shortly afterwards.

Separation of the orbiter from the core stage took place at $T + 8$m02.8s at an altitude of roughly 150 km. The core stage was scheduled to re-enter the atmosphere, with fragments coming down in the Pacific some 19,500 km from the launch point. After firing its thrusters to move to a safe distance from the core stage, Buran now positioned itself for a critical burn of one of its two DOM orbital maneuvering engines to impart the 66.7 m/s of additional velocity needed to reach orbit. The burn, monitored by the easternmost Soviet ground stations as Buran headed for the Soviet–Chinese border, got underway at $T + 11$m28s and lasted 67 seconds.

About thirty-five minutes later, at $T + 46$m07s, as Buran came within range of the tracking ships *Dobrovolskiy* and *Nedelin* in the South Pacific, one of the DOM engines burned for another 40 seconds (delta-V of 41.7 m/s) to place the orbiter into its final 247×255 km orbit. Inclination was 51.6°, the same as that of the Mir space station, but the two were in different orbital planes. Since this was a conservative two-orbit test flight, there was no need for Buran to further increase its orbital altitude.

Although Soviet media did not carry the launch live, both Radio Moscow World Service and the Soviet domestic *Mayak* radio station reported the launch at the very beginning of their 3:00 GMT newscasts. The World Service even optimistically said Buran had been placed into orbit, although orbital insertion was still at least ten minutes away. At 4:10 GMT *Mayak* broadcast a recorded live report of the launch from its reporters both at Baykonur and in Mission Control in Kaliningrad. Moscow television showed the first footage of the launch 1.5 hours after blast-off. The TASS news agency issued the following official statement on the launch:

"On 15 November 1988 at 6.00 Moscow Time the Soviet Union launched the universal rocket space transportation system Energiya with the reusable ship

Buran. At 6.47 the orbital ship went into the planned orbit. The test program envisages a two-orbit flight of the orbital ship around the Earth and a landing in automatic mode at the Baykonur cosmodrome at 9.25 Moscow time."

Orbital operations

After the second DOM burn, Buran was placed into its nominal orbital orientation, with the left wing pointing towards Earth, which was considered the most favorable position from a thermal standpoint of view. The correct orientation was confirmed both by telemetry and television views of the Earth from the camera mounted in the cockpit.

Artist's conception of Buran in orbit with left wing pointed towards the Earth (*source*: *www.buran.ru*).

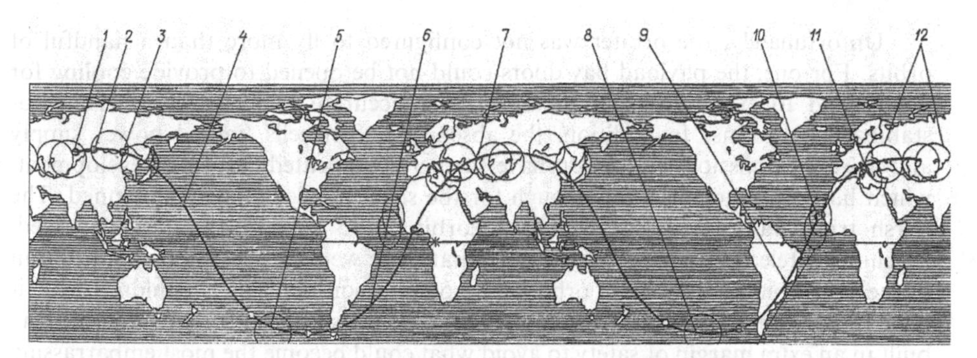

Main phases of Buran's mission: circles represent coverage zones of Soviet tracking means; 1, launch; 2, Buran separation; 3, first DOM burn; 4, flight in transfer orbit; 5, second DOM burn; 6, flight in final orbit; 7, preparation for deorbit burn; 8, deorbit burn; 9, descent; 10, entry interface; 11, approach and landing; 12, coverage zones of tracking stations (*source*: Yuriy Semyonov/Mashinostroyeniye).

Barely had Buran settled into its final orbit when preparations began for re-entry. As Buran came back within range of Soviet tracking means, it sent the computed deorbit parameters to TsUP, while TsUP in turn uplinked the latest data on wind speed and direction at the runway to Buran's computers. At 4:31 GMT the on-board software switched to the re-entry and landing sequence. Propellant was transferred from the forward to aft reaction control system to meet center-of-gravity requirements for re-entry and landing. Finally, the automatic systems commanded the orbiter to maneuver its tail toward the direction of flight in preparation for retrofire.

The deorbit burn began at a mission elapsed time of 2h20m07s as Buran flew over the South Pacific, where the burn was monitored by both the *Dobrovolskiy* and the *Nedelin*. It was an unusual retrofire location for the Russians, whose ballistic manned space capsules had always deorbited off the southeast coast of South America. However, Buran, being a winged vehicle, afforded a gentler and less steep re-entry trajectory. The 2m38s burn reduced Buran's speed by 162.4 m/s, enough for the vehicle to begin the one-hour descent back to the Baykonur cosmodrome.

Some Western space analysts were surprised that the mission was planned to last just two instead of three orbits. The ground track for the second orbit was well west of the launch site, meaning the orbiter had to initiate a lengthy right bank eastward to reach Baykonur as it flew on a northern trajectory up to the west coast of Africa. Had Buran stayed aloft for three revolutions, its ground track would naturally have taken it almost directly to the launch site.

Possibly, the Russians were eager to demonstrate Buran's cross-range capability and reduce the stress on the energy management system. However, a more likely explanation is that if something went wrong with the deorbit burn on the second orbit, there would still be time to correct the problem for three more deorbit attempts later that day. Any delays beyond that would have required Buran to stay in orbit for an additional day until its ground track brought it back within range of the Baykonur cosmodrome.

Unfortunately, the orbiter was not configured to fly more than a handful of orbits. For one, the payload bay doors could not be opened to provide cooling for Buran and to expose navigation sensors for accurate realignment of the gyro-stabilized platforms. In addition, the absence of fuel cells limited power supply and a 24-hour mission would also have further complicated software development, which had proved challenging enough for the short mission that was planned. The harsh truth was that if Buran didn't deorbit on 15 November, it was doomed. Documents released in recent years show that there were even plans to dump Buran in the ocean on the 4th or 5th orbit if for some reason the vehicle couldn't make it back to Baykonur [52]. By scheduling re-entry for the second orbit, flight controllers built in an extra margin of safety to avoid what could become the most embarrassing setback in the history of the Soviet space program.

Re-entry and landing

Some thirty minutes after the deorbit burn, having turned around with its nose pitched up, Buran hit the upper layers of the atmosphere at an angle of attack of 39° just off the west coast of Central Africa. Telemetry was relayed via the *Volkov* and *Belyayev* in the South Atlantic and through the Kosmos-1897 Luch/Altair satellite until Buran entered its communications black-out at 5:53 GMT at an altitude of roughly 90 km. The black-out, caused by a sheath of plasma enveloping the spacecraft, was expected to last between 16 and 19 minutes, more than three times longer than during a normal Soyuz re-entry.

Right within the expected timeframe, at 6:11 GMT, Buran came out of the black-out, with telemetry now being received by the ground station in Dzhusaly. Uplink from TsUP was via the Luch satellite, with Dzhusaly acting as back-up. At this point Buran, still flying ten times the speed of sound, was about 500 km from the runway and 10 minutes away from landing. Soon the vehicle was also picked up by the radars of the Vympel system at Baykonur.

Taking into account the strong southwest winds, Buran had already been programmed before the deorbit burn to approach the runway from the east. At an altitude of 20 km it was now up to the orbiter's on-board computers themselves to determine whether the vehicle would intercept the southern or northern Heading Alignment Cylinder (TsVK), two imaginary cylinders that help the vehicle line up with the runway. Later calculations showed there was a 97 percent chance Buran would pick the southern cylinder, but completely unexpectedly, Buran's computers decided to dissipate the remaining energy by flying around the northern cylinder. The result was that Buran initially flew at an angle almost perpendicular to the runway axis before making a right turn to line up with the runway. Many flight controllers and onlookers were in a state of shock until they realized what was going on. For a few brief moments, thought was reportedly given to activating the self-destruct system that was on board Buran in case it deviated from its landing path.

Meanwhile, Magomed Tolboyev and his cameraman had once again taken off aboard the MiG-25-SOTN to shoot video of the orbiter, among other things to determine the state of the heat shield before touchdown. A MiG-25RBK with Ural

Buran touches down.

Sultanov behind the controls had been on stand-by in case Buran was forced to return to Baykonur after a single orbit, but that did not turn out to be necessary. Buran's unexpected approach path made it difficult for Tolboyev to intercept the descending orbiter. As he later recalled:

> "... it was almost as if Buran felt it was being attacked ... It was only afterwards that everyone realized that the ship had picked the most optimal entry point for the final approach and had carried out a complex maneuver. But, for me it was unexpected, we hadn't even trained for it ... When Buran picked the complex descent trajectory, I had a feeling as if it was being controlled by a person. Of course, no one was on board, just equipment. And it did an outstanding job. The software for the Buran flight was just impeccable" [53].

TV transmissions from the MiG showed Buran's speed brake fully open at one point to adjust air speed as the vehicle went subsonic. The orbiter then entered a steep glideslope toward the approach end of the runway, breaking through the low cloud deck only seconds before touchdown. Finally, at 6:24.42 GMT, just one second earlier than planned, Buran landed at a speed of 263 km/h, with the left main gear touching down slightly ahead of the right main wheels. The vehicle then rocked briefly to the right before settling down on both main landing gears. Battling a strong head and crosswind of up to 20 m/s, the ship deployed its drag chutes as Tolboyev's MiG-25 triumphantly flew by. Wheels stop came at 6:25.24 GMT at a mission elapsed time of 3 hours 25 minutes 24 seconds.

One of the officials in the Baykonur control room later recalled the triumphant atmosphere:

> "A second of silence and then a storm of applause. Controllers and leaders jumped from their seats, embraced, and kissed each other. Many cried and bashfully dried their tears, both civilians and people in uniform. A feeling of achievement literally poured out of everyone. People congratulated the chief designers, Y.P. Semyonov and G.E. Lozino-Lozinskiy, and they, with their pinched faces, could not restrain their joy either and also thanked and congratulated everyone. We boarded some buses and rushed to the aerodrome. Stepping onto the runway, we could experience for ourselves that the warnings of the meteorologists had not been in vain. The wind blew away hats and literally blew us off our feet. And she, that beauty, was standing there right in the middle of the runway" [54].

According to official data Buran had overshot the planned touchdown point by 15 m and come down 3 m off the runway centerline, coming to a stop just 80 cm off the centerline after a landing roll-out of 1,620 m. However, a closer analysis shows it landed about 190 m *short* of the planned touchdown point and some 9 m off the runway centerline. During the landing roll-out Buran crossed the centerline and finally came to a standstill with its nose gear about 5 m off the other side of the centerline [55]. After landing, it took about 10 minutes to shut down on-board systems, with TsUP sending the final power-down command via the Luch relay satellite.

As expected, Soviet media did not carry the landing live, but both Radio Moscow World Service and *Mayak* reported the safe return only minutes after touchdown. Half an hour after landing, *Mayak* aired the recorded live report. One of the reporters couldn't contain his enthusiasm, shouting "Hurray!" and "Victory!" as Buran rolled to a stop on the runway. TASS, on the other hand, was typically brief and businesslike in its landing announcement:

> "On 15 November 1988 at 9.25 Moscow time the orbital ship Buran, having conducted a two-orbit flight around the Earth, touched down at the landing strip of the Baykonur cosmodrome. For the first time in the world a landing in automatic mode has been conducted. The program of the test launch of the universal rocket and space transportation system Energiya and the reusable orbital ship Buran has been carried out in full" [56].

POST-FLIGHT OPERATIONS

Post-landing operations on the runway included removal of residual LOX from the ODU propulsion system. After that, Buran was wheeled back to the MZK building,

Post-landing operations underway. Note heavy scorching on aft fuselage and elevons (*source*: *www.buran.ru*).

where—among other things—residual kerosene in the ODU system and hydrazine for the Auxiliary Power Units were drained from the vehicle's tanks. Buran was still in the MZK at the end of the month, when a French delegation headed by President François Mitterand visited the cosmodrome to watch the launch of "spationaut" Jean-Loup Chrétien aboard Soyuz TM-7 on 26 November.

After Buran was towed back to its MIK OK processing building, engineers got down to a close inspection of the vehicle. Much attention was focused on the ship's heat shield. Several dozen tiles were damaged, showing cracks or signs of erosion or melting, and seven were lost altogether (compared with sixteen on Columbia during STS-1). These were one black tile each on the vertical stabilizer, rudder/speed brake, and body flap, three black tiles on the underside of the left wing and one white tile near one of the overhead windows. The three black tiles were in an area bordering on one of the reinforced carbon–carbon panels on the leading edge of the wing. This is the only area where the underlying surface suffered major damage, fortunately without catastrophic consequences. There were also two missing blankets of flexible thermal insulation on the upper left wing and several gapfillers were missing on the vehicle's underside.

With the launch having taken place in cold and wet conditions, much of the damage sustained by the thermal protection system is believed to have been caused by chunks of ice falling from the launch tower, Energiya's core stage, and the orbiter itself. There was also some significant scorching of tiles on the vertical stabilizer and the aft fuselage of the vehicle. This was attributed not only to the thermal effects of re-entry, but also to exhaust gases impinging on the vehicle from the separation motors of Energiya's strap-on boosters [57].

Little more has been revealed about post-flight analysis of Buran. Before thorough checks could be completed, the orbiter had to be readied for a series of test flights atop the new Mriya carrier aircraft in May 1989 in preparation for a flight to the Paris Air Show in June 1989 (see Chapter 4). By the time Buran returned to its hangar in Baykonur, there were already growing doubts about the program's future. Moreover, since the second mission was to be flown by the second orbiter, there was no urgency in preparing Buran for its next flight.

POST-FLIGHT REACTIONS

Only hours after the mission the Central Committee of the Communist Party sent the obligatory congratulatory message to the Energiya–Buran team.

> "The launching of the Buran craft ... and its successful return to Earth open up a qualitatively new stage in the Soviet space research program and substantially extend our opportunities for space exploration. From now on, Soviet cosmonautics possesses not only the means of placing large payloads into various orbits, but also the ability to return them to Earth. The use of the new space transportation system in conjunction with expendable carrier rockets and with permanent manned orbital complexes makes it possible to concentrate the principal efforts and means on those areas of space exploration that will ensure the maximum economic return to the national economy and will advance science towards higher frontiers ... The new success of Soviet cosmonautics has once again convincingly demonstrated to the whole world the high level of our homeland's scientific and technical potential."

The message may not have sounded so convincing to the Energiya–Buran officials who had heard the private comments of Mikhail Gorbachov during his visit to the Baykonur cosmodrome in May 1987 (see Chapter 6). Gleb Lozino-Lozinskiy didn't spare Gorbachov in one of his final interviews many years later:

> "... we already felt that there probably wouldn't be any more flights ... Buran had flown. You'd think that ... the General Secretary of the CPSU, responsible for the country, its prestige, should show some interest. But when that General Secretary ... was told that Buran had landed, he just said: "OK, fine". He displayed absolutely no understanding, interest in the country's ... successes and achievements in the field of technology and science ...Gorbachov ...was notable for an exceptional ability to display inability. I later called him, tried to meet him to explain things, but to no avail" [58].

The Soviet press generally hailed the flight in a style typical of the pre-*glasnost* days, although that would gradually change in the following months as the space program in general increasingly became a target for public criticism. However, even amid the initial flush of excitement over the successful completion of the mission, there were voices of dissent, surprisingly from the space community itself. Just days after the flight, Roald Sagdeyev, the head of the Institute of Space Research (IKI) and a science adviser to Gorbachov, termed the Soviet shuttle "an outstanding technological achievement but a costly mistake." Sagdeyev, who was visiting the United States with dissident scientist Andrey Sakharov, said: "It went up and it came down. But it had absolutely no scientific value. My personal view is that American experience with the Shuttle indicates that from the point of view of cost efficiency, the shuttle is in deep trouble. It is much simpler and cheaper to fly a payload with any kind of expendable vehicle ... We have put too much emphasis on manned flight at the expense of unmanned efforts that produced more scientific information at lower cost" [59].

International reaction to the flight was largely positive, with many observers admiring the pinpoint precision of Buran's automatic landing system. "[The flight] shows that the Russians' boldness and ambition is matched by their ingenuity," said Soviet space expert James Oberg in an interview for *Time* magazine. "It blows us out of our last space-operations monopoly" [60].

Inevitably omnipresent in the Western reactions were comments on the similarity of Energiya–Buran to the US Space Shuttle. US specialists generally questioned the official Soviet explanation that the laws of aerodynamics require similar designs, pointing out that American engineers considered several quite distinct designs, including some markedly different wing and fuselage shapes, before settling on the one adopted in the early 1970s. Nicholas Johnson, another respected American observer of the Soviet space program, said:

> "The fact that the Soviets picked a design identical to ours can't be coincidental. There's no doubt they took advantage of a vast amount of engineering development that went into ours. I don't think stealing was necessary. A lot of the information was unclassified and open, if you knew where to look for it" [61].

In an editorial only days after the flight, *Aviation Week* downplayed the significance of the similarity between the two systems, focusing instead on the implications the mission had for America's place on the international space scene:

> "There is validity in the contention by Western observers that much of the technology embodied in the Buran has been gleaned from the data base generated by the U.S.'s development of the space shuttle. But concentrating on that issue misses the point and gives small comfort to U.S. officials concerned with maintaining a position of space leadership. The advanced materials, computers, software, aerodynamics and propulsion in the Soviet shuttle system and ability of the Soviet team to integrate multiple fault-tolerant computers and manage them effectively is something they have never before demonstrated ... The USSR has joined the reusable shuttle club and will not be turned back. The Soviets can be expected to aggressively exploit their shuttle's potential. Major applications they see are to build a large permanently manned space station and then prepare a springboard for a manned mission to Mars ... The Buran/Energiya mission is to be hailed as a success. It also should be taken as another reminder that an aggressive, broadly based space program is an integral

part of the Soviet Union's national policy. It cannot be considered any less than that by the U.S." [62].

Little did anyone know at the time that Buran was destined to remain on the ground forever.

REFERENCES

[1] Radio Moscow World Service, 30 January 1986.

[2] P. James, *Soviet Conquest from Space*, New Rochelle: Arlington House Publishers, 1974, pp. 125–142.

[3] M. Houtman, "Albatros: De Sowjet Shuttle", *Spaceview*, July 1976.

[4] G. Sinyaryov *et al.*, "Space transportation system Albatros" (in Russian), in: *K.E. Tsiolkovskiy i nauchno-tekhnicheskiy progress*, Moscow: Nauka, 1982, pp. 38–41; S. Reznik, "Student design organizations and interdisciplinary projects in the field of rocket and space technology" (in Russian), paper presented at the *31st Academic Readings on Cosmonautics in Moscow, January–February 2007*.

[5] *The Soviet Space Program. National Intelligence Estimate*, NIE 11-1-83, *Volume II : The Estimate*, 18 July 1983, p. V-10.

[6] C. Covault, "Soviets Build Reusable Shuttle", *Aviation Week & Space Technology*, 20 March 1978, pp. 14–15.

[7] *Soviet Military Capabilities and Intentions in Space. National Intelligence Estimate*, NIE 11-1-80, 6 August 1980, pp. 32–34.

[8] T. Furniss, "Soviet shuttle claims doubted", *Flight International*, 24 May 1986.

[9] J. Oberg, "Soviet shuttle mysteries", *Aerospace America*, June 1987, pp. 24–28.

[10] *Soviet Reusable Space Systems Program: Implications for Space Operations in the 1990s (An Intelligence Assessment)*, 1 September 1988.

[11] K. Vasilchenko, G. Lozino-Lozinskiy, G. Svishchev, "The Road to Buran" (in Russian), *Pravda*, 24 November 1988.

[12] N. Dombrovskiy, "Space orbits of a canard" (in Russian), *Sovetskaya Rossiya*, 17 May 1989.

[13] P. Pesavento, "Russian Space Shuttle Projects, 1957–1994. Part 2", *Spaceflight*, June 1995, p. 194.

[14] *Soviet Capabilities and Intentions for Permanently Manned Space Stations. An Intelligence Assessment*, 1 November 1982, p. 7.

[15] *Soviet Military Power 1983*, Washington, D.C.: US Department of Defense, 1983; C. Covault, "Soviets Building Heavy Shuttle", *Aviation Week & Space Technology*, 14 March 1983, pp. 256–258.

[16] *The Soviet Space Program. National Intelligence Estimate*, NIE 11-1-83, *Volume II: The Estimate*, 18 July 1983, p. V-8.

[17] "Soviet Shuttle, Heavy Booster in Serious Development Trouble", *Aviation Week & Space Technology*, 27 May 1985, pp. 21–22.

[18] "Soviet Shuttle", *Aviation Week & Space Technology*, 1 December 1986.

[19] "Soviets Demonstrate Flight Readiness with Firing of Heavy-Lift Booster", *Aviation Week & Space Technology*, 6 March 1987.

[20] C. Covault, "USSR's Reusable Orbiter Nears Approach, Landing Tests", *Aviation Week & Space Technology*, 3 December 1984, pp. 18–19.

[21] C. Covault, "Soviets Begin Orbiter Tests Following Engine Installation", *Aviation Week & Space Technology*, 14 April 1986, pp. 16–17.
[22] J. Oberg, *op. cit.*
[23] D. Fink, J. Lenorowitz, "Soviets Fly Jet-Powered Space Shuttle Testbed", *Aviation Week & Space Technology*, 12 October 1987, p. 23.
[24] *The Soviet Space Program. National Intelligence Estimate*, NIE 11-1-83, *Volume II: The Estimate*, 18 July 1983, p. V-8; R. McKie, "Space 'breakout' race", *The Observer*, 2 September 1984; E. Stevens, J. Witherow, "Russian Snowstorm to Blast US in Space Race", *The Sunday Times*, 11 January 1987.
[25] *Soviet Reusable Space Systems Program: Implications for Space Operations in the 1990s (An Intelligence Assessment)*, 1 September 1988.
[26] Y. Golovanov, "Just where are we flying to?" (in Russian), *Izvestiya*, 12 December 1991.
[27] Y. Semyonov, *Mnogorazovyy orbitalnyy korabl Buran*, Moscow: Mashinostroyeniye, 1995, p. 45.
[28] S. Grachov, "Energiya–Buran: The swan song of Soviet cosmonautics or the thorny path to space?" (in Russian), *Aviatsiya i vremya*, 2/2005.
[29] "Buran Expands Soviet Ability to Orbit Large Payloads", *Aviation Week & Space Technology*, 21 November 1988, p. 22.
[30] Y. Semyonov, *Raketno-kosmicheskaya korporatsiya Energiya 1946–1996*, Moscow: RKK Energiya, 1996, p. 385.
[31] N. Kidger, "The Soviet Shuttle Story", *Spaceflight*, January 1990, pp. 4–6.
[32] "Government Production Decision Awaited in Additional Space Shuttle Orbiters", *Aviation Week & Space Technology*, 5 June 1989, p. 95.
[33] "The turbojet engine installation" (in Russian), on-line at *http://www.buran.ru/htm/vrdu.htm*; letter by former Energiya–Buran chief designer I. Sadovskiy at *http://www.buran.ru/htm/sadovsky.htm#vrdu*
[34] V. Mokhov, "Module for Buran" (in Russian), *Novosti kosmonavtiki*, 23–24/1998, pp. 69–70.
[35] B. Gubanov, *Triumf i tragediya Energii (tom 3)*, Nizhniy Novgorod: Izdatelstvo Nizhegorodskogo instituta ekonomicheskogo razvitiya, 1998, pp. 292–293.
[36] B. Gubanov, *op. cit.*, pp. 404–405.
[37] E. Vaskevich archives.
[38] Y. Semyonov, *Mnogorazovyy orbitalnyy korabl Buran*, *op. cit.*, pp. 376–379; Y. Semyonov, *Raketno-kosmicheskaya korporatsiya Energiya 1946–1996*, *op. cit.*, pp. 383–384, 390, 591; B. Gubanov, *op. cit.*, pp. 404, 416–417.
[39] B. Gubanov, *Triumf i tragediya Energii (tom 4: Polyot v nebytiye)*, Nizhniy Novgorod: Izdatelstvo Nizhegorodskogo instituta ekonomicheskogo razvitiya, 1999, p. 190.
[40] B. Gertz, "Soviet shuttle might upstage U.S. comeback", *The Washington Times*, 20 September 1988.
[41] Radio Moscow World Service, 26 October 1988.
[42] A. Tarasov, "At the last minute" (in Russian), *Pravda*, 29 October 1988.
[43] James Oberg Cosmogram, 8 November 1988.
[44] C. Covault, "Soviet Space Shuttle Prepared for Second Launch Attempt", *Aviation Week & Space Technology*, 7 November 1988, p. 26.
[45] Memoirs of Andrey Lavrentyev, on-line at *http://www.buran.ru/htm/memory10.htm*
[46] B. Gubanov, (volume 3), *op. cit.*, pp. 416–432; V. Filin, *Put k Energii*, Moscow: Logos, 2001, pp. 179–184; V. Gudilin, "Take-off and landing of Buran" (in Russian), on-line at *http://www.buran.ru/htm/gudilin.htm*
[47] "Soviets Pushed Shuttle Launch in Bad Weather", *Soviet Aerospace*, 5 December 1988.

[48] Memoirs of Andrey Lavrentyev, *op. cit*. Apparently, for reasons that are not entirely clear, each scrub required a relatively lengthy turnaround time, especially after the tanks had been loaded with propellant.
[49] B. Gubanov, (volume 4), *op. cit*., pp. 6–7 ; V. Filin, *op. cit*., pp. 185–187; V. Gudilin, *op. cit.*
[50] Y. Semyonov, *Mnogorazovyy orbitalnyy korabl Buran*, *op. cit*., pp. 404–425; B. Gubanov, (volume 3), *op. cit*., p. 422; N. Johnson, *The Soviet Year in Space 1988*, Colorado Springs: Teledyne Brown Engineering, 1989, pp. 39–43, 110–112; B. Hendrickx correspondence with Chris van den Berg.
[51] A. Tarasov, "First launch of Buran" (in Russian), *Pravda*, 16 November 1988.
[52] "Flight programme for Buran on 29 October 1988" (in Russian), on-line at *http://www.buran.ru/htm/program.htm*
[53] A. Andryushkov, "First and last meeting" (in Russian), *Vestnik aviatsii i kosmonavtiki*, 28 March 2001.
[54] V. Filin, *op. cit*., p. 190.
[55] "The descent and landing trajectories of the orbital ship Buran" (in Russian), on-line at *http://www.buran.ru/htm/algoritm.htm*
[56] Mission description compiled from: Y. Semyonov, *Mnogorazovyy orbitalnyy korabl Buran*, *op. cit*., pp. 432–435; B. Gubanov, (volume 4), *op. cit*., pp. 9–11; S. Grachov, "Energiya–Buran: The swan song of Soviet cosmonautics or the thorny path to space?" (in Russian), *Aviatsiya i vremya*, 4/2005; various press reports.
[57] V. Lukashevich, "First flight" (in Russian), on-line at *http://www.buran.ru/htm/mtkkmain.htm*
[58] V. Lukashevich, "Last interview with G.E. Lozino-Lozinskiy" (in Russian), on-line at *http://www.buran.ru/htm/lastin.htm*
[59] J. Wilford, "Soviet Shuttle Faces U.S.-Style Criticism, *International Herald Tribune*, 23 November 1988.
[60] J. Langone, "Sunny Debut for Snowstorm", *Time*, 28 November 1988.
[61] J. Wilford, "U.S. Experts Say Soviet Shuttle Strongly Recalls U.S. Design", *International Herald Tribune*, 17 November 1988.
[62] "Soviet Shuttle Success", *Aviation Week & Space Technology*, 21 November 1988, p. 9.

8

Shattered dreams, new beginnings

CHANGING SCHEDULES

The original goal for the Buran program was to fly a total of ten test flights using five orbiters. Although considered test flights, most or all of these missions were to be flown to the Mir space station and carry out what would usually be considered operational tasks. All missions would carry 37KB modules in the cargo bay. Three of those were supposed to be built by the Khrunichev factory (serial nrs. 37070, 37071, and 37072). On the early test flights these modules would mainly carry instrumentation and remain in the cargo bay, but eventually at least two of them were to be turned into small scientific laboratories (renamed 37KBI, "I" standing for "research") that would be left behind at the lateral docking port of Mir's Kristall module to be picked up on a subsequent mission [1].

In September 1988 officials of the Energiya–Buran program reported to the Council of Ministers that the plan still was to fly ten test flights, with the first two to be flown unmanned [2]. Internal LII planning documents show that by the end of the year the second unmanned flight was scheduled for late 1989, to be followed by the first manned mission in late 1990. There would be one manned flight in 1991 and two in 1992. All missions would be flown by two-man LII crews, except for the first 1992 mission, which was to be piloted by an Air Force GKNII crew. All these initial missions would use either vehicle 1K or 2K [3].

An improvised payload for Energiya 2L

Even as Buran was undergoing initial post-flight servicing at the MIK OK, assembly of Energiya rocket 2L was nearing completion in the nearby MIK RN. Apparently, this rocket had been configured from the beginning for launching an unmanned payload canister rather than a Buran orbiter. With no Energiya-tailored payloads or upper stages ready to fly at this stage, a plan was devised to launch 2L with two satellites that would usually be orbited separately by the Proton rocket, an unidentified geostationary communications satellite and an Uragan navigation satellite for the Glonass network. Few details have been released about this configuration, known as GK-199: only that the satellites would have been housed in a Polyus-type vehicle with the payload shroud of the Proton rocket. Two Blok-DM type upper stages were probably required to inject the satellites into their proper orbits. One other objective of the launch was to test the parachute recovery of Energiya's strap-on boosters. A "draft plan" for the GK-199 mission was approved at meetings of the Council of Chief Designers in March and May 1989. The 2L vehicle was expected to be ready for roll-out to the pad by March 1990. However, the project received only lukewarm support from the Ministry of General Machine Building, which argued there was no room in its budget for such a flight [4].

Defense Council meeting

On 6 May 1989 the Energiya–Buran program was again on the agenda of the Defense Council, chaired by Gorbachov. Appearing before the Council, leaders of the Energiya–Buran program outlined future plans for the system, including the GK-199 mission, the creation of fully reusable versions of the Energiya rocket, and the development of derived launch vehicles such as Energiya-M, Groza, and Vulkan. While acknowledging the success of Buran's mission and praising the work of the people involved, the Council expressed dissatisfaction with the progress made on devising payloads and missions for the Soviet shuttle. The Council also made some cost-cutting moves, ordering the number of operational shuttle vehicles to be reduced from five (as planned since 1977) to three and curtail Buran's test flight program to just five missions by combining some of the objectives of the earlier planned missions. At the same time it called for speeding up work on Buran payloads and Energiya-derived launch vehicles.

The plan was now to fly Energiya 2L with the GK-199 payload in 1990, giving the team an extra opportunity to man-rate the rocket for future Buran missions. The second unmanned orbiter mission was now delayed to the first quarter of 1991 and apart from a docking with Mir would feature a link-up with a manned Soyuz "rescue vehicle" (see Chapter 5). This mission, designated 2K1 (the first flight of orbiter 2K) had already been approved by the Military Industrial Commission on 22 February 1989. The first manned Buran flight was now scheduled for the first half of 1992. LII internal planning documents drawn up around this time show the unmanned mission was scheduled for April/May 1991 and the manned flight for May 1992, with crew training to begin in December 1990. The decisions of the Council were consolidated by a government decree in June 1989, which laid out plans for the use of Buran until the year 2000 [5].

CHANGE OF LEADERSHIP AT NPO ENERGIYA

On 10 January 1989 NPO Energiya general designer Valentin P. Glushko passed away at age 80. On 8 April 1988 Glushko had suffered a stroke in his office, but wasn't found until four hours later. He underwent complex neurological surgery the following day, but never made a full recovery. Glushko spent most of the final months of his life in hospital, watching the Energiya–Buran mission on television rather than witnessing it first hand [6].

Glushko's death set in motion an internal battle within NPO Energiya to name his successor. On 23 January 1989 leading officials at NPO Energiya sent a letter to the Central Committee, VPK, and MOM, recommending Yuriy P. Semyonov as Glushko's successor. After a six-year stint at Yangel's OKB-586, Semyonov had

Glushko's grave at Novodevichi cemetery (B. Vis).

Yuriy Semyonov (*source*: RKK Energiya).

joined Korolyov's OKB-1 in 1964 and had quickly risen through the ranks of the design bureau, possibly helped by the fact that he was the son-in-law of the influential Politburo member Andrey Kirilenko, who also was the *de facto* head of the Soviet space program in his capacity as Central Committee Secretary for Defense Matters from 1979 to 1983. Semyonov began his career at OKB-1 as a leading designer of the Soyuz spacecraft and the L-1 ("Zond") circumlunar vehicles, going on to become the chief designer of Soyuz and Salyut in 1972. After the split of the Energiya and Buran offices within NPO Energiya in 1981 he also became chief designer of Buran.

It wasn't until 21 August 1989, after another appeal from leading NPO Energiya officials the month before, that Semyonov was officially named general designer of NPO Energiya, following in the footsteps of Korolyov, Mishin, and Glushko. One also wonders if there wasn't unequivocal support from Minister of General Machine Building Vitaliy Doguzhiyev, a former classmate of Semyonov, although he left the post to Oleg Shishkin in July 1989. The official history of NPO Energiya (edited by Semyonov!) largely attributes the 7-month power vacuum at NPO Energiya to Boris Gubanov's attempts to split off his rocket design department from the bureau and incorporate it into an independent design bureau for the creation of heavy-lift launch vehicles and upper stages. After the death of "rocket man" Glushko, Gubanov had evidently become worried about the future of his department within NPO Energiya, which did not only work on Energiya itself, but also on various derived launch vehicles that had no immediate relevance to the piloted space programs that were NPO Energiya's main focus.

The June 1989 government decree resulting from the May meeting of the Defense Council had basically given the go-ahead for further development of such systems, but according to Gubanov's memoirs the plans were scrapped by the so-called Scientific Technical Council of NPO Energiya on 18 August 1989 (three days before Semyonov's official appointment). The only exception was Energiya-M, a lightweight version of Energiya. Gubanov describes this move as the "initial castration" of the Energiya program. According to the official NPO Energiya history the Council divided the company's space-related activities into five levels of priority:

(1) Energiya–Buran and Mir.
(2) Heavy payloads for Energiya, including a geostationary communications platform.
(3) The Mir-2 space station and the further modification of Soyuz.
(4) Work on future air-launched systems (including reusable ones), spaceplanes, "reusable multipurpose space systems", piloted Mars missions, further improvement of Energiya–Buran (including work on a reusable strap-on booster).
(5) Other work, including that on the Blok-D upper stage.

In September, Semyonov canceled plans for the GK-199 mission, ordering instead preparation of the Energiya vehicle 2L for the launch of a massive geostationary communications platform by the end of 1992, even though the development of such a platform and the upper stages to place it into the required orbit were only in an embryonic stage.

On 28 August 1989 Gubanov wrote a letter to Gorbachov, warning him that the Energiya program was to suffer the same fate as the N-1 unless action was taken to make it economically viable. He once again outlined plans for Energiya-M, cargo versions of the standard Energiya, and fully reusable versions of Energiya, arguing that such systems could save costs by orbiting heavier satellites with more built-in redundancy and hence longer lifetimes. Effective development of such rocket systems, Gubanov once again stressed, could only be performed by a specialized design bureau. Gorbachov directed the task of looking into that possibility to Oleg Baklanov, who was now the Central Committee Secretary for Defense Matters after having served as Minister of General Machine Building from 1983 until 1988. One option considered was a merger of three organizations based in Kuybyshev—namely, the Volga Branch of NPO Energiya, the Central Specialized Design Bureau (TsSKB), and the Progress factory. The idea met with stiff opposition from Semyonov and TsSKB chief Dmitriy Kozlov, the latter having already refused to become involved in Energiya in the mid-1970s.

On 29 September 1989 a new structure was officially approved for NPO Energiya. Responsibility for the orbiter was now in the hands of Department 351 under the leadership of V.N. Pogorlyuk. Gubanov remained in his function as chief designer of the entire Energiya–Buran system, but the sections working under him on future versions of Energiya were abolished. A final decision on the creation of a new launch vehicle design bureau was to be made at a meeting of the Central Committee in March 1990, but no consensus was reached, leaving the issue unresolved. At a meeting on 7 May 1990 the Scientific Technical Council of NPO Energiya decided that the formation of such a bureau was "inexpedient". Gubanov was eventually dismissed from NPO Energiya on 5 March 1992 for his involvement in a deal between the Progress factory and an organization called Kazakhobshchemash to sell Soyuz rockets to Kazakhstan, although Gubanov himself saw it as just an excuse to get rid of him. With that move the post of "chief designer of Energiya–Buran" was officially abolished. Gubanov retired and passed away in 1999 [7].

MOUNTING CRITICISM

While Gorbachov's policy of openness or *glasnost* had enabled unprecedented media coverage of Buran's maiden flight, it also exposed the program to severe and often sound criticism, especially as the launch date for the next mission kept slipping and few realistic missions for the Soviet shuttle were announced. All this was against the backdrop of increasing public skepticism about the cost and purpose of the space program in general.

As in the US, much of the criticism came from space scientists, who saw little scientific value in Buran. In a rarely seen op-ed on the space program published in the official Communist Party newspaper *Pravda* in March 1989, the Institute of Space Research's K. Gringauz called Energiya–Buran a remarkable engineering feat, but pointed out that just like the Space Shuttle in the US it had caused significant cuts in scientific space research, the difference being that the USSR had a smaller economic potential and—unlike the US—already *had* a permanently manned space station and adequate space transportation systems available. Gringauz continued:

> "Rockets of the Energiya type can apparently be used not only to launch Buran, but also for manned flights to Mars. However, the beginning of such flights is planned for 2015, and in a quarter century's time the control systems and all its special components will have become obsolete. In my opinion, it cannot be ruled out that the main reason for developing the Energiya–Buran system was the industry's striving for self-confirmation and not the real needs of the country and science" [8].

The following month *Pravda*'s science editor lamented the lack of progress in the program and its high cost:

> "Much has been said and written about Buran and all kinds of hopes were pinned on it. But after its unmanned test flight it has got stuck in the hangars of Baykonur. Can't it be incorporated into the well-established system of space stations and expendable spacecraft? Have technical difficulties been discovered? As usual, one can only guess, because no news is leaking out from those hangars. But even without such news, it is clear that billions of rubles so badly needed for the national economy have been withdrawn from circulation for a long time to come" [9].

In late 1991 Yaroslav Golovanov, one of the country's most respected space journalists, published a damning analysis of Buran's intended missions in the prestigious *Izvestiya* newspaper. For one, he said, Buran was not an effective satellite launcher:

> "The cost of a single [Space Shuttle] flight is some 10–20 times higher than people initially thought it would be. Of course, our Buran ... is infected with the same disease ... The Buran launch cost roughly 170 million rubles. Calcula-

tions show that that puts the cost of lifting one kilogram of payload on Buran at 6,000 rubles. If that payload were launched on a Soyuz rocket, it would cost only one-sixth of that."

As for returning satellites from space, Golovanov wrote that "not a single of our puny satellites is so valuable that its return via Buran wouldn't be wasteful," adding that no such satellites were going to appear in the foreseeable future either. He also questioned the need to use Buran for servicing space stations, quoting Soviet space officials themselves as saying that their expendable transport ships were more effective. Turning to the military uses of Buran, Golovanov noted its ineffectiveness as a quick-response weapon because of the lengthy launch preparations and the limited number of launch azimuths. Concluding his analysis, Golovanov wrote:

> "Who can explain to me and to the millions of my countrymen—whose money has been used to build that star plane—why we need it if none of the space systems that has been created or is actually under development has been adapted to be put into orbit by Buran or Energiya or brought back down from orbit [by Buran]?" [10]

Faced with ever more penetrating questions from journalists relishing their newly found freedom, Soviet space officials had little choice but to disclose the true motives behind the creation of Energiya–Buran. In a television interview on 12 April 1991 Yuriy Semyonov said:

> "I have to say frankly that Buran was developed to counter the Shuttle. It's only now that everyone, including [Defense Minister] Marshal Yazov, is repudiating it: they say Buran is unnecessary ... The project was originated by the Defense Ministry, although they are now disowning it. All this took place before my very eyes. It was designed to counter or parry, as it were, the work that was being done in the United States" [11].

Responding to criticism about the lack of payloads, officials were quick to point out that Buran should not be primarily seen as a system to launch and retrieve ordinary satellites. In an interview in late 1989 Aleksandr Dunayev, the head of Glavkosmos, said:

> "The irony is that we have always said that the Energiya–Buran system should not be regarded as a transportation system (what will it carry?); it will be considerably more costly than conventional launch vehicles, and now these very arguments are being used against us: We have made a mistake, they say. We have made no mistake. The Energiya–Buran system was conceived primarily for defense purposes and it was deemed quite essential, and all other issues ... were to be secondary. Does this mean that the system has no peaceful applications? It is impossible to imagine that" [12].

When elaborating on those "peaceful applications", Buran's designers were hardly able to make a convincing case. Speaking in an interview shortly after Buran's flight, Semyonov said Buran's primary task would be:

> "to launch costly facilities outfitted with unique scientific instruments, for example, large optical telescopes, with sophisticated electronic equipment. Other uses could include the creation in orbit of big radio telescopes, aerial systems, solar power stations, and interplanetary complexes. These are extremely expensive constructions, each of which is the only one of its kind and needs to be serviced by manipulators, robots, and qualified personnel." [13].

However, as Semyonov was probably all too well aware himself, such plans existed only on paper and would take many years if not decades to come to fruition. The harsh truth was that Buran was slowly turning into a relic of the Cold War and its developers were having a difficult time concealing it.

THE USSR BREAKS UP

Despite all the criticism, preparations continued at Baykonur for future Buran missions. In the summer of 1990 the OK-MT full-scale test orbiter spent a month on pad 37 (6 July–7 August) for crew boarding and evacuation exercises and also for tests in which the fuel cells were loaded with liquid oxygen and liquid hydrogen.

By the first half of 1991 more than two years had elapsed since the first flight, making many wonder if a second flight was going to take place at all. Space officials kept stressing that the 2K1 mission to Mir was still on and would be flown sometime in 1992. One glimmer of hope was a test roll-out of the 2K orbiter to the launch pad in May 1991.

However, it wasn't long before Buran's future was further thrown into doubt by events that shook the very foundations of the Soviet Union. On 19 August 1991 tanks rolled into Moscow as a group of Communist Party hardliners calling themselves the State Emergency Committee attempted to take control of the country while Gorbachov was vacationing in the Crimea. The coup was timed to prevent the signing of a new Union Treaty which would have fundamentally recast the relationship between the center and the republics in favor of the latter. Although the putsch collapsed in only three days, it accelerated the events that would lead to the disintegration of the USSR at the end of the year. Adding to the growing unpopularity of the space program was the fact that one of the coup plotters had been Oleg Baklanov, who had been a strong supporter of the Energiya–Buran program in particular.

In the wake of the failed coup the Russian government took over the union government, ministry by ministry. In the autumn of 1991 the Ministry of General Machine Building was dissolved. The rocket and space enterprises located on Russian territory were transferred to the Russian Ministry of the Industry. Many of the enterprises were expected to merge into specialized conglomerates that would be

2K vehicle on the pad in the spring of 1991. Note missing tiles (*source*: Luc van den Abeelen).

subordinate to an organization called Rosobshchemash. Established in October 1991 on the vestiges of the Ministry of General Machine Building, it would act as a middleman between the Russian government and other nations for space and defense project orders. It was headed by outgoing MOM minister Oleg Shishkin, with Yuriy Koptev acting as his deputy for space matters. However, several leading companies, including NPO Energiya, refused to join Rosobshchemash. As Koptev later recalled, the organization was ineffective in bringing together the Russian space industry. In December 1991 leading space officials requested the government to set up a Russian Space Agency, in response to which a special commission was created led by Yegor Gaydar, the Minister of Economy and Finance [14].

The committee's findings were presented to President Boris Yeltsin during a key meeting at the Kremlin on 18 February 1992. It was attended among others by Yuriy Semyonov, Gleb Lozino-Lozinskiy, TsNIIMash director Vladimir Utkin, Vice-President of the Academy of Sciences Yevgeniy Velikhov and Koptev, who had been Gaydar's deputy in the committee and was the leading candidate to head the new agency. While the formation of the agency topped the agenda, the meeting also addressed the future of specific programs. Opening the meeting, Yeltsin spoke out against the continuation of the Energiya–Buran program. Semyonov countered the President by saying that its cancellation would be a repeat of the flawed decision to terminate the N-1 program in the 1970s and would deal an irreparable blow to the country's scientific, technical, military, and industrial potential. Semyonov was supported by Koptev and Utkin, while Velikhov echoed Yeltsin's sentiments, calling for an immediate shutdown of the program. The official minutes of the meeting said the future of the Energiya–Buran program would require further analysis, but according to the official history of NPO Energiya "all present at the meeting felt that the fate of the program had been sealed." On 25 February 1992 Yeltsin issued an edict approving the establishment of the Russian Space Agency (RKA) [15].

BURAN'S SWAN SONG

Later that year, RKA, the Ministry of Defense, the Academy of Sciences, and several other organizations drew up a "State Space Program up to the Year 2000", which did not include any plans for continued use of Buran [16]. This was a clear sign that, as far as RKA was concerned, Buran had no place in the new political and economical environment following the collapse of the USSR. In fact, some sources say the agency decided that same year to cancel further work on Buran [17]. The only more or less optimistic statements on the future of Buran in 1992 came from NPO Energiya officials themselves. Early in the year Vladimir Nikitskiy, Energiya's director of international affairs, said funding for Buran was being maintained on a low level and that the program had not yet been canceled outright, although it would be expensive to keep the already built orbiters in flyable storage [18]. In the summer Semyonov said nearly 4 billion rubles would be directed to continuation of the program. He noted, however, that the launch complex needed to be restored because no routine inspection and maintenance work had been done on it for nearly a year and a half. Semyonov held out hope that the 2K1 mission would fly in 1993 [19].

It appears Semyonov's words were no more than wishful thinking. As the months progressed, it was becoming ever clearer that the program was in its death throes. In May 1993 the Council of Chief Designers issued the following statement, which confirmed what had been obvious all along:

> "The two successful launches of the Energiya rocket ... have confirmed the correctness of the design decisions and the reliability of all elements of this new rocket and space system, unmatched in its capabilities by anything in the world. Taking into consideration that the government is not in a position not only to ensure the continuation of work, but also to take measures to maintain the cooperation between the designers and the acquired scientific and technical potential, the Council of Chief Designers is forced to conclude with deep regret that further work on the orbital vehicle Buran and the Energiya rocket carrier, [once] destined to provide our country a leading position in the exploration of space, is not considered possible" [20].

This statement is the closest that the Russians ever came to officially announcing the end of Energiya–Buran. There was no single day when the program was canceled.

Since the project had been sanctioned by a government and Communist Party decree in 1976, the only way to *officially* terminate it was by another government decree or by a presidential edict ("ukase"). This also meant that no funds were allocated to mothball, demolish, or reuse surviving hardware, something which companies had to pay for out of their own pockets. It wasn't until 2005, after numerous pleas from the Russian Space Agency, that the Russian government began to settle outstanding debts with companies involved in the Energiya–Buran program and also to provide funds to destroy or reuse surviving hardware [21].

There are few hard figures on the exact cost of the Energiya–Buran program, but there can be little doubt that it gobbled up a significant portion of the annual Soviet space budget, especially during the 1980s. This was even to the detriment of ongoing piloted space programs. According to the official NPO Energiya history so many funds had been diverted to Buran that by early 1984 work on the Mir space station had come to a virtual standstill [22].

In 1989 Soviet space officials for the first time released details of the budget. The 1989 space budget amounted to 6.9 billion rubles (about $10 billion according to the official exchange rates at the time), of which 3.9 billion went to military space programs, 1.7 billion to "economic and scientific programs" and 1.3 billion to Energiya–Buran [23]. Speaking at a Cosmonautics Day meeting on 12 April 1993, Koptev said that wielding the axe on the program had freed up 40–45 percent of the resources spent on the entire civilian space program [24]. As for the overall cost of the program, at the end of 1989 Glavkosmos chief Dunayev said that 14 billion rubles had been spent during thirteen years of development and testing [25]. Boris Gubanov says that by 1 January 1991 the program had cost a total of 16.4 billion rubles, of which 12.3 billion had gone to design and testing and 4.1 billion to "capital construction" [26].

Soviet officials regularly made optimistic statements along the lines that the numerous technological spin-offs from the Energiya–Buran program would eventually pay back its cost. Dunayev said in late 1989 that 581 proposals had been made to other industrial sectors to introduce those spin-offs, adding that the expected savings from proposals already adopted amounted to hundreds of millions of rubles and that the total 14 billion rubles invested in research and development would be returned by the year 2000 [27]. Korolyov bureau veteran Boris Chertok even claimed that the spin-offs would more than pay for the expenditures on creating the system, even if it was never launched into space again [28].

MISSIONS FOR BURAN

The goals outlined for Buran in the original February 1976 government and party decree were primarily military in nature, although civilian satellite deployment and retrieval missions as well as space station servicing missions were seen as additional objectives (see Chapter 2). At any rate, unlike the Space Shuttle in the US, Buran was not supposed to replace the entire expendable launch vehicle fleet, but merely complement that fleet by flying missions that it was uniquely designed to perform. However, strange as it may seem, all indications are that the primary emphasis throughout the history of the program was on the development of the rocket and orbiter themselves, not so much on the type of missions they would eventually fulfill. The Russians were so blindly focused on building a system matching the capabilities of the Space Shuttle that this became almost a goal in itself.

This is not to say that no thought was given to payloads at all. Specific orders to design payloads and work out flight programs for the Soviet orbiter came in party and government decrees issued in December 1981, August 1985, and August 1987. In 1981–1982 the military space R&D institute TsNII-50 conducted studies of possible military uses of Buran until 1995 under the name "Complex". In January 1984 the Ministry of Defense, MOM, and the Academy of Sciences jointly approved a program of Buran missions until the year 1995, and a concrete program of Buran operations up until 2000 seems to have been included in the earlier mentioned government/party decree of June 1989. Unfortunately, details of all those decrees and studies remain classified.

What *is* known is that all the major Soviet space design bureaus were asked to come up with ideas: NPO Energiya in Kaliningrad (manned spacecraft), NPO Mashinostroyeniya in Reutov (manned spacecraft, military satellites), KB Salyut in Fili (manned spacecraft, space combat means), KB Yuzhnoye in Dnepropetrovsk (military/scientific satellites), TsSKB in Kuybyshev (photoreconnaissance, remote sensing, materials-processing spacecraft), NPO Lavochkin in Moscow (deep-space probes and early-warning satellites), and NPO PM in Krasnoyarsk (communications and navigation satellites). The chief designers of those organizations were asked to take into account not only the significant payload capacity of the Soviet orbiter, but also its ability to repair satellites in orbit and return them back to Earth.

Despite the repeated calls to formulate ideas for Buran payloads, the response was meager. Presumably, all the new satellites being designed at those organizations could easily be accommodated by existing launch vehicles, and their chief designers must have felt as if they were asked to invent payloads to fit a space transportation system with which they had little affinity. It also made little sense to launch *existing* satellites on Buran, because (unlike the Space Shuttle) the intention of the Buran program had never been to replace the expendable launch vehicle fleet. Moreover, there had always been a tradition in the Soviet space industry that the design bureaus that developed rockets also needed to design payloads tailored to fly on those rockets. As far as the chief designers were concerned, Buran would be no exception and it was up to NPO Energiya to think up missions for its shuttle system.

Another factor that probably discouraged the satellite chief designers from becoming involved in Buran was that there was little confidence that the Soviet shuttle would ever fly. Already made wary by the N-1 debacle, they became even more skeptical when the RD-170 development problems brought the Energiya–Buran program to the verge of collapse in the early 1980s, and the original 1983 launch date kept slipping ever further. It wasn't really until the first missions of the Zenit rocket in 1985 and, ultimately, the first flight of Energiya in 1987 that the program to many became a credible undertaking, but by that time the political constellation that would eventually lead to its downfall was already beginning to take shape [29].

The result of all this was that Buran was never seriously considered for routine satellite deployment missions. NPO Energiya's predecessor OKB-1 had moved out of satellite construction back in the early 1960s, farming out the development of communications satellites, photoreconnaissance satellites, and deep-space probes to other organizations. Instead, Buran's primary mission would be to support NPO Energiya's core business—namely, space station operations. Unlike the situation in the US, where the Space Shuttle had to wait until the mid-1990s to perform its originally planned role of a space station ferry, the Russians had space stations readily available from the outset. Most of the missions that were *seriously* planned beyond the first flight were related either to Mir or its planned successor Mir-2, with the primary payload (the 37KB modules) being developed by KB Salyut, a branch of NPO Energiya between 1981 and 1988.

While military missions were expected to become Buran's main goal, few such missions have ever been identified. Ironically, by the time Buran was ready to fly, the Pentagon was withdrawing from the Shuttle program in the aftermath of the Challenger disaster, reorienting its heavy payloads to unmanned rockets.

In the end, the relative dearth of payloads and missions backfired on the program as the Soviet empire collapsed and became one of the main arguments to justify its cancellation.

The Buran/Mir/Soyuz mission

The only other Buran mission that ever came close to flying was 2K1, in which vehicle 2K would have been launched unmanned to the Mir complex, and after undocking would have been briefly boarded by a Soyuz crew before returning to Earth unmanned. This mission, along with crew assignments, has been described in detail in Chapter 5.

Building Mir-2

By mid-1991 the 2K1 mission had slipped to 1992 from its original launch date in the first quarter of 1991. Beyond that Buran was now scheduled to take part in the assembly and operation of the Mir-2 complex, where the emphasis would be on the industrial production of ultra-pure medicines and semiconductor materials and also on remote sensing. The plans were presented in detail by Yuriy Semyonov at the congress of the International Astronautical Federation in Montreal in October 1991.

First, the 2K orbiter would go up again in 1993 on an unmanned solo flight (2K2) to test some of the biotechnological installations to be flown under the Mir-2 program. Then in 1994 the 1K vehicle would fly the first manned mission (1K2) as part of a plan sometimes light-heartedly referred to as "Mir-1.5", in which Mir would gradually be replaced in orbit by Mir-2. After the launch of the Mir-2 core module by a Proton rocket, Buran would rendezvous with the module, grab it with its two remote manipulator arms, and dock it to a bridge in the cargo bay. Buran would then

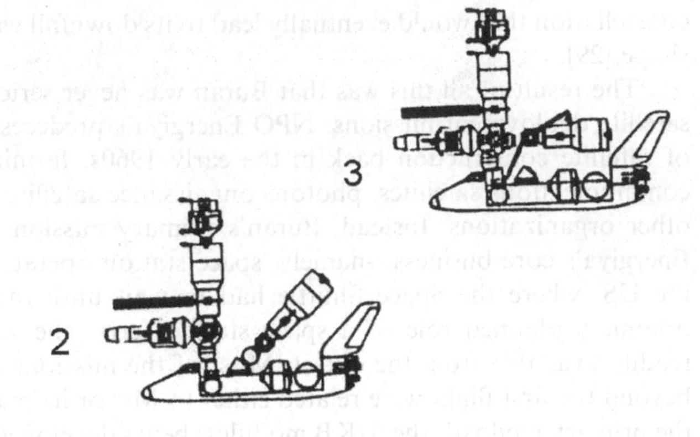

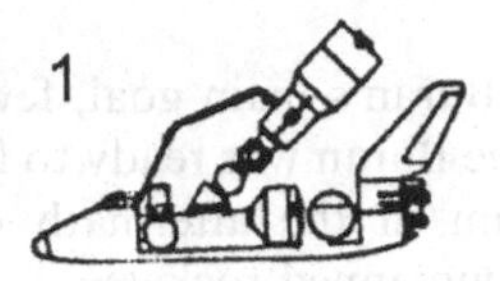

1K2 mission as planned in late 1991: 1, Buran picks up Mir-2 core module; 2, Buran docks with Mir; 3, Buran mechanical arm transfers Mir-2 core module to Mir lateral docking port (*source*: Yuriy Semyonov).

link up with a small docking module on Mir's multiple docking adapter and again use its manipulator arms to transfer the Mir-2 core module to a lateral docking on Mir previously occupied by the Spektr module. The two modules would remain docked for about two years. After the transfer of the Priroda Earth resources module to the Mir-2 core, Mir and its remaining add-on modules would then have been undocked and discarded, setting the stage for the four-year assembly of the Mir-2 complex (1996–2000).

Before that, in 1995, vehicle 2K would be launched on another autonomous flight (2K3) to test a biotechnological module called 37KBT, based on the original 37KB instrumentation modules. With the emphasis having shifted from fundamental scientific research to biotechnological production, the original plans for the 37KBI scientific add-on modules had been scrapped in late 1989. Buran would now regularly fly two biotechnological modules (37KBT nr. 1 and nr. 2), carrying one up and bringing the other down.

Between 1996 and 2000 there would be two missions annually, one using vehicle 2K to swap out the 37KBT biotechnological modules (2K4, 2K5, 2K6, 2K7, and 2K8) and another using the 1K orbiter for assembly and logistics missions (1K3, 1K4, 1K5, 1K6, 1K7). Planned for addition to Mir-2 was a 37KBE "power module" equipped with extra solar panels. Further Buran missions would have been required to add a large 85 m truss structure to Mir-2 and outfit it with solar arrays, large radiators, and an array of scientific instruments [30].

The "Mir 1.5" plan was dropped in 1992, when it was decided that Mir-2 would

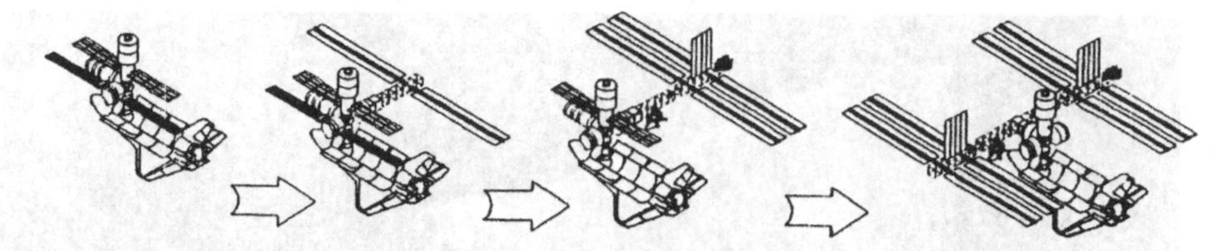

Build-up of Mir-2 using Buran orbiters (*source*: Yuriy Semyonov).

only be launched after Mir had outlived its usefulness. This would also allow the new station to be placed into a higher inclination orbit (65° vs. 51.6° for Mir) for better remote-sensing coverage. At this point the big Buran-launched 37KB-type modules were abandoned in favor of smaller modules based on the Zenit-launched Progress-M2 cargo ship. The new Mir-2 concept was approved by the Council of Chief Designers in November 1992. Although it left open the option of launching the add-on modules and the station's truss structure with Buran, Zenit was clearly the preferred option. By the time Mir-2 was merged with Freedom to become the International Space Station in late 1993, work on Buran had been suspended.

Buran's role in space station operations

Although Buran figured prominently in plans for both Mir and Mir-2, there are no indications it was ever supposed to replace traditional transportation systems such as Soyuz and Progress. The idea was that it would be used in parallel with those systems for missions requiring its unique capabilities, such as assembly of large structures, swapping out of modules and delivery and return of large pieces of equipment. While Buran could have made it possible to reduce the number of Soyuz and Progress missions, these vehicles would have continued to play a crucial role in Soviet space station operations. This also explains why the Russians never stopped improving Soyuz and Progress during the development of Buran.

In fact, the International Space Station (ISS) is now pretty much operated as the Russians had set out to do with Mir and Mir-2, being serviced by a combination of large shuttles and smaller capsule-type vehicles. The ISS itself is clear proof that it is impossible to operate a space station with large shuttle vehicles alone. Although such vehicles can deliver larger crews and more supplies than capsule-type spacecraft, it is not economically justified to use them for dedicated crew rotation and resupply missions. Ideally, these tasks should be combined with shuttle-unique assignments and not be seen as mission objectives in themselves.

The biggest problem with Shuttle/Buran-type vehicles is that they can only stay docked to a space station for several weeks at most until their consumables run out. Vehicles like Soyuz and Progress can be largely deactivated after docking to a station and remain attached to it for months on end. This means they are always available for reboost and refueling operations when needed and—crucially for crew safety—can always immediately return a resident crew back home if an emergency situation arises. NASA had originally planned to service Space Station Freedom solely with

Artist's conception of Buran docked to Mir space station.

the Space Shuttle and leave crews on board in between Shuttle missions. Only after the 1986 Challenger disaster did it dawn on the agency that it would be dangerous to have crews on the station without a lifeboat attached. NASA then found itself scrambling to find a US contractor capable of building a station lifeboat at short notice. Fortunately enough for NASA, political changes in the USSR allowed the agency to adopt Soyuz as a lifeboat for Freedom in 1992 and the vehicle continued to serve in that role as part of the ISS.

The simultaneous operation of large shuttles and capsules also provides redundancy. One vehicle can continue to service the station in case the other is grounded. This was vividly demonstrated by the 2003 Columbia accident, after which Soyuz and Progress vehicles served as a lifeline for the station. One can only imagine what things would have been like if *both* the Space Shuttle and Buran had been around for ISS operations. Having been built to the same specifications as the Shuttle, Buran could have continued ISS assembly work during the Shuttle's standdown. Of course, this is no more than wishful thinking, because the very conditions that lay at the foundation of Buran's downfall enabled the creation of the ISS.

Buran's ISS legacy

Despite the fact that the Soviet orbiters had long been mothballed by the time ISS construction began, one piece of Buran technology does play a vital role in station

APAS-95 docking port (*source*: NASA).

Russian Pirs module. Central and aft parts are derived from Buran's Docking Module (*source*: NASA).

operations. This is the Androgynous Peripheral Docking System (APDS—Russian acronym APAS), a Russian-built docking mechanism that allows Space Shuttles to dock with the US-built Pressurized Mating Adapters (PMAs) on the ISS "node" modules. Built at RKK Energiya under the leadership of Vladimir Syromyatnikov, the first APAS (APAS-75) was developed back in the 1970s for the Apollo–Soyuz Test Project. A modified version (APAS-89) appeared in the 1980s to enable Soviet orbiters to dock with the axial APAS docking port of Mir's Kristall module. In the end, Buran never flew to Mir and the Kristall APAS docking port was used only once by Soyuz TM-16 in 1993.

In July 1992 NASA initiated the development of the Orbiter Docking System (ODS) to support Shuttle flights to Mir. Mounted in the forward end of the payload bay, the ODS consists of an external airlock, a supporting truss structure, and an APAS docking port. While the first two elements were built by Rockwell, the APAS was manufactured by RKK Energiya. Although Energiya's internal designator for the Shuttle APAS is APAS-95, it is essentially the same as Buran's APAS-89. While the ODS was slightly modified for Shuttle missions to ISS, APAS remained unchanged. There was even a suggestion to launch Buran to Mir to test the docking system prior to the beginning of the Shuttle–Mir flights [31].

The APAS consists of a three-petal androgynous capture ring mounted on six interconnected, ball screw shock absorbers that arrest the relative motion of the two vehicles and prevent them from colliding. The APAS-89 differs from APAS-75 in several key respects. It is much more compact (although the inner egress tunnel diameter is more or less the same), has twelve structural latches rather than eight, the guide ring and its extend/retract mechanism are packaged inside rather than outside the egress tunnel, and the three guide petals are pointed inboard rather than outboard [32].

Russian plans to sell their *entire* Buran Docking Module to NASA fell through, but its design did serve as the basis for the construction of the Russian Pirs airlock module, docked to ISS in September 2001. This retains the central part of the adapter's spherical section (2.55 m in diameter). Mounted to its aft end is a small section of the Buran airlock's cylindrical tunnel (without the extendable part) and attached to the front end is the forward part of a Soyuz-TM/Progress-M orbital module. The design was more complex than that of the Docking Module of Mir (316GK), which did not have to be used as an airlock and was merely an extension to the Kristall module to facilitate Shuttle dockings [33].

Spacelab-type missions

Among the autonomous missions would have been flights with Spacelab-type modules installed in Buran's cargo bay. NPO Energiya had plans for a so-called "Laboratory Compartment" (LO) (index 14F33) that was reportedly based on the 37KB design. This would be connected to the crew compartment by a special tunnel. Such flights would have lasted anywhere from 9 to 30 days and be devoted to scientific, materials-processing, and biotechnological experiments, which take a relatively long time to produce the necessary results. The longest missions would have required the installation of an extra cryo kit for the fuel cells. Buran was also seen as an ideal platform for long-duration *unmanned* materials-processing and biotechnology missions, benefiting from the undisturbed microgravity environment and the ample power provided by the fuel cells. A military version of the LO known as the Undetachable Useful Payload (NPG) was also considered, but its index 17F32 indicates that it was to be built on the basis of a different design and no further information on it is available [34].

Deployment, servicing, and retrieval missions

While Buran was never seriously considered for routine satellite deployment missions, the Russians did look at the possibility of placing big payloads in the cargo bay. Among these were spacecraft developed as part of a Soviet "Star Wars" program in which NPO Energiya was given the leading role in 1976. It would have seen the use of space-based assets to destroy enemy satellites, ballistic missiles, and ground-based targets. Making maximum use of existing technology, NPO Energiya tabled proposals for "battle stations" that would be based on Salyut and Mir technology.

For anti-satellite operations the idea was to develop two types of Salyut look-alike space stations, one equipped with missiles (Kaskad) and the other with laser weapons (Skif). The stations carried much larger propellant supplies than their

Mir-type "battle station" with Buran-based combat modules (*source*: RKK Energiya/ *www.buran.ru*).

progenitors, but had man-tended capability, being able to house two-man crews for up to seven days. Kaskad stations would target high-orbiting satellites, while the Skif stations were to knock out satellites in low orbits. Experimental versions of these stations would be orbited by the Proton rocket, but the operational ones were designed to go up in the cargo bay of Buran. The Soviet orbiter would also be responsible for refueling missions to these stations. In 1981 work on Skif/Kaskad was transferred to Energiya's new KB Salyut branch, which dropped the Salyut-based design in favor of 100-ton Energiya-launched spacecraft (see Chapter 6). There are no indications Buran still had any role to play in Skif/Kaskad from that moment on.

For destruction of ground-based targets the NPO Energiya planners came up with a Mir-type core module with four specialized modules docked to a ball-shaped multiple docking adapter. Attached to the axial front port was a module with an additional multiple docking adapter that served as the berthing place for so-called "combat modules" resembling Buran orbiters without wings or other aerodynamic surfaces. After undocking from the station, the unmanned combat modules would maneuver to the proper location and then deploy small vehicles tipped with (unspecified) weapons that could re-enter the atmosphere. These could be either ballistic-type vehicles or lifting bodies. One design studied for these re-entry vehicles was based on the BOR-4 lifting bodies. Presumably, the idea was that after deploying the weapons the Buran-based combat modules would return to base to be reloaded with new ones [35].

One big military satellite intended for launch by Buran was Sapfir ("Sapphire"), a 24-ton optical reconnaissance satellite developed by TsSKB in Kuybyshev. This was equipped with a 3 m diameter telescope to photograph targets of interest in great detail. The idea was that Buran crews would regularly visit Sapfir for servicing. Although the telescope for the first such satellite was nearly finished, the project was discontinued after the cancellation of Buran, since the Proton rocket was not capable of orbiting the satellite [36].

Another big payload eyed for launch by Buran was ROS-7K ("Radiotechnical Orbital Station"), a man-tended Salyut-derived space station equipped with a 30 m diameter dish antenna called KRT-30. Capable of serving as a radio telescope and a radar, the KRT-30 was to be used for all-weather remote-sensing, astrophysical, and geophysical observations and target localization for the Soviet Navy. Together with the ground-based components needed to receive, process, and distribute data from the station, the system was called Gals ("Tack", in the nautical meaning), an indication that its observations in support of the Soviet Navy were seen as its primary mission.

Flying in a circular 600 km orbit inclined 64.8° to the equator, ROS-7K could house two-man crews up to seven days for maintenance operations and could be refueled in orbit. The complete ROS-7K with the stowed KRT-30 fitted in the cargo bay of Buran, although launch by the Proton rocket was studied as an alternative. Buran was also supposed to fly a technology demonstration mission in support of ROS-7K/Gals called "Karat", but no further details on this are available. Gals was studied at NPO Energiya from 1978 until 1987 [37].

Buran (along with Proton) was also considered to launch a giant space tug powered by a nuclear electric engine. Called Gerkules (Russian for "Hercules"), the tug was to be stationed in a 200 km orbit and one of its tasks was to maneuver 100-ton spacecraft launched by Energiya to geostationary orbit. Given the 35 m length of the tug, several missions would have been required to assemble it in orbit. Gerkules studies at NPO Energiya began in 1978 and lasted until at least 1986 [38].

Another exotic payload studied for launch by Buran or Proton was an experimental, orbiting solar power station, consisting of a solar tug and a dish antenna (based on the KRT-30). Deployment of the experimental solar power station would have required two Proton or Buran launches [39].

NPO Energiya also looked at so-called "Experimental Space Apparatuses" (EKA) that appear to have been prototypes of expensive new satellites that would be thoroughly checked out in orbit by Buran. The crew would, for instance, check if vital systems (such as various appendages) worked and carry out repair work if necessary. The EKA could then later be revisited for maintenance operations or the retrieval of valuable parts for analysis on Earth or reuse on later satellites [40].

Another future assignment for Buran occasionally mentioned by Russian sources was the retrieval of satellites from space. While this may have sounded attractive, such missions usually require that satellites are designed to be picked up by an orbiter—that is, have grapple fixtures for the orbiter's remote manipulator system and be small enough to fit in the payload bay—and, above all, circle the Earth in orbits that can be reached by it. In practice, that would have virtually limited such

The Salyut-7 space station.

missions to satellites deployed by the orbiter itself and not equipped with a kick motor to be boosted to high orbits. The original 1976 government/party decree on Buran had called for the development of a reusable space tug (11F45) to operate between low and high orbits, but that was never developed. In one interview Yuriy Semyonov mentioned the possibility of retrieving nuclear-powered satellites that threatened to fall back to Earth [41]. The only such satellites operated by the Soviet Union were the US-A radar ocean reconnaissance satellites and it looks unlikely they could ever have been retrieved by Buran, if only because of the radiation threat to the crew.

One other mission studied for Buran was to retrieve elements of the Salyut-7 space station. Launched in 1982, Salyut-7 played host to its final crew in May 1986 before definitively passing the torch to Mir. However, rather than deorbiting it, as had been the usual practice with earlier Salyuts, the Russians boosted the station and the attached Kosmos-1686 spacecraft (a Transport Supply Ship or TKS) to a 474 × 492 km storage orbit in August 1986 to see how well their systems would stand up to a prolonged stay in space and use that experience in designing future spacecraft. Some two weeks after the maneuvers Yuriy Semyonov said in an interview that "in a few years a group of cosmonauts could be sent to Salyut to study the state of the orbital complex" [42].

In December 1988, with Buran no longer a state secret, Semyonov acknowledged that the idea was to send a Buran crew to Salyut-7 in 1995–2000 and retrieve parts of the complex for detailed analysis on Earth, adding this would provide invaluable data on prolonged exposure of materials to space conditions [43]. Some reports at the time suggested the plan was to retrieve the *entire* Salyut-7 space station, but given the technical complexity of such a mission, that never seems to have been the intention.

However, Salyut's orbit decayed much faster than predicted due to unexpectedly high solar activity in the late 1980s/early 1990s that caused the upper layers of the atmosphere to expand considerably. On top of that, Kosmos-1686 suffered a failure of its electrical systems in December 1989, making it impossible to use the vehicle's thrusters to keep the station in a gravity-gradient mode. With little fuel left in Salyut's own tanks, the complex eventually made an uncontrolled re-entry on 7 February 1991, showering debris over South America.

FATE OF THE SOVIET ORBITERS

BTS-002

After the completion of the Horizontal Flight Test program in December 1989, BTS-002 was kept in storage at the Flight Research Institute in Zhukovskiy, where it was put on display during the biennial MAKS aerospace shows in 1997 and 1999.

In 1999 the vehicle was leased to an Australian company called Buran Space Corporation. Chaired by Australian-born astronaut Paul Scully-Power, it planned to put the vehicle on display during the 2000 Summer Olympic Games in Sydney. Since the VM-T and Mriya carrier aircraft were no longer available, BTS-002 had to be transported to Australia by water. In order to ease the transport, the vehicle was stripped of its landing gear, vertical stabilizer, wings, and the two side-mounted AL-31F turbojet engines, which would then later be reassembled after arrival in Sydney.

The first leg of the cumbersome journey took BTS-002 from Zhukovskiy to St. Petersburg. The vehicle left Zhukovskiy on 30 October 1999 on a submersible flat pontoon owned by the British company Brambles Project Services. Later that day Muscovites were treated to the unusual view of two orbiters side by side, when

BTS-002 arrives in Sydney (*source*: *www.buran.ru*).

BTS-002 sailed past a full-scale Buran test model serving as an attraction in Gorkiy Park on the banks of the Moscow River. After arriving in St. Petersburg, it took BTS-002 two weeks to be cleared by customs and continue its journey to Göteborg in Sweden, where it remained stuck for another six weeks until an appropriate container ship (the *Tampa*) was found for the long trip to Australia. The atmospheric shuttle made a stop in New Brunswick, Canada, before proceeding through the Panama Canal on to its final destination "down under". BTS-002 arrived at Darling Harbor on 9 February 2000, where it was welcomed with much fanfare in a ceremony broadcast live by several Australian television stations and attended among others by Andrew Thomas, another Australian-born NASA astronaut.

BTS-002 was on display in Sydney under a temporary structure for several months. BSC, which had taken out a nine-year lease on the vehicle, had ambitious plans to take the BTS-002 on an extensive tour of cities throughout Australia and Southeast Asia, but poor ticket sales forced the company into bankruptcy. BTS-002 then spent the following months in a fenced-in parking lot in Sydney, protected by nothing more than a large tarp. It was subject to repeated vandalism, with some sections becoming covered in graffiti.

With Buran Space Corporation unable to complete its payments, ownership of the vehicle reverted back to NPO Molniya, which then sought a new owner because it lacked the resources to bring the craft back home. NPO Molniya approached an American company called First FX that arranged for the auction of BTS-002 through a radio station in Los Angeles in May 2002, but the $6 million minimum asking price turned out to be too high. Somewhat later NPO Molniya did find a buyer for the vehicle, a Singapore-based company called Space Shuttle World Tours (SSWT), which shipped it to Bahrain to be displayed at the 2002 Summer Festival. With that exhibition not successful either, SSWT planned to move the vehicle to Thailand as a tourist attraction. However, the company had apparently defaulted on its payments to Molniya, which then brought a lawsuit against SSWT to prevent the transfer to Thailand. Pending the outcome of the legal dispute, SSWT negotiated to place BTS-002 at a junkyard in Bahrain.

BTS-002 finally seemed to have a lucky break in 2004, when a group of German journalists stumbled on it while covering a Formula-1 Grand Prix race in Bahrain. Their articles generated quite some interest back in Germany, where the Auto & Technik Museum in Sinsheim offered a large sum to NPO Molniya to add the vehicle to its collection. Unfortunately, ongoing legal battles between Molniya and SSWT have so far blocked the potential deal and BTS-002 remains stuck in Bahrain [44].

Orbiter 1K (Buran)

After returning from its mission on 15 November 1988, the 1K orbiter was sent back to the MIK OK orbiter-processing facility for post-flight inspections. Those were interrupted in early May 1989 for test flights with the An-225 Mriya carrier aircraft, staged from the Yubileynyy runway. This was in preparation for the long flight to the Paris Air Show the following month (see Chapter 4). In late June 1989 Mriya returned Buran back to Baykonur. The pair made one last demonstration flight over

Buran at the MIK OK in 1997 (B. Vis).

Baykonur on 12 April 1991 as part of an air parade to mark the 30th anniversary of the flight of Yuriy Gagarin. Later that year Soviet space officials discussed the possibility of bringing Mriya and Buran to the International Aerospace Convention in Huntsville, Alabama in July 1992, but those plans were never realized [45].

As it turned out, the demonstration flight in April 1991 had been the last time that 1K ever took to the skies. In early 1990 Western reports claimed vehicle 1K had been retired from flight status. As it had been built without a life support system, full avionics systems, and a fuel cell electrical system, it was considered too costly and difficult to modify the vehicle for a future useful unmanned or manned flight [46]. Another reason given for the grounding of vehicle 1K was that design changes had been introduced in the newer airframes under construction at the Tushino Machine Building Factory that would be too expensive to incorporate into the older vehicle. Moreover, the Soviet policy apparently was that each airframe had a lifetime of 10 years, no matter how many flights it made. With construction of 1K having started in 1983, the ship was nearing the end of its warranty [47]. However, given the astronomical cost of building an orbiter, it is hard to imagine that the Russians would have strictly adhered to this policy. In fact, vehicle 1K still figured prominently in the Mir-2 assembly plans outlined in late 1991. The reported retirement of the first flight vehicle may have had more to do with the increasingly bleak prospects for the program than anything else.

For years, Buran languished in the thermal protection system bay of the MIK OK, only to be shown to occasional visitors. In 1994 the idea arose to save it from oblivion by mating it with an Energiya rocket for permanent display at the MIK RN Energiya assembly building. Since it was considered that the stack should resemble the real thing as much as possible, the intention was to attach Buran to flightworthy Energiya hardware mothballed at the MIK RN—namely, the core stage of vehicle 2L and the strap-on boosters originally earmarked for vehicle 3L. However, in the mid-1990s a decision was made to remove the RD-170 engines from the strap-on boosters for vehicles 3L to 6L and ship them back to Energomash to be modified for use on Zenit-3SL rockets being built under the Sea Launch program. Finally, Buran was moved to high bay 4 of the MIK RN in July 1998 and assembled the following month with a non-flight rated Energiya consisting of the core stage of vehicle 5S1 and the strap-ons of vehicle 4M-KS [48].

As if its fate wasn't sad enough, Russia's only flown shuttle was reduced to scrap when the roof covering the three high bays of the MIK RN collapsed on 12 May 2002. The accident claimed the lives of seven men conducting repairs on the leaky roof. It was blamed on a combination of factors that had made the roof 1.5 times heavier than specified. First, the roof insulation material used during the construction of the MIK back in the 1960s was heavier than it should have been. Second, the roof had absorbed several days' worth of heavy rains, and, third, more than 10 tons of new roofing material had accumulated on the building in preparation for the repairs. The collapse destroyed not only orbiter 1K, but also the other Energiya hardware remain-

Remains of Buran after the roof collapse (*source*: *www.buran.ru*).

ing in the MIK: three core stages and two Blok-Ya launch pad adapters in high bay 3, eight strap-on boosters in high bay 4, and another eight strap-ons in high bay 5 [49].

Orbiter 2K ("Buran-2")

The second spaceworthy orbiter 2K (sometimes called "Buran-2", airframe nr. 1.02) arrived at Baykonur atop a VM-T carrier aircraft on 23 March 1988. Because of the relatively limited lifting capability of the VM-T, vehicle 2K arrived at the cosmodrome underweight and in a relatively "raw" state, with a lot of work still remaining to be done before its first flight. The Mir/Soyuz docking mission planned for 2K was

2K orbiter (front) and OK-MT mock-up in storage inside the MZK (*source*: Sergey Kazak/ Novosti kosmonavtiki).

much more challenging than the conservative two-orbit test flight of vehicle 1K. Because of that the ship needed to be outfitted with systems that had not been on the original orbiter, such as a partially operational life support system, fuel cells, a fully operational payload bay door actuation system, a docking mechanism, and a remote manipulator arm. The vehicle was also equipped with a single ejection seat and several non-ejection seats.

On 16 May 1991 vehicle 2K, attached to the 4M core stage and strap-ons, was rolled out to Energiya–Buran pad 37 for two weeks of tests. The roll-out was witnessed by the crew of Soyuz TM-12 (Artsebarskiy, Krikalyov, Sharman), who were making final preparations for their mission. One obvious external sign that the ship wasn't quite ready for flight yet was that many of its heat-resistant tiles were still missing. Among the work done at the pad were load tests of the fuel cells and orbiter evacuation exercises. On 30 May the stack trundled back to the MIK RN to be prepared for three days of tests at the Dynamic Test Stand (7 June–10 June 1991). After that it once again returned to the MIK RN, where vehicle 2K was detached from the core stage and returned to the MIK OK [50].

Speaking at the International Astronautical Federation Congress in Montreal in October 1991, Yuriy Semyonov told reporters that the Energiya rocket for the Buran-2 mission had been placed on the launch pad in preparation for a 10–15 second static test firing of its main engines, and there is some evidence that Energiya 3L did undergo such a test that year [51].

By the time the Energiya–Buran program was canceled, the 2K orbiter was said to be 95–97 percent ready. The vehicle was later moved from the MIK OK to the defunct Assembly and Fueling Facility (MZK), where it still resides today.

Other flight vehicles

In the mid-1980s NPO Molniya began building three more airframes intended for use in spaceflight-qualified vehicles (3K, 4K, and 5K). Talking about vehicle 3K in early 1990, Gleb Lozino-Lozinskiy said it would be lighter and more reliable than the earlier orbiters thanks to the use of composite materials and an improved thermal protection system. He expected the spacecraft to be ready in 1992 [52].

Actually, vehicle 3K (airframe nr. 2.01) was only about 30 percent ready when the Buran program was canceled in 1993. It consisted of a complete fuselage, but apparently had very few internal systems installed. Pictures of the vehicle show that the crew compartment and the aft compartment were virtually empty. The fuselage was only partly covered with tiles and the payload bay doors were missing. Vehicle 3K remained at the Tushino Machine Building Factory near Moscow until October 2004, when the fuselage, wings, and vertical stabilizer were transported separately to a nearby berth on the Moscow River, the same one where earlier orbiters were loaded onto a barge for transportation to Zhukovskiy. It will either be turned into scrap metal or sold to a museum if anyone displays interest [53].

The airframes for vehicles 4K and 5K (nrs. 2.02 and 2.03) never reached completion. Construction work was presumably halted after the Defense Council's May 1989 decision to reduce the orbiter fleet from five to three vehicles. One report

Partially assembled 3K orbiter (*source*: *www.buran.ru*/Aleksey Mikheyev).

suggests there were plans to turn one of the airframes into an underwater training mock-up for the neutral buoyancy facility at Star City, but that never happened [54]. Some elements of these airframes still lie in storage at the Tushino Machine Building Factory, but most parts have been turned to scrap.

Full-scale test articles

Currently, five full-scale Buran test articles still reside at various locations in the former Soviet Union, four of them in a fully assembled state. Two vehicles, OK-MT and OK-ML1, remain at the Baykonur cosmodrome. OK-MT is situated in the MZK building together with flight vehicle 2K and is said to be in relatively good condition. The same cannot be said of OK-ML1, which for a long time sat exposed to the elements at the orbiter test-firing stand. It is in a sorry state, with many of its parts having been stripped by visiting tourists. In January 2007 it was parked next to the Baykonur museum on Site 2 of the cosmodrome and there were plans to turn it into an exhibit.

Any plans to ferry the remaining vehicles to Russia or other countries are compounded by the fact that all the remaining Energiya–Buran hardware at Baykonur is now the property of Kazakhstan. Moreover, the mate–demate device needed to lift the orbiters onto the Mriya aircraft near the Yubileynyy runway has not been maintained in an operational state.

OK-ML1 sits outside at Baykonur (B. Vis).

Probably the most famous full-scale test article is OK-M, which sits as a tourist attraction in Gorkiy Park on the banks of the Moscow River. The cargo bay has been turned into what looks like the passenger section of an aircraft, with visitors being able to watch images of the Earth projected on screens and being treated to space food. Before entering the vehicle they get a symbolic medical check-up and a certificate clearing them for an imaginary flight in space.

The electrical test model OK-KS remains at the facilities of RKK Energiya in Korolyov, while the partially disassembled test model OK-TVI is in storage at NIIkhimmash near Sergiyev Posad. RKK Energiya has been trying to somehow get rid of OK-KS, which occupies valuable space at its facilities, but to no apparent avail, mainly due to the absence of a presidential edict officially canceling the Energiya–Buran program. However, something may happen with them after all, now that the Russian government has begun allocating at least some funds for dealing with remaining Energiya–Buran hardware [55].

ENERGIYA CARGO MISSIONS

The decision to launch Buran as a passive payload on Energiya made it possible to use the same rocket in various configurations to orbit heavy unmanned payloads. Although this was one of the main advantages of the Soviet system as compared with the Space Shuttle, the development of these cargo versions of Energiya always took a backseat to that of the main Energiya–Buran system. One of the main reasons for this must have been that there were few payloads in the given mass range that stood any chance of flying soon. The final go-ahead seems to have been given by the government decree on Buran issued on 21 November 1977.

Cargo Transport Container

For cargo missions the orbiter would have been replaced by a so-called Cargo Transport Container (GTK or 14S70) that could house a variety of payloads. This configuration was known as Buran-T (T standing for "transport") before the name Energiya was adopted in 1987. The interfaces between the rocket and the payload would have been virtually identical to those on Energiya–Buran. Two diameters were considered for the GTK—namely, 5.5 m and 6.7 m—with the final choice falling on the latter, which turned out to be the most favorable in terms of aerodynamic and other characteristics. The container was 42 m long and had an internal volume of about 1,000 m^3. The two main sections of the container were to be jettisoned after the rocket passed through the thickest layers of the atmosphere. The GTK was *not* used on the maiden flight of Energiya with the Skif-DM/Polyus payload, which flew the launch profile unprotected, except for a shroud on the upper FSB section. Strictly speaking, this was not a standard Buran-T configuration [56].

Upper stages

Since the core stage was suborbital, another element that needed to be developed for Buran-T besides the GTK were the upper stages to place payloads into orbit. One of these was a modification of the Proton rocket's Blok-DM upper stage. Having a diameter of 3.7 m and a length of 5.56 m, it was to carry between 11 and 15 tons of LOX/kerosene. Its engine was to have a thrust of up to 8.5 tons and have the capability of being ignited up to seven times. It could also act as a retro- and correction stage for long-duration deep-space missions, in which case it would need a special propellant-cooling system.

The other upper stage, known as 14S40 or Smerch ("Tornado"), was to use liquid oxygen and hydrogen. It was only one in a family of cryogenic upper stages that the KB Salyut design bureau (part of NPO Energiya in the 1980s) had been tasked to develop by a government decree in December 1984. The others were Shtorm ("Gale") for the Proton rocket, Vikhr ("Whirlwind") for Groza (an Energiya with two strap-ons), and the 11K37 (a "heavy Zenit") and Vezuviy ("Vesuvius") for Vulkan (an Energiya with eight strap-ons). Manufacturing was to take place at the Krasnoyarsk Machine Building Factory.

By late 1985 KB Salyut came up with a plan for using the cryogenic 11D56M engine, an improved version of the 11D56 engine developed back in the 1960s by KB Khimmash for the N-1 rocket. With its thrust of 7.1 tons and specific impulse of 461 s, it was well suited for KB Salyut's own Proton, but did not meet the requirements that NPO Energiya had laid down for Smerch. In July 1988 Minister of General Machine

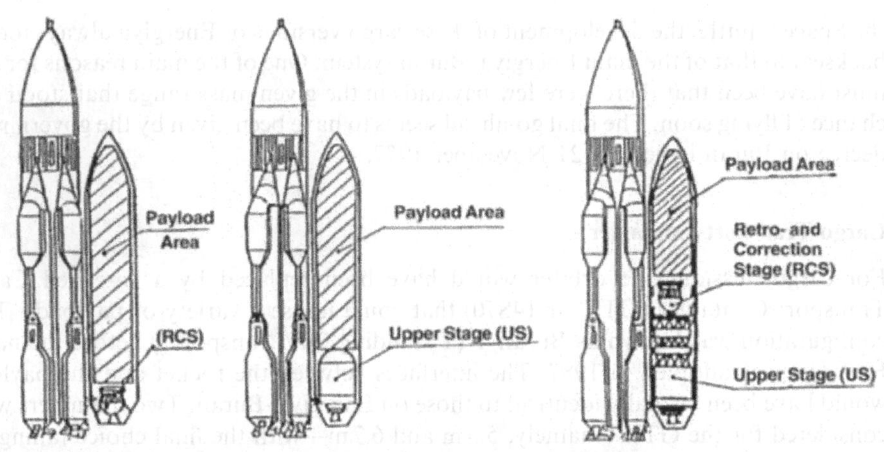

Buran-T configurations (*source*: RKK Energiya).

Building Vitaliy Doguzhiyev directed NPO Energiya and its Volga Branch to propose its own upper stages for Buran-T and Vulkan. NPO Energiya set its sights on the RO-95, an open-cycle LOX/LH_2 engine under development at KBKhA in Voronezh.

With a thrust of 10 tons and a specific impulse of 475 s, the RO-95 outperformed the 11D56M by a considerable margin and was also optimized for use in Vulkan's Vezuviy upper stage. Unlike the upper stage that KB Salyut had proposed, NPO Energiya's Smerch had the LOX tank on top, which was more favorable in terms of center-of-gravity requirements and also made it easier to ignite the engine in zero gravity. In this configuration Smerch was 5.5 m wide and 16 m long with a propellant mass of up to 70 tons. The engine could be re-ignited up to ten times. Technical requirements for the RO-95 were sent to KBKhA in December 1988 and test firings of the engine were expected to begin in 1991–1992. Yet in February 1989 Doguzhiyev seems to have turned around his earlier decision by limiting work on cryogenic upper-stage engines to KB Khimmash's 11D56M, arguing that there were no payloads in the pipeline for Buran-T and Vulkan that justified the development of an entirely new engine.

Initially, three upper-stage configurations were studied for Buran-T: only the Blok-DM derived stage for low-orbiting payloads (up to 1,000 km), only the Smerch for payloads destined for geostationary orbit, lunar libration points, and lunar orbit, and the two stages combined for lunar-landing missions, flights to Mars and Jupiter. Payload capacity would have been about 88 tons to low Earth orbit, 18–19 tons to geostationary orbit, 21.5–23 tons into lunar orbit, 9–10 tons to the lunar surface, and 10–13 tons into Martian orbit [57].

Globis

Although Energiya never came anywhere close to flying in the Buran-T configuration, there was no lack of ideas for payloads. As mentioned earlier, in September 1989

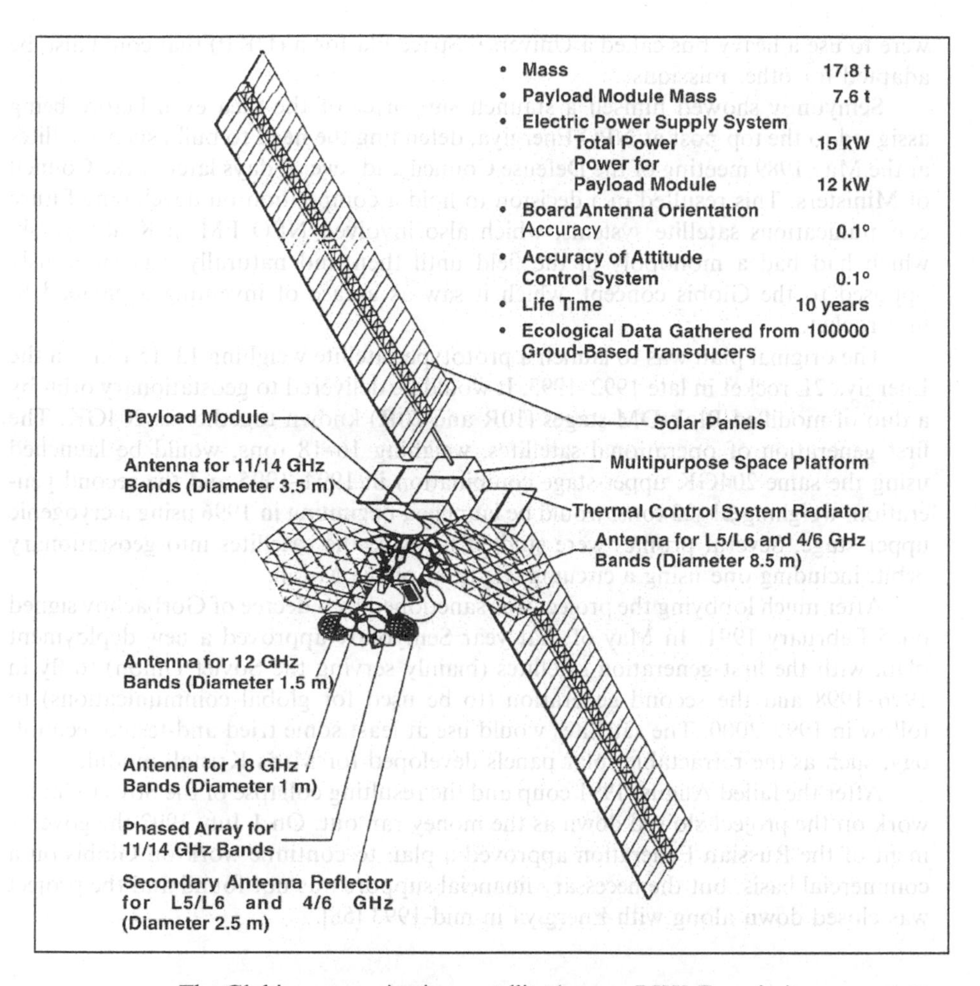

The Globis communications satellite (*source*: RKK Energiya).

NPO Energiya's new chief Yuriy Semyonov canceled plans to launch two Proton payloads on the Energiya 2L vehicle and ordered the use of that rocket instead to orbit a heavy geostationary communications platform called Globis. The idea to build these heavy communications satellites had originated in 1988 as a result of efforts to find useful payloads for Energiya, the existence of which could no longer be justified on the basis of Buran alone. Studies showed that a small network of such platforms could vastly improve communication links over the vast territory of the Soviet Union and eliminate the need to regularly launch smaller communications satellites, thereby preventing overcrowding of the geostationary belt. It was estimated that three such satellites could replace 32 conventional communications satellites. The Globis satellites, sometimes referred to in the Soviet press as the "czar satellites", were to use a heavy bus called a Universal Space Platform (UKP) that could also be adapted for other missions.

Semyonov showed himself a staunch supporter of the idea even before being assigned to the top post at NPO Energiya, defending the need to build such satellites at the May 1989 meeting of the Defense Council and several days later at the Council of Ministers. This resulted in a decision to hold a competition on developing future communications satellite systems, which also involved NPO PM in Krasnoyarsk, which had had a monopoly in the field until then and naturally was vigorously opposed to the Globis concept, which it saw as a case of inventing a payload to fit a rocket.

The original plan was to launch a prototype satellite weighing 13–15 tons on the Energiya 2L rocket in late 1992–1993. It would be delivered to geostationary orbit by a duo of modified Blok-DM stages (10R and 20R) known together as 204GK. The first generation of operational satellites, weighing 16–18 tons, would be launched using the same 204GK upper-stage combination in 1994–1995 and the second generation, weighing 21–23 tons, would be launched beginning in 1996 using a cryogenic upper stage. Several profiles were studied to place the satellites into geostationary orbit, including one using a circumnavigation of the Moon.

After much lobbying the project was sanctioned by a decree of Gorbachov signed on 5 February 1991. In May of that year Semyonov approved a new deployment plan, with the first-generation satellites (mainly serving the Soviet Union) to fly in 1996–1998 and the second generation (to be used for global communications) to follow in 1999–2000. The satellites would use at least some tried-and-tested technology, such as the retractable solar panels developed for Mir's Kristall module.

After the failed August 1991 coup and the resulting collapse of the Soviet Union, work on the project slowed down as the money ran out. On 1 July 1992 the government of the Russian Federation approved a plan to continue work on Globis on a commercial basis, but the necessary financial support was not found and the project was closed down along with Energiya in mid-1993 [58].

Skif

After the failed launch of the Skif-DM/Polyus payload by Energiya 6SL on 15 May 1987 (see Chapter 6), KB Salyut continued work on the Skif project, albeit at a slow pace. Original plans to launch Skif-D1 (without a laser payload) and Skif-D2 (with a laser payload) in 1987 and 1988 soon turned out to be unattainable. Several major components of Skif-D1 (both for the FSB and the Payload Module) were finished at the Khrunichev factory by early 1987, but serious problems with the development of the vehicle's acquisition, tracking, and pointing system delayed the final assembly of the spacecraft. That, combined with the declining support from the Gorbachov administration for Star Wars programs, led to the suspension of all Skif-related work at KB Salyut and the Khrunichev factory in September 1987.

While Skif was supposed to use laser-type weapons to destroy low-orbiting satellites, KB Salyut concurrently also developed the Kaskad system, armed with conventional missiles to destroy satellites in medium and geostationary orbits. Virtually nothing has been revealed about this project, but it was almost certainly also supposed to be launched by Energiya [59].

Mir-2 modules

In response to the announcement of America's Freedom space station in 1984, the Soviet Union devised plans for a massive version of the Mir-2 space station using giant modules launched by the Energiya rocket. There were two competing proposals for such modules, one put forward by the NPO Energiya central design bureau and another by its KB Salyut branch. In the NPO Energiya design, the modules weighed roughly 75 tons and were delivered to the station by a space tug known as GTA-S (Cargo Transport Supply Ship), which would be detached from the module after docking. Orbit insertion of the module/GTA-S combination after separation from the Energiya rocket would have been achieved with a Blok-DM derived upper stage. This particular configuration of Energiya was known as 14A10.

Little is known about the KB Salyut proposal, only that it was based on the Skif design and would have used an FGB section, presumably *both* for orbit insertion and subsequent maneuvering to the space station.

There were ambitious plans to gradually assemble a station consisting of at least eight such giant modules, but in 1991 budget realities forced these plans to be shelved in favor of a downsized Mir-2 with a 20-ton core module and smaller add-on modules. In 1993 this version of Mir-2 was united with Freedom to become the International Space Station [60].

TMP

Using the experience gained in the Skif program, KB Salyut proposed a heavy Energiya-launched space factory to produce ultra-pure semiconductor alloys and crystals. First announced in 1990, the factory was called TMP ("Technological Production Module") and had a launch mass of 102 tons and in-orbit mass of 88 tons. It was about 35 m long with a main diameter of just over 4 m. The spacecraft consisted of a Laboratory Compartment based on the main core cylinder of the Proton rocket's first stage, and an Instrument Cargo Compartment derived from the FGB. Solar arrays extended from both compartments, producing 60 kW for a mission exceeding five years.

The on-board production complex, derived from that of Mir's Kristall module, would weigh a total of 25 tons. The finished products would return to Earth in ballistic or gliding-type return capsules that could each hold up to 140 kg of materials. Robotic manipulator arms would be used to remove a capsule from storage, load it, and then transfer it to a small airlock for ejection. The TMP had two docking ports to receive Progress resupply ships and Soyuz spacecraft or air-launched MAKS spaceplanes, with crews being able to spend up to 10 days aboard the facility to unload supply ships and perform maintenance work. In KB Salyut's vision, the TMP was only the final step in a phased program for space-based materials processing, which also included the launch of small 1.2-ton capsules and bigger 20-ton vehicles [61].

Other payloads

Bolstered by the success of the maiden Energiya launch in 1987, NPO Energiya worked out a series of ambitious plans for future use of the rocket. Taking into account the changing international climate, those missions focused not so much on national, but global needs. Some of these projects bordered on the realm of science fiction and were way beyond even the generous budgets of the Soviet days, which is why the Russians were clearly counting on international partners to join them. The following missions were studied in 1987–1993:

- A constellation of 30 to 40 satellites to restore the depleted ozone layer by aiming laser beams at the stratosphere, causing excited oxygen molecules to break up under the influence of solar radiation and to recombine into ozone molecules. Weighing 60 to 80 tons each, the satellites would have flown in Sun-synchronous orbits at an altitude of 1,600 km, using electric propulsion systems to maneuver from their initial insertion orbits. Using this satellite constellation, it would have taken an estimated 30 years to solve the ozone depletion problem.
- Containers with radioactive waste to be placed into heliocentric graveyard orbits between Earth and Mars at a distance of approximately 1.2 astronomical units from the Sun. Weighing 50 tons each, the hardened containers could house 6 to 9 tons of radioactive waste. It was estimated that 10 to 15 Energiya missions would be required annually to dispose of the 100 tons of high-level radioactive waste produced around the world each year. Each container was to be boosted to an 800 km parking orbit by a conventional upper stage before being sent on an escape trajectory by a nuclear electric propulsion system.
- A constellation of solar reflector satellites to illuminate the polar regions, provide energy from space, and improve crop yields by stimulating photosynthesis. With each of the satellites weighing 5–6 tons, a single Energiya was capable of placing a cluster of 10 to 12 such satellites into a low parking orbit with the help of an upper stage. A reusable, solar electric interorbital space tug would have boosted the satellites to a 1,700 km polar orbit inclined 103° to the equator. Each satellite had a 10 year lifetime and would be usable 8 hours daily, illuminating a 17 km diameter circular area on the Earth's surface.
- An Earth-to-Moon shuttle service to collect helium-3 on the lunar surface for use in nuclear fusion reactors.
- 20-ton environmental monitoring satellites in geostationary orbit. Using the same UKP platform as the Globis satellites, they would monitor the Earth with optical, infrared, and microwave remote-sensing instruments, study Sun–Earth relations with ultraviolet spectrometers and particle detectors, and relay data from low-orbiting satellites in radio and optical wavelengths.
- 30-ton UKP-based satellites in 600 km polar orbits to monitor observance of international disarmament treaties and perform remote-sensing tasks such

as studies of natural resources and environmental monitoring. The 12-ton payload would have included a videospectrometer, optical electronic cameras, and phased-array antennas.

- Satellites to clear the geostationary belt of space debris. Equipped with an engine unit and grappling devices, they would each spend about half a year in 0° to 14° inclination orbits at geostationary altitude, moving defunct satellites and debris to graveyard orbits.
- A 27-ton space-based radio telescope to provide Very Long Baseline Interferometry (VLBI) in concert with ground-based radio telescopes. Called IVS (International VLBI Satellite), this was a joint Soviet–European project put forward in response to a 1989 Call for Mission Proposals for the second medium-size mission under ESA's Horizon-2000 program. The IVS was to consist of NPO Energiya's UKP bus and a European-built 20 m diameter radio telescope. With an inclination of 65° and a perigee of 6,000 km, the apogee would be varied from an initial height of 20,000 km to 40,000 km and 150,000 km over the satellite's five-year operational lifetime. IVS was picked along with five other projects for further assessment in 1991, but was not approved for further development. Had it been selected, it could have flown in 2001 [62].

Even though the Skif-DM launch had demonstrated that Energiya was capable of being used as a heavy cargo carrier, Buran-T failed to gain impetus, mainly due to a lack of interest from the military, who were supposed to be the system's main customers. A government decree in August 1985 had ordered the Ministry of Defense to work out "technical requirements" for Buran-T and Vulkan in a three-month period and NPO Energiya to prepare a draft government decree on these systems in the first quarter of 1986, outlining their objectives and setting a timeline for their development. The draft was sent for review to the VPK by July 1986 and called for starting Buran-T flights in 1988, with the introduction of the Smerch cryogenic upper stage expected in 1995. It was not until December 1987, one and a half years later, that the VPK responded by rejecting the draft, claiming it had not been agreed upon with the military. For the military a rocket could only be declared operational if there was a concrete payload for it, which was hardly the case for Buran-T. Eventually, the military even withdrew their "technical requirements" for Buran-T [63].

THE ENERGIYA FAMILY

As had been Valentin Glushko's original intention with the RLA rockets, the modular design of the Energiya rocket allowed it to be transformed into a wide variety of lighter and heavier launch vehicles. If all had gone according to plan, Russia could now have had an entire family of heavy-lift launch vehicles capable of orbiting payloads from just 30 tons to a mind-boggling 500 tons.

Vulkan

The Energiya version with eight strap-on boosters was known as Vulkan ("Volcano"). With the lower section of the core stage completely surrounded by strap-ons, the payload and upper stage were mounted atop the core stage as on a conventional rocket. The tanks of the strap-on boosters and core stage were stretched and the strap-ons did not have the parachute recovery systems of the standard Energiya. Vulkan would have required the development of a new adapter platform to place it on the launch table. Launches would only have taken place from the UKSS pad, which was built from the beginning with a view to supporting Vulkan launches in the future.

Two slightly different versions of Vulkan have been described in Russian literature. One used the same RD-0120 and RD-170 engines as the standard Energiya and was capable of placing 170 tons into a low 50.7° orbit. Equipped with an 11D57M cryogenic upper-stage engine of the KB Saturn "Lyulka" design bureau (vacuum thrust 42 tons, specific impulse 460 s), it could inject a 28-ton payload into geostationary orbit [64].

The other version carried upgraded first and second-stage engines and the Vezuviy cryogenic upper stage, probably outfitted with the RO-95 engine of KBKhA. The upgraded first-stage engines were known as RD-172 or 14D20 and had a sea-level thrust of 784 tons (as compared with 740 tons for the RD-170). Also mentioned has been an even more powerful version called RD-179 with a reported sea-level thrust of 860 tons. The core stage engines retained the RD-0120 designator and had a vacuum thrust of 200 tons (as compared with 190 tons for the standard RD-0120). The following payload capacities are given: 200 tons to a low 50.7° orbit, 172 tons to a 97° orbit, 36 tons to geostationary orbit, 43 tons into lunar orbit, and 52 tons to Mars. Possibly, the first version was an early proposal that was later superseded by the more capable one [65].

The Vulkan rocket (*source*: *www.buran.ru*).

The development of Vulkan seems to have been set in motion by a government decree released in July 1981, which called for making "technical proposals" for the rocket within the next five years. The "technical requirements" that formed the basis for these proposals were issued in July 1982. With a payload capacity of around 200 tons, Vulkan was seen by the Russians as a rocket that could play a crucial role in future manned missions to Mars and other planets of the solar system. It was the subject of further government decrees between 1983 and 1986, but timelines for its development remained vague as no concrete payloads were ever defined for it.

Groza and Energiya-M

In the course of Energiya's history several studies were made of configurations in which the core stage was flanked by just two strap-ons, providing payload capacities of between about 30 and 60 tons. The first version, called RLA-125, was proposed in 1976 and another one known as Groza ("Thunderstorm") appeared in the mid-1980s. Groza, using a standard core stage with four RD-0120 engines and two strap-ons, had a reported payload capacity to low orbit of up to 63 tons. The Cargo Transport Container strapped to the side would be a downsized version of that developed for Buran-T. Groza required virtually no modifications to the existing Energiya pads. All that needed to be done was to bolt the strap-ons more firmly to the pad because the rocket would be more susceptible to high winds. Because of this, launch weather rules were also tightened.

On 25 December 1984 the Soviet government released a major decree on rocket and space systems to be developed in the period 1986–1995. One of these was planned to be a series of rockets with payload capacities between 30 and 60 tons, although it is not clear what payloads exactly were being considered. Three systems were adopted for parallel studies: a modernized version of the Proton rocket, several heavier versions of the Zenit (11K37), and Groza. As mentioned earlier, these boosters were to use a standardized series of cryogenic upper stages, Shtorm for Proton and Vikhr for the 11K37 and Groza.

The preliminary design for Groza was completed in December 1985. However, on 18 August 1988 the Ministry of General Machine Building ordered NPO Energiya to modify the rocket in order to make it compatible with more realistic payloads of between 25 and 40 tons. This made it necessary to reduce the number of RD-0120 engines to one or two and hence make the core stage smaller. The first idea was to reduce the diameter of the core stage to 4.1 m or 5.5 m and lower the propellant mass to between 200 and 450 tons. However, since this would have required different manufacturing techniques, it was decided to retain the standard Energiya core stage diameter of 7.7 m. By late 1989 engineers were focusing on a version with one RD-0120 engine and a propellant mass of around 240 tons. With the core stage (called Blok-V) only about half as high as that of Energiya, the payload had to be stacked on top rather than strapped to the side. At the intersection between the core stage and the payload bay the rocket would taper off to a diameter of 6.7 m, the same as that of the 14S70 Cargo Transport Container of Buran-T. With a length of 25 m,

Energiya-M on the UKSS pad (*source*: *www.buran.ru*).

the payload bay was probably almost identical in dimensions to the Cargo Transport Container for Groza. The concept was approved by the Council of Chief Designers on 19 July 1990. Initially called Neytron ("Neutron"), the new rocket eventually became known as Energiya-M.

Four configurations were considered for the payload section, one in which the satellite would occupy the entire bay and have its own engine system (N-11) and three where the satellite would be attached to various upper stages (N-12, N-14, and N-15).

The N-12 was a Blok-DM modification with an engine known as the 11D58MF and was also planned for use on Zenit, Proton, and the original Angara. It allowed the rocket to place 29 tons into low Earth orbit or up to 3 tons into geostationary orbit. The N-14 was a Blok-DM modification with the standard 11D58M engine and was identical to the second stage of the 204GK upper-stage combination planned for Buran-T. It was capable of delivering a 5.5-ton payload to geostationary orbit. The N-15, able to launch 6.5 tons into geostationary orbit, was a LOX/LH_2 upper stage but no further information on this is available. It is known that in 1992 work got underway on a LOX/LH_2 upper stage known as Yastreb carrying the RO-97 engine of KBKhA. This stage was primarily intended for Proton, Zenit, and Angara, but with slight modifications could also be mounted on Energiya-M. However, it was smaller than the N-15 and also had its propellant tanks configured differently.

As early as 1990 a mock-up of Energiya-M was ready for tests at Baykonur. It was placed both on the UKSS pad and Energiya–Buran pad 37. It was only afterwards, on 8 April 1991, that the government issued a decree ordering NPO Energiya, KB Yuzhnoye, and KB Salyut to come up with competing proposals for boosters in the 25 to 40-ton payload range. This basically was a repeat of the order given in the 25 December 1984 decree, although in a somewhat lighter payload class. KB Salyut and KB Yuzhnoye had apparently also been optimizing their Proton and 11K37 designs. Eventually, on 6 July 1991 the Ministry of General Machine Building opted for Energiya-M. Between 1991 and 1993 preparations were made for starting production of flight models.

During that period, NPO Energiya worked out plans to launch a 30-ton spaceplane (OK-M2) atop the rocket and also to turn the two strap-ons into reusable flyback boosters, something which appears to have been studied as early as 1989. Another idea was to launch the rocket from an ocean-based platform near the equator. This would not only allow Energiya-M to loft heavier payloads, but would also resolve the political problems associated with flying it from Baykonur, which became foreign territory after the collapse of the Soviet Union. One exotic mission considered for the ocean-launched Energiya-M was to deposit radioactive waste into heliocentric orbits, eliminating the risks involved in launching such dangerous payloads over populated territories. These studies formed the basis for the creation of the international Sea Launch venture, which would eventually use the three-stage Zenit rocket.

Despite the fact that Energiya-M used existing hardware and infrastructure and outperformed rockets like the Titan-4 and Ariane-5, it was ahead of its time. At the time there was simply no demand for the types of satellites that the rocket could place into orbit. On 15 September 1992 the Russian government started yet another competition to develop a family of even lighter rockets, which would eventually evolve into the Angara series. By late 1993 government funding for Energiya-M was stopped, with Russian Space Agency officials stating there was no demand for the rocket on the market. The following year NPO Energiya made an ultimate attempt to attract Western customers to Energiya-M and other Energiya variants, but to no avail. The prototype Energiya-M still stands today inside the Dynamic Test Stand at Baykonur [66].

Making Energiya reusable

The ultimate dream of the Energiya designers was to develop a rocket that would be fully reusable (Energiya-2 or GK-175). The plan was to achieve full reusability in various steps, first of all by having the strap-on rockets parachute back to Earth for recovery. In the next step the core stage was to be turned into a reusable winged stage with three RD-0120 engines and a payload compartment in the upper section. Despite the lower amount of propellant, the overall dimensions of the core stage remained the same, freeing up some 610 m^3 of volume in the payload section, which compared favorably with the 350 m^3 offered by Buran's cargo bay.

The massive nose fairing would not separate during ascent, but open in space, somewhat like the forward cargo door of a Lockheed C-5 transport aircraft, allowing it to be reused on subsequent flights. After deployment of the payload (30 to 40 tons), the fairing would slide down over the LOX tank so that the core stage would shrink in size from 60 to 44 m to prevent stability problems during re-entry. In order to cut costs, the core stage would inherit as many systems as possible from Buran (wings, vertical stabilizer, landing gear, avionics, and hydraulic systems). It was not considered expedient though to cover the stage with Buran's heat-resistant tiles and efforts focused instead on using innovative non-ablative and active cooling thermal protection systems.

The next phase was to replace the standard strap-ons by the same type of flyback boosters being envisaged for Energiya-M. The parachute recovery system imposed a

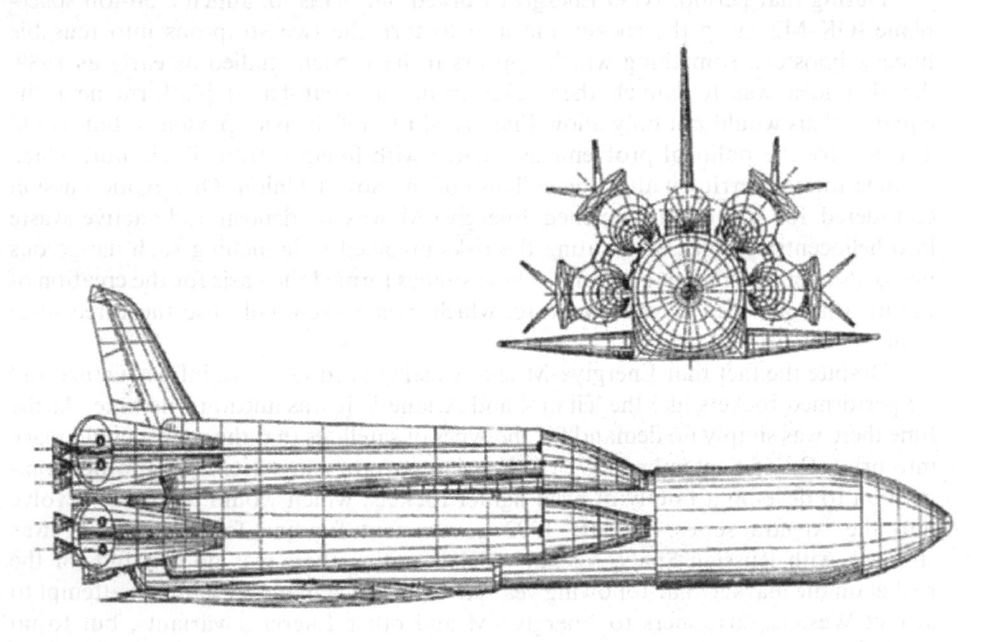

Fully reusable Energiya with flyback strap-on boosters and winged core stage (*source*: RKK Energiya).

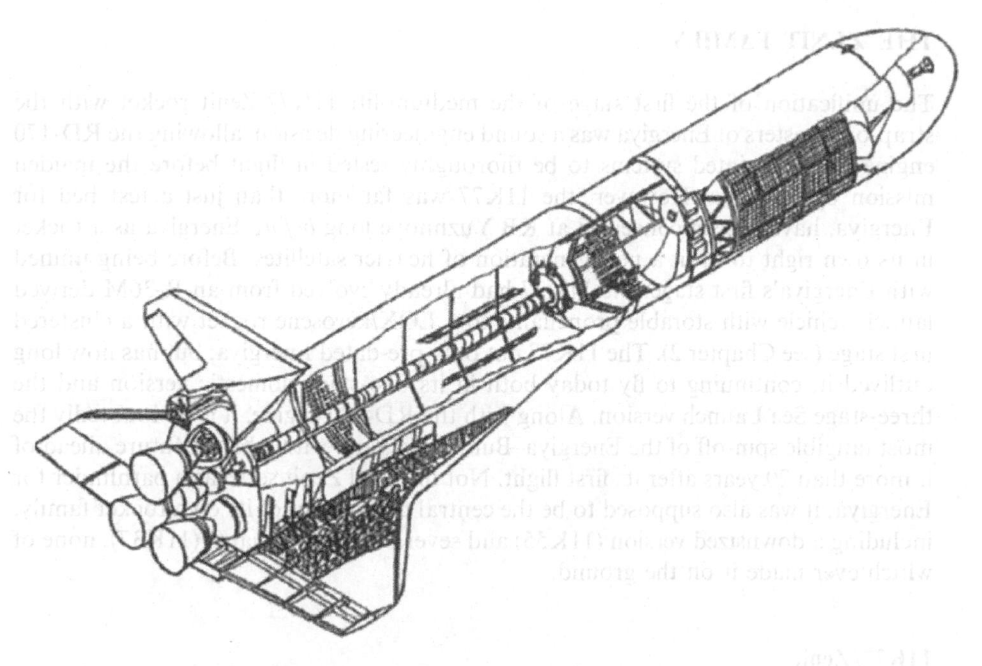

Winged core stage (*source*: RKK Energiya).

heavy weight penalty on the rocket, the impact zones limited the number of launch azimuths, and recovery from those distant impact zones would have been a laborious and costly undertaking. The flyback strap-ons would be equipped with long foldout wings, a V-shaped tail, and a small jet engine enabling them to fly back to the launch site after separation from the core stage. Although the idea was tempting, the landing of four strap-ons in quick succession on the single Baykonur runway would probably have caused tremendous logistical problems.

In the final step the four flyback strap-ons would be replaced by a first stage equal in size to the second stage and with similar landing systems, but without thermal protection, and equipped with four RD-170 engines. This vehicle would have a payload capacity of between 30 and 50 tons. By using two such first stages it would be possible to increase payload capacity to 200 tons, about the same as the Vulkan with its eight strap-ons. There was even an idea to use four of the large first stages and lengthen the second stage to reach a phenomenal payload capacity of 500 tons. Despite the impressive prospects, any of those three variants would probably have required major modifications to the existing Energiya launch facilities.

Not surprisingly, all these bold proposals faced an uphill battle as the budgets for the space program became ever tighter towards the end of the 1980s. The only spin-off from the Energiya-2 studies is Baykal, a reusable flyback booster now being proposed as a first stage for the Angara rocket family. This incorporates many ideas that had originally been conceived for Energiya-2's flyback strap-ons [67].

THE ZENIT FAMILY

The unification of the first stage of the medium-lift 11K77/Zenit rocket with the strap-on boosters of Energiya was a sound engineering decision, allowing the RD-170 engine and associated systems to be thoroughly tested in flight before the maiden mission of Energiya. However, the 11K77 was far more than just a test bed for Energiya, having been conceived at KB Yuzhnoye long *before* Energiya as a rocket in its own right to orbit a new generation of heavier satellites. Before being unified with Energiya's first stage, the 11K77 had already evolved from an R-36M derived launch vehicle with storable propellants to a LOX/kerosene rocket with a clustered first stage (see Chapter 2). The 11K77 not only pre-dated Energiya, but has now long outlived it, continuing to fly today both in its two-stage domestic version and the three-stage Sea Launch version. Along with the RD-180 engine, it is undoubtedly the most tangible spin-off of the Energiya–Buran program, with a bright future ahead of it more than 20 years after its first flight. Not only did Zenit serve as a pathfinder for Energiya, it was also supposed to be the central component of its own rocket family, including a downsized version (11K55) and several heavier variants (11K37), none of which ever made it off the ground.

11K77/Zenit

The final version of the 11K77 was approved by a government decree released on 16 March 1976, which set the maiden launch for the second quarter of 1979.

Zenit on the pad at Baykonur. Crew access tower is still in place (*source*: Russian Space Agency).

However, the project soon ran into substantial delays, mainly due to development problems with the RD-170/171 engines, highlighted by the explosion of a Zenit first stage at the test stand of NIIkhimmash in June 1982. The switch to the single-chamber MD-185 engines considered for Energiya was also weighed for Zenit. Eventually, Zenit made its maiden flight on 13 April 1985, almost six years later than originally planned (see Chapter 6).

Original specifications for the 11K77 were to launch payloads into orbits with inclinations between 46° and 98° from both Baykonur and Plesetsk. The Zenit used highly automated launch facilities developed by the Design Bureau of Transport Machine Building (KBTM). These enabled several rockets to be placed on stand-by and be launched in quick succession. The idea was that the Zenit could swiftly replenish constellations of military satellites in case of an impending conflict or if some of them were knocked out by the enemy. Two launch pads were built at Baykonur. Construction of a Zenit pad at Plesetsk got underway in 1986, but the work was suspended in 1994 and the pad is being rebuilt for the Angara rocket family.

The vast majority of Zenit launches have carried KB Yuzhnoye's Tselina-2 electronic intelligence satellites, placed into 850 km circular orbits inclined 71° to the equator. Actually, the Tselina-2 satellites are far underweight for Zenit, having been originally developed for launch by the lighter Tsiklon-3 rocket and then reoriented to Zenit because of slight increases in dimensions and mass. This was also the case for the Resurs-O1 and Meteor-3M satellites, originally built for launch by the Soyuz and Tsiklon-3 rockets. Most of the payloads *really* tailored for Zenit never flew as a result of the break-up of the Soviet Union in 1991, which not only led to a major economic crisis but also turned Zenit into a Ukrainian booster. Exceptions were two heavy photoreconnaissance satellites launched in 1994 and 2000 and the Okean-O ocean-monitoring satellite orbited in 1999.

From the outset Zenit was also developed as a man-rated launch vehicle with the necessary built-in redundancy and safety features. In the late 1980s NPO Energiya designed a Zenit-launched vehicle called Zarya ("Dawn"), which outwardly re-sembled an enlarged Soyuz descent capsule. A relic of those plans is a crew access tower still in place at the Baykonur Zenit pad. Zenit was also supposed to launch a variety of cargo ships and modules to Mir-2 and later to the International Space Station, but those plans were abandoned in 1996.

Compounding the problems for Zenit were three back-to-back launch failures that the rocket suffered in the 1990–1992 timeframe. The first of these resulted in the rocket crashing back seconds after lift-off, completely devastating one of the two Baykonur Zenit pads, which still lies in ruins today. But, while the end of the Cold War spelled bad news for Zenit as a domestic launch vehicle, it opened up new frontiers for its use in international programs. The first such opportunity arose in 1989, when Glavkosmos signed a deal to launch Zenits with Blok-DM upper stages from Cape York in Australia. Located on the east coast of Australia's northernmost peninsula just 12 degrees south of the equator, Cape York was ideal for due east launches over the Pacific Ocean to place communications satellites into geostationary orbits.

Zenit-3SL lifts off from its ocean launch pad (*source*: Sea Launch).

A three-stage version of the 11K77 had already been envisaged for Soviet domestic missions by the original March 1976 government decree on Zenit. Although the Blok-DM had been considered from the outset, KB Yuzhnoye had preferred an upper stage with storable propellants (nitric acid and dimethyl hydrazine) plus an additional solid-fuel apogee kick motor, together capable of placing 1.3-ton payloads into geostationary orbit from Baykonur (compared with 1 ton for the Blok-DM). However, the use of the toxic storable propellants was considered unacceptable for launches from Australian territory, leaving Yuzhnoye no choice but to revert to NPO Energiya's Blok-DM. Eventually, the Cape York plan fell through because of a lack of investor support.

The big break for the Zenit came in May 1995 with the official establishment of Sea Launch, a joint venture between KB Yuzhnoye, RKK Energiya, Boeing, and Kvaerner to launch three-stage Zenit rockets (Zenit-3SL) with Blok-DM upper stages on commercial satellite deployment missions from a converted Norwegian oil rig near the equator. Initial studies of sea-launched versions of Energiya, Energiya-M, and Zenit had been conducted at NPO Energiya in November 1991–December 1992 because of uncertainty over the future use of the Baykonur cosmodrome and rocket stage impact zones in independent Kazakhstan. Realizing that such a venture would require foreign investors, NPO Energiya officials pitched the idea of a sea-launched Zenit or Energiya-M to Boeing during a visit to the company's Seattle headquarters in March 1993, with the final choice falling on Zenit in July 1993. Unknown to most of the parties involved (even Energiya), KB Yuzhnoye *itself* had studied sea-launched versions of Zenit together with KBTM in 1976–1980 under a research program known as Plavuchest ("Buoyancy"). This would have seen the use of two catamaran-type vehicles, one acting as a launch pad and the other as a command center and storage facility for as many as five Zenit rockets with hypergolic upper stages. Many of the ideas worked out under Plavuchest were later incorporated into Sea Launch.

Sea Launch saw its inaugural mission on 27 March 1999 and has since averaged three launches per year, securing a solid place in the international commercial launch market. The company did suffer a significant setback on 31 January 2007, when one of its rockets exploded during lift-off. Although the launch platform escaped relatively unscathed, the commercial implications of this accident are as of yet unclear.

Significant differences between the heritage Zenit and the Sea Launch version were a new navigation system, a next-generation flight computer, and increased performance by mass reductions. The propulsion system remained essentially unchanged. Originally, the hope was to use an improved first-stage engine called RD-173 on which Energomash had begun work in the second half of the 1980s. This engine delivered 5 percent more thrust than the RD-171, had an improved turbopump assembly, and a modernized guidance and control system. Experimental versions of the engine underwent static test firings between 1990 and 1996, but further testing was suspended for financial reasons.

With the production line for the standard RD-171 closed due to a lack of state orders, Energomash had no other option but to modify existing RD-170 Energiya engines for use in the Sea Launch program. In 1996–1997 a total of fourteen RD-170 engines were "cannibalized" from mothballed Energiya strap-on boosters and shipped back to Energomash for modification. This batch was enough to ensure several years of Sea Launch operations, but eventually Energomash returned to its RD-173 plans. The modified engine, now redesignated RD-171M, has the same thrust as the RD-171, but is 200 kg lighter and has an improved guidance and control system. Testing started in 2004 and the engine made its debut in February 2006. In May 2004 Sea Launch also introduced a slightly improved RD-120 engine for the second stage (93 tons of thrust vs. 85 tons for the earlier version). Further performance improvements may be achieved by adding suspended propellant tanks to the first stage.

In late 2003 the Sea Launch Board of Directors resolved to go forward with plans to offer launch services from Baykonur in Kazakhstan, in addition to its sea-based launches at the equator. An earlier attempt by Yuzhnoye to commercialize the two-stage Zenit from Baykonur had ended with an embarrassing launch failure in 1998 in which 12 Globalstar satellites came tumbling back to Earth minutes after lift-off. The new offering, Land Launch, is based on the collaboration of the Sea Launch Company and Space International Services (SIS) of Russia to meet the launch needs of commercial customers with medium-weight satellites. The Land Launch Zenits will have the same modifications as the Sea Launch version and can fly in a two-stage configuration for launches to low and elliptical orbits and with three stages to geostationary orbits [68].

11K55

The 11K55 was a lighter version of Zenit conceived jointly by KB Yuzhnoye and the Omsk-based PO Polyot in 1976. In its original design it had a modified Zenit first stage with a smaller propellant load and a two-chamber version of the RD-170. The newly developed second stage would be powered by a cluster of three LOX/kerosene 11D58M engines of NPO Energiya's Blok-DM upper stage. With a launch mass of 210 tons, the rocket had a payload capacity of roughly 5 tons. The 11K55 was seen as an environmentally clean replacement for the Kosmos-3M and Tsiklon launch vehicles, rockets with storable propellants built at Polyot and Yuzhnoye, respectively. Launches were to take place from Plesetsk and Kapustin Yar. Other engines considered for use on the 11K55 in the late 1980s/early 1990s were a cluster of three or four RD-120K engines on the first stage and two different re-ignitable LOX/kerosene engines for the second stage, the RD-133 and RD-134, both with a vacuum thrust of 35 tons. The 11K55 is known to have been the subject of a government decree in 1986 [69].

11K37

In 1976 KB Yuzhnoye also began looking at heavier versions of Zenit. One idea was to build a 5.4 m diameter first stage with two RD-171 engines, but this was deemed unrealistic because it required new manufacturing techniques and untested transportation methods by road, water, or air. Instead, planners concentrated on rockets with two, three, or four Zenit first stages clustered together, capable of putting between about 30 and 60 tons into low orbit. The common designator for these boosters was 11K37.

The most thoroughly studied version was the one with three Zenit first stages. Mounted above that would have been a second stage with a 214-ton thrust single-chamber version of the RD-170 engine known as RD-141 as well as an RD-8 vernier engine. A later idea was to equip the stage with three 90-ton thrust RD-142 engines and several verniers. The RD-142 was apparently an improved version of the RD-120 engine used in the second stage of the 11K77. With the latter second stage, the 11K37 would have been capable of putting 40 tons into low orbit, 35 tons into polar orbit, and about 5 tons into geostationary orbit. Upper stages studied by Yuzhnoye for the 11K37 were an interorbital space tug originally developed for deployment from Buran as well as the Vikhr cryogenic upper stage.

As mentioned earlier, the 11K37 was proposed by KB Yuzhnoye in a competition started in 1984 to develop boosters in the 30–60 ton range, with the other candidates being Groza and an upgraded Proton rocket. A major disadvantage of the 11K37 was that it required the construction of a new launch facility. Therefore, the Ministry of General Machine Building recommended in the late 1980s to make the rocket compatible with Energiya's UKSS launch pad at Baykonur. When a new competition was launched in the early 1990s to develop rockets in the 20–40 ton payload range, the 11K37 was still in the running, but eventually lost out to Energiya-M in 1991 [70].

THE RD-180 ENGINE

With state orders for RD-170 and RD-171 engines running out, NPO Energomash began looking at international marketing opportunities for its engines in the early 1990s, setting its sights on America in particular. Realizing that the RD-170/171 thrust levels were beyond what was needed on American launch vehicles, the company designed a two-chamber version of the engine called RD-180 that was tailored to the US market and was probably similar to a first-stage engine studied for the 11K55. In October 1992 Pratt & Whitney started working with Energomash to draw American customers to the RD-180 and a tripropellant engine known as the RD-701, and also to advise the Russian company on ways to implement a cost-accounting system [71]. The RD-180 was considered for use on a new two-stage Martin Marietta booster as well as an upgraded version of General Dynamics' Atlas-II rocket [72].

In 1994 General Dynamics Space Systems was sold to Martin Marietta, which in turn merged with Lockheed in 1995 to become Lockheed Martin. The company continued looking at new engines to power its new Atlas-IIAR rocket (later renamed Atlas III) as well as a new generation of Atlas vehicles (Atlas V) being developed under the Air Force's Evolved Expendable Launch Vehicle (EELV) competition. In January 1996 Lockheed Martin's choice fell on the RD-180, which beat the Aerojet-sponsored NK-33, a Russian engine originally developed for the N-1 Moon rocket,

The RD-180 engine (*source*: Lockheed Martin).

and a derivative of Rocketdyne's venerable MA-5A called the MA-5D. Just one nozzle of the RD-180 generates as much thrust as all three MA-5A nozzles combined on the older Atlas configuration.

The RD-180 essentially is an RD-170 "cut in half" with a new, less powerful turbopump driven by a single gas generator. About 75 percent of parts are identical to those of the RD-170. It has a sea-level thrust of 390 tons and a specific impulse of 311 s. The engine has several features that made it attractive to Lockheed Martin. It operates at much higher pressures than most other expendable booster engines, allowing the deep throttling capacity critical to effective engine use. The RD-180 can throttle over a 40–100 percent range, yet it remains flat in specific impulse throughout this range (losing just about a second of Isp), which is very important to fly the engine on both light and heavy-lift launch vehicles. The RD-180's single-shaft turbine, liquid-oxygen pump, and single-stage propellant pump are all on one shaft, which cuts overall parts count, reduces cost, and translates to excellent reliability. Furthermore, adoption of Russian seal and flange technologies virtually eliminated cryogenic system leaks that were accepted as normal on US boosters.

In early 1997 Energomash and Pratt & Whitney expanded their cooperation on the RD-180 into a joint venture called RD Amross LCC to build and market the engine. At the time the RD-180 accounted for 75 percent of Energomash's business. In June 1997 Lockheed Martin announced it would purchase 101 RD-180 engines from Amross under a contract expected to be worth 1 billion dollars. In a move to allay concerns about relying on Russian technology for placing military and intelligence satellites into orbit, Lockheed Martin vowed that the US would set up its own production line at a new Pratt & Whitney facility in West Palm Beach, Florida, but those plans have run into numerous delays.

A prototype version of the RD-180 underwent an initial test firing at Energomash's test facilities in Khimki in November 1996. The first test firing of a full-fledged engine followed in April 1997. Also applicable to the RD-180 were test firings of the RD-173, which had several new features that were incorporated into the RD-180. An RD-180 mated to an Atlas III thrust structure and tank simulator was first test-fired at the Marshall Space Flight Center in Huntsville, Alabama in July 1998. The Atlas III debuted in a spectacular launch from Cape Canaveral on 24 May 2000, successfully placing into orbit a Eutelsat communications satellite. It was a landmark event in US–Russian space cooperation, very illustrative of the new, post Cold War atmosphere. A US rocket that had evolved from an ICBM conceived to level Soviet cities was now powered by a Russian rocket engine that itself had its origins in a program once seen as a crucial part of the military space race.

The Atlas III was retired in 2005 after seven successful missions, clearing the way for the new Atlas V generation. In November 1997 the Air Force had decided to modify its procurement plans for the EELV program, splitting the work between a pair of finalists rather than going for a single winner-take-all award. One of the major reasons given for the redirection was to enhance US space launch competitiveness by keeping two rocket builders in business. The work would now be divided between Lockheed Martin with its Atlas V family and McDonnell Douglas (later Boeing) with its Delta-4 family.

Atlas-V launch from Cape Canaveral (*source*: Lockheed Martin).

The Atlas V family uses a Common Core Booster (CCB) first stage fitted with an RD-180 engine and flanked by up to five solid rocket boosters. The Centaur second stage is powered by either a single or two RLA-10A-4-2 engines and the payload is protected by either a 4 or 5 m diameter payload fairing. There were also plans for an

Atlas V Heavy featuring three CCBs coupled together, but Lockheed Martin is no longer actively pursuing development of this version.

Given the slightly different flight modes for the medium-lift and heavy-lift Atlas V versions, the RD-180 had to undergo separate certification programs for the two versions, although it is exactly the same engine as flown on the Atlas III. The impressive RD-180 test-firing program was completed in early 2002. Since the first test in 1996, the RD-180 averaged a full flight duration firing every 10 days, encompassing 135 total development and certification tests in Khimki, comprised of 91 Atlas III class tests, 30 Atlas V Medium class tests, and 14 Atlas V Heavy class tests. All totaled, the RD-180 racked up an impressive 25,449 seconds of development and certification test firing in Khimki alone, equivalent to 110 nominal Atlas V missions. The inaugural flight of the Atlas V took place on 21 August 2002. The RD-180 may also fly on the first stage of a Japanese rocket called Galaxy Express, which is expected to use the first stage of the Atlas III [73].

THE RD-191 ENGINE

Another offshoot of the Energiya propulsion system is the RD-191, a single-chamber version of the RD-170/171 designed to power Russia's new Angara family of launch vehicles.

Originally, Angara (named after a Siberian river) was conceived as a heavy launch vehicle to replace the Proton. Its history began with a government decree (nr. 716-53) issued on 15 September 1992 calling for the development of a launch vehicle capable of lifting 24 tons to low orbit and 3.5 tons to geostationary orbit from the Plesetsk cosmodrome in northern Russia. Proton could only fly from Baykonur in Kazakhstan, the future of which was uncertain after the collapse of the USSR. The new launch vehicle was to be built exclusively by Russian enterprises and make maximum use of Energiya–Zenit technology.

Responding to the decree, the Russian Space Agency and the Ministry of Defense, the two organizations that had ordered the vehicle, launched a tender between three design bureaus: NPO Energiya, the Khrunichev Center, and the Makeyev bureau, the latter having specialized for years in sea-launched intercontinental ballistic missiles. Rather than proposing separate projects, NPO Energiya and Makeyev joined forces in January 1994 to design a launch vehicle provisionally called Energiya-3. This had a first stage comprised of three modules each powered by an RD-180 engine, and a second stage with an RD-146 engine, a re-ignitable version of the Zenit second stage's RD-120.

The Khrunichev vehicle had a first stage employing the RD-174, yet another modification of the RD-170/171, and a second stage using one of the Energiya core stage's RD-0120 LOX/LH_2 engines. Both the first and second stages had suspended propellant tanks, giving the rocket an odd external appearance. For geostationary missions the rocket would have used a LOX/LH_2 upper stage. Other plans were to turn the first stage into a reusable flyback booster and eventually to launch the vehicle

Khrunichev's original Angara configuration (*source*: Khrunichev).

from the Svobodnyy cosmodrome in the Russian Far East. Both proposed launch vehicles would have been able to fly from modified Zenit pads.

In September 1994 the Russian Space Agency and the Ministry of Defense selected Khrunichev as the winner of the competition, although development of the second stage was subcontracted to RKK Energiya. The decision was consolidated by another government decree on 26 August 1995, which set the maiden flight of Angara for 2005. However, during the following two years it became apparent that Khrunichev's design was basically flawed. Among the problems were the rocket's very low thrust-to-weight ratio (1.09) and the challenges associated with igniting the RD-0120 at altitude, not to mention the fact that the engine's production line at KBKhA in Voronezh had been closed [74].

Another problem with Khrunichev's design was that it was primarily geared to replacing the Proton and left little room for building derived launch vehicles. The best Khrunichev had come up with were two rockets called Yenisey and Neva. Yenisey basically was an Angara first stage topped by a Zenit second stage, giving an 18-ton capacity to low orbit, and a 2.5-ton capacity to geostationary orbit with a KVRB cryogenic upper stage. Neva was an all-cryogenic vehicle, using Angara's second stage as its first stage and the KVRB as second stage, optimized to place 4.1-ton payloads into polar orbit [75]. However, the range of payloads that could be orbited by these boosters was minimal and, moreover, Neva required a dedicated launch pad.

All this was at a time when the Russians realized that they needed a replacement not only for Proton, but also for other venerable 1960s launch vehicles such as the Tsiklon and Kosmos-3M, whose production lines were expected to be closed. The problem could at least temporarily be solved by switching to converted ballistic missiles such as Rokot and Dnepr, declared excess in accordance with international disarmament agreements. However, under those agreements all redundant Russian ICBMs had to be destroyed by 2007.

In 1997 several Russian companies devised strategies to modernize the Russian rocket fleet, aiming to lower costs by relying on modular designs and allay environmental concerns by using ecologically clean propellants rather than the toxic storable propellants employed by the old ICBM-derived rockets. In many ways this was a return to the plans for standardized launch vehicles in the early 1970s that had given rise to the Energiya and Zenit families. Ironically, the Energiya family was now dead because of a lack of affordable payloads and the Zenit family had remained restricted to the 11K77 because its design bureau was now situated in independent Ukraine.

In RKK Energiya's vision the light payload range (1–5 tons) would be covered by rockets known as Kvant and Diana, medium-size payloads (7–15 tons) would continue to be launched by Soyuz-derived rockets and Zenit rockets, and heavy payloads (20–25 tons) would go up on RKK Energiya's originally planned Angara version. For geostationary missions the Angara would carry a Blok-DM-SL upper stage of the Sea Launch program with an increased propellant load. The company also proposed to launch the same Angara vehicle from Baykonur's UKSS pad under the name Sodruzhestvo ("Commonwealth") with equal financial input from Russia, Kazakhstan, and Ukraine. The Kvant rocket would be redesigned to have a first stage with an RD-180 engine and a Blok-DM-SL second stage, thereby paving the way to Angara, much like Zenit had served as a pathfinder for Energiya. Other contenders were the Makeyev design bureau with a family of both land and sea-launched lightweight boosters (Riksha) burning liquid oxygen and methane, the Kompomash corporation, TsNIIMash, the Keldysh Center, and TsNII-50 [76].

Khrunichev also changed its strategy in accordance with the new requirements. In March 1997 the company scrapped plans for its original heavy-lift Angara and decided to turn Angara into a family of launch vehicles accommodating a broad range of payloads. The core element of the Angara family became a so-called Universal Rocket Module (URM) acting as the first stage. Powered by the RD-191, the URM could fly as a single unit or in clusters of three to six. Mounted above it would be one or two stages, depending on the mass of the payload. Propellant combinations considered for these upper stages were nitric acid/UDMH, LOX/kerosene, LOX/methane, and LOX/LH_2. Khrunichev retained its earlier idea of turning the first stage into a reusable flyback booster, now called Baykal, in cooperation with NPO Molniya.

Baykal flyback booster (*source*: Timofey Prygichev).

The Khrunichev plan seems to have been given the nod in late 1997/early 1998 without much consideration for the other proposals, which was possibly a result of the fact that Khrunichev became subordinate to the Russian Space Agency in 1998. This sparked off angry reactions from RKK Energiya, whose Angara proposal had been turned down in 1994 largely on the basis of the fact that the RD-180 was an unflown engine. However, by now the RD-180 was undergoing successful (US-sponsored) test firings in support of the Atlas program, and it was the RD-191 that was the untested engine.

By late 1998 Khrunichev had more than 10 Angara configurations on the drawing boards, some looking more exotic than others. Payload capacity would have been between 2 to 4 tons to low orbit on the lower end of the payload scale, and between 13 and 24 tons on the heavier end of the payload spectrum, leaving the 7–10 ton niche to the Soyuz family. Geostationary orbit capability would have been between about 1 and 5 tons [77]. However, economic realities soon forced Khrunichev to scale down its ambitions to just four rockets, two light versions (Angara 1.1 and 1.2) with a single URM and two heavier versions (Angara-A3 and Angara-A5) with three and five URM modules, respectively. Payload mass to low orbit for the first two rockets is 2.0 and 3.7 tons and 14.6 and 24.5 tons for the latter two.

Although the RD-191 is an unflown engine, the idea to build a single-chamber version of the RD-170/171 is not at all new. In the late 1970s/early 1980s Energomash had already done detailed design work on such an engine (the MD-185) for the

Angara-A3 (left) and Angara-A5 (*source*: Timofey Prygichev).

first stage of Zenit and Energiya's strap-ons to safeguard against problems with the RD-170/171 (see Chapter 6). A similar engine (the RD-141) had been considered for the second stage of the 11K37. The RD-191 delivers 196 tons of thrust at sea level and has a specific impulse of 310 s. It features a new turbopump unit driven by a single gas generator and a new system of mixture ratio control. Control of the thrust vector is provided through gimbaling of the engine about two axes.

On 31 December 1998 Khrunichev signed a deal with Energomash for the design, manufacture, and delivery of RD-191 engines for the Angara family. Mock-up versions of the engine were built for full-scale models of Angara 1.1 and

The RD-191 engine (*source*: NPO Energomash).

the Baykal flyback booster shown at the Paris Air Show in 1999 and 2001, respectively. The RD-191 test-firing program began with a short 5 s test at Energomash's Khimki facilities on 27 July 2001. About 70 test firings are required to certify the engine for flight [78].

Launch dates for the Angara rockets have continued to slip, partly due to slow progress in the construction of launch facilities at Plesetsk. Although these are located on the same site once intended to support launches of Zenit rockets, they essentially had to be built from scratch. In a plan reminiscent of RKK Energiya's 1997 Sodruzhestvo proposal, Angara is also expected to be launched from a modified Proton complex at Baykonur called Bayterek ("Poplar") under a joint venture between Khrunichev and the Kazakh Finance Ministry's State Property and Privatization Committee. In addition to that, Khrunichev is developing a modified version of Angara's URM (also carrying the RD-191) that will act as the first stage of a South Korean launch vehicle called KSLV-1. The first Angara is now not expected to fly until after 2010, but whenever it goes up, it will carry with it the legacy of the Energiya–Buran program.

STATUS OF COSMODROME INFRASTRUCTURE

After several years of uncertainty over the status of the Baykonur cosmodrome, the Russian and Kazakh governments agreed in December 1994 that Russia would lease the cosmodrome for 20 years for an annual rent of $115 million. Several months earlier the Russian government had issued a decree calling for the Military Space Forces to transfer authority over a large part of its facilities to the Russian Space Agency. This included all the Energiya–Buran facilities, authority over which was divided by the agency among several design bureaus. While some of the Energiya–Buran facilities stand rusting in the steppes without any prospects for future use, others have been modified for other programs and are once again buzzing with activity.

MIK RN/MIK 112

The Energiya assembly building is now run by the Samara-based TsSKB/Progress organization. As mentioned earlier, the three high bays were rendered useless by the roof collapse in May 2002, which among other hardware destroyed the only flown Buran orbiter. In 2004 the Russian Space Agency earmarked funds to refurbish high Bay 3 for processing future satellites built by TsSKB/Progress, but there are no signs that any work is being done. No plans have been announced for repairing the other two high bays and these may be demolished. By the end of 2006 the debris from the roof collapse had reportedly still not been cleared.

The collapsed roof of the MIK RN (*source*: *www.buran.ru*).

Soyuz rocket assembly in low bay 1 (*source*: RKK Energiya).

Low bay 1 now houses the processing area for Soyuz launch vehicles flying from pad 5 (the "Gagarin pad") on Site 1, which is now used for Soyuz and Progress launches to the International Space Station and also for occasional launches of civilian satellites built by TsSKB/Progress (Foton, Resurs). Soyuz rockets destined for the Gagarin pad used to be assembled in the old MIK-2B building in the center of the cosmodrome. Although the new assembly hall is much roomier than MIK-2B, it is much farther away from the Gagarin pad, making the roll-out procedure more cumbersome. The roll-out now takes about 2 hours, compared with 30 minutes in the old days. The first vehicle to be rolled out from the modified low bay 1 was Progress M1-10 in June 2003.

Low bay 2 was modified in the late 1990s to process payloads for Starsem, a Russian/European company founded in 1996 by Arianespace, EADS SPACE, TsSKB/Progress, and the Russian Space Agency to commercially market the Soyuz launch vehicle. There are three Starsem clean rooms here: a 286 m^2 Payload Processing Facility (PPF) with two independent control rooms to permit parallel operations, a 285 m^2 Hazardous Processing Facility (HPF) to fill satellites with toxic hypergolic propellants, and a 587 m^2 Upper Composite Integration Facility (UCIF) for integration with the Fregat upper stage and upper fairing encapsulation. Mating of the upper composite with the first three stages of the Soyuz launch vehicle takes place in the Soyuz assembly building on Site 31 on the "right flank" of the cosmodrome. Starsem exclusively flies from pad 6, situated on Site 31 [79].

MIK OK/MIK 254

Now operated by RKK Energiya, the MIK OK Buran processing building has once again become a very active facility, processing hardware for both manned and unmanned programs. The dilapidated building was even given a fresh coat of paint in 2004, making it look as good as new, at least from a distance.

In the early 1990s, bay 104, where orbiters used to undergo electrical tests, was modified to enable final launch preparations of Proton-launched space station modules, which before that used to be processed in the Proton area of the cosmodrome. The first module to be processed here was Mir's Spektr module in 1995, followed later by Priroda and the Russian ISS modules Zarya and Zvezda. At least one more Russian ISS module, the Multipurpose Laboratory Module (the original Zarya back-up vehicle), will pass through the MIK OK before its expected launch in 2009. After processing is completed, the modules are transported to the MIK 92A-50 Proton assembly building for integration with the launch vehicle.

Bay 104 now also houses the processing area for Soyuz and Progress vehicles, which used to be situated in the old MIK-2A assembly building in the center of the cosmodrome. This is where the vehicles are placed after arriving from RKK Energiya for final outfitting and testing. Four vehicles can be prepared here simultaneously. Less than two weeks before a Soyuz launch, the prime and back-up crews go to bay 104 to perform fit checks aboard the spacecraft, giving them a chance to see how the flight vehicle is configured. Once tests are completed, the spacecraft are sent to Site 31 of the cosmodrome for fueling and then return to the MIK OK for installation of the launch vehicle adapter and payload shroud. Next the combination is moved to the nearby MIK RN low bay 1 for mating with the Soyuz launch vehicle. The first vehicles to undergo launch preparations in Bay 104 were Progress M-40 and Soyuz TM-29 in late 1998. Also used for Soyuz/Progress preparations are bays 103 and 105. Bay 105 continues to serve as an anechoic chamber for compatibility tests of radio systems, and bay 103, the former Buran assembly bay, now houses a vacuum chamber.

Finally, bay 103 and bay 102 (the former a thermal protection system bay) are used to prepare RKK Energiya's Blok-DM upper stages and Yamal communications satellites. Iridium satellites were processed here as well [80].

MZK and SDI

The Assembly and Fueling Facility (MZK) and Dynamic Test Stand (SDI) on Site 112A, both turned over to RKK Energiya, are no longer being used and have not been refurbished for years. The MZK now serves a storage facility for two full-scale orbiters (the 2K flight vehicle and the OK-MT engineering test model). The SDI still houses a mock-up of the Energiya-M launch vehicle. The Russian Space Agency is considering dismantling these facilities [81].

Abandoned pad 38 (*source*: *www.buran.ru*).

Energiya–Buran launch pads

The two Energiya–Buran launch pads 37 and 38 on Site 110 of the cosmodrome (the former N-1 pads), now owned by the NIIkhimmash organization, are in poor shape. Maintenance work was discontinued in 1993 and many parts were stolen by marauders. The underground levels of the pads are flooded with an estimated 50,000 m^3 of water. All that NIIkhimmash does is to guard the pads and ensure that any useful remaining parts can be used in other programs. Six boxcars' worth of equipment was dismantled for use in the Sea Launch program. If this site is ever reactivated, the cheapest option will probably be to tear down the existing infrastructure and build new launch pads from scratch [82].

UKSS

The Universal Test Stand and Launch Pad, used for Energiya fueling tests, test firings, and also for the first launch of Energiya in May 1987, is in relatively good condition. Also run by NIIkhimmash, it is being maintained by a 110-man strong team. Key systems such as the sound suppression water system are still intact. The huge "tank farm" situated at some distance from the pad is now used to store liquid oxygen, liquid hydrogen, nitrogen, and helium for other programs [83].

There have been several proposals to revive the UKSS for new rocket programs. One suggestion around the turn of the century was to use it for test launches of the Avrora rocket, a much upgraded version of the Soyuz rocket that would be launched on commercial missions from Christmas Island in the Indian Ocean under a contract between the Asia Pacific Space Center, RKK Energiya, and several other Russian organizations. Unfortunately, the deal to build the rocket and the island launch pad fell through [84]. The UKSS has also been eyed to serve as a launch pad operated and financed jointly by CIS countries for launches of Angara rockets. In the late 1990s there was RKK Energiya's short-lived Sodruzhestvo proposal and more recently the UKSS was also considered for the Russian/Kazakh Bayterek complex. However, it was later decided that Bayterek will be built on an old Proton site. In late 2004 a group of US experts visited the UKSS and supporting facilities to study its possible use as "an international spaceport", but nothing has been heard of such plans since [85].

At any rate, UKSS' designers say the pad can be quite easily modified to accommodate launch vehicles other than Energiya. Against Soviet/Russian tradition, it would even be possible to assemble rockets on the UKSS vertically. That may eventually become a necessity, because at least part of the railroad track that used to connect the UKSS with the MIK RN has reportedly been removed and reused to connect that assembly building with Site 31 for Starsem missions. The two crawler transporters remain parked outside the MIK RN [86].

Landing facilities

The Yubileynyy runway was rarely used in the early 1990s and gradually deteriorated as a result. One idea put forward in 1994 was to use it as a refueling base for cargo planes on intercontinental flights [87]. What really saved Yubileynyy in the end was the establishment of international launch vehicle organizations like International Launch Services, Starsem, and Land Launch, which need a well-equipped airfield to deliver their customers' satellites to the cosmodrome. The advantages that Yubileynyy has over the older Krayniy airfield near the town of Baykonur are that it can receive heavy cargo airplanes and is situated much closer to the launch facilities.

Conforming to International Civil Aviation Organization standards for Class 1 airports, Yubileynyy handles aircraft of all classes for both freight and charter flights, including Boeing 747s and Antonov 124s. The airfield can operate year-round at any time of day. It has cranes, forklifts, and other equipment needed for offloading satellites. The payloads are transferred to railcars that are located approximately 50 to 80 m from the aircraft. The airport is connected by rail and road to all major cosmodrome facilities [88].

Playing a crucial role in commercial launch operations, Yubileynyy is continuously being maintained in a good state. It is also the place where most visiting delegations now arrive at the launch site.

REFERENCES

[1] V. Mokhov, "Module for Buran" (in Russian), *Novosti kosmonavtiki*, 23–24/1998, pp. 69–70.

[2] B. Gubanov, *Triumf i tragediya Energii (tom 3: Energiya-Buran)*, Nizhniy Novgorod: Izdatelstvo Nizhegorodskogo instituta ekonomicheskogo razvitiya, 1998, p. 408.
[3] E. Vaskevich archives.
[4] B. Gubanov, *Triumf i tragediya Energii (tom 4: Polyot v nebytiye)*, Nizhniy Novgorod: Izdatelstvo Nizhegorodskogo instituta ekonomicheskogo razvitiya, 1999, pp. 18–20.
[5] B. Gubanov, (volume 4), *op. cit.*, pp. 23–28; E. Vaskevich archives.
[6] Y. Semyonov, *Raketno-kosmicheskaya korporatsiya Energiya 1946–1996*, Moscow: RKK Energiya, 1996, p. 434.
[7] Y. Semyonov, *op. cit.*, pp. 435–441, 648; B. Gubanov, *op. cit.*, pp. 26–28, 47, 178–194.
[8] K. Gringauz, "Loss of cosmic speed" (in Russian), *Pravda*, 25 March 1989.
[9] A. Pokrovskiy, "The living soul of cosmonautics: Isn't it being affected by the virus of departmentalism?" (in Russian), *Pravda*, 8 April 1989.
[10] Y. Golovanov, "Just where are we flying to?" (in Russian), *Izvestiya*, 12 December 1991.
[11] Semyonov interview on Moscow Central Television, 12 April 1991, as translated by *JPRS Report*, 26 June 1991, pp. 56–57.
[12] O. Moroz, "Can we economize on space?" (in Russian), *Literaturnaya gazeta*, 20 December 1989.
[13] B. Konovalov, "Why do we need Buran?" (in Russian), *Izvestiya*, 5 December 1988.
[14] *Cosmonautics 1991*, Moscow: Mashinostroyeniye/Matson Press, 1992, p. 56; I. Marinin, "10th anniversary of Rosaviakosmos" (interview with Koptev) (in Russian), *Novosti kosmonavtiki*, 2/2002, pp. 2–3.
[15] Y. Semyonov, *op. cit.*, p. 448.
[16] Y. Koptev, "On Russia's Space Effort in the Period up to the Year 2000," *Aviation and Space News*, 1/1993, pp. 5–11.
[17] B. Gubanov, (volume 4), *op. cit.*, p. 191.
[18] J. Lenorovitz, "Russia to Upgrade Mir 1 Space Station, Prepares for New Orbital Facility," *Aviation Week & Space Technology*, 4 May 1992, pp. 84–85.
[19] Y. Meshkov, "Buran could have perished, but it has been saved" (in Russian), *Nezavisimaya gazeta*, 12 August 1992, p. 6.
[20] S. Konyukhov, *Prizvany vremenem: rakety i kosmicheskiye apparaty konstruktorskogo byuro Yuzhnoye*, Dnepropetrovsk: Art Press, 2004.
[21] V. Rybakov, "ISS could repeat the fate of Mir" (in Russian), *Zavtra*, 24 October 2002, p. 6; I. Lisov, "2005 space budget passed" (in Russian), *Novosti kosmonavtiki*, 2/2005, pp. 6–8.
[22] Y. Semyonov, *op. cit.*, p. 307.
[23] O. Moroz, *op. cit.*; Y. Koptev, "Space fantasies" (in Russian), *Ekonomika i zhizn*, September 1990, p. 19.
[24] A. Zak, "Chaos in the American space program" (in Russian), *Nezavisimaya gazeta*, 6 May 1993, p. 6.
[25] O. Moroz, *op. cit.*
[26] B. Gubanov, (volume 4), *op. cit.*, p. 25.
[27] O. Moroz, *op. cit.*
[28] B. Olesyuk, "Cosmic heights and gaping abysses: Two views on the development of cosmonautics" (in Russian), *Trud*, 5 April 1991.
[29] B. Gubanov, (volume 4), *op. cit.*, pp. 13–16, 25; V. Favorskiy, I. Meshcheryakov, *Voyenno-kosmicheskiye sily, kniga 2*, Moscow: Izdatelstvo Sankt-Peterburgskoy tipografii, 1998, pp. 127, 201.
[30] Y. Semyonov presentation at *1991 IAF Congress*; L. van den Abeelen, "Mir and Buran to Become Integrated," *Spaceflight*, September 1992, pp. 293–294.

[31] M. Tarasenko, "Buran Rises from the Ashes", *Space News*, 24–30 May 1993, p. 15.
[32] D. Jenkins, *The History of the National Space Transportation System*, Hinckley: Midland Publishing, 2001, p. 381.
[33] K. Lantratov, "A new Russian element of ISS launched" (in Russian), *Novosti kosmonavtiki*, 11/2001, pp. 6–10.
[34] Y. Semyonov, *Mnogorazovyy orbitalnyy korabl Buran*, Moscow: Mashinostroyeniye, 1995, pp. 55; V. Mokhov, *op. cit.*
[35] Y. Semyonov, *Raketno-kosmicheskaya korporatsiya Energiya 1946–1996, op. cit.*, pp. 419–420.
[36] V. Lukashevich, "Applications of Buran" (in Russian), on-line at *http://www.buran.ru/htm/spirit.htm#sapfir*
[37] Y. Semyonov, *Raketno-kosmicheskaya korporatsiya Energiya 1946–1996, op. cit.*, pp. 416–418.
[38] *Ibid.*, pp. 412–413.
[39] *Ibid.*, pp. 421–422.
[40] *Ibid.*, pp. 403–404.
[41] A. Yuskovets, "On the launch pad: A conversation with Yuriy Pavlovich Semyonov" (in Russian), *Leninskoye znamya*, 16 February 1989.
[42] Radio Moscow World Service, 8 September 1986.
[43] B. Konovalov, *op. cit.*
[44] V. Lukashevich, "Buran in Sydney", *Novosti kosmonavtiki*, 4/2000, p. 71; various Internet sites.
[45] "Buran visit", *Aviation Week & Space Technology*, 9 September 1991, p. 15.
[46] "First Soviet Shuttle Orbiter Grounded from Future Flights", *Aviation Week & Space Technology*, 12 March 1990, p. 20.
[47] J. Lenorovitz, "Soviets' Second Buran Orbiter Undergoes Launch Pad Tests," *Aviation Week & Space Technology*, 12 August 1991, pp. 38–40.
[48] Y. Semyonov, *Raketno-kosmicheskaya korporatsiya Energiya 1996–2001*, Moscow: RKK Energiya, 2001, p. 789.
[49] L. Osadchaya, "Accident at Baykonur" (in Russian), *Novosti kosmonavtiki*, 7/2002, pp. 52–53; L. Osadchaya, "What happened at Baykonur?" (in Russian), *Novosti kosmonavtiki*, 8/2002, p. 60.
[50] J. Lenorovitz, "Soviets' Second Buran Orbiter Undergoes Launch Pad Tests," *op. cit.*; "Assembled versions of the Energiya rocket" (in Russian), on-line at *http://www.buran.ru/htm/rocrt3.htm.* The complete stack had the designator 3D.
[51] "Soviets Eye Air-Launch ICBM to Put Civil Satellites in Orbit", *Aviation Week & Space Technology*, 14 October 1991, pp. 72–73; "Assembled versions of the Energiya rocket", *op. cit.*
[52] "Improvements to Third Soviet Shuttle", *Spaceflight*, May 1990, p. 147.
[53] "Orbital ship of the second series (Product 2.01)," on-line at *http://www.buran.ru/htm/2-01.htm*
[54] J. Lenorovitz, "Soviets' Second Buran Orbiter Undergoes Launch Pad Tests, *op. cit.*
[55] O. Urusov, "Berth for 'birds'" (in Russian), *Novosti kosmonavtiki*, 3/2004, p. 72; A. Chernoivanova, N. Polyakov, "Quiet before the snowstorm" (in Russian), 25 October 2004, on-line at *http://www.gazeta.ru/2004/10/14/oa_136521.shtml*
[56] B. Gubanov, (volume 3), *op. cit.*, p. 329.
[57] *Ibid.*, pp. 330–331, 388–393; D. Vorontsov, "About Deytron and others: Notes of a rank-and-file engineer," *Novosti kosmonavtiki*, 10/2006, pp. 60–62.

[58] Y. Semyonov, *Raketno-kosmicheskaya korporatsiya Energiya 1946–1996*, *op. cit.*, pp. 477–478; B. Gubanov, (volume 4), *op. cit.*, pp. 47–55, 180.

[59] K. Lantratov, "The 'Star Wars' which never happened" (in Russian), detailed history of Skif published on-line at *http://www.buran.ru/htm/str163.htm*

[60] *Ibid.*; K. Lantratov, "Zvezda: Road into space" (in Russian), *Novosti kosmonavtiki*, 9/2000, pp. 3–7.

[61] V. Pallo, "KB Salyut's program: The reaches of space or just space mirages?" (in Russian), *Zemlya i vselennaya*, March–April 1992, pp. 18–25; N. Johnson, *Europe and Asia in Space 1993–1994*, Colorado Springs: Kaman Sciences Corporation, pp. 237–239.

[62] N. Kidger, "ESA/Soviet radio telescope proposed", *Spaceflight*, April 1991, p. 112; Y. Semyonov, *Raketno-kosmicheskaya korporatsiya Energiya 1946–1996*, *op. cit.*, pp. 400–403.

[63] B. Gubanov, (volume 3), *op. cit.*, pp. 393–395.

[64] Y. Semyonov, *Raketno-kosmicheskaya korporatsiya Energiya 1946–1996*, *op. cit.*, pp. 366.

[65] B. Gubanov, (volume 3), *op. cit.*, pp. 336–337.

[66] Y. Semyonov, *Raketno-kosmicheskaya korporatsiya Energiya 1946–1996*, *op. cit.*, pp. 407–408, 470–474, 486–488; B. Gubanov, (volume 3), *op. cit.*, pp. 268, 334–336, 400; B. Gubanov, (volume 4), pp. 55–57.

[67] C. Covault, "USSR Conducts Extensive Design Work on Large, Unmanned Flyback Booster," *Aviation Week & Space Technology*, 19 November 1990, pp. 23–24; B. Gubanov, (volume 3), *op. cit.*, pp. 367–388.

[68] S. Konyukhov, *op. cit.*; B. Katorgin, *NPO Energomash imeni akademika V.P. Glushko. Put v raketnoy tekhnike*, Moscow: Mashinostroyeniye/Polyot, 2004, pp. 125–126, 209–210, 216.

[69] S. Konyukhov, *op. cit.*; B. Katorgin, *op. cit.*, pp. 203–204, 427.

[70] B. Gubanov, (volume 3), *op. cit.*, pp. 58–59, 335; S. Konyukhov, *op. cit.*; B. Katorgin, *op. cit.*, pp. 427–428.

[71] A. Lawler, "Energomash Seeks RD-180 Sales in West," *Space News*, 19–25 April 1993, p. 12.

[72] J. Lenorovitz, "US Industry Offers New ELV Proposals," *Aviation Week & Space Technology*, 21 March 1994, pp. 24–25.

[73] B. Katorgin, *op. cit.*, pp. 134–137.

[74] Y. Semyonov, *Raketno-kosmicheskaya korporatsiya Energiya 1946–1996*, *op. cit.*, pp. 491–492.

[75] A. Ovchinnikov, S. Sergeyev, "Cosmodrome Plesetsk: The basis for an independent space policy of Russia," *Novosti kosmonavtiki*, 24/1995, pp. 46–52.

[76] Y. Semyonov, *Raketno-kosmicheskaya korporatsiya Energiya 1996–2001*, *op. cit.*, pp. 703–721.

[77] Y. Zhuravin, "The 'sea' of plans for Angara" (in Russian), *Novosti kosmonavtiki*, 3/1999, pp. 48–50.

[78] B. Katorgin, *op. cit.*, pp. 137–138, 312–314.

[79] Starsem website at *http://www.starsem.com/services/sppf.htm*

[80] I. Marinin, O. Tverskoy, "Baykonur today and tomorrow" (in Russian), *Novosti kosmonavtiki*, 2/2005, pp. 58–63.

[81] *Ibid.*

[82] *Ibid.*

[83] *Ibid.*

[84] O. Urusov, "Avrora will not fly from Baykonur" (in Russian), *Novosti kosmonavtiki*, 1/2002, p. 48.

[85] I. Marinin, "Russian cosmonautics: Results of the year" (in Russian) (interview with Russian space agency chief Perminov), *Novosti kosmonavtiki*, 2/2005, p. 2.
[86] O. Urusov, "Baykonur preparing for Soyuz-2 operations" (in Russian), *Novosti kosmonavtiki*, 4/2002, p. 49.
[87] I. Marinin, "Contract signed on leasing Baykonur" (in Russian), *Novosti kosmonavtiki*, 26/1994, p. 14.
[88] *Land Launch User's Guide.*

9

Beyond Buran

Although Buran was the focus of attention between the mid-1970s and the early 1990s, in the background the Russians continued working on other spaceplane concepts to either complement Buran or succeed it in the future. Many of these efforts concentrated on smaller spaceplanes that were considered to be more efficient for space station support. At the same time, looking further into the future, considerable research has been done into single-stage-to-orbit spaceplanes that may one day significantly reduce the cost of Earth-to-orbit transportation.

CHELOMEY'S LKS

Vladimir Chelomey's Central Design Bureau of Machine Building (TsKBM), already engaged in spaceplane research in the early 1960s, resumed work on reusable spacecraft in the mid-1970s in response to an order by the Military Industrial Commission on 27 December 1973 to formulate proposals for reusable space transportation systems of different sizes (see Chapter 2). That work continued even after the February 1976 Energiya–Buran decree, with Chelomey enjoying the support of Yakob Ryabov, who was Central Committee Secretary for Defense Matters from late 1976 to 1979, succeeding Dmitriy Ustinov [1].

By 1978, after having studied numerous configurations, launch, and landing techniques for a small reusable spaceplane, Chelomey's engineers had settled on a Light Spaceplane (LKS) to be launched by the bureau's Proton rocket. Weighing 20 tons, it would be a delta-wing vehicle capable of carrying four tons of cargo, two tons of fuel and a crew of two. The pressurized compartment of the LKS had a volume of 16 m^3 and consisted of two decks, an upper deck with the cockpit in front and a living compartment in the back, and a lower deck with support equipment. The ship would not be protected from the heat of re-entry by tiles, but rely on a different type of heat shield material said to be rated for 100 missions. It was similar to the

Mock-up of LKS spaceplane (*source*: Timofey Prygichev).

material used on the return capsules of Chelomey's Transport Supply Ships (TKS), the reusability of which was demonstrated during a number of test flights in the late 1970s and early 1980s. The LKS would land on a runway at a speed of 300 km/h using a nose wheel and aft skid landing gear. The relatively low landing speed meant that it could use a wide variety of runways. The LKS had emergency escape systems allowing the crew to be saved during virtually any phase of the flight.

TsKBM worked out plans for both unmanned and manned versions of the LKS, capable of staying in orbit for up to one year and 10 days, respectively. Its missions would range from crew transport and cargo delivery to space stations to a broad array of military missions. By the end of the 1970s the LKS was seen by Chelomey as a key element in a "Star Wars" plan to deploy a space-based missile defense shield to protect the entire territory of the Soviet Union from nuclear attack. All this was several years *before* President Ronald Reagan announced the Strategic Defense Initiative (SDI) in 1983.

Chelomey considered Energiya–Buran as a vastly expensive undertaking that the country could barely afford. One of his associates once quoted him as saying:

> "Whereas for the Americans the expenditures on [the Space Shuttle] are [serious], but bearable, for us [such expenditures] will plunge us into bankruptcy. I won't be surprised if our cosmonauts will have to fly on our shuttle naked."

Convinced that Buran would not be ready to fly for many more years, Chelomey pitched the LKS as a vehicle that could be ready in four years' time.

In 1980 Chelomey took the risky move of "going public" with the LKS. He sent his LKS proposals directly to Leonid Brezhnev, who in response set up an inter-departmental commission headed by deputy Defense Minister V.M. Shabanov and consisting of representatives of the major design bureaus and research institutes. The commission turned down Chelomey's proposal, calling it "cumbersome", "difficult to realize", and "expensive" (terms that could have very well been applied to Buran as well).

Nevertheless, Chelomey ordered his team to clandestinely build a full-scale mock-up of the LKS in just a month's time, a task that was successfully accomplished. However, news of Chelomey's underground initiative was soon leaked to the Ministry of General Machine Building, which strongly reprimanded Chelomey for having illegally spent 140,000 rubles of government money. Still undeterred, Chelomey pressed on with his LKS work and was finally stopped dead in his tracks with an official reprimand from the Communist Party.

In the political constellation of those days Chelomey stood no real chance of mustering the support needed to get LKS off the ground. His star had been waning ever since his lifelong enemy (and Glushko supporter) Dmitriy Ustinov had become Minister of Defense in 1976. In 1978 Ustinov had already been instrumental in canceling Chelomey's Almaz military space station program. The LKS followed suit. Seated next to Chelomey in the very cockpit of the LKS mock-up during a visit to TsKBM, Ustinov made it clear to him that the LKS had no future given the amount of effort and money already invested in Energiya–Buran. Things got much worse for Chelomey in December 1981, when the Central Committee and the Council of Ministers issued a decree that banned TsKBM from any further involvement in the Soviet ballistic missile and space program, essentially ending Chelomey's career as a missile and spacecraft designer. In a cruel twist of fate, Ustinov and Chelomey passed away only days apart in December 1984—moreover, in the same hospital. Even the LKS mock-up did not survive. It was demolished in what has been described as "an act of sabotage" in 1991.

With hindsight, if the Soviet space program required any type of reusable spacecraft in the 1970s/1980s, the LKS probably would have been the way to go. Leaving aside its potential military applications, it could have played an important role in ferrying crews and cargo to space stations, reducing the number of Soyuz and Progress missions. Having said that, one wonders if it would have been an economically advantageous system, since it relied on the expensive and expendable Proton rocket and on a heat shield that possibly required long turnaround times [2].

NPO MOLNIYA'S MAKS

Even as the newly created NPO Molniya got down to Buran development in 1976, the Mikoyan bureau contingent in the organization seemingly had a hard time parting with the air-launched Spiral concept. In fact, one NPO Molniya veteran recalls that

Lozino-Lozinskiy was never overly enthusiastic about Buran, which had been forced upon him from above, and that his real passion remained with air-launched systems [3]. Realizing that one of the major drawbacks of Spiral had been the need to develop a futuristic hypersonic aircraft, the Mikoyan designers began drawing up plans for spaceplanes launched from existing subsonic transport planes. The aim was to expand their missions beyond military reconnaissance and offensive operations to satellite deployment/retrieval and space station support. Unlike Buran, such spaceplanes would be suited to launch payloads usually orbited by expendable launch vehicles and had many other advantages such as quicker turnaround, more launch flexibility, and a wider range of attainable orbits. The new air-launched concept benefited heavily from experience gained in the Spiral, BOR, and Buran programs.

System 49 and Bizan

Studies of new air-launched systems began at NPO Molniya in 1977 under a research program known as Rosa ("Dew") and initially focused on the use of the Antonov-124 Ruslan as the carrier aircraft. By 1981 this resulted in the so-called System 49, in which the Ruslan would carry a single-person 13-ton spaceplane attached in tandem to a two-stage rocket. The rocket had two Kuznetsov NK-43 LOX/kerosene engines in the first stage and a single Lyulka 11D57M LOX/LH_2 engine in the second stage. With an overall take-off mass of 430 tons, System 49 allowed the spaceplane to place about 4 tons into a low 51° inclination orbit. Payloads could be launched into orbits with altitudes between 120 and 1,000 km and with inclinations between 45° and 94°.

In 1982 System 49 was superseded by a modified system called Bizan ("Mizzen"). Having the same performance as System 49, it differed from the latter in that the spaceplane was placed on top of a single-stage rocket and had main engines itself. The advantage of the single-stage rocket was that it would burn up over the ocean across the world from the launch point. In the two-stage System 49 the first stage would have crashed in a zone about 2,000 km from the launch point, requiring that area to be cleared for impact. Bizan's rocket was fitted with a single NK-43A, while the spaceplane itself had two 11D57M engines, which could now be reused on subsequent missions. Also considered was a cargo version known as Bizan-T, where the spaceplane was replaced by an unmanned cargo canister [4]. Bizan was also

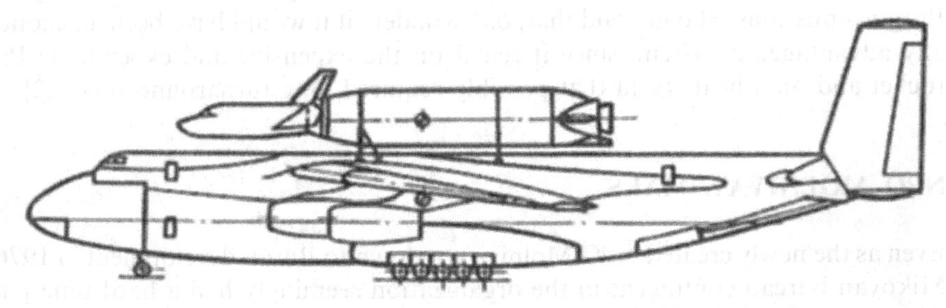

System 49 (*source*: NPO Molniya/Moscow Aviation Institute).

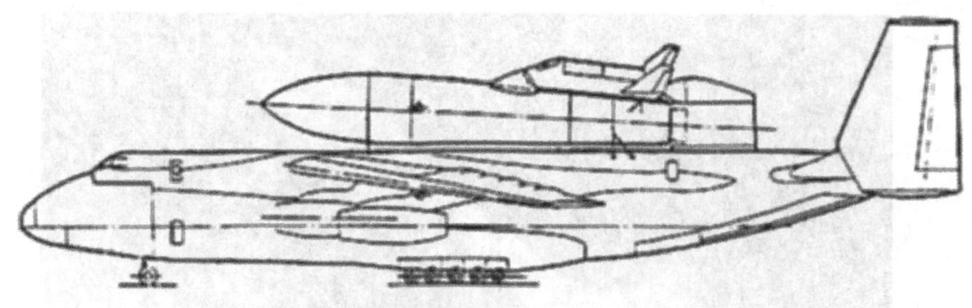

Bizan (*source*: NPO Molniya/Moscow Aviation Institute).

the name of an unmanned rocket system launched from the An-225 Mriya that was studied by the Volga Branch of NPO Energiya in 1984–1988 [5].

MAKS design features

The plans underwent further changes with the inception in the mid-1980s of the more capable An-225 Mriya carrier aircraft. Although conceived in the first place to transport Buran and elements of the Energiya rocket from the manufacturers to the Baykonur cosmodrome, designers may have had air-launch capability in the back of their minds from the outset.

The Mriya-based system was dubbed the Multipurpose Aerospace System (*Mnogotselevaya aviatsionno-kosmicheskaya sistema* or MAKS). The rocket was now replaced by an expendable external fuel tank (VTB), perched on top of which was either a reusable spaceplane (MAKS-OS) or an expendable unmanned cargo canister (MAKS-T). Also envisaged was a fully reusable unmanned winged cargo carrier with integrated propellant tanks (MAKS-M).

The OS was a 26-ton, two-man spaceplane with a length of 19.3 m, a height of 8.6 m, and a wingspan of 8.6 m. As on Spiral and BOR, the wings could be folded back for re-entry. The thermal protection system was the same as that of Buran, although a different material was needed for the much thinner wing leading edges. Behind the crew compartment was a 2.8 × 6.8 m payload bay. The original plan was for the spaceplane to have three Kuznetsov NK-45 LOX/LH_2 main engines with a vacuum thrust of 90 tons each. That idea was turned down in favor of a tripropellant LOX/LH_2/kerosene engine called RD-701, developed at NPO Energomash on the basis of the RD-170. Although this lowered the mean specific impulse, it still resulted in better performance because the external tank became much lighter by reducing the amount of liquid hydrogen, which is a low-density fuel taking up a lot of volume.

The RD-701 is a twin-chambered, staged, combustion cycle engine. Each chamber has a pair of turbopumps. One pump processes liquid oxygen and kerosene, which is turned into an oxygen-rich gas at 700 atmospheres after passing through a preburner. The other pump feeds liquid hydrogen to the main combustion chamber at ambient temperatures. The RD-701 has two modes of operation, combining first and second-stage engine characteristics in one package. During the initial phase it burns 81.4 percent liquid oxygen, 12.6 percent kerosene, and 6 percent liquid hydro-

MAKS launch.

gen, producing a total thrust of 400 tons with a specific impulse of 415 s. Then it switches to a combination of just liquid oxygen and hydrogen, with the thrust decreasing to 162 tons, but the specific impulse climbing to 460 s, helped by the deployment of a nozzle extension.

A typical MAKS-OS launch profile would see Mriya climb to an altitude of 9 km and assume the proper pre-launch attitude. The spaceplane would then ignite its main engine while still riding piggyback on the aircraft, making it possible to check its performance before separation. Some ten seconds later the 275-ton combination of spaceplane and external tank would be released from the Mriya to begin the trip to orbit. The engines would shut down before the spaceplane reached orbital velocity, allowing the external tank to burn up over the ocean across the world some 19,000 km from the launch point. The OS would then perform two burns of its two hydrogen peroxide/kerosene orbital maneuvering engines to place itself into orbit.

MAKS versions

The basic version of the OS was designed to launch and retrieve small and medium-size satellites. Payload capacity was 8.3 tons to a 200 km orbit with a 51° inclination and 4.6 tons back to Earth. For space station missions there were two configurations. In one of them (TTO-1) the payload bay would house a small pressurized module capable of carrying four passengers plus cargo. This would be used for crew rotation or rescue missions, although the latter required additional fuel supplies for quick maneuvering. In the other (TTO-2) the payload bay would remain unpressurized and carry structures such as solar panels, antennas, or propellant tanks for refueling a space station. In both configurations a docking adapter was installed just behind the crew compartment. Also considered was an unmanned OS without a crew compartment and with a slightly enlarged cargo bay to fly heavier payloads (9.5 tons into a 200 km, 51° inclination orbit). Before committing MAKS-OS to flight, NPO Molniya planned to fly a suborbital unmanned demonstrator (MAKS-D). This would have the same size and shape as the OS, but would be equipped with a single RD-120 Zenit second-stage engine fed by propellant tanks in the payload bay.

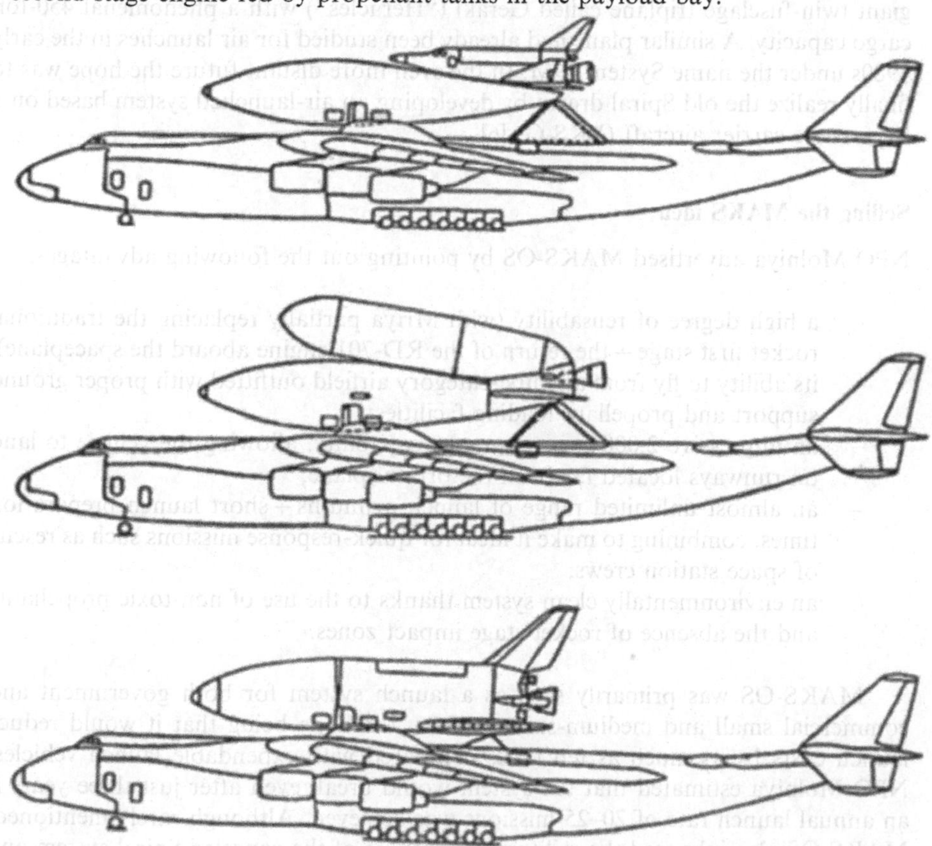

MAKS-OS, MAKS-T, and MAKS-M (*source*: NPO Molniya/Moscow Aviation Institute).

In the MAKS-T configuration the OS was replaced by an unmanned cargo canister equipped with an RD-701 tripropellant engine and an upper stage for inserting payloads into the proper orbit. Maximum payload capacity was 18 tons to a 200 km, 51° inclination orbit, and 4.8 tons to geostationary orbit. For geostationary missions the Mriya would fly to the equator to make maximum use of the Earth's eastward rotation and be refueled in flight.

MAKS-M was a fully reusable, unmanned, winged cargo container with integrated propellant tanks, designed to deliver payloads to low orbits (5.5 tons to a 200 km, 51° orbit). Situated in between the propellant tanks was a cargo bay slightly larger than that of the OS. An earlier version of this (VKS-D) had the cargo bay on top of the propellant tanks. NPO Molniya designers even floated the idea of transforming VKS-D into a suborbital intercontinental passenger plane capable of carrying 52 passengers to any point on the globe within 3 hours at a price of $40,000 per ticket.

NPO Molniya had plans to further upgrade the MAKS system by fitting the Mriya with more powerful NK-44 engines and eventually by replacing Mriya with a giant twin-fuselage triplane called Gerakl ("Heracles") with a phenomenal 450-ton cargo capacity. A similar plane had already been studied for air launches in the early 1980s under the name System 49M. In the even more distant future the hope was to finally realize the old Spiral dream by developing an air-launched system based on a hypersonic carrier aircraft (VKS-G) [6].

Selling the MAKS idea

NPO Molniya advertised MAKS-OS by pointing out the following advantages:

- a high degree of reusability (with Mriya partially replacing the traditional rocket first stage + the return of the RD-701 engine aboard the spaceplane);
- its ability to fly from any first-category airfield outfitted with proper ground support and propellant loading facilities;
- an impressive 2,000 km cross-range capability, allowing the vehicle to land on runways located far from the orbital plane;
- an almost unlimited range of launch azimuths + short launch preparation times, combining to make it ideal for quick-response missions such as rescue of space station crews;
- an environmentally clean system thanks to the use of non-toxic propellants and the absence of rocket stage impact zones.

MAKS-OS was primarily seen as a launch system for both government and commercial small and medium-size satellites, the hope being that it would reduce launch costs by as much as ten times compared with expendable launch vehicles. NPO Molniya estimated that the system would break even after just three years if an annual launch rate of 20–25 missions was achieved. Although rarely mentioned, MAKS-OS also inherited the military advantages of the canceled Spiral system and was considered for reconnaissance, inspection, and attack missions [7]. Vladimir

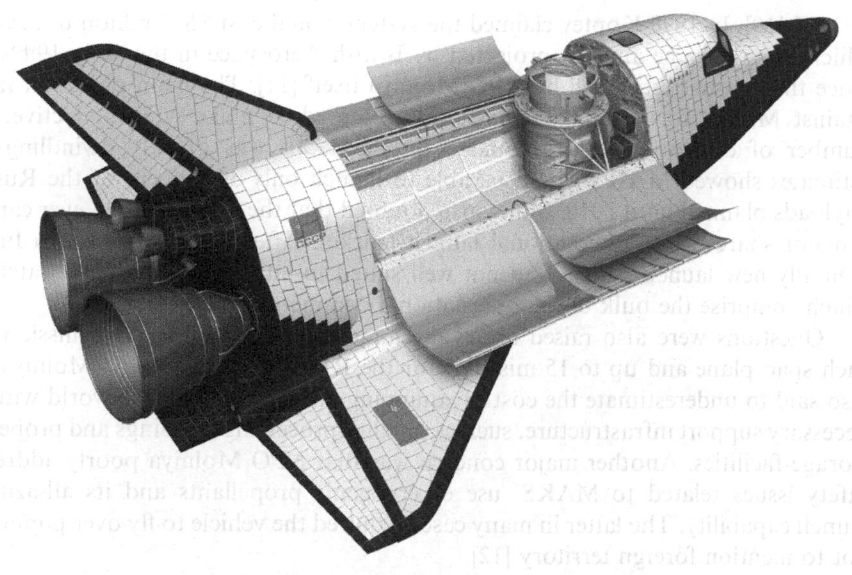

MAKS spaceplane (*source*: *www.buran.ru*).

Skorodelov, a deputy chief designer at NPO Molniya, later acknowledged that MAKS was conceived with both civilian and military goals in mind and that the civilian applications came to the foreground only as the Cold War drew to a close [8]. One may even wonder if MAKS wasn't at least partially inspired by the US Air Force's Space Sortie system, a quick-response military spaceplane studied in the early 1980s that would be launched with an external fuel tank from the back of a modified Boeing 747.

The preliminary design for MAKS was finished in 1988 and the system was first publicly presented by Gleb Lozino-Lozinskiy at the 40th Congress of the International Astronautical Federation in Malaga, Spain in October 1989. Realizing that MAKS stood little chance as an exclusively government-funded project, NPO Molniya sought international partners to join the project. There was considerable European interest in the early 1990s. British Aerospace saw MAKS as a possible intermediate step towards its own Interim HOTOL, an An-225 launched version of the original British HOTOL single-stage-to-orbit spaceplane. ESA displayed interest in MAKS as an alternative to its own Hermes spaceplane. In 1993–1994 ESA sponsored a joint study by British Aerospace, NPO Molniya, TsAGI, and the Antonov design bureau on a MAKS look-alike Rocket Ascent Demonstrator Mission (RADEM) to prove the technology for a possible European/Russian/Ukrainian reusable air-launched system. However, the results of the study were never implemented as ESA lost interest in a European-funded space transportation system. The study did result in NPO Molniya's later MAKS-D proposal [9].

MAKS received little support within the Russian Space Agency, with Yuriy Koptev having spoken out against it even before becoming the head of the agency in 1992 [10]. In 1998 Koptev claimed the system would cost $6–7 billion to develop, which was the same amount projected by British Aerospace in the early 1990s and twice the amount estimated by NPO Molniya itself [11]. The main objection raised against MAKS was the high launch rate required to make it cost-effective. The number of domestic satellite launches in the 1990s was quickly dwindling and estimates showed MAKS would be able to launch only 30 percent of the Russian payloads planned until 2010. Many also doubted that the system would ever capture a major share of the international launch market. After all, MAKS was a fundamentally new launch system and not well suited to launch geostationary satellites, which comprise the bulk of the international commercial payloads.

Questions were also raised about the announced reusability (100 missions for each spaceplane and up to 15 missions for the RD-701 engine). NPO Molniya was also said to underestimate the cost of equipping airfields all over the world with the necessary support infrastructure, such as satellite-processing buildings and propellant storage facilities. Another major concern was that NPO Molniya poorly addressed safety issues related to MAKS' use of cryogenic propellants and its all-azimuth launch capability. The latter in many cases required the vehicle to fly over populated, not to mention foreign territory [12].

Despite all the objections, NPO Molniya continued low-level research on MAKS using shoestring government funds (at least partially thanks to continued support from the military) and other financial means, some provided by the Moscow city government. Full-scale mock-ups were built of the OS spaceplane and the external fuel tank. A crude experimental version of the RD-701 began testing at NPO Energomash in 1994. Mriya re-entered service in 2001 after having been grounded for seven years.

MAKS was given a new chance in late 2005, when it competed with proposals by RKK Energiya and the Khrunichev Center in a tender to develop a successor for the Soyuz spacecraft. The spaceplane proper now had a slightly differently shaped fuselage and no longer had foldable wings. However, the Russian Space Agency canceled the tender in July 2006, preferring to develop a capsule-type vehicle in collaboration with ESA. One of the main drawbacks cited for MAKS was the considerable Ukrainian involvement—namely, the Antonov bureau's Mriya aircraft. One of the requirements in the tender had been to limit foreign contributions. MAKS is now destined to go down in history as yet another unrealized Russian spaceplane project.

NPO ENERGIYA'S OK-M SPACEPLANES

Between 1984 and 1993 NPO Energiya studied several relatively small spaceplanes that were primarily intended to replace Soyuz and Progress for space station support. These had the general designator OK-M (Small-Size Orbital Ship).

The basic OK-M was a 15-ton spaceplane launched by the Zenit rocket. The interface between the vehicle and the rocket was an adapter equipped with four 25-ton solid-fuel motors that could be used to either pull the ship away from the

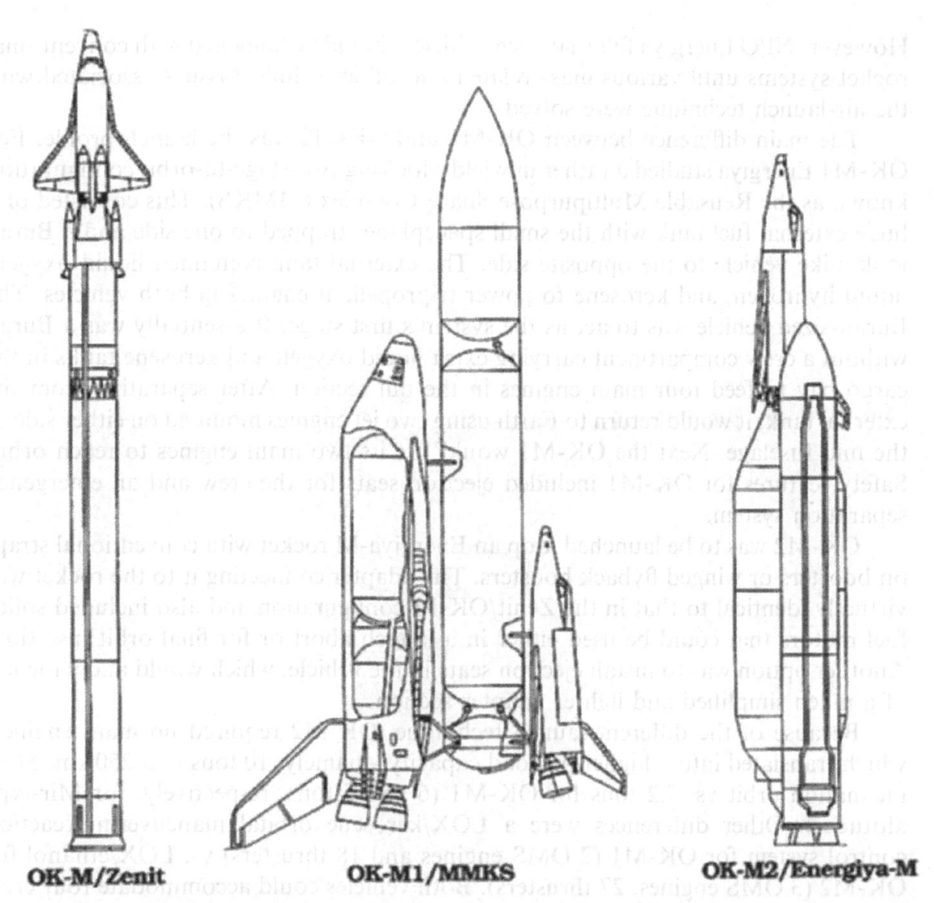

OK-M, OK-M1, and OK-M2 (*source*: RKK Energiya).

rocket in an emergency or to provide extra thrust during launch by being activated shortly after second-stage ignition. Aerodynamically, OK-M was a mini-version of Buran, having delta wings with elevons and a vertical stabilizer with rudder/speed brake. The outer surface was covered with Buran-type heat-resistant tiles. The ship could carry a crew of two in the cabin and—if required—four more cosmonauts in a pressurized module inside its 20 m^3 cargo bay. The nosecap of the vehicle would be retracted to expose an androgynous docking port. OK-M had two orbital maneuvering engines and 34 thrusters, all burning nitrogen tetroxide/UDMH. Power was to be provided by 16 batteries, although solar panels were also considered. Payload capacity was 3.5 tons to a 51.6°, 200 km orbit and just 2 tons to a 450 km Mir-type orbit.

Better satisfying the logistics requirements of space stations were two heavier spaceplanes called OK-M1 (31.8 tons) and OK-M2 (30 tons). Jointly developed with NPO Molniya, they were very similar to the air-launched MAKS-OS spaceplane.

However, NPO Energiya felt that such vehicles should be launched with conventional rocket systems until various mass-related and other technical issues associated with the air-launch technique were solved.

The main difference between OK-M1 and OK-M2 was the launch profile. For OK-M1 Energiya studied a rather unwieldy-looking two-stage-to-orbit configuration known as the Reusable Multipurpose Space Complex (MMKS). This consisted of a huge external fuel tank with the small spaceplane strapped to one side and a Buran look-alike vehicle to the opposite side. The external tank contained liquid oxygen, liquid hydrogen, and kerosene to power tripropellant engines in both vehicles. The Buran-sized vehicle was to act as the system's first stage. It essentially was a Buran without a crew compartment carrying extra liquid oxygen and kerosene tanks in the cargo bay to feed four main engines in the tail section. After separating from the external tank, it would return to Earth using two jet engines mounted on either side of the mid fuselage. Next the OK-M1 would fire its two main engines to reach orbit. Safety features for OK-M1 included ejection seats for the crew and an emergency separation system.

OK-M2 was to be launched atop an Energiya-M rocket with conventional strap-on boosters or winged flyback boosters. The adapter connecting it to the rocket was virtually identical to that in the Zenit/OK-M configuration and also included solid-fuel motors that could be used either in a launch abort or for final orbit insertion. Another option was to install ejection seats in the vehicle, which would allow the use of a much simplified and lighter adapter section.

Because of the different launch technique, OK-M2 required no main engines, which translated into a higher payload capacity—namely, 10 tons to a 250 km, 51.6° inclination orbit vs. 7.2 tons for OK-M1 (6 and 5 tons, respectively, for Mir-type altitudes). Other differences were a LOX/kerosene orbital maneuvering/reaction control system for OK-M1 (2 OMS engines and 18 thrusters) vs. LOX/ethanol for OK-M2 (3 OMS engines, 27 thrusters). Both vehicles could accommodate four crew members in the crew compartment and another four in a pressurized module in the 40 m^3 cargo bay. The power supply system relied on a combination of fuel cells and batteries. Just like MAKS-OS, both ships had foldable wings [13]. In 1994 a proposal was made to launch OK-M2 with a European Ariane-5 booster outfitted with Energiya strap-on boosters [14].

Concurrently with the OK-M studies, NPO Energiya worked out plans for a ballistic reusable spacecraft called Zarya ("Dawn"). This looked like an enlarged Soyuz descent capsule with a small expendable instrument section attached to it. Weighing 15 tons, it would be launched by Zenit and make a vertical landing using a cluster of 24 liquid-fuel braking engines rather than parachutes. The heat shield would be similar to that of Buran. Zarya was mainly intended for space station support, but also was to fly autonomous missions in the interests of the Ministry of Defense and the Academy of Sciences. Maximum crew capacity was eight.

Indications are that Zarya was considered a much more likely contender to replace or complement Soyuz/Progress than the OK-M spaceplanes. While OK-M was no more than a conceptual study, Zarya development was sanctioned by a government decree in January 1985 and even went beyond the "preliminary design" phase. Zarya was eventually canceled in January 1989 due to a lack of financing, although Valentin Glushko's death that same month may have contributed to the decision [15].

RKK ENERGIYA'S KLIPER

Original plans

For most of the 1990s the hard economic times forced RKK Energiya to limit manned spacecraft development to upgrading the existing Soyuz spacecraft. Work on a Soyuz successor didn't resume in earnest until the turn of the century. By 2002 Energiya designers had settled on a lifting body borrowing technology from Soyuz and Buran. This was publicly announced as Kliper ("Clipper") in February 2004 and described as a 12–14 ton spacecraft with a reusable return capsule. However, in November 2004 company officials revealed an alternative winged design for Kliper that was ultimately preferred over the lifting body.

In the 2004 plans Kliper had a reusable "Return Vehicle" (VA), made up of a crew cabin embedded in an unpressurized fuselage which could be either a lifting

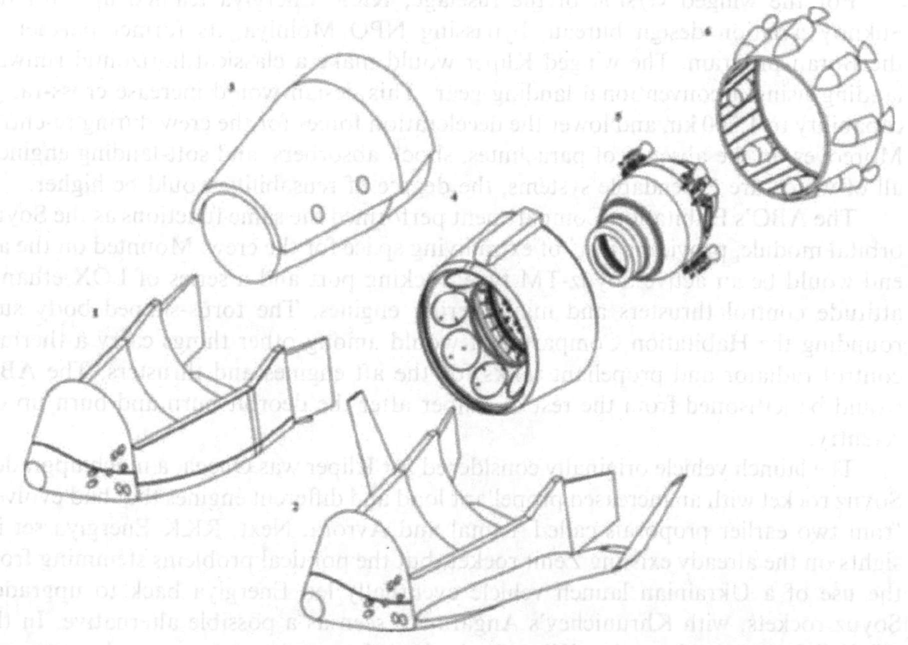

Exploded view of Kliper: 1, fuselage shaped as lifting body; 2, winged fuselage; 3, crew cabin; 4, ABO body; 5, ABO Habitation Compartment; 6, launch vehicle adapter with emergency escape rockets (*source*: RKK Energiya).

body or a winged design. Attached to the aft of that was an expendable "Aggregate/ Habitation Compartment" (ABO), consisting of a Soyuz orbital module surrounded by a torus-shaped body. About half of the habitation compartment protruded from the aft of the body and had a docking port to link up with the ISS or other spacecraft.

The Return Vehicle's blunt-shaped crew cabin offered 20 m^3 of working space (five times as much as Soyuz) and could house a maximum crew of six (minimum crew of two). The independently developed fuselage would protect the front and lower part of the crew cabin during re-entry, descent, and touchdown. The fuselage and the upper part of the crew cabin had a heat shield consisting of 60 × 60 cm thermal covers made from the same material as Buran's tiles. Also installed in the fuselage were LOX/ethanol attitude control thrusters and electricity-producing fuel cells derived from those developed for Buran.

The lifting-body configuration was probably inherited from so-called Recoverable Maneuverable Capsules (VMK) studied by Energiya in the early 1990s as return capsules for a series of vehicles intended for autonomous microgravity missions or space station servicing missions. The lifting-body fuselage was fitted with two rudders and two body flaps and would have a cross-range capability of up to 500 km (10 times more than Soyuz). Kliper would descend on parachutes, stowed in a container in the top section of the crew cabin, with pneumatic shock absorbers and small solid-fuel engines softening the touchdown.

For the winged version of the fuselage, RKK Energiya teamed up with the Sukhoy aviation design bureau, bypassing NPO Molniya, its former partner in the Buran program. The winged Kliper would make a classical horizontal runway landing using a conventional landing gear. This design would increase cross-range capability to 1,200 km and lower the deceleration forces for the crew during re-entry. Moreover, in the absence of parachutes, shock absorbers, and soft-landing engines, all of which are expendable systems, the degree of reusability would be higher.

The ABO's Habitation Compartment performed the same functions as the Soyuz orbital module, providing 8 m^3 of extra living space for the crew. Mounted on the aft end would be an active Soyuz-TM type docking port and a series of LOX/ethanol attitude control thrusters and maneuvering engines. The torus-shaped body surrounding the Habitation Compartment would among other things carry a thermal control radiator and propellant tanks for the aft engines and thrusters. The ABO would be jettisoned from the rest of Kliper after the deorbit burn and burn up on re-entry.

The launch vehicle originally considered for Kliper was Onega, a much upgraded Soyuz rocket with an increased propellant load and different engines that had evolved from two earlier proposals called Yamal and Avrora. Next, RKK Energiya set its sights on the already existing Zenit rocket, but the political problems stemming from the use of a Ukrainian launch vehicle eventually led Energiya back to upgraded Soyuz rockets, with Khrunichev's Angara-A3 seen as a possible alternative. In the original Onega configuration Kliper had a launch escape tower mounted on its nose section, but in the later configurations this was replaced by eight solid-fuel rocket motors installed on an adapter between the launch vehicle and the spacecraft. As in the earlier Zenit/OK-M and Energiya-M/OK-M2 plans, these motors could be used either in abort scenarios or to augment thrust in a nominal launch [16].

The manned spacecraft tender

Throughout 2000–2005 RKK Energiya funded Kliper exclusively with its own means. Prospects for government funding emerged in October 2005 when an advanced manned transportation system was approved for development under Russia's Federal Space Program for 2006–2015. Under new government rules, the Russian Space Agency launched a tender between three companies in November 2005 to build the ship. Requirements were for the spacecraft to be at least 80 percent reusable and fly at least 20 missions, carry up to six people, and have a cargo capacity of at least 500 kg up and down. The maiden flight should take place in 2013. Vying for the contract—apart from RKK Energiya—were NPO Molniya with an adapted version of its air-launched MAKS system and the Khrunichev Center with Angara-launched capsule-type vehicles derived from its TKS spacecraft.

Kliper spacecraft (*source*: *www.buran.ru*).

In its tender proposal RKK Energiya portrayed Kliper as one step in a multi-phased architecture for Russia's piloted space program until 2030. In this vision Kliper would be preceded by a cardinally modernized Soyuz incorporating many elements of Kliper and succeeded by piloted Moon and Mars ships using a combination of Soyuz and Kliper technology. By now only the winged version of Kliper was eyed, because that would offer more comfortable re-entry and landing conditions for the many paying passengers (including tourists) that were expected to fly on the vehicle to return at least some of the development cost. The VA's crew module was now cylindrically shaped with different seating arrangements.

Under the new plans Kliper was to be developed in two versions, one primarily intended for long solo missions and the other for flights to ISS. The first would be a partially reusable vehicle consisting of the VA and ABO plus the launch escape adapter. The other would be a fully reusable spacecraft made up of just the VA and the launch escape adapter with its solid-fuel engines, which after a nominal launch would stay attached to the VA to perform the deorbit burn. The ABO would be made redundant by a reusable space tug called Parom ("Ferry") that would pick up Kliper after launch, tow it to ISS, and later return it back to a lower orbit for retrofire. Parom is a 12.5-ton space tug permanently stationed in orbit that can tow both Kliper vehicles and unmanned cargo containers to ISS. Differing in mass (14 tons vs. 12.5 tons), the two versions of Kliper would be launched by different versions of the upgraded Soyuz launch vehicle known as Soyuz-3 (for the solo version) and Soyuz-2-3 (for the ISS version). Payload capacity up and down would be 500 kg for both [17].

Kliper on the backburner

The requirements stipulated for the new piloted spacecraft in the Federal Space Program were so obviously tailored to RKK Energiya's Kliper that many wondered if the tender was no more than a formality. However, in January 2006 the Russian Space Agency decided to extend the tender, asking the three companies to bring their proposals in closer agreement with the tender specifications. Finally, in a rather embarrassing move, the agency's head Anatoliy Perminov announced at the Farnborough air show in England in July 2006 that the tender had been canceled without a winner. Instead, Russia would join forces with ESA to build an Advanced Crew Transportation System (ACTS), with RKK Energiya serving as the prime contractor on the Russian side. This is expected to become a much upgraded version of Soyuz incorporating European technology. An earlier invitation to ESA to join the development of Kliper had been turned down at an ESA ministerial meeting in December 2005.

Despite cancellation of the tender, RKK Energiya is continuing work on Kliper using its own resources. It has the full backing of Nikolay Sevastyanov, who succeeded Yuriy Semyonov as head of RKK Energiya in May 2005. Sevastyanov holds out hope Kliper will eventually receive government funding and be ready to fly in 2015. However, there are signs of growing rifts between RKK Energiya and the Russian Space Agency, which considers Energiya's plans overly ambitious and way beyond affordable limits. Only time will tell if the differences can be resolved and if Kliper will become the country's first new piloted space transportation system since Buran.

SINGLE-STAGE-TO-ORBIT SPACEPLANES

While the main focus over the past forty years has been on winged spacecraft launched with conventional rockets or from airplanes, the Russians have never abandoned the idea of eventually fielding a single-stage-to-orbit (SSTO) spaceplane that can take off horizontally like an ordinary aircraft. Although the development of an SSTO system remains a distant dream (even in the West), plenty of research has been done in the field in the past decades.

One of the first Soviet SSTO spaceplanes was put on the drawing board by Yevgeniy S. Shchetinkov in 1966 at the Scientific Research Institute of Thermal Processes (NII TP, the former NII-1 and the later Keldysh Research Center). Shchetinkov, a veteran of the GIRD and RNII rocket research institutes of the 1930s, had formulated ideas for scramjet engines as early as 1957. Using a combination of ramjet, scramjet, and liquid-fuel engines, his proposed spaceplane had a take-off mass of between 150 and 250 tons and was capable of placing between 6 and 11 tons into orbit [18].

The M(G)-19 "Gurkolyot"

Another idea for an SSTO spaceplane emerged at the Strategic Rocket Forces' NII-4 research institute. It was the brainchild of Oleg Gurko, who had been working out ideas for such systems since the late 1940s. The novelty in Gurko's plan was a hydrogen scramjet that would suck in air heated by the exhaust of a liquid-fuel rocket engine placed in front of it. In the 1960s he approached both Mikoyan and Myasishchev, who both showed interest in building a vehicle using such a propulsion system. However, the barriers between the Strategic Rocket Forces and the Ministry of the Aviation Industry proved too high.

It was not until after the approval of the US Space Shuttle in the early 1970s that Soviet top brass started showing some interest in Gurko's ideas. On 10 October 1974 the Minister of the Aviation Industry and the Air Force Commander-in-Chief signed an order allowing Myasishchev's Experimental Machine Building Factory (EMZ) to work out "technical proposals" for an SSTO using Gurko's propulsion system under a program called Kholod-2 ("Cold-2"). Placed in charge of the project within EMZ was A. Tokhunts, one of Myasishchev's deputies, and the leading engineer was I. Plyusnin. Development of the propulsion system was entrusted to the Kuznetsov design bureau in Kuybyshev, the same bureau that had built the engines for the lower stages of the N-1 rocket. Gurko, now employed by the TsNII-50 R&D institute that had split off from NII-4 in April 1972, continued to provide technical support. Within EMZ the project was known as "Theme 19" and the SSTO itself was designated

Oleg Gurko poses in front of an M-19 model in his apartment in Moscow in 1999 (*source*: Asif Siddiqi).

M-19. It has also been referred to as MG-19 ("G" for Gurko) and was affectionately known as "Gurkolyot".

In its final design the M-19 was a 69 m long triangular-shaped lifting body with small aft and front-mounted wings and two fins. Having a take-off mass of 500 tons, it was capable of inserting a 40-ton payload into low Earth orbit. There was also an alternative plan for a Buran look-alike vehicle with big delta wings and a single vertical stabilizer. Having the same take-off mass as the primary design, it would have a payload capacity of just 30 tons.

The M-19 had an impressive cross-range capability of 4,500 km and could significantly change its inclination by making dips into the atmosphere down to 50–60 km. Thermal protection was provided by carbon–carbon material and tiles. Situated in the nose was the crew compartment, which could be jettisoned from the vehicle in case of an emergency. It consisted of a flight deck and living compartment and was designed to carry a crew of between three and seven. Behind the crew compartment was a 15 × 4 m payload bay, equipped with an airlock/docking system and a remote manipulator arm.

Installed behind the payload bay was a big, removable tank containing liquid hydrogen for the vehicle's propulsion system. The latter was made up of a nuclear engine in the aft section of the vehicle, a pair of two-spool turbojet engines, and a set of hypersonic scramjet engines mounted on the underside of the aft fuselage. The propulsion system was adapted from Gurko's original proposal by the introduction of an on-board nuclear reactor that would heat up the air entering the turbojet and scramjet engines to very high temperatures. This would allow the air to escape from the nozzles at very high speeds with little combustion taking place, making it possible to save hydrogen for later stages of the orbit insertion process. The turbojet and scramjet engines would be used to accelerate the M-19 to a speed of Mach 16 and carry it to an altitude of 50 km. At that point the nuclear engine would kick in to place the ship into orbit.

The M-19 was billed as a multi-purpose system, capable of performing routine space transportation tasks, missions in the interests of science and the national economy, as well as reconnaissance and offensive missions. One big advantage of the SSTO was that it required no staging during launch, meaning that it had an almost unlimited number of launch azimuths.

EMZ was aiming for a step-by-step approach in the development of the M-19. This would include test flights of several "flying laboratories" to test the nuclear and scramjet engines, drop tests and re-entry tests of M-19 models, and the construction of an experimental hypersonic airplane that could act as a long-range bomber with a range of up to 12,000 km or a launch platform to place into orbit 40-ton payloads. The SSTO itself was expected to be ready for its first flight in 1987–1988.

Myasishchev perfectly understood the technical challenges posed by such a system and was well aware it wouldn't be ready to fly until many years after the Space Shuttle. However, since the Soviet Union was already several years behind in the development of a Space Shuttle response, he reasoned it would be better to start work on a much more advanced and capable system straightaway rather than build a copy that *itself* would be upstaged by the Space Shuttle by several years.

Despite the futuristic nature of the M-19, Myasishchev was not hampered by his "boss" Pyotr Dementyev, the Minister of the Aviation Industry, albeit more for political reasons than anything else. Just as he did with Spiral, Dementyev seems to have considered the M-19 a convenient tool in his arguments with MOM. Dementyev was wary of getting involved in NPO Energiya's (read: MOM's) Space Shuttle "copy", fearing that if his aviation design bureaus were assigned to the project, some of them would eventually be transferred to MOM. By tacitly supporting the M-19, he hoped to drag out the decision-making process leading to the approval of a Space Shuttle response.

Work on the project continued even after Myasishchev's EMZ was absorbed by NPO Molniya to work on Buran in February 1976. On 25 May 1976 the Military Industrial Commission decided to continue basic research on the SSTO spaceplane. Research in support of the M-19 was conducted by leading aviation institutes such as TsAGI, TsIAM, and ITPM. EMZ drew up plans to fly an experimental Lyulka liquid-hydrogen engine on an Ilyushin Il-76 airplane, mainly to test the techniques required to store liquid hydrogen at cryogenic temperatures. After Myasishchev's death in October 1978, this work was transferred to the Tupolev bureau, where it was successfully completed using the Tu-155.

Myasishchev's death was a major setback for the Gurkolyot. Nevertheless, work on the project seems to have continued at some level until the collapse of the USSR. Between 1978 and 1988 it was mentioned in three more VPK decisions and even in two government/party decrees. While the M-19 may have been considered a serious contender to counter the Space Shuttle before 1976, it quickly moved into the background once work on Energiya–Buran got underway in earnest. From then on it was probably seen as no more than a promising design for a second-generation shuttle vehicle.

One problem with further research on the M-19 was that it had to be done by organizations already preoccupied with Buran. Although there were government orders for NPO Molniya and TsNIIMash to conduct research on the M-19, there was very little enthusiasm for it. There does seem to have been at least some support for it from Boris Gubanov after he was appointed chief designer of the Energiya–Buran system in 1982. The M-19 was also hampered by interdepartmental squabbling between MOM and MAP, on the one hand, and the Strategic Rocket Forces and the Air Force, on the other hand.

Another problem with the M-19 was the use of a nuclear reactor and propulsion system, which posed safety risks both to the crew and the general public, even though designers went to great lengths to make it as safe as possible. However, the biggest showstopper for the M-19 must have been that few believed it was technically feasible, and perhaps rightly so, because even today, thirty years on, a vehicle of this type remains no more than a distant dream [19].

The Soviet response to NASP

Research on aerospace planes in the Soviet Union got a fresh impetus in the mid-1980s, presumably in response to similar work started in the US in 1982 at the Defense Advanced Research Projects Agency (DARPA) under the name Copper Canyon and then transferred to NASA and the Air Force as the National Aerospace Plane (NASP) in 1986. President Ronald Reagan mentioned the project in his State of the Union speech on 4 February 1986, calling it:

> "a new Orient Express that could, by the end of the next decade, take off from Dulles Airport, accelerate up to 15 times the speed of sound, attaining low Earth orbit or flying to Tokyo within two hours."

Although touted by the Reagan Administration for its civilian commercial applications and as a possible follow-on to the Space Shuttle for NASA, the 80–20 split of funding between the Air Force and NASA clearly indicates NASP was first and foremost a military program. The objective of the program was to develop a prototype SSTO vehicle taking off with turbojets, then switching to hydrogen-fueled scramjets at subsonic and hypersonic speeds, with a LOX/LH_2 rocket engine performing orbit insertion.

The go-ahead for the Soviet response came in two government decrees on 27 January and 19 July 1986, followed by the release of technical specifications by

The Tu-2000.

the Ministry of Defense on 1 September 1986. Three organizations were tasked to come up with proposals: NPO Energiya, the Yakovlev bureau, and the Tupolev bureau. While nothing is known about the Yakovlev concept, NPO Energiya's aerospace plane was a 71 m long vehicle with a wingspan of 42 m and a maximum height of 10 m. With a take-off mass of approximately 700 tons (dry mass 140 tons), the vehicle would use a combination of turbojets, scramjets, and rocket engines to reach orbit. It was designed for the deployment of payloads into low orbits (at least 25 tons into a 200 km, 51° orbit), servicing of orbital complexes, intercontinental passenger transport and also for military operations "in and from orbit". The project was headed by veteran designer Pavel Tsybin [20].

The project eventually selected for further development was the Tupolev bureau's Tu-2000. Actually, the bureau was no newcomer to SSTO vehicles, having already performed low-priority studies of horizontal take-off and landing spaceplanes with a take-off mass of up to 300 tons in 1968–1971. Overall Tu-2000 was very similar in design to NASP, relying on the same combination of engines to go into orbit. It had a vertical stabilizer and small delta wings, with much of the lift provided by the flat-shaped underside of the fuselage. A huge hydrogen tank occupied most of the mid and aft fuselage and would feed both the scramjet and rocket engines. The oxygen tank for the rocket engine was located in the tail section.

The Tupolev bureau proposed to carry out the project in two stages. First, it would develop a 55–60 m long two-man suborbital demonstrator (Tu-2000A) to reach a maximum velocity of Mach 5/6 and an altitude of up to 30 km. With a take-off mass between 70 and 90 tons, the vehicle would be equipped with four turbojet engines, two scramjets, and two liquid-fuel rocket engines. Then the project would move on to an experimental 71 m long two-man orbital version with a take-off mass between 210 and 280 tons and six rather than four turbojet engines. Payload capacity was 6–10 tons to low orbits between 200 and 400 km. Unconfirmed reports suggest the Tupolev bureau also planned a long-range bomber (Tu-2000B) and a hypersonic passenger plane based on the Tu-2000 design.

By the early 1990s the Tupolev bureau had reportedly built a wing torque box made of a nickel alloy, elements of the fuselage, cryogenic fuel tanks, and composite fuel lines. Estimates made in 1995 showed that Tu-2000 related R&D would cost at least $5.29 billion, a high price-tag even if Russia had a healthy economy. Budget realities had also forced NASA and the Air Force to cancel NASP in 1993. Although low-level research on the Tu-2000 may have continued for several more years, this project obviously stands no chance of being realized any time soon [21].

The Oryol program

In 1993 the Russian Space Agency initiated a research and development program called Oryol ("Eagle") to devise a strategy for the development of reusable space transportation systems in the 21st century. While the program was mainly aimed at technology development, several design bureaus were also invited to work out possible schemes for a Russian Aerospace Plane (RAKS), although it is hardly likely the intention was to actually build one. The focus was both on SSTO and two-stage-to-orbit (TSTO) concepts.

Schemes for vertically launched, partially reusable TSTO systems were devised by RKK Energiya, KB Salyut (which became part of the Khrunichev Center in 1993), and TsNIImash. All these revolved around the use of winged flyback boosters and expendable second stages, capable of placing about 25 tons into low 51° inclination orbits. All the concepts relied on the use of LOX/LH_2 engines, with the RD-0120 figuring prominently in three of the four schemes. The payloads could either be traditional satellites placed under a payload fairing or spaceplanes. Primarily intended for space station support, these Reusable Orbital Ships (MOK) would have an expendable instrument and cargo compartment.

Attention was also given to air-launched systems. It would seem that NPO Molniya got some funding under Oryol to continue work on its air-launched MAKS versions. Meanwhile, the Mikoyan bureau studied a fully reusable TSTO system called MiGAKS, consisting of a turbojet/ramjet powered hypersonic carrier aircraft and a spaceplane with rocket engines. The aircraft would propel the spaceplane to Mach 6 before releasing it and would then return either to its home base or to a runway downrange. The Mikoyan bureau studied hypersonic planes burning a combination of kerosene and hydrogen (total take-off mass 420 tons) or hydrogen alone (take-off mass 350 tons). Payload capacity to a low 51° inclination orbit was 12.3 tons for the first version and 10 tons for the second version.

In the SSTO area, the Mikoyan bureau came up with an unmanned spaceplane called MiG-2000. Weighing 300 tons at take-off, the 54 m long vehicle would be accelerated to Mach 0.8 by a liquid-fueled rocket sled, with ramjets propelling it to Mach 5 before rocket engines burning LOX and subcooled liquid hydrogen took

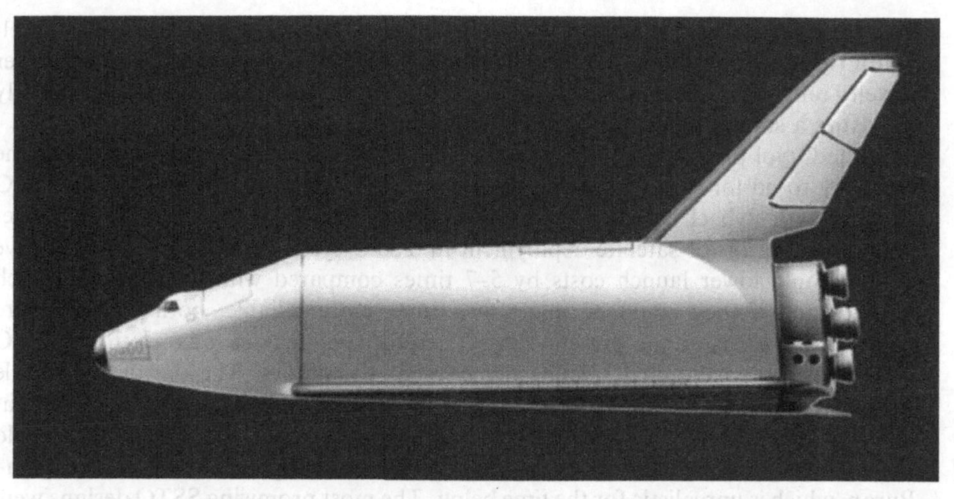

RKK Energiya's MKR spaceplane (*source*: RKK Energiya).

over to boost it to orbit. Payload capacity was 9 tons to a low 51° inclination orbit and cross-range capability was up to 3,000 km.

RKK Energiya proposed a 1,400-ton SSTO spaceplane called MKR (Reusable Space Rocket Plane). This would be launched on its own vertically, powered by seven tripropellant LOX/LH_2/kerosene engines with a sea-level thrust of 250 tons each. Externally resembling a Buran orbiter, most of the mid and aft fuselage was occupied by propellant tanks, leaving room only for a 8.0 × 4.5 m payload bay. Payload capacity was anywhere from 10 to 18 tons to low 51° orbits, depending on whether the vehicle was manned (maximum crew of three) or unmanned. Missions would last no longer than seven days. Cross-range capability was 2,000 km [22].

There was other SSTO research in the 1990s apparently not funded under Oryol. Khrunichev's KB Salyut worked on a vertical take-off/horizontal landing system reminiscent of America's VentureStar, and the Makeyev bureau designed a vertical take-off/vertical landing system called Korona similar to the American DC-X and its Delta Clipper prototype [23]. Finally, NPO Molniya did paper studies of sled-launched SSTOs (VKS-R) as well as vertical take-off/horizontal landing systems (VKS-O) [24].

Perhaps the most exotic SSTO concept was Ajax, originally conceived in the late 1980s by Vladimir L. Frayshtadt at the holding concern Leninets in Leningrad, but not made public until the 1990s. The basic principle is that Ajax turns the kinetic energy produced by the incoming airflow into chemical energy and power. Hydrocarbon fuel circulating under the skin is decomposed into several constituents by aerodynamic heating ("endothermic fuel conversion") and routed to a so-called magnetohydrodynamics (MHD) propulsion system, consisting of an MHD generator, a scramjet, and an MHD accelerator. The MHD generator extracts energy and thereby slows down the airflow before it enters the combustion chamber, circumventing the problems associated with mixing fuel and air at high Mach numbers.

Subsequently, the extracted energy is re-injected into the system by the MHD accelerator (located *behind* the combustion chamber) which speeds up the airflow. Another novelty on Ajax is the creation of plasma at the leading and trailing edges of its body to ensure a smoother air flow across the fuselage [25].

The Oryol program was finished in 2001. The general conclusion was that the best way to go forward in the near future was to develop partially reusable TSTO systems with flyback boosters and conventional rocket engines. Including spaceplanes as a means of satellite deployment in TSTO systems would only be effective if they could lower launch costs by 5–7 times compared with expendable launch vehicles and if they could be made five times more reliable, both of which are unattainable goals at the present time. Therefore, preference was given to TSTO systems with conventional satellite deployment techniques. The partially reusable Angara rockets using the Baykal flyback stage were seen as a first step in that direction. SSTOs were considered worth developing only if their dry mass could be made 30 percent lower than that of systems like the Space Shuttle or Energiya–Buran, which is unrealistic for the time being. The most promising SSTO designs were considered to be vertical take-off/horizontal landing systems [26].

Under the Federal Space Program for 2001–2005 Oryol was followed by another research program called Grif ("Vulture"), focusing among other things on studies of new, heat-resistant materials, construction materials, and air-breathing engines [27]. The latest Federal Space Program (2006–2015) only envisages the development of a partially reusable TSTO system with a flyback booster, an indication that SSTO has been shelved for many years to come. A tender to develop the TSTO is to be held in 2009 and the system is supposed to be fielded in 2016, although this is subject to further review. Payload capacity should be 25–35 tons to low orbit and launch costs should be reduced 1.5 times by avoiding the expenditures associated with clearing first-stage impact zones.

A possible contender is the RN-35, a TSTO system designed by the Keldysh Research Center in 2001–2003. Having a payload capacity of 35 tons, it would have a winged flyback booster burning liquid oxygen and methane. This may eventually be followed around 2030 by the RN-70, a similar system with a 70-ton payload capacity. There may be cooperation with the French CNES space agency under a program known as Ural [28]. At any rate, given the conclusions of the Oryol studies, it is unlikely that spaceplanes will be part of the TSTO program.

Hypersonic scramjet research

Most of the SSTO designs studied thus far require the use of air-breathing hypersonic scramjets, thereby limiting the amount of fuel that needs to be carried on board. The Russians did extensive research on such scramjets, which have applications not only in aerospace planes but also in aviation and long-range missiles. They were the first to test hypersonic scramjets in flight under a research program called Kholod ("Cold"). Initiated by the Military Industrial Commission in March 1979, Kholod was actually a wide-ranging research program to study the use of cryogenic fuels such as hydrogen and methane in aviation. This included studies of efficient ways of producing such

Scramjet ready for launch on S200 missile.

propellants, but also the study of hydrogen-fueled scramjets. The latter component of the Kholod program was entrusted to the Baranov Central Institute of Aviation Engine Building (TsIAM), while actual construction of the scramjet took place at the Soyuz design bureau (the former OKB-300), already involved earlier in developing propulsion systems for Spiral's hypersonic aircraft.

The 1 m long scramjet was configured with an asymmetrical three-shock fixed intake and a coaxial combustion chamber. It was launched by the S200 surface-to-air missile from the Saryshagan test site near Lake Balkhash in Kazakhstan and remained attached to the rocket throughout the flight, although at least some of the vehicles were recovered. Five missions were flown with mixed success between November 1991 and February 1998. The French space agency CNES took part in missions 2 and 3 and NASA was involved in the final two missions. The last mission, using a modified version of the scramjet developed at KBKhA in Voronezh, reached a record velocity of Mach 6.47 [29].

NASA used the acquired experience for its own hypersonic scramjet test program called Hyper-X, in which small X-43A experimental research aircraft with an airframe-integrated scramjet were launched from a B-52 bomber using modified single-stage Pegasus rockets. After an initial launch failure in 2001, the X-43A reached record speeds of Mach 6.83 and Mach 9.6 during two test flights in 2004. In Russia, the Soyuz design bureau also had plans for an air-launched, hydrogen-powered scramjet test bed (GLL-31) capable of reaching Mach 10. This would be launched from the belly of a MiG-31 jet with the help of a modified S-300 surface-to-air missile and be separated from the missile for later parachute recovery (unlike the non-recoverable X-43A). Soyuz also worked on a ground-launched, kerosene-fueled scramjet capable of reaching Mach 4.5, but the status of the latter two projects is unclear [30].

In 1993 TsIAM together with the Flight Research Institute (LII) and NPO Mashinostroyeniya (the former Chelomey bureau) started work on a new hypersonic scramjet test effort under the Oryol program, more specifically under the propulsion component of the program designated Oryol-2-1. This resulted in a new scramjet test bed called Igla ("Needle"), intended to conduct free-flight tests with the scramjet configured to operate in a usable flight vehicle. Measuring 8 m long, the vehicle has a three-module scramjet engine powered by liquid hydrogen and is launched by a UR-100N (SS-19 "Stiletto") intercontinental ballistic missile (also used as the basis for the Rokot launch vehicle). Capable of operating up to Mach 14.0, it parachutes down back to Earth after separation from the launch vehicle [31].

There is some speculation that Igla may have made its first test flight as part of a major military exercise staged by Russia's armed forces on 18 February 2004. This saw the launch of several rockets from Plesetsk and Baykonur, with President Vladimir Putin on hand at Plesetsk to witness the launches there. After the Baykonur launch, which involved a UR-100N, Putin said:

> "An experiment has been conducted and successfully concluded. Very soon we will have in service the most up-to-date technical systems, which are able to hit intercontinental targets at hypersonic speed and with great precision and have the ability to carry out a deep maneuver both in altitude and direction."

Putin was apparently referring to maneuverable hypersonic warheads that are extremely difficult to counter with missile defense systems. NPO Mashinostroyeniya started work on such a system under the name Albatros in 1987, which was then abandoned at the end of the Cold War but may have been resurrected after the United States' withdrawal from the Anti Ballistic Missile treaty in 2002 [32].

All indications are that scramjet research in Russia is now mainly being performed in the interests of the military, but the experience will still come in handy if Russia ever decides to build an SSTO aerospace plane in the distant future.

SPACE TOURISM

Quite possibly, the next Russian winged spacecraft to make its appearance will not be a state-sponsored vehicle, but one financed by the private sector for suborbital tourist missions. Suborbital space tourism got a major boost in the 1990s with the initiation of the X-Prize (later renamed Ansari X-Prize), a $10 million prize designed to jumpstart the space tourism industry through competition between entrepreneurs and rocket experts around the world. The cash prize would be awarded to the first team that privately built and launched a spaceship capable of carrying three people to 100 km altitude and repeat that launch with the same ship within two weeks.

The M-55X and the C-XXI suborbital rocket plane (*source*: EMZ).

One of the 26 contenders for the X-Prize was a consortium consisting of the Experimental Machine Building Factory (EMZ) in Zhukovskiy, the Russian Suborbital Corporation, and the Virginia-based company Space Adventures, which also brokers deals for millionaires wishing to fly to the International Space Station. On 14 March 2002 the consortium unveiled plans for a system called Constellation XXI, consisting of the M-55X carrier aircraft and the C-XXI suborbital vehicle, both designed by EMZ, which played a leading role in the Buran program as part of NPO Molniya. The M-55X is a modified version of the M-55 "Geofizika", a high-altitude research aircraft that made its debut in 1988. Journalists invited to EMZ's facilities were shown one of the M-55 aircraft with a wooden mock-up of the rocket plane suspended above it. The C-XXI was described as a 7.7 m long and 2.02 m high vehicle capable of carrying one pilot and two passengers. It was made up of a crew module and a jettisonable engine unit.

The plan was for the M-55X to carry the C-XXI to an altitude of 17 km, where the pair would separate at a speed of 750 km/h. Shortly afterwards, the C-XXI would ignite a solid-fuel rocket engine that would accelerate it to a speed of 1,600 km/h and take it to an altitude of 50 km. After engine burnout, the engine unit would be separated, while the rocket plane continued to an altitude of over 100 km, allowing the passengers to experience 3 to 5 minutes of weightlessness. The C-XXI would then make a 360° turn to glide to a landing on an ordinary runway at a speed of 220 km/h. All three crew members were supposed to wear pressure suits and could be ejected from the vehicle during the entire piggyback ride on the M-55X as well as during the early and final stages of the ship's autonomous flight [33].

The Ansari X-Prize was eventually won by Mojave Aerospace Ventures/Scaled Composites, the team led by the famed US aerospace designer Burt Rutan and sponsored by financier Paul Allen. After several powered test flights earlier in the year, the team's SpaceShipOne, dropped from the White Knight One carrier aircraft, made two successful suborbital flights in September–October 2004 less than two weeks apart. Building upon the success of SpaceShipOne, Rutan teamed up in July 2005 with the British business tycoon Richard Branson to form a new aerospace production company (the "Spaceship Company") that will build a fleet of commercial suborbital spaceships (SpaceShipTwo) and launch aircraft (White Knight Two). Owned and operated by a company called Virgin Galactic, at least five ships will be carrying two pilots and up to six paying passengers on suborbital flights reaching an altitude of 140 km.

Although Constellation XXI lost out in the X-Prize competition, its design now serves as the basis for a new suborbital tourist project that may eventually compete with Virgin Galactic. Space Adventures has again joined forces with EMZ to build an advanced version of the C-XXI that will use the same M-55X as its parent aircraft. Dubbed Explorer, the rocket plane will be able to haul five people to the edge of space and have emergency rescue systems similar to those of its predecessor. Also part of the partnership is Texas-based Prodea, a firm founded by the Ansari family, which put up the $10 million prize money for the X-Prize competition. Space Adventures intends to sell Explorer vehicles to operator companies to conduct the actual missions. It has deals in place to fly the Explorer vehicles from spaceports near major airports in the United Arab Emirates and Singapore [34].

REFERENCES

[1] "Yakob Ryabov", *Rossiyskiy kto est kto*, 5/1999.

[2] M. Rudenko, "Star Wars" (in Russian), *Trud*, 26 August 1993; A. Kirpil, O. Okara, "Designer of spaceplanes" (in Russian), *Nezavisimaya gazeta*, 5 July 1994, p. 6; M. Rudenko, "The rocket planes of designer Chelomey" (in Russian), *Vozdushnyy Transport*, 1996; A. Zavalashin, "That unknown and outstanding Chelomey" (in Russian), on-line at *Aerokosmicheskiy portal Ukrainy*; G. Yefremov (ed.), *60 let samootverzhennogo truda vo imya mira*, Moscow: Oruzhiye i tekhnologii, 2004, pp. 140–142.

[3] A. Gribelyuk, "Who created Buran?" (in Russian), *Zapad Vostok*, 7–13 March 2006.

[4] E. Kutyakin, "Two-stage horizontally launched aerospace systems" (in Russian), in: G. Lozino-Lozinskiy, A. Bratukhin, *Aviatsionno-kosmicheskiye sistemy*, Moscow: Izdatelstvo MAI, 1997, pp. 292–295.

[5] See: D. Vorontsov, "About Deytron and others" (in Russian), *Novosti kosmonavtiki*, 10/2006, pp. 60–62.

[6] E. Dudar, "Analysis of different concepts for reusable space transportation systems" (in Russian); V. Skorodelov, "Multipurpose Aerospace System MAKS" (in Russian); E. Dudar, T. Lobzova, "Flying and technical characteristics of MAKS" (in Russian), in: G. Lozino-Lozinskiy, A. Bratukhin, *op. cit.*, pp. 279–291, 303–307, 334–342.

[7] V. Kucherenko, L. Batalov, "No RAKS without MAKS" (in Russian), *Rossiyskaya gazeta*, 11 March 1999.

[8] TV documentary *Udarnaya Sila*, ORT television, 20 March 2007.

[9] V. Skorodelov, "Experimental aerospace system MAKS-D" (in Russian), in: G. Lozino-Lozinskiy, A. Bratukhin, *op. cit.*, pp. 308–311.

[10] B. Konovalov, "Into space on wings: Project for launches to orbit from Mriya airplanes" (in Russian), *Izvestiya*, 24 November 1989, p. 3.

[11] A. Bakina, "General director of RKA against MAKS project" (in Russian), *Novosti kosmonavtiki*, 3/1998, p. 41.
[12] E. Devtyarov, "MAKS project raises questions" (in Russian), *Novosti kosmonavtiki*, 5/1999, pp. 48–49.
[13] Y. Semyonov, *Raketno-kosmicheskaya korporatsiya Energiya 1946–1996*, Moscow: RKK Energiya, 1996, pp. 404–408.
[14] V. Filin, "Small Multipurpose Reusable Orbiter (OK-M) Vehicle," paper presented at *30th AIAA/ASME/SAE/ASEE Joint Propulsion Conference in Indianapolis, 27–29 June 1994*.
[15] Y. Semyonov, *op. cit.*, pp. 423–424.
[16] S. Shamsutdinov, "Project Kliper" (in Russian), *Novosti kosmonavtiki*, 7/2005, pp. 1–7.
[17] S. Shamsutdinov, "RKK Energiya: A concept for the development of Russian piloted cosmonautics" (in Russian), *Novosti kosmonavtiki*, 7/2006, pp. 6–13; I. Marinin, "N. Sevastyanov: We are realizing the ideas of Korolyov" (in Russian), *Novosti kosmonavtiki*, 1/2007, pp. 4–7.
[18] V. Fomin *et al.*, "Jet planes of the future" (in Russian), *Nauka iz pervykh ruk*, March 2005, pp. 146–155.
[19] A. Bruk *et al.*, *Illyustrirovannaya entsiklopediya samolyotov EMZ im. V.M. Myasishcheva (tom 3, chast 1)*, Moscow: Aviko Press, 1999, pp. 26–31; A. Bruk *et al.*, *Illyustrirovannaya entsiklopediya samolyotov EMZ im. V.M. Myasishcheva (tom 8)*, Moscow: Aviko Press, 2005; A. Zuzulskiy, *Vperedi svoego vremeni*, Moscow: SIP RIA, 2000.
[20] Y. Semyonov, *op. cit.*, p. 408.
[21] "Single-stage aerospace plane Tu-2000" (in Russian), on-line at *http://airbase.ru/sb/russia/tupolev/2000/*; I. Chornyy, "The taming of hypersonics" (in Russian), *Novosti kosmonavtiki*, 17–18/1998, pp. 46–48.
[22] S. Umanskiy, *Rakety-nositeli. Kosmodromy*. Moscow: Restart+, 2001, pp. 112–120; Y. Semyonov, *Raketno-kosmicheskaya korporatsiya Energiya 1996–2001*, Moscow: RKK Energiya, 2001, pp. 787–788.
[23] I. Chornyy, *op. cit.*
[24] E. Dudar, *op. cit.*
[25] I. Chornyy, "Skylon and Ajax: Dissimilar twins" (in Russian), *Novosti kosmonavtiki*, 23–24/1998, pp. 50–51.
[26] I. Chornyy, *op. cit.*; "Reusable carrier rockets will become the basis for future space research, say participants in an international symposium" (in Russian), ITAR-TASS report, 1 November 2001.
[27] ITAR-TASS report, 1 November 2001.
[28] I. Chornyy, V. Dmitriyev, "Course of work on the Ural theme" (in Russian), *Novosti kosmonavtiki*, 12/2006, p. 60; V. Golovachov, "It is time to build factories on the Moon" (in Russian), *Trud*, 12 December 2006.
[29] "The Kholod hypersonic flying laboratory" (in Russian), on-line at *http://www.sergib.agava.ru/russia/tsiam/holod/holod.htm*
[30] I. Afanasyev, "Flight tests with scramjets" (in Russian), *Novosti kosmonavtiki*, 12/1999, pp. 48-49.
[31] I. Afanasyev, "Lessons of NASP in the background of Kholod" (in Russian), *Novosti kosmonavtiki*, 7/2002, pp. 40–43.
[32] K. Platt, "Russia's Hypersonic Engine Development—A World Leader," *Space Chronicle*, 2005, pp. 93–96; A. Rotkin, "Kuzkina mat-M" (in Russian), *Russkiy Newsweek*, 19–25 June 2006.

[33] A. Kopik, "Russian rocket plane for space hops" (in Russian), *Novosti kosmonavtiki*, 5/2002, pp. 58–59.
[34] T. Malik, "Suborbital Rocketship Fleet to Carry Tourists Spaceward in Style," article on *space.com* website, 22 February 2006.

Appendix A

Key Energiya–Buran specifications

A.1 ENERGIYA ROCKET

Overall specifications

Total launch mass	
Without payload	2,270 t
With Buran	2,375–2,419 t
Rocket mass prior to orbiter separation	178.5 t
Core stage	
Wet mass	776.2 t
Liquid oxygen mass	602.775 t
Liquid hydrogen mass	100.868 t
Strap-on booster	
Wet mass	372.6 t
Liquid oxygen mass	221.7 t
Kerosene mass	85.3 t
Dimensions	
Core stage length	58.765 m
Core stage tank diameter	7.75 m
Strap-on booster length	39.46 m
Strap-on booster tank diameter	3.92 m
Total lift-off thrust	3,550 t

Engine specifications

Parameter	RD-170	RD-0120
Propellants	LOX/kerosene	LOX/LH_2
Oxidizer/fuel mixture ratio	2.6:1	6:1
Combustion cycle	Closed	Closed
Sea-level thrust	740 t	147.6 t
Vacuum thrust	806.4 t	200 t
Sea-level specific impulse	308.5 s	353.2 s
Vacuum specific impulse	336.2 s	454.7 s
Chamber pressure	250 kg/cm^2	223 kg/cm^2
Turbopump power	257,360 hp	92,700 hp
Turbopump rotation	13,850 rpm	32,500 rpm
Throttle range	50–100%	45–100%
Nozzle area ratio	36.87:1	85.7:1
Gimbal capability	8°	7–11°
Nominal burn time	140–150 s	500 s
Dry mass	9,755 kg	3,450 kg
Length	4.0 m	4.550 m
Diameter	3.800 m	2.420 m

Energiya data collected from: Y. Baturin (ed.), *Mirovaya pilotiruemaya kosmonavtika*, Moscow: RTSoft, 2005, p. 443; Fact sheet of Voronezh Machine Building Factory; "The RD-170 and RD-171" (in Russian), on-line at *http://www.lpre.de/energomash/RD-170/index.htm*

A.2 BURAN

Maximum launch mass	105 t
Mass on first mission	79.4 t
Landing mass	
Nominal	82 t
Maximum	87 t
Dry mass	62 t
Maximum payload to orbit	
For 200 km, 50.7° orbit	30 t
For 200 km, 97° orbit	16 t
Mass of returned payload	
Nominal	15 t
Maximum	20 t
Dimensions	
Length	36.37 m
Height (on runway)	16.35 m
Maximum width of fuselage	5.60 m
Wingspan	23.92 m
Wing area	250 m^2
Tail area	39 m^2
Body flap area	10.3 m^2
Payload bay length	18.55 m
Payload bay diameter	4.70 m
Crew	
Minimum (with ejection seats)	2
Maximum (without ejection seats)	10
Volume of crew compartment	73 m^3
Flight duration	
Nominal	7 days
Maximum (with extra tanks)	30 days
Range of orbital inclinations	50.7–110°
Orbital altitude	
Nominal (circular)	250–500 km
Maximum (with extra tanks)	1,000 km
Maximum g-forces	
Launch (nominal trajectory)	2.95 g
Re-entry (nominal trajectory)	1.6 g
Lift-to-drag ratio	
At hypersonic speeds	1.3
At subsonic speeds	5.6
Crossrange capability	
Maximum	1,700 km
Demonstrated during 1st flight	550 km
Landing speed	
Average (for 82 ton landing mass)	312 km/h
Maximum	360 km/h
On first flight	263 km/h
Landing rollout distance	
Minimum/maximum	1,100–2,000 m
On first flight	1,620 m
Maximum number of flights	100

Buran data taken from: Y. Baturin, *op. cit.*, p. 438.

Appendix B

Short biographies of Buran cosmonauts

(Listed here are only those cosmonauts who originally were selected *specifically* for the Buran program)

Afanasyev, Viktor Mikhaylovich was born in Bryansk on 31 December 1948. He graduated in 1970 from the Kachinskoye Higher Military Aviation Pilot School, where Yuriy Sheffer, Aleksandr Puchkov, and Aleksandr Shchukin, all future Buran cosmonauts, had been his classmates. Following graduation, he served as an Air Force pilot until he enrolled in the Air Force's test pilot school in 1976, graduating the following year. He subsequently worked as a test pilot in the flight test center in Akhtubinsk and in 1985 he and two colleagues were selected to join GKNII's cosmonaut team. From 1985 until 1987 the three underwent OKP training at the Gagarin Cosmonaut Training Center in Star City. After they had graduated and qualified as cosmonaut-tester, TsPK chief Vladimir Shatalov offered them a transfer to the TsPK cosmonaut detachment, which all three accepted. Afanasyev would make three long-duration space flights to Mir: Soyuz TM-11 in 1990–1991 (EO-8), Soyuz TM-18 in 1994 (EO-15), and Soyuz TM-29 (EO-27) in 1999. He also made a short mission to the International Space Station aboard Soyuz TM-33 in 2001. In total, Viktor Afanasyev logged over 545 days in space, during which he conducted seven EVAs, totalling over 38 hours. He has since retired from the cosmonaut team.

Artsebarskiy, Anatoliy Pavlovich was born on 9 September 1956 in the village of Prosyanaya, Dnepropetrovsk Region in the Ukraine. In 1977 he graduated from the Kharkov Higher Military Aviation Pilot School and remained there as instructor. In 1982 he transferred to the Air Force's test pilot school in Akhtubinsk and graduated as a Test Pilot 3rd Class the following year. In 1985 he was selected as one of three new cosmonauts in the GKNII Buran cosmonaut group and sent to Star City for OKP. However, upon graduating in 1987, Artsebarskiy and his two colleagues accepted the offer to transfer to the TsPK cosmonaut detachment and remained in Star City. After having been back-up commander for Soyuz TM-11, he went on to command Soyuz TM-12 and Mir expedition EO-9. It would be his only spaceflight. The official reason that he didn't fly again is unknown, but it has been said that he was grounded for attending the 1992 Planetary Congress of the Association of Space Explorers without explicit personal permission from TsPK head Pyotr Klimuk. In September 1993 he was detached to the Russian Academy of Sciences and considered a member of their cosmonaut group. When the government decided to limit the number of institutions to which military personnel could be detached, the Academy of Sciences was not among them. In July 1994 Artsebarskiy was sent to the Academy of the General Staff to study. In 1998 he retired from the Air Force.

Bachurin, Ivan Ivanovich was born on 29 January 1942 in Berestovenka in the Kharkov Region. In 1959 he entered the Orenburg Higher Military Aviation Pilot School and following graduation in 1963 he remained there as an instructor. In 1967 he enrolled in the Soviet Air Force's Chkalov test pilot school, graduating the following year and beginning flight testing at the flight test center in Akhtubinsk. In 1973, he also graduated from the Moscow Aviation Institute. Five years later, Bachurin was selected as one of the GKNII cosmonaut candidates to eventually fly on Buran. As senior officer, he was also named the group's commander. In 1980, he completed OKP, becoming a cosmonaut-tester. Bachurin then began Buran-related test flying in Akhtubinsk, and in 1987–1988 was involved in the approach and landing test program on BTS-002. Together with Aleksey Boroday, Bachurin flew BTS six times, three times in the commander's seat. He also trained as one of three GKNII cosmonauts for a Soyuz docking mission with an unmanned Buran, but that was never flown. Shortly thereafter, Bachurin was medically disqualified and left the cosmonaut group in 1992. He is retired and lives in the town of Chkalovskiy, near Star City.

Boroday, Aleksey Sergeyevich was born on 28 July 1947 in the village of Borodayevka, near Volgograd (then still called Stalingrad). After finishing school he worked in a factory but at the same time took flying lessons in a DOSAAF air club. In 1969 he graduated from the Kachinskoye Higher Military Aviation Pilot School and subsequently served as a fighter pilot in the Air Force. In 1977 he graduated from the Air Force's test pilot school in Akhtubinsk and began test pilot work in GKNII. Soon, he was selected to become one of the GKNII Buran cosmonauts and together with the other candidates he was sent to Star City for the OKP basic cosmonaut training course, finishing that in 1980. In 1981, he graduated from the Akhtubinsk branch of the Moscow Aviation Institute. Together with Ivan Bachurin, Boroday took part in the approach and landing test flight program on Buran's analog BTS-002, flying a total of six missions in 1987 and 1988. Later he trained as commander of one of three crews that was preparing for a Soyuz docking mission with an unmanned Buran. He left the GKNII team in 1993. Boroday returned to flying on heavy transport planes, including the Antonov An-225 Mriya while it was transporting the Buran orbiter. On 8 October 1996, he commanded an Antonov An-124 Ruslan on a cargo flight to Turin, Italy. During landing the plane lost engine thrust and hit the ground with a wingtip. The plane cartwheeled and crashed in a field near the airport, killing the co-pilot and injuring the other crew members. Boroday, who regained consciousness in a hospital after five days, lost both his legs. He still lives in Star City.

Sovetskiye i rossiyskiye kosmonavty

Chirkin, Viktor Martynovich was born on 13 July 1944 in Barnaul. He graduated from the Armavir Higher Military Aviation Pilot School in 1971 with the qualification of pilot-engineer. From 1973 he studied in Akhtubinsk at the Air Force's test pilot school, becoming a Test Pilot 3rd Class in 1974 and a Test Pilot 2nd Class in 1977. The following year Chirkin was one of the GKNII pilots selected for Buran and was sent to Star City for OKP. He graduated in November 1980 and received his cosmonaut-tester certificate, but by then Chirkin had growing doubts that Buran had a future and decided to resign from that program and return to full-time test flying in Akhtubinsk. Eventually, he would rise to the rank of Major-General and become both a Merited Test Pilot and a Hero of the Russian Federation.

Grekov, Nikolay Sergeyevich was born in Kalinin in Kirgizia on 15 February 1950. After graduating from the Armavir Higher Military Aviation Pilot School in 1971, he served with the Soviet Air Defense Forces in Belorussia and in the city of Gorkiy. In May 1978 Grekov (representing the Air Defense Forces) was selected to join the 1976 Buran cosmonaut group of TsPK. As had been the case with the 1976 candidates, he was first sent to the Air Force's test pilot school in Akhtubinsk, from which he graduated as Test Pilot 3rd Class in July 1979. He then went on to undergo OKP basic cosmonaut training and finished that in February 1982. Shortage of qualified commanders for Soyuz and delays in the Buran program then led to the decision to transfer all members of the TsPK Buran team to the Soyuz–Salyut training group. However, in spite of this transfer, Grekov would not fly in space. After Vladimir Vasyutin had been forced to return to Earth due to illness in November 1985, all cosmonauts were given an extra physical examination. Grekov was found to have a chronic form of hepatitis and was forced to end his cosmonaut career in December 1986. He did stay on at TsPK, however, eventually becoming the head of the Search and Recovery Department. In that capacity, he was responsible not only for the recovery of crews after landing, but also for splashdown and winter survival training of cosmonauts. Grekov retired in 2004.

Ivanov, Leonid Georgyevich was born on 25 June 1950 in Safonovo near Smolensk. He attended the Kachinskoye Higher Military Aviation Pilot School and following graduation in 1971 he served in an Air Force unit in the town of Mukachevo in the Prikarpat military district. After having been selected by TsPK in 1976 to become a cosmonaut, Ivanov and his fellow cosmonaut candidates were sent to the Air Force's test pilot school in Akhtubinsk. Having graduated as Test Pilot 3rd Class in 1977, the group underwent OKP in Star City until September 1978. Ivanov then became one of seven pilots to return to Akhtubinsk to obtain the title of Test Pilot 2nd Class. During this training course on 24 October 1980 Ivanov's MiG-27 fighter went into a spin and crashed, killing the pilot. Ivanov was buried in the village of Leonikha, near Star City.

Kadenyuk, Leonid Konstantinovich was born on 28 January 1951 in what is now the city of Chernovtsi in the Ukraine. In 1971 he graduated from the Chernigov Higher Military Aviation Pilot School, becoming an Air Force pilot until he was selected as a cosmonaut candidate in 1976. Together with the other pilots from his selection group, he was sent to the Air Force's test pilot school in Akhtubinsk, where he graduated as a Test Pilot 3rd Class in June 1977. He then took the OKP basic cosmonaut training course at TsPK, qualifying as a cosmonaut-tester in September 1978. Next, Kadenyuk and a number of other cosmonauts from his group returned to Akhtubinsk and continued their test pilot training. In 1981 he finished that and became a Test Pilot 2nd Class. The group returned to TsPK and began Buran-related training, but in March 1983 Kadenyuk's divorce resulted in his dismissal from the cosmonaut detachment. He subsequently returned to GKNII in Akhtubinsk to work as a test pilot. In December 1988 he managed to be included in the GKNII Buran cosmonaut team and was assigned as one of three commanders to train for a Soyuz mission that was to dock with an unmanned Buran. Despite cancellation of that flight, Kadenyuk didn't give up his dream of flying in space. After the break-up of the Soviet Union he moved to the Ukraine and became a pilot in the Air Force. When the United States signed a contract with the Ukraine to fly a Ukrainian astronaut on the Space Shuttle, Kadenyuk applied for the Ukrainian cosmonaut team. In November 1996 Kadenyuk was assigned as payload specialist for Space Shuttle mission STS-87. He flew a 15 day 16 hour mission on the Shuttle Columbia from 19 November until 5 December 1987. All in all, he had waited 21 years and 3 months since he had first been selected as a cosmonaut. With that Kadenyuk holds the record for time elapsed between initial selection and the first spaceflight.

Kononenko, Oleg Grigoryevich was born on 16 August 1938 in the village of Samarskoye in the Rostov Region. In 1958 he graduated from the DOSAAF school in Saransk and became a pilot instructor. In 1965 he entered the Ministry of the Aviation Industry's test pilot school in Zhukovskiy, graduating a year later from the helicopter branch. In addition, in 1975 he graduated from the Zhukovskiy branch of the Moscow Aviation Institute. Kononenko was selected by LII for Buran training in 1977 and began his OKP at TsPK in 1979. Kononenko was in the middle of his final exams when his Yak-38 jet crashed in the South China Sea on 8 August 1980. The vertical take-off and landing jet lost engine power shortly after take-off from the aircraft carrier *Minsk*. Kononenko, who had been a Merited Test Pilot of the Soviet Union, didn't manage to eject in time and was killed instantly. Although he received the Order of Lenin (his second) posthumously, he did not get the title of cosmonaut-tester posthumously, as he had not yet passed all his final exams at that time.

Levchenko, Anatoliy Semyonovich was born on 21 May 1941 in Krasnokutsk, near Kharkov in the Ukraine. Eager to become a pilot, he enrolled in the Kremenchug Higher Air Force Pilot School, but it was closed a year later and he finished his pilot education at the Chernigov Higher Military Aviation Pilot School, where he was a classmate of future cosmonaut Pyotr Klimuk. Upon graduation he served for five years as a MiG-21 pilot in Turkmenistan, and then left the Air Force to enroll in the test pilot school of the Ministry of the Aviation Industry in Zhukovskiy. He graduated in 1971, becoming a test pilot at LII. In 1977 Levchenko was one of the pilots selected to undergo cosmonaut training for the Buran flight test program. He was also one of the pilots who flew approach and landing tests on BTS-002, conducting four flights. As he was the designated back-up commander for Buran's first manned orbital mission, Levchenko first acted as back-up to Igor Volk for the Soyuz T-12 mission to Salyut-7 in 1984 and then went on to fly an eight-day mission to the Mir space station on Soyuz TM-4 in December 1987. However, several months after his flight he was diagnosed with a brain tumor, from which he died on 6 August 1988.

Maksimenko, Valeriy Yevgenyevich was born on 16 July 1950 in Tyumen. He graduated from secondary school in 1967 and went to the Kharkov Higher Military Aviation Pilot School, where he studied until 1971. Upon graduating, he remained there as a pilot instructor. In 1977 he enrolled in the Air Force's test pilot school in Akhtubinsk, becoming a test pilot the following year. After having been selected to become a cosmonaut in the GKNII group, he followed the OKP basic cosmonaut training course at TsPK from 1989 until 1991. At the same time, he continued test flight work for the Air Force, almost exclusively in high-performance fighter aircraft like the MiG-29 and Su-27. When he came to the conclusion that his future was not in the Buran program, he requested to be allowed to return to full-time test flying, a request that was granted. In January 1993 he became the head of the GKNII test pilot school (TsPLI).

Manakov, Gennadiy Mikhaylovich was born on 1 June 1950 in Yefimovka, in the Orenburg Region. He graduated from the Armavir Higher Military Aviation Pilot School in 1973 and remained there for two years as an instructor. After that he served in Kamchatka and in the Moscow military district. In 1979 he was admitted to the Chkalov test pilot school of the Soviet Air Force in Akhtubinsk, graduating in 1980. He was assigned as a test pilot of fighter planes, at the same time taking a course at the Akhtubinsk branch of the Moscow Aviation Institute, from which he graduated in 1985. In August of that year, Manakov was one of three test pilots who were assigned to GKNII's Buran cosmonaut group. They underwent OKP training in Star City until 1987, but upon graduation were offered to stay in Star City as members of TsPK's cosmonaut detachment. Manakov accepted the offer and began mission training for Soyuz flights to the Mir space station. Eventually, he would fly to Mir on Soyuz TM-10 in 1990 (EO-7) and Soyuz TM-16 in 1993 (EO-13), logging a total of 310 days in space. In addition, he conducted three EVAs, spending more than 12 hours outside the station. Manakov was training for the Soyuz TM-24/EO-22 mission in 1996, when he was medically disqualified and grounded. In July 2000 he retired from the Air Force.

Sovetskiye i rossiyskiye kosmonavty

Moskalenko, Nikolay Tikhonovich was born on 1 January 1949 in the village of Goragorskiy, in the Chechen-Ingush Republic of the Russian Federation. From 1966 until 1970 he attended the Yeysk Higher Military Aviation School, and upon graduating he served in the Air Force, until he was selected as part of the 1976 TsPK intake. Having first been trained as military test pilots until June 1977, the group took OKP between October 1977 and September 1978. Moskalenko was then sent back to Akhtubinsk for further test pilot training and, after becoming a Test Pilot 2nd Class in 1981, he returned to TsPK for mission training as a cosmonaut. He was assigned to the third crew of what eventually became Soyuz T-14, but he would never fly in space himself. His divorce resulted in his expulsion from the cosmonaut detachment in June 1986. Moskalenko returned to test flying in Akhtubinsk until he left the Air Force in June 1990. He died after a long illness on 26 November 2004.

Mosolov, Vladimir Yemelyanovich was born in Kaliningrad (now Korolyov) near Moscow on 22 February 1944. He enrolled in the Tambov Higher Military Aviation Pilot School, from which he graduated in 1967. After that he served in long-range aircraft units. In 1976 Mosolov graduated from the Soviet Air Force's test pilot school in Akhtubinsk and became a test pilot at GKNII. When the Air Force began looking for a group of test pilots to fly on Buran, Mosolov was one of the eight candidates selected. In 1979 the group went to the Gagarin Cosmonaut Training Center to undergo OKP, with five members graduating in 1980. The following year Mosolov also graduated from the Akhtubinsk branch of the Moscow Aviation Institute. For a number of years, Mosolov did Buran-related test flying until he was dismissed from the cosmonaut group in 1987 because of his divorce. He returned to regular test flying and left the Air Force in 1995 to work for private aviation companies.

Sovetskiye i rossiyskiye kosmonavty

Polonskiy, Anatoliy Borisovich was born on 1 January 1956 in the village of Pogranichnik in Kazakhstan. In 1977 he graduated from the Orenburg Higher Military Aviation Pilot School and served as a pilot in units of the Baltic Fleet until 1985, when he enrolled in the test pilot school in Akhtubinsk. He graduated in 1986 and went to work as a test pilot for GKNII. In February 1988 Polonskiy was selected as one of the new GKNII cosmonaut candidates to undergo OKP training in TsPK, qualifying as a cosmonaut-tester in April 1991. However, as Buran never flew in space again, Polonskiy became occupied full-time with test flying and operational flying of heavy transport aircraft, becoming a squadron commander in GKNII. Among the planes Polonskiy flew was the largest aircraft in the world, the Antonov An-225 Mriya. He lives in Chkalovskiy near Star City.

Prikhodko, Yuriy Viktorovich was born on 15 November 1953 in Dushanbe, the capital of the former Soviet republic of Tadzhikistan in Central Asia. After graduating from secondary school he worked as a laboratory assistant for a short time, before enrolling in the Kachinskoye Higher Military Aviation Pilot School in 1971, where he was a classmate of Sergey Tresvyatskiy. Upon graduation in 1975, he remained at the school as a pilot instructor until he resigned his commission from the Air Force and enrolled in the Ministry of the Aviation Industry's test pilot school in Zhukovskiy. Upon graduation in 1986 he became a Test Pilot 3rd Class and began test flying different types of aircraft. At the same time, in the evenings, he studied at the Zhukovskiy branch of the Moscow Aviation Institute, graduating in 1989. The previous year, he had been selected to join the LII cosmonaut team and in 1989 he began OKP, passing his final exam on 28 March 1991 and earning the qualification of cosmonaut-tester. By that time, however, it was becoming clear that Buran would probably not fly again. After having worked as a test pilot for twelve years, he left LII in 1998 and went to the United States, where he worked as an exchange pilot in California. His dream was to earn a green card and stay in the US, possibly even as a test pilot for NASA, but on 27 July 2001, he died from cancer, only 47 years old. He is buried in the town of Ostrovtsy, not far from Zhukovskiy.

Protchenko, Sergey Filippovich was born on 3 January 1947 in the village of Senitskiy in the Bryansk Region. In 1969 he graduated from the Chernigov Higher Military Aviation Pilot School and then served as a pilot in the Air Force until 1976, when he was one of nine pilots selected to become cosmonauts for the TsPK cosmonaut detachment. Protchenko and his eight fellow pilots were sent to Akhtubinsk to be trained as test pilots at GKNII. Upon earning the qualification of Test Pilot 3rd Class, Protchenko proceeded to take the OKP basic cosmonaut training course at TsPK, which he successfully concluded in 1978. He was then one of the seven group members who were sent back to Akhtubinsk for further training as test pilots. It was during this second course in Akhtubinsk in 1978 that Protchenko failed a medical and was dismissed from the cosmonaut team. In August 1986 he also retired from the Air Force with the rank of Lieutenant-Colonel.

Puchkov, Aleksandr Sergeyevich was born on 15 October 1948 in the town of Medyn in the Kaluga Region. In 1966, Puchkov enrolled in the Kachinskoye Higher Military Aviation Pilot School in Volgograd, where three more future Buran cosmonauts, Viktor Afanasyev, Yuriy Sheffer, and Aleksandr Shchukin, were his classmates. Upon graduation Puchkov and Sheffer both worked as pilot instructors. In 1977, Puchkov finished his test pilot course at the Air Force's test pilot school in Akhtubinsk and stayed there to work as a test pilot. In 1989 he was selected to undergo OKP training in order to become a cosmonaut in GKNII's Buran cosmonaut group. Puchkov graduated in 1991, although he had continued his test pilot work during OPK. His Buran career ended in November 1996 when the GKNII cosmonaut team was officially disbanded. In June 1997, he retired from the Air Force and went to work for VPK MAPO, a company in which a number of design bureaus had merged to produce MiG fighter aircraft. Puchkov became a department head in VPK MAPO.

Pushenko, Nikolay Alekseyevich was born on 10 August 1952 in the village of Povalikha in the Altay Region. He graduated from the Barnaul Higher Military Aviation Pilot School in 1974, and subsequently served in Air Force units until 1982, when he was admitted to the Air Force test pilot school in Akthubinsk. He graduated in 1983 and became a GKNII test pilot. In 1989 he was selected as one of six new Buran cosmonaut candidates for the GKNII team and from 1989 until April 1991 he underwent OKP in Star City. When it became clear that Buran would never fly again, Pushenko requested a transfer to the TsPK cosmonaut detachment, but, although his commander had promised him the transfer, this never took place for unknown reasons. After the GKNII cosmonaut team was disbanded in November 1996, Pushenko returned to test flying until he retired from the Air Force in 1998. In 2000 he began working for the State Research Institute for Civil Aviation at Sheremetyevo Airport outside Moscow.

Saley, Yevgeniy Vladimirovich was born on 1 January 1950 in Tavda in the Sverdlovsk Region. He studied at the Kachinskoye Higher Military Aviation Pilot School from 1967 until 1971 and after graduation worked there as an instructor. Later, he was stationed at Air Force units in Poland and Uzbekistan. In 1975 he applied for enrollment in the Gagarin Air Force Academy in Monino, but instead was offered to undergo the selection procedure to become a cosmonaut. In 1976, Saley was selected by TsPK and together with the other group members spent the next nine months in test pilot school in Akhtubinsk, graduating as a Test Pilot 3rd Class. After an additional parachute course, the group then took OKP in Star City, and upon graduating as cosmonaut-testers in September 1978, seven group members, including Saley, returned to Akhtubinsk. Saley graduated as Test Pilot 2nd Class in June 1981. After their return to Star City, they were moved over to the Soyuz–Salyut group, given the shortage of commanders for Soyuz and the delays Buran was facing. Saley trained as a back-up crew member for Soyuz T-14 in 1984–1985. He would undoubtedly have been assigned as prime crew member for a subsequent Soyuz mission, but reportedly had a falling-out with cosmonaut training chief Vladimir Shatalov over crew assignments, after which Shatalov immediately grounded him. He left the cosmonaut team in October 1987. He went on to become deputy director of the Chkalov Central Flying Club in Moscow. Saley still lives in Star City.

Sattarov, Nail Sharipovich was born on 23 December 1941 in the village of Kabakovo in Bashkiria. In 1967 he graduated from the Orenburg Higher Aviation Military Pilot School, and stayed on there as an instructor until he was invited to train at the Air Force test pilot school in Akhtubinsk. In 1978 Sattarov was one of the eight pilots selected by GKNII to fly Buran. In April 1980, before ending his OKP training, he left the training group, reportedly because he had violated safety rules by making a roll maneuver in a Tupolev Tu-134 passenger jet. Sattarov was also grounded for a short period but eventually went back to test flying in Akhtubinsk. He rose to the rank of Colonel, and in 1993 left GKNII and the Air Force, becoming a test pilot for the Tupolev design bureau in Zhukovskiy.

Shchukin, Aleksandr Vladimirovich was born on 19 January 1946 in Vienna, where his father was serving in the Soviet army when it was occupying Austria after the war. In 1966 Shchukin enrolled in the Kachinskoye Higher Military Aviation Pilot School in Volgograd, where Viktor Afanasyev, Yuriy Sheffer, and Aleksandr Puchkov were among his classmates. After graduating in 1970 he served for five years in the former German Democratic Republic. In 1975 Shchukin, then a Major, left the Air Force and enrolled in the Ministry of the Aviation Industry test pilot school in Zhukovksiy. In June 1977 he graduated as a Test Pilot 3rd Class and began working at LII in Zhukovskiy. Soon afterwards he was included in the first selection group to train as cosmonauts for the Buran program. He underwent basic cosmonaut training at TsPK without interrupting his test pilot work and upon graduation was awarded the title of cosmonaut-tester. During the approach and landing test program from 1985 until 1988 Shchukin flew seven times on Buran's analog BTS-002. He also served as back-up to Anatoliy Levchenko on Soyuz TM-4. On 18 August 1988 Shchukin was killed when his Sukhoy Su-26 sports plane crashed during a test flight.

Sheffer, Yuriy Petrovich was born on 30 June 1947 in Chelyabinsk. He attended the Kachinskoye Higher Military Aviation Pilot School in Volgograd, where he studied in the same class as Aleksandr Shchukin, Viktor Afanasyev, and Aleksandr Puchkov, graduating in 1970. He remained as an instructor, and had another future colleague, Sergey Tresvyatskiy, as one of his students. Sheffer left the Air Force in 1975 and enrolled in the civilian test pilot school of the Ministry of the Aviation Industry in Zhukovskiy. Upon graduation in 1977 he became a test pilot at the Tupolev design bureau, moving to the Flight Research Institute in 1985. In 1980 he completed a graduate degree at the Zhukovskiy branch of the Moscow Aviation Institute. Sheffer was selected as a member of the LII cosmonaut team in 1985 and underwent OKP, graduating in 1987. He conducted Buran-related flight testing, but in the end, no manned missions would take place. While the cosmonaut team dispersed, Yuriy Sheffer remained at LII. He retired as a cosmonaut in early 2001, but became a department head while continuing his duties as a test pilot. On 5 June 2001 he died of a heart attack while in his office.

Sovetskiye i rossiyskiye kosmonavty

Sokovykh, Anatoliy Mikhaylovich was born on 12 January 1944 in Skovorodino in the Amur Region. He graduated from the Kachinskoye Higher Military Aviation Pilot School in 1966, and subsequently served as a pilot in the former German Democratic Republic. In 1973 he enrolled in the Air Force's test pilot school in Akhtubinsk, graduating as a Test Pilot 3rd Class the following year. In 1978 he was selected as one of the pilots to form the GKNII Buran cosmonaut team. He underwent OKP in Star City in 1979 and 1980, qualifying as cosmonaut-tester. However, he didn't get involved in any Buran mission training. In 1985 he was involved in an accident for which he was blamed. As a consequence, he left the cosmonaut group, although he remained in Akhtubinsk, test flying for GKNII, until his retirement in 1999. In 1994 he was awarded the title of Merited Test Pilot of the Russian Federation.

Solovyov, Anatoliy Yakovlevich was born in Riga (Latvia) on 16 January 1948. He graduated from the Chernigov Higher Military Aviation Pilot School in 1972 and then served as a pilot in the Far East until 1976, when he was selected as a TsPK cosmonaut. He trained in Akhtubinsk to become a Test Pilot 3rd Class. After his OKP he returned to Akhtubinsk and became a Test Pilot 2nd Class. He was subsequently transferred to the space station training groups. His first spaceflight came in 1988 as commander of Soyuz TM-5, a Soviet–Bulgarian visiting mission to the Mir space station. He subsequently flew four more missions, all long-duration expeditions to Mir: Soyuz TM-9 (EO-6) in 1990, Soyuz TM-15 (EO-12) in 1992–1993, STS-71 (EO-19) in 1995, and Soyuz TM-26 (EO-24) in 1997–1998. For the third of these four, Solovyov and his flight engineer Gennadiy Strekalov were launched on the Space Shuttle Atlantis in 1995 on its first docking mission with the station under the joint Shuttle–Mir program. Solovyov was slated to fly the first expedition mission to the International Space Station, but when it became clear to him that not he, but American astronaut Bill Shepherd would be the commander, he declined the assignment. With five flights as commander, a flight experience of 651 days in orbit, and 16 EVAs totaling more than 82 hours under his belt, he refused to be subordinate to an American who had flown only three missions, none as commander, with a total of fewer than 19 days and no EVA experience at all. He reportedly was offered to become the commander of TsPK after Pyotr Klimuk's retirement but declined the offer. Solovyov has since retired from the Air Force and gone into business.

Stankyavichus, Rimantas Antanas-Antano was born in Mariyampole in Lithuania on 26 July 1944. In 1966 he graduated from the Chernigov Higher Military Aviation Pilot School and then served in Soviet Air Force units in Poland, Central Asia, and Egypt, where he flew 25 combat missions during the Six-Day Israeli–Arab war of 1967. In 1972 he was awarded the Order of the Red Star. The next year Stankyavichus left the Air Force and enrolled in the Ministry of the Aviation Industry test pilot school in Zhukovskiy, graduating in 1975. Three years later, he was selected as one of five LII pilots to undergo cosmonaut training for the Buran program. Stankyavichus became one of the principal members of the LII cosmonaut team, flying a total of 13 approach and landing test flights on BTS-002 between 1985 and 1988. He was also LII's prime candidate to occupy the co-pilot seat on Buran's first manned orbital mission. In 1989 he was assigned to a Soyuz mission to Mir in order to prepare him for his future Buran mission, but later that year he was dropped from the crew because of changes in the Mir flight schedule. Stankyavichus returned to test flying and was killed on 9 September 1990 when his Sukhoy Su-27 crashed during a demonstration flight in Italy.

Sultanov, Ural Nazibovich was born on 18 November 1948 in the village of Nikifarovo in Bashkiria. He attended the Kharkov Higher Military Aviation Pilot School, and after graduating in 1971 remained there as a pilot instructor. In 1983 he joined LII's cosmonaut group and undertook OKP from 1985 until 1987, when he was awarded his cosmonaut-tester certificate. From April until November 1988 Sultanov joined several other LII cosmonauts at Baykonur, flying MiG-25 jets and Tupolev Tu-154LL flying laboratories in Buran approach and landing profiles. When Buran made its only spaceflight, he was Magomed Tolboyev's back-up as MiG-25 chase pilot. At the same time, he continued his other test pilot duties at LII. Sultanov left LII in March 2002, becoming an Ilyushin Il-18 pilot for an aviation company named after Valentina Grizodubova, a famous female pilot in the Soviet Union. Currently, he is the deputy chief of Bashkir Airlines.

Titov, Vladimir Georgyevich was born on 1 January 1947 in Sretensk in the Chita Region. After graduating from secondary school, he enrolled in the Chernigov Higher Military Aviation Pilot School. Upon graduating he remained there as an instructor until 1974. From 1974 until his selection to the TsPK cosmonaut detachment in 1976, he served in the Seryogin Regiment, TsPK's air wing at Chkalovskiy air base near Star City. After selection in 1976 Titov and his fellow selectees were first sent to the Air Force test pilot school in Akhtubinsk to become Test Pilots 3rd Class. They underwent OKP from October 1977 until September 1978. When it became clear that Buran was suffering delays and there was a growing shortage of Soyuz commanders, some of the group were transferred to the Soyuz–Salyut program. Titov would fly four missions. His first, on Soyuz T-8 in April 1983, had to be cut short because his ship failed to dock with Salyut-7. The next attempt almost ended in disaster in September 1983 when his launch vehicle caught fire and exploded on the launch pad. Titov's Soyuz was pulled away to safety by the launch escape system with seconds to spare. Following this almost fatal accident, Titov and his flight engineer on both occasions, Gennadiy Strekalov, were separated. They were considered an unlucky crew. Paired with a new engineer, Musa Manarov, and with LII cosmonaut Anatoliy Levchenko in the third seat, Titov flew Soyuz TM-4 to the Mir station in 1987 for the first mission that would spend over a year in orbit: 365 days and 22 hours. His third and fourth missions were both on the Space Shuttle. He was a mission specialist on STS-63 in 1995, the dress rehearsal rendezvous mission with Mir. The Shuttle didn't dock with the station on that occasion, but on Titov's second shuttle mission, STS-86 in 1997, Atlantis did dock. During the docked phase, Titov and NASA astronaut Scott Parazynski performed a five-hour EVA. Following the flight, Titov retired from the cosmonaut detachment and the Russian Air Force to become Boeing's representative in Moscow. On his four missions, Vladimir Titov logged a total of 387 days in space. He performed four EVAs, totaling a little under 19 hours.

Tokarev, Valeriy Ivanovich was born on 29 October 1952 in Kapustin Yar, where his father was serving on the missile launch base. In 1969 he finished secondary school in Rostov and in 1973 he graduated from the Stavropol Higher Military Aviation Pilot School, after which he began service in the Air Force. In 1981 he began studying at the Air Force's test pilot school in Akhtubinsk, graduating in 1982 and becoming a test pilot at GKNII. Selected for the Buran program, Tokarev underwent OKP from 1989 until 1991, without interrupting his test pilot work. Like Leonid Kadenyuk, Valeriy Tokarev had set his mind on flying in space, if not on Buran, then on another spacecraft. When it became clear that Buran would not fly, he requested a transfer to the TsPK cosmonaut team. On 29 July 1997 the State Interdepartmental Commission agreed to the transfer and as early as December 1998, Tokarev was assigned as the Russian mission specialist for Space Shuttle mission STS-96. As such, he flew on Discovery to the International Space Station between 27 May and 6 June 1999. From August 2001 he trained as flight engineer for ISS Expedition 8, but then lost his prime crew assignment as a result of the crew reshuffling that took place after the February 2003 Columbia accident. Instead, he was reassigned as Expedition 8 back-up commander. His next assignment was as Soyuz commander and ISS flight engineer for Expedition 10, but when his commander Bill McArthur was temporarily grounded for medical reasons, the crew was replaced and moved down the line. Once McArthur was returned to flight status, he and Tokarev were teamed up again and assigned as back-up crew for Expedition 10. The two were eventually launched as the Expedition 12 crew from Baykonur on 30 September 2005. After a successful mission of almost 190 days, during which he performed two EVAs, Tokarev and his crew landed their Soyuz safely in Kazakhstan. In late 2006, he was preparing for new flight assignments.

Tolboyev, Magomed Omarovich was born on 20 January 1951 in the Dagestan Soviet autonomous republic in the Caucasus. In 1969 he enrolled in the Yeysk Higher Military Aviation Pilot School, from which he graduated in 1973. After that he served in Air Force units until 1980. In 1976 he tried to get selected as a cosmonaut in the TsPK cosmonaut detachment but didn't pass the medical commission because of a spine trauma he had suffered in an accident with a Sukhoy Su-7B. He made another attempt in 1979 but again failed the medical commission, this time because of a foot injury that was the result of another ejection from a plane. In 1980 Tolboyev left the Air Force and enrolled in the Ministry of the Aviation Industry test pilot school in Zhukovskiy. Upon graduation in 1981 he began working for LII. In 1983 he was selected to the LII cosmonaut group for Buran, and the following year he graduated from the Zhukovskiy branch of the Moscow Aviation Institute. Tolboyev followed the OKP basic cosmonaut training course between 1985 and 1987, qualifying as cosmonaut-tester. From April until November 1988 Tolboyev flew Buran approach and landing profiles on the Tupolev Tu-154LL flying laboratory and on MiG-25 jets. He acted as a chase plane pilot during the launch and landing of Buran on its one and only mission on 15 November 1988. His MiG-25 is the one visible in the well-known video of the orbiter's roll-out after landing. For some time he was slated to become Igor Volk's co-pilot on Buran's first manned mission. After resigning from LII, Tolboyev entered politics in 1994, becoming a representative for the Republic of Dagestan in the State Duma. In 1997 he became president of the biennial MAKS air show in Zhukovskiy.

Tresvyatskiy, Sergey Nikolayevich was born in the town of Nizhne-Udinsk in the Irkutsk Region on 6 May 1954. After graduating from secondary school in 1971, he enrolled in the Kachinskoye Higher Military Aviation Pilot School. There his future colleague Yuriy Sheffer was one of his instructors and Yuriy Prikhodko was a classmate. From 1975 until 1980 he was assigned as a pilot in the former German Democratic Republic and later he was stationed in the Far East. In 1981 Tresvyatskiy left the Air Force and enrolled in the Ministry of the Aviation Industry test pilot school in Zhukovskiy. Upon graduation in 1983, he became a test pilot at the Flight Research Institute. In 1985 he graduated from the Zhukovskiy branch of the Moscow Aviation Institute, where he had attended evening classes. That same year he was included in the LII cosmonaut team that was preparing for spaceflights on Buran. He followed the OKP training at TsPK from 1985 until 1987, when he received the qualification of cosmonaut-tester. From April until November 1988 Tresvyatskiy worked at the Baykonur cosmodrome, rehearsing pre-landing maneuvers, runway approaches, and landings using the Tupolev Tu-154LL and MiG-25 jet aircraft. While it became clear that the Buran program was dying, Sergey Tresvyatskiy stayed on as a test pilot at LII. Besides that, he also performed demonstration flights on the MiG-29. He became world famous when his MiG collided in mid-air with his colleague's during an air show at RAF Fairford on 24 July 1993. Both pilots were able to eject and survived unharmed. Especially Tresvyatskiy's cool reaction after he had landed was the talk of the town. After he had removed his parachute, he simply lit up a cigarette and walked away to meet the emergency services that were hurrying toward the place where he had come down. Tresvyatskiy later was placed in charge of LII's Ramenskoye airfield in Zhukovskiy. In 2004 he became the last of the Buran pilots to leave LII and went on to become general director of the Samara Scientific and Technical Complex (the former Kuznetsov design bureau), which develops the NK series of aircraft and rocket engines.

Vasyutin, Vladimir Vladimirovich was born on 8 March 1952 in Kharkov in the Ukraine. He studied at the Kharkov Higher Air Force School from 1970 until 1974, and upon graduation was assigned as pilot instructor there until he was selected as a TsPK cosmonaut candidate in 1976. He was first sent to the test pilot school in Akhtubinsk to become a Test Pilot 3rd Class. Subsequently, he took the basic cosmonaut training course at TsPK to qualify him as a cosmonaut-tester. Unlike the other cosmonauts from his group, Vasyutin and Vladimir Titov were not sent back to Akhtubinsk for further test pilot training, but were transferred to the Soyuz–Salyut program. After three back-up assignments (Soyuz T-7, Soyuz T-10, and Soyuz T-12), Vasyutin was finally launched aboard Soyuz T-14 in September 1985. However, after a few weeks aboard the Salyut-7 space station Vasyutin became ill and, when treatment on board didn't help, Mission Control was forced to cut his mission short. Upon his return, Vasyutin was hospitalized, but, although he fully recovered, he never flew in space again. He retired from the cosmonaut team and went back to the Air Force, where he rose to the rank of Lieutenant-General and became Deputy Commander of the Gagarin Air Force Academy in Monino. Vasyutin passed away on 19 July 2002.

Viktorenko, Aleksandr Stepanovich was born on 29 March 1947 in Olginka in Kazakhstan. He graduated from the Orenburg Higher Aviation Pilot School in 1969, after which he served as an Ilyushin Il-28 pilot in the Baltic Fleet. In May 1978 Viktorenko (representing the Navy) was one of two additional candidates selected to join the 1976 Buran cosmonaut group of TsPK. Like the 1976 candidates, he was first sent to Akhtubinsk in order to qualify as a military test pilot. He graduated as Test Pilot 3rd Class in July 1979 and then went on to OKP basic cosmonaut training, finishing that in February 1982. Shortage of qualified commanders for Soyuz and delays in the Buran program then led to the decision to transfer all members of the TsPK Buran team to the Soyuz–Salyut training group. Eventually, Viktorenko would fly four missions to the Mir space station (Soyuz TM-3, Soyuz TM-8, Soyuz TM-14, Soyuz TM-20), logging a total of 489 days in orbit. He also spent almost 18 hours outside the spacecraft during six EVAs. In July 1997 Viktorenko retired from both the cosmonaut team and the Air Force.

Volk, Igor Petrovich was born in the city of Zmiyev in the Kharkov Region in the Ukraine on 12 April 1937. He attended the Military Aviation Pilot School in Kirovograd in the Ukraine, graduating as a bomber pilot. He was then stationed in Baku where he flew Tupolev Tu-16 and Ilyushin Il-28 bombers. In 1962 Volk left the Air Force and from 1963 underwent test pilot training at the Ministry of the Aviation Industry test pilot school in Zhukovskiy, graduating in 1965. Volk's first involvement in the manned space program was in that same year, when he was a pilot on the Tupolev Tu-104 in which cosmonauts Pavel Belyayev and Aleksey Leonov flew parabolic flights to train for Leonov's EVA on Voskhod-2. In 1969 he graduated from the Zhukovskiy branch of the Moscow Aviation Institute, and in 1975 he was a member of the examination board that passed his future crewmate Svetlana Savitskaya as a test pilot. In 1976 Volk performed a brief flight on the 105.11 atmospheric test bed of the Spiral spaceplane. In 1977 Volk was selected as one of the pilots to undergo cosmonaut training in preparation for the initial test flight program for Buran. He flew 12 times on Buran's analog BTS-002. Volk was the designated commander for Buran's first manned orbital mission. In order to get the mandatory spaceflight experience and see what the influence of spaceflight on his piloting skills would be, he made a spaceflight on Soyuz T-12 in July 1984. Igor Volk became a deputy head of LII in 1995, but left the institute in February 2002 for a position in private industry. He lives in Moscow.

Volkov, Aleksandr Aleksandrovich was born on 27 April 1948 in Gorlovka in the Donetsk Region in the Ukraine. He enrolled in the Kharkov Higher Military Aviation Pilot School in 1966, and, upon graduation in 1970, he stayed as an instructor until he was selected as a TsPK cosmonaut candidate in 1976. After his selection Volkov and the other eight candidates were sent to Akhtubinsk to qualify as Test Pilots 3rd Class. After finishing OKP in September 1978, Volkov returned to Akhtubinsk and in 1981 he became a Test Pilot 2nd Class. After that he was transferred to the Soyuz–Salyut program. Volkov subsequently made three spaceflights, one to Salyut-7 in 1985 (Soyuz T-14) and two to Mir in 1988–1989 (Soyuz TM-7) and 1991–1992 (Soyuz TM-13). In total he logged 391 days in orbit. He also conducted two EVAs, spending more than 10 hours outside the spacecraft. Volkov has held various management positions at TsPK, among them commander of the cosmonaut team from 1990 until 1998.

Yablontsev, Aleksandr Nikolayevich was born on 3 April 1955 in Warsaw (Poland) where his father was stationed. After graduating from secondary school in 1972, he enrolled in the Armavir Higher Military Aviation Pilot School, where he studied until graduating in 1976. He subsequently served in various Air Force units before entering the Air Force's test pilot school in Akhtubinsk. After graduation in 1985 he worked there as a test pilot. In January 1989 the State Interdepartmental Commission confirmed Yablontsev and five colleagues as cosmonauts in the GKNII team. They underwent OKP in Star City without interrupting their test flight work, and in 1991 they passed their exams, qualifying as cosmonaut-testers. In 1989 he had also successfully concluded a course at the Akhtubinsk branch of the Moscow Aviation Institute. In September 1996 TsPK's commander Pyotr Klimuk offered him a transfer to the TsPK cosmonaut team, but Yablontsev declined, feeling that flying on Soyuz wasn't really flying. In 1997 he retired from the Air Force and, like Nikolay Pushenko, became a civilian test pilot at the State Research Institute for Civil Aviation.

Zabolotskiy, Viktor Vasilyevich was born on 19 April 1946 in Moscow. He began flying in the First Moscow Flying Club in 1964, and went on to study for two years (1967–1969) in DOSAAF's Central Joint Flying and Technical School. After graduating, he worked in Kaluga as a pilot instructor on the MiG-15 and MiG-17. In 1973 he was admitted to the test pilot school in Zhukovskiy. He graduated two years later and began working as a test pilot for LII. In 1981 Zabolotskiy graduated from the Academy of Civil Aviation in Leningrad. He joined the LII Buran team in 1984 and underwent OKP in Star City between 1985 and 1987. In 1988 Zabolotskiy was scheduled to fly a MiG-25 chase plane during the landing phase of Buran's only spaceflight, but this assignment was canceled when he was assigned to the commission that investigated the Su-26 crash that had killed Aleksandr Shchukin. In 1989 Zabolotskiy became a Merited Test Pilot of the Soviet Union. It was also in that period that he was assigned as Rimantas Stankyavichus's backup for a familiarization spaceflight on a Soyuz, but this flight would never take place. Viktor Zabolotskiy left LII in late 1996 to become a test pilot for the Khrunichev Center and also became head of the Russian Federation for Aviation Amateurs.

Appendix C

List of Russian abbreviations

ABO — *Agregatno-Bytovoy Otsek* (Aggregate/Habitation Compartment: aft compartment of Kliper)

AO — *Agregatnyy Otsek* (Aggregate Compartment: Buran lower deck)

APAS — *Androginnyy Periferiynyy Agregat Stykovki* (Androgynous Peripheral Docking System)

BB — *Bazovyy Blok* (Base Unit: pod housing orbital maneuvering engines and tanks for Buran's propulsion system)

BDP — *Blok Dopolnitelnykh Priborov* (Unit for Additional Instruments: payload on first Buran mission)

BDU — *Blok Dvigateley Upravleniya* (Control Engine Unit: Buran unit housing thrusters and verniers)

BO — *Bytovoy Otsek* (Habitation Compartment: Buran mid-deck)

BOR — *Bespilotnyy Orbitalnyy Raketoplan* (Unmanned Orbital Rocket Plane: scale models of the Spiral spaceplane and Buran flown on suborbital and orbital missions)

BTS (002) — *Bolshoy Transportnyy Samolyot* (Big Transport Plane: full-scale Buran model for atmosheric tests)

DO — *Dvigatel Orientatsii* (Orientation Engine: Buran vernier)

DOM — *Dvigatel Orbitalnogo Manevrirovaniya* (Orbital Maneuvering Engine: Buran engine for on-orbit maneuvers)

DOS — *Dolgovremennaya Orbitalnaya Stantsiya* (Long-Term Orbital Station: the civilian space stations officially launched as Salyut-1, 4, 6, 7 and Mir)

DOSAAF — *Dobrovolnoe Obshchestvo Sodeystviya Armii, Aviatsii i Flotu* (Voluntary Society of Assistance to the Army, the Air Force and the Navy: Soviet paramilitary society)

DP — *Dalniy Planiruyushchiy* (Long-Distance Glider: Tupolev unmanned glider)

EKA — *Eksperimentalnyy Kosmicheskiy Apparat* (Experimental Space Apparatus: prototype satellites for launch by Buran)

EMZ — *Eksperimentalnyy Mashinostroitelnyy Zavod* (Experimental Machine Building Factory: one of the organizations making up NPO Molniya, situated in Zhukovskiy)

EO — *Ekspeditsiya Osnovnaya* (Main Expedition: space station resident crew)

EO — *Ekstrennoye Otdeleniye* (Emergency Separation: Buran launch abort profile)

EPOS — *Eksperimentalnyy Pilotiruyemyy Orbitalnyy Samolyot* (Experimental Piloted Orbital Plane: orbital and atmospheric test beds for the Spiral spaceplane)

FGB — *Funktsionalno-Gruzovoy Blok* (Functional Cargo Block: main part of the TKS spacecraft)

FSB — *Funktsionalno-Sluzhebnyy Blok* (Functional Service Block: modified FGB section for the Skif-DM/Polyus spacecraft)

GDL — *Gazodinamicheskaya Laboratoriya* (Gas Dynamics Laboratory: rocket research group 1921–1933)

GIRD — *Gruppa Izucheniya Reaktivnogo Dvizheniya* (Group for the Investigation of Reactive Motion: rocket research group 1931–1933)

GKNII — *Gosudarstvennyy Krasnoznamennyy Nauchno-Ispytatelnyy Institut* (State Red Banner Scientific Test Institute: Air Force test and training site in Akhtubinsk)

GLI — *Gorizontalnye Lyotnye Ispytaniya* (Horizontal Flight Tests: the approach and landing tests flown with the BTS-002 vehicle)

GMK — *Glavnaya Meditsinskaya Komissiya* (Chief Medical Commission: medical board for cosmonaut selection)

GMVK — *Gosudarstvennaya Mezhvedomstvennaya Komissiya* (State Interdepartmental Commission: leading body for cosmonaut selection)

GOR — *Gruppa Operativnogo Rukovodstva* (Operations Control Group: management body for Energiya–Buran program)

GSP — *Girostabilizirovannaya Platforma* (Gyro-Stabilized Platform: Buran inertial measurement unit)

GSR — *Giperzvukovoy Samolyot-Razgonshchik* (Hypersonic Boost Aircraft: hypersonic aircraft of the Spiral system)

GTA-S — *Gruzovoy Transportnyy Apparat Snabzheniya* (Cargo Transport Supply Ship: space tug for 75-ton space station modules)

GTK — *Gruzovoy Transportnyy Konteyner* (Cargo Transport Container: payload container strapped to the side of Energiya)

GUKOS	*Glavnoye Upravleniye Kosmicheskikh Sredstv* (Chief Directorate of Space Assets: the "space branch" of the Strategic Rocket Forces, forerunner of the Military Space Forces)
IMBP	*Institut Mediko-Biologicheskikh Problem* (Institute of Medical and Biological Problems: space medicine institute in Moscow)
IPM	*Institut Prikladnoy Matematiki* (Institute of Applied Mathematics, Moscow)
IPMekh	*Institut Problem Mekhaniki* (Institute of Mechanical Problems, Moscow)
IPRIM	*Institut Prikladnoy Mekhaniki* (Institute of Applied Mechanics, Moscow)
ISZhO	*Individualnye Sredstva Zhizneobespecheniya* (Personal Life Support System: part of Buran life support system)
ITPM	*Institut Teoreticheskoy i Prikladnoy Mekhaniki* (Institute of Theoretical and Applied Mechanics, Novosibirsk)
KBKhA	*Konstruktorskoye Byuro Khimicheskoy Avtomatiki* (Design Bureau of Chemical Automatics: bureau in charge of developing Energiya's RD-0120 engine, situated in Voronezh)
KBOM	*Konstruktorskoye Byuro Obshchego Mashinostroyeniya* (Design Bureau of General Machine Building: launch pad design bureau in Moscow)
KBTM	*Konstruktorskoye Byuro Transportnogo Mashinostroyeniya* (Design Bureau of Transport Machine Building: launch pad design bureau in Moscow)
KhChF	*Khvostovaya Chast Fyuzelyazha* (aft fuselage)
KIK	*Komandno-Izmeritelnyy Kompleks* (Command and Measurement Complex: Soviet tracking station network)
KO	*Komandnyy Otsek* (Command Compartment: Buran flight deck)
KTOK	*Korpus Trenazhorov Orbitalnogo Korablya* (Orbiter Simulator Building: building with Buran simulators at Star City)
KV	*Kosmicheskiye Voyska* (Space Troops)
KVRB	*Kislorodno-Vodorodnyy Raketnyy Blok* (Oxygen/Hydrogen Rocket Stage: cryogenic upper stage)
LII	*Lyotno-Issledovatelskiy Institut* (Flight Research Institute: civilian flight test center in Zhukovskiy)
LKS	*Lyogkiy Kosmicheskiy Samolyot* (Light Spaceplane: Chelomey spaceplane)
LL	*Letayushchaya Laboratoriya* (Flying Laboratory: aircraft used for training Buran test pilots)
LO	*Laboratornyy Otsek* ("Laboratory Compartment": Spacelab-type payload for Buran)
LOK	*Lunnyy Orbitalnyy Korabl* (Lunar Orbital Ship: Soyuz-derived vehicle for the N-1/L-3 piloted lunar-landing program)

MAKS	*Mnogotselevaya Aviatsionno-Kosmicheskaya Sistema* (Multipurpose Aerospace System: air-launched spaceplane system developed by NPO Molniya)
MAP	*Ministerstvo Aviatsionnoy Promyshlennosti* (Ministry of the Aviation Industry)
MEK	*Mezhvedomstvennaya Ekspertnaya Komissiya* (Interdepartmental Expert Commission)
MIK OK	*Montazhno-Ispytatelnyy Korpus Orbitalnogo Korablya* (Orbiter Assembly and Test Facility: Buran processing facility at Baykonur)
MIK RN	*Montazhno-Ispytatelnyy Korpus Rakety-Nositelya* (Rocket Assembly and Test Facility: Energiya assembly building at Baykonur)
MKBS	*Mnogotselevaya Kosmicheskaya Baza Stantsiya* (Multipurpose Space Base Station: N-1 launched space station studied by TsKBEM in the late 1960s/early 1970s)
MKR	*Mnogorazovyy Kosmicheskiy Raketoplan* (Reusable Space Rocket Plane: single-stage-to-orbit spaceplane studied by RKK Energiya)
MKS	*Mnogorazovaya Kosmicheskaya Sistema* (Reusable Space System: the official name of the Energiya–Buran system and its cosmodrome infrastructure)
MOK	*Mnogorazovyy Orbitalnyy Korabl* (Reusable Orbital Ship: spaceplanes studied in the 1990s as part of two-stage-to-orbit systems)
MOK	*Mnogotselevoy Orbitalnyy Kompleks* (Multipurpose Orbital Complex: constellation of space stations, spacecraft, and satellites studied by TsKBEM in the late 1960s/early 1970s)
MOM	*Ministerstvo Obshchego Mashinostroyeniya* (Ministry of General Machine Building: Soviet "space and missile ministry")
MRKK	*Mnogorazovyy Raketno-Kosmicheskiy Kompleks* (Reusable Rocket Space Complex: name for the combination of Energiya rocket and Buran orbiter)
MTKVP	*Mnogorazovyy Transportnyy Korabl Vertikalnoy Posadki* (Reusable Vertical Landing Transport Ship: lifting body studied by NPO Energiya in the mid-1970s)
MV	*Manyovr Vozvrashcheniya* (Return Maneuver: Buran launch abort profile)
MVKS	*Mezhvedomstvennyy Koordinatsionnyy Sovet* (Interdepartmental Coordinating Council: leading body overseeing Energiya–Buran program)
MZK	*Montazhno-Zapravochnyy Korpus* (Assembly and Fueling Facility: hangar at Baykonur for hazardous operations)
NChF	*Nosovaya Chast Fyuzelyazha* (forward fuselage)

NII *Nauchno-Issledovatelskiy Institut* (Scientific Research Institute)

NIIkhimmash *Nauchno-Issledovatelskiy Institut Khimicheskogo Mashinostroyeniya* (Scientific Research Institute of Chemical Machine Building: major rocket engine and spacecraft test center near Zagorsk/Sergiyev Posad)

NIIKhSM *Nauchno-Issledovatelskiy Institut Khimicheskikh i Stroitelnykh Mashin* (Scientific Research Institute of Chemical and Building Machines: test center in Zagorsk/Sergiyev Posad)

NIIMash *Nauchno-Issledovatelskiy Institut Mashinostroyeniya* (Scientific Research Institute of Machine Building: rocket engine test center near Nizhnyaya Salda)

NII TP *Nauchno-Issledovatelskiy Institut Teplovykh Protsessov* (Scientific Research Institute of Thermal Processes: the name of the current Keldysh Research Center between 1965 and 1995. Originally RNII/NII-3/NII-1)

NIVS *Navigatsionnaya Izmeritelnaya Vizualnaya Sistema* (Visual Navigation Measurement System: Buran navigation aid)

NPG *Neotdelyayemyy Poleznyy Gruz* (Undetachable Useful Payload: classified military payload for Buran)

NPO AP *Nauchno-Proizvodstvennoye Obyedineniye Avtomatiki i Priborostroyeniya* (Scientific Production Association of Automatics and Instrument Building: design bureau in charge of Buran computers and software, situated in Moscow)

NPO PM *Nauchno-Proizvodstvennoye Obyedineniye Prikladnoy Mekhaniki* (Scientific Production Association of Applied Mechanics: leading design bureau of communications and navigation satellites near Krasnoyarsk)

ODU *Obyedinyonnaya Dvigatelnaya Ustanovka* (Combined Engine Installation: general name for the Buran propulsion system)

OE *Otsek Energeticheskiy* (Energy Compartment: section of Skif-DM/Polyus)

OK *Orbitalnyy Korabl* (Orbital Ship: general name for Russian shuttle orbiters)

OKB *Opytno-Konstruktorskoye Byuro* (Experimental Design Bureau)

OKP *Obshchekosmicheskaya Podgotovka* (General Space Training: basic training course for cosmonauts at Star City)

OKPD *Obyedinyonnyy Komandno-Dispetcherskiy Punkt* (Combined Command and Air Control Center: control building near Baykonur runway)

OKPKI *Otraslevoy Kompleks Podgotovki Kosmonavtov-Ispytateley* (Departmental Training Complex for Cosmonaut-Testers: cosmonaut training center of the Flight Research Institute)

ONA *Ostronapravlennaya Antenna* (Narrow-Beam Antenna: Buran high-gain antenna)

ORT *Otsek Rabochikh Tel* (Gas Compartment: section of Skif-DM/Polyus)

OS *Orbitalnyy Samolyot* (Spaceplane: orbital element of the Spiral system)

OSA *Otsek Spetsialnoy Apparatury* (Special Equipment Compartment: section of Skif-DM/Polyus)

OT *Odnovitkovaya Trayektoriya* (Single-Orbit Trajectory: Buran abort profile)

OZEM *Opytnyy Zavod Energeticheskogo Mashinostroyeniya* (Experimental Factory of Energetic Machine Building: production facility aligned with KB Energomash in Khimki)

PDST *Pilotazhno-Dinamicheskiy Stend-Trenazhor* (Piloting Dynamic Test Stand/Simulator: Buran simulator at NPO Molniya)

PKA *Planiruyushchiy Kosmicheskiy Apparat* (Gliding Space Apparatus: spaceplane studied by Pavel Tsybin in the late 1950s)

PMV *Postroitel Mestnoy Vertikali* (Local Vertical Sensor: Buran navigation aid)

PRSO *Polnorazmernyy Stend Oborudovaniya* (Full-Scale Equipment Test Stand: Buran simulator at NPO Molniya)

PRZS *Pribor Registratsii Voskhoda i Zakhoda Solntsa* (Sunrise/Sunset Detection Instrument: Buran navigation aid)

PSS *Pilotazhno-Staticheskiy Stend* (Piloting Static Test Stand: Buran test stand at NPO Molniya)

RD *Raketnyy Dvigatel* (Rocket Engine: designator for Soviet rocket engines)

RDS *Radiodalnomernaya Sistema* (Radio Rangefinder System: Buran landing navigation aid)

RKA *Rossiyskoye Kosmicheskoye Agentstvo* (Russian Space Agency: official name of the agency from 1992 to 1999, later RAKA and FKA)

RLA *Raketno-Letatelnyy Apparat* (Rocket Flying Apparatus: family of heavy-lift launch vehicles studied by NPO Energiya in the mid-1970s)

RM *Rabochee Mesto* (Workstation: workstation in Buran cockpit)

RMS *Radiomayachnaya Sistema Posadki* (Radio Beacon Landing System: Buran landing navigation aid)

RNII *Reaktivnyy Nauchno-Issledovatelskiy Institut* (Reactive Scientific Research Institute: rocket research group in the 1930s)

ROS *Radiotekhnicheskaya Orbitalnaya Stantsiya* (Radiotechnical Orbital Station: man-tended space station studied by NPO Energiya in the 1970s/1980s)

RSBN *Radiotekhnicheskaya Sistema Blizhney Navigatsii* (Radiotechnical Short-Range Navigation System: Buran landing navigation aid)

RSU *Reaktivnaya Sistema Upravleniya* (Reaction Control System: Buran thrusters and verniers)

RVB *Radiovysotomer Bolshikh Vysot* (High-Altitude Radio Altimeter: Buran navigation aid)

RVM *Radiovysotomer Malykh Vysot* (Low-Altitude Radio Altimeter: Buran navigation aid)

RVSN *Raketnye Voyska Strategicheskogo Naznacheniya* (Strategic Rocket Forces: branch of the armed forces responsible for missile launches)

RVV *Radiovysotomer-Vertikal* (Vertical Radio Altimeter: Buran navigation aid)

SBI *Sistema Vzaimnykh Izmereniy* (Mutual Measurement System: Buran navigation aid)

SBM *Sistema Bortovykh Manipulyatorov* (On-Board Manipulator System: Buran mechanical arm)

SChF *Srednyaya Chast Fyuzelyazha* (mid fuselage)

SDI *Stend Dinamicheskikh Ispytaniy* (Dynamic Test Stand: vibration test facility at Baykonur)

SEP *Sistema Elektropitaniya* (Electric Power System: Buran power supply system)

SGK *Sovet Glavnykh Konstruktorov* (Council of Chief Designers: coordinating group made up of chief designers and other space officials)

SGS *Sistema Gazovogo Sostava* (Gas Composition System: part of Buran life support system)

ShLI *Shkola Lyotchikov-Ispytateley* (Test Pilot School: test pilot school aligned with LII in Zhukovskiy)

SibNIA *Sibirskiy Nauchno-Issledovatelskiy Institut Aviatsii* (Siberian Scientific Research Institute of Aviation, situated in Novosibirsk)

SM *Stykovochnyy Modul* (Docking Module: Buran docking adapter and external airlock)

SNiR *Sistema Nadduva i Razgermetizatsii* (Pressurization and Depressurization System: part of Buran life support system)

SNVP *Sistema Nadduva i Ventilyatsii Planera* (Airframe Pressurization and Ventilation System: Buran ventilation system)

SPK *Sredstvo Peredvizheniya Kosmonavta* (Cosmonaut Maneuvering Unit: jet-powered backpack for spacewalks)

SPV *Sistema Pityevoy Vody* (Potable Water System: part of Buran life support system)

STR *Sistema Termoregulirovaniya* (Thermal Control System: part of Buran environmental control system)

STV *Sistema Tekhnicheskoy Vody* (Process Water System: part of Buran life support system)

SVO *Sistema Vodoobespecheniya* (Water Supply System: part of Buran life support system)

SVSP *Sistema Vysotno-Skorostnykh Parametrov* (Altitude/Velocity Parameter System: Buran landing navigation aid)

TK *Tekhnicheskiy Kompleks* (Technical Zone: area of assembly and support buildings at Baykonur)

TKS *Transportnyy Korabl Snabzheniya* (Transport Supply Ship: 20-ton crew and cargo vehicle for Almaz military space station, eventually only flown to civilian Salyut-6 and Salyut-7 space stations)

TMP *Tekhnologicheskiy Modul Proizvodstva* (Technological Production Module: 100-ton Energiya-launched space factory)

TMZ *Tushinskiy Mashinostroitelnyy Zavod* (Tushino Machine Building Factory: main production facility of NPO Molniya in Tushino)

TsAGI *Tsentralnyy Aerogidrodinamicheskiy Institut* (Central Aerohydrodynamics Institute: leading R&D institute of the Ministry of the Aviation Industry in Zhukovskiy)

TsIAM *Tsentralnyy Institut Aviatsionnogo Motorostroyeniya* (Central Institute of Aviation Engine Building, situated in Moscow)

TsKBEM *Tsentralnoye Konstruktorskoye Byuro Eksperimentalnogo Mashinostroyeniya* (Central Design Bureau of Experimental Machine Building: the name of the "Korolyov bureau" between 1965 and 1974)

TsKBM *Tsentralnoye Konstruktorskoye Byuro Mashinostroyeniya* (Central Design Bureau of Machine Building: the name of the "Chelomey bureau" from 1965 to 1983)

TsM *Tselevoy Modul* (Payload Module: payload section of Skif-DM/Polyus)

TsNII-50 *Tsentralnyy Nauchno-Issledovatelskiy Institut 50* (Central Scientific Research Institute 50: military space R&D institute in Yubileynyy, Moscow area)

TsNIIMash *Tsentralnyy Nauchno-Issledovatelskiy Institut Mashinostroyeniya* (Central Scientific Research Institute of Machine Building: leading space R&D institute of the Ministry of General Machine Building in Kaliningrad/Korolyov)

TsNII RTK *Tsentralnyy Nauchno-Issledovatelskiy Institut Robototekhniki i Tekhnicheskoy Kibernetiki* (Central Scientific Research Institute of Robotic Technology and Technical Cybernetics: design bureau in charge of Buran's mechanical arms, situated in Leningrad/St. Petersburg)

TsPK *Tsentr Podgotovki Kosmonavtov* (Cosmonaut Training Center: "Star City" near Moscow)

TsPLI *Tsentr Podgotovki Lyotchikov-Ispytateley* (Test Pilot Training Center: test pilot school aligned with GKNII in Akhtubinsk)

TsSKB *Tsentralnoye Spetsializirovannoye Konstruktorskoye Byuro* (Central Specialized Design Bureau: design bureau for Soyuz rockets and spy satellites in Kuybyshev/Samara)

TsUP *Tsentr Upravleniya Polyotom* (Mission Control Center in Kaliningrad/Korolyov)

TsVK *Tsilindr Vyverki Kursa* (Heading Alignment Cylinder: imaginary cylinder helping Buran to line up with the runway)

TUA *Transportnyy Ustanovochnyy Agregat* (Transportation and Erection Aggregate: crawler transporter for Energiya at Baykonur)

TZP *Teplozashchitnoye Pokrytiye* (Thermal Protection System: Buran heat shield)

UD *Upravlyayushchiy Dvigatel* (Control Engine: Buran thruster)

UER *Upravleniye Eksperimentalnoy Raboty* (Directorate of Experimental Work: directorate of MOM in charge of Energiya–Buran)

UKP *Universalnaya Kosmicheskaya Platforma* (Universal Space Platform: "bus" for large Energiya-launched payloads)

UKSS *Universalnyy Kompleks Stend-Start* (Universal Test Stand and Launch Pad: combined test-firing stand and launch pad for Energiya at Baykonur)

UNKS *Upravleniye Nachalnika Kosmicheskikh Sredstv* (Directorate of the Commander of Space Assets: name of the Military Space Forces between 1986 and 1992)

URKTS *Universalnaya Raketno-Kosmicheskaya Transportnaya Sistema* (Universal Rocket and Space Transportation System: general name for the Energiya rocket family)

VA *Vozvrashchayemyy Apparat* (Return Apparatus: return capsule of Kliper)

VDK *Vizir-Dalnomer Kosmonavta* (Cosmonaut Visual Rangefinder: Buran navigation aid)

VIAM *Vsesoyuznyy Institut Aviatsionnykh Materialov* (All-Union Institute of Aviation Materials, situated in Moscow)

VKK *Vozdushno-Kosmicheskiy Korabl* (aerospace ship: general word for winged spacecraft, usually used for single-stage-to-orbit spaceplanes)

VKS *Voyenno-Kosmicheskiye Sily* (Military Space Forces)

VMK *Vozvrashchaemaya Manevriruyushchaya Kapsula* (Recoverable Maneuverable Capsule: lifting body studied by NPO/RKK Energiya in the 1990s)

VM-T Modified Myasishchev bomber for transporting Buran and elements of Energiya's core stage

VMZ *Voronezhskiy Mashinostroitelnyy Zavod* (Voronezh Machine Building Factory: production facility aligned with the KBKhA design bureau in Voronezh)

VNIIRA	*Vsesoyuznyy Nauchno-Issledovatelskiy Institut Radioapparatury* (All-Union Scientific Research Institute of Radio Equipment: design bureau in charge of the Vympel landing navigation system)
VPK	*Voyenno-Promyshlennaya Komissiya* (Military Industrial Commission: leading government body for implementation of space policy)
VSU	*Vspomogatelnaya Silovaya Ustanovka* (Auxiliary Power Unit)
VTB	*Vneshniy Toplivnyy Bak* (External Tank: external tank for the MAKS air-launched spaceplane)
ZEM	*Zavod Eksperimentalnogo Mashinostroyeniya* (Factory of Experimental Machine Building: production facility aligned with NPO Energiya)
ZSP	*Zvyozdno-Solnechnyy Pribor* (Stellar–Solar Instrument: Buran navigation aid)

Appendix D

Bibliographical notes and selected bibliography

BIBLIOGRAPHICAL NOTES

Russian-language sources

Any researcher wishing to gain a good understanding of the Energiya–Buran program will inevitably have to consult Russian-language sources. The bulk of the present book is based on the Russian sources listed below.

Dedicated books

Dedicated books on Energiya–Buran did not start appearing until well after the cancellation of the program. The first substantial work in Russian on Buran was *Mnogorazovyy orbitalnyy korabl Buran* ("Reusable Orbiter Buran") (Y. Semyonov *et al.*, 1995). Authored by several leading designers of Buran, it is a detailed and indispensable nuts-and-bolts description of the Soviet shuttle orbiter, but virtually ignores the Energiya rocket. It also lacks historical and organizational background on the program, apparently because it was originally written in the late 1980s, at a time when much of that information was still considered sensitive. Unfortunately, the original text was not updated when the book finally appeared in 1995.

Another detailed technical description of the orbiter can be found in *Aviatsionno-kosmicheskiye sistemy* ("Aerospace Systems") (G. Lozino-Lozinskiy, A. Bratukhin, eds., 1997), which contains over 40 specialized technical articles on Buran and about 20 on the MAKS air-launched spaceplane. Written by specialists of NPO Molniya, the articles deal only with those elements of Buran that NPO Molniya was involved in. Readers looking for historical background will again be largely disappointed.

Another landmark work on the Energiya–Buran program are the memoirs of Energiya–Buran chief designer Boris Gubanov. Called *Triumf i tragediya Energii* ("Triumph and Tragedy of Energiya") (1998–2000), they were published in four volumes, the last two of which are dedicated largely to Energiya–Buran. While Buran is given scarce attention, Gubanov's memoirs are a goldmine of information on the Energiya rocket and planned derivatives, filling a gap left by the aforementioned works. Gubanov covers both the historical and technical aspects of the Energiya program, clearly relying not only on memory, but on a wealth of documentary material as well. Another important memoir of the Energiya–Buran program is *Put k Energii* ("The Road to Energiya") (2001) by RKK Energiya's Vyacheslav Filin. Like Gubanov's work, it mainly concentrates on the development of the Energiya rocket, but is far less voluminous and more anecdotal in style. Vladimir Trofimov, the first deputy director of NPO Energomash from 1974 to 1993, tracks the development of the Energiya rocket's RD-170 engine in his book *Osushchestvleniye mechty* ("Realization of a Dream") (2001).

Compensating for the lack of decent pictures in these books is a coffee table photo album called *Mnogorazovaya kosmicheskaya sistema Energiya-Buran* ("The Reusable Space System Energiya–Buran") (A. Kuznetsov, 2004). This contains over 1,000 high-quality photographs not only of Energiya–Buran hardware, but also of the people and the ground facilities, most of them never published before. The accompanying text adds little new information, but particularly interesting for the historian is the near complete text of the 17 February 1976 government and party decree that gave the official go-ahead for the Energiya–Buran program. This is one of the very few primary documents on the program released thus far.

Notably absent so far in the Russian-language Buran bibliography are memoirs of the test pilots involved in the program and histories written by independent analysts. The best summary of our current knowledge of the program is given in *Mirovaya pilotiruyemaya kosmonavtika* ("World Piloted Spaceflight") (Y. Baturin, ed., 2005). Written by Russia's leading space historians, this is the most comprehensive and up-to-date history of the world's piloted space projects.

Organization histories

Many Russian space enterprises have published histories of their spacecraft and/or rocket programs. Although inevitably biased, they are an indispensable source of information nevertheless. RKK Energiya, the "prime contractor" for the Energiya–Buran program, published a copious history on the occasion of its 50th anniversary called *Raketno-kosmicheskaya korporatsiya Energiya imeni S.P. Korolyova 1946–1996* ("Rocket and Space Corporation Energiya Named after S.P. Korolyov") (Y. Semyonov, ed., 1996). This has over thirty pages of information on the Energiya–Buran program.

Specialists of the Experimental Machine Building Factory (EMZ) have put out a series of lavishly illustrated books covering the airplanes, missiles, and space vehicles designed under the leadership of Vladimir Myasishchev. One series, focusing on Myasishchev projects of the 1950s and 1960s, is called *Illyustrirovannaya entsiklopediya samolyotov OKB V.M. Myasishcheva* ("Illustrated Encyclopedia of Airplanes of the Experimental Design Bureau of Myasishchev"), and a second series is devoted to projects developed by Myasishchev's EMZ in the 1970s and 1980s and called *Illyustrirovannaya entsiklopediya samolyotov EMZ im. V.M. Myasishcheva* ("Illustrated Encyclopedia of Airplanes of the Experimental Machine Building Factory named after Myasishchev") (A. Bruk *et al.*, 1999–2001). One volume in the first series details early spaceplane concepts studied at Myasishchev's OKB-23 in the late 1950s/early 1960s, and one book in the second series elaborates in particular on the VM-T carrier aircraft used to transport Buran and elements of the Energiya rocket. More recently, several more books have come out in the second series, one covering Buran, but these were unfortunately not yet available to the authors when this book went to press.

The Chelomey design bureau published a history with the rather bombastic title *60 let samootverzhennogo truda vo imya mira* ("60 Years of Selfless Work in the Name of Peace") (G. Yefremov, ed., 2004). Published to mark the bureau's 60th anniversary, it offers disappointingly little new insight into Chelomey's spaceplane projects.

Valuable information on Energiya's strap-on booster rockets and on the Zenit rocket is provided in two histories of the Yuzhnoye design bureau titled *Rakety i kosmicheskiye apparaty konstruktorskogo byuro Yuzhnoye* ("Rockets and Space Apparatuses of the Yuzhnoye Design Bureau") (S. Konyukhov, ed., 2000) and *Prizvany vremenem: rakety i kosmicheskiye apparaty konstruktorskogo byuro Yuzhnoye* ("Summoned by the Times: Rockets and Space Apparatuses of the Yuzhnoye Design Bureau") (S. Konyukhov, ed., 2004). The engines of the Energiya rocket are described in detail in histories of NPO Energomash (*NPO Energomash imeni akademika V.P. Glushko: Put v raketnoy tekhnike*) ("NPO Energomash Named after Academician V.P. Glushko: A Road in Rocket Technology") (B. Katorgin, ed., 2004) and the Chemical Automatics Design Bureau (*KBKhA: stranitsy istorii*) ("KBKhA: Pages of History") (Voronezh, 1995).

Magazine articles

Before the first dedicated books were published, most of the revelations on early Soviet spaceplane projects and the Buran program came in newspaper and magazine articles in the early 1990s. Among these were aerospace magazines such as *Kosmonavtika, astronomiya* ("Spaceflight, Astronomy") (in the *Znaniye* series) and *Aviatsiya i kosmonavtika* ("Aviation and Spaceflight"). Many of these articles were translated into English at the time by the US Foreign Broadcast Information Service (FBIS) in *JPRS Report: USSR Space* and *JPRS Report: Eurasia Space*.

Russia's leading space magazine now is *Novosti kosmonavtiki* ("News of Cosmonautics"), considered by many to be the best spaceflight journal in the world. The magazine first appeared in 1991, at a time when the Energiya–Buran program was already in its death throes, but has carried several historical articles on the program throughout the years. Particularly helpful are a series of articles on the origins of Buran by Vadim Lukashevich published in 2006.

On-line sources

The richest on-line source of information on the Energiya–Buran program is the website of Russian spaceflight historian Vadim Lukashevich at *www.buran.ru*

This contains a wealth of information on all aspects of the Energiya–Buran program and also has extensive sections on the Spiral program (where Lukashevich has done ground-breaking research using primary documents) and MAKS. The website has an extensive bibliography of Soviet/Russian books and articles on Buran and other Soviet spaceplane projects, providing electronic versions of many of these. For instance, it has significant extracts from many of the Russian-language books mentioned above—namely, Semyonov (1995 and 1996), Lozino-Lozinskiy/Bratukhin (1997) (with English translations of most articles), Gubanov (1998–2000), and Filin (2001). This is a blessing for the researcher, since hard copies of most of these books and articles are extremely difficult to obtain. Also noteworthy are several personal recollections of the Energiya–Buran program that have never appeared in print before as well as a handful of primary documents. The site has a big photo gallery and also contains some excellent artwork produced by the webmaster himself, some examples of which he has kindly allowed to be reproduced in this book. All of the information, pictures, and artwork on the website plus additional material (including documentaries) are also available on a set of commercially available CD-ROMs.

English-language sources

The only book in English devoted solely to the Energiya–Buran program was Henry Matthews' *The Secret Story of the Soviet Space Shuttle* (Beirut, 1994), which provided a useful summary of the information then available, but unfortunately contained gaps and inaccuracies that were avoidable even at the time of writing. Most of the information on the Energiya–Buran program released by Russian sources in subsequent years has never been summarized in Western books. Asif Siddiqi's monumental *Challenge to Apollo* (2000), without question the most elaborate and best-documented history of the Soviet piloted space program to date, was the first book to give a detailed account of the Soviet spaceplane programs of the 1950s and 1960s. However, the Energiya–Buran program is beyond the book's scope, since it covers events only until 1974.

Good contemporary reports on the Energiya–Buran program can be found in the leading US trade magazine *Aviation Week & Space Technology*, in the British Interplanetary Society's *Spaceflight* magazine, the short-lived British space journal *Spaceflight News*, and Nicholas Johnson's annual *Soviet Year in Space* reports (published by Teledyne Brown Engineering in Colorado Springs). Unfortunately, surprisingly little was written on the program in subsequent years, exactly when most of the crucial information came out. A few exceptions are "Russian Space Shuttle Projects 1957–1994" by Peter Pesavento (*Spaceflight*, May/June/July/August 1995), "A Cold Snow Falls: The Soviet Buran Space Shuttle" by Stephen Garber (*Quest*, 5/2002), "The Origins and Evolution of the Energiya Rocket Family" (*Journal of the British Interplanetary Society*, July/August 2002), and "The Soviet BOR-4 Spaceplanes and Their Legacy" (*Journal of the British Interplanetary Society*, March 2007), both by Bart Hendrickx.

Researchers interested in US intelligence assessments of the Energiya–Buran program can find a handful of declassified CIA "National Intelligence Estimates" of the Soviet space program on the CIA Electronic Reading Room at *http://www.foia.cia.gov* and on Mark Wade's Encyclopedia Astronautica at *http://www.astronautix.com/data/index.htm* However, the number of declassified documents is still too small to form a good picture of what was really known. Public Defense Department assessments were given in the annual *Soviet Military Power* reports (1983–1991), most of which are now on-line on the website of the Federation of American Scientists at *http://www.fas.org/irp/dia/product/smp_index.htm*

Final note

This book is an attempt to synthesize the wealth of information released on Energiya–Buran in the past twenty years or so, but does not claim to be a definitive history of the program. While Russian books and articles have provided a good technical understanding of the system, key questions remain about such things as the motives behind the decision to go ahead with Buran in the mid-1970s, efforts to collect information on the US Space Shuttle, the financial aspects of the program, the payloads Buran was supposed to fly, and the eventual demise of the project. Acutely missing is primary source material, the holy grail of any historian. Government archives related to developments in the defense industry in the 1970s and 1980s are still classified and design bureau archives remain largely off-limits to researchers as well. Unfortunately, that situation is not likely to change anytime soon in an environment where the Russian government is again increasingly clamping down on the release of sensitive information. In light of the fact that Buran was to a large extent a military program, the future may not bode well for researchers wishing to delve deeper into the project's history.

SELECTED BIBLIOGRAPHY

Extensive bibliographical references can be found at the end of each chapter. Only a selection of important works used in the book is listed here.

Baturin, Y. (ed.), *Mirovaya pilotiruyemaya kosmonavtika*, Moscow: RTSoft, 2005.

Bruk, A., Udalov, K., Smirnov, S. *et al.*, *Illyustrirovannaya entsiklopediya samolyotov EMZ im. V.M. Myasishcheva (tom 3, chast 1)*, Moscow: Aviko Press, 1999.

Bruk, A., Udalov, K., Smirnov, S., Brezginova, N. *et al.*, *Illyustrirovannaya entsiklopediya samolyotov OKB V.M. Myasishcheva (tom 2, chast 1)*, Moscow: Aviko Press, 2001.

Favorskiy, V., Meshcheryakov, I., *Voyenno-kosmicheskiye sily, kniga 1*, Moscow: Izdatelstvo Sankt-Peterburgskoy tipografii, 1997.

Favorskiy, V., Meshcheryakov, I., *Voyenno-kosmicheskiye sily, kniga 2*, Moscow: Izdatelstvo Sankt-Peterburgskoy tipografii, 1998.

Filin, V., *Put k Energii*, Moscow: Logos, 2001.

Gladkiy, V., "How the Energiya–Buran project was born" (in Russian), *Aviatsiya i kosmonavtika*, 1/2002.

Gubanov, B., *Triumf i tragediya Energii (tom 3: Energiya-Buran)*, Nizhniy Novgorod: Izdatelstvo Nizhegorodskogo instituta ekonomicheskogo razvitiya, 1998.

Gubanov, B., *Triumf i tragediya Energii (tom 4: Polyot v nebytiye)*, Nizhniy Novgorod: Izdatelstvo Nizhegorodskogo instituta ekonomicheskogo razvitiya, 1999.

Jenkins, D., *The History of the National Space Transportation System*, Hinckley: Midland Publishing, 2001.

Katorgin, B. (ed.), *NPO Energomash imeni akademika V.P. Glushko. Put v raketnoy tekhnike*, Moscow: Mashinostroyeniye/Polyot, 2004.

KB Khimavtomatiki, stranitsy istorii, tom 1, Voronezh: KBKhA, 1995.

Konyukhov, S. (ed.), *Prizvany vremenem: rakety i kosmicheskiye apparaty konstruktorskogo byuro Yuzhnoye*, Dnepropetrovsk: Art Press, 2004.

Kuznetsov, A., *Mnogorazovaya kosmicheskaya sistema Energiya-Buran*, Moscow: OmV-Luch, 2004.

Lantratov, K., "The 'Star Wars' which never happened" (in Russian), detailed history of Skif/Polyus, published on-line at *http://www.buran.ru/htm/str163.htm*, translated into English in *Quest* magazine, 1/2007, 2/2007.

Lozino-Lozinskiy, G., Bratukhin, A. (ed.), *Aviatsionno-kosmicheskiye sistemy*, Moscow: Izdatelstvo MAI, 1997.

Lukashevich, V., *www.buran.ru* website and accompanying CD-ROMs.

Lukashevich, V., Trufakin, V., Mikoyan, S., "The aerospace system Spiral"/"Spaceplanes of the Spiral System"/"Spiral in national cosmonautics" (in Russian), *Aerokosmicheskoye obozreniye*, 3/2005, 4/2005, 5/2005, 6/2005, 1/2006, 2/2006.

Lukashevich, V., Trufakin, V., Mikoyan, S., "The aerospace system Spiral," *Aviatsiya i kosmonavtika*, 10/2006, 11/2006, 12/2006, 1/2007, 2/2007.

Lukashevich, V., "The OK-92 that became Buran" (in Russian), *Novosti kosmonavtiki*, 3/2006, 4/2006.

Lukashevich, V., "A Soviet copy of the Shuttle: The orbital ship OS-120" (in Russian), *Novosti kosmonavtiki*, 8/2006.

Marinin, I., Shamsutdinov, S., Glushko, A., *Sovetskiye i rossiyskiye kosmonavty 1960–2000*, Moscow: Novosti kosmonavtiki, 2001.

Semyonov, Y. *et al.*, *Mnogorazovyy orbitalnyy korabl Buran*, Moscow: Mashinostroyeniye, 1995.

Semyonov, Y. (ed.), *Raketno-kosmicheskaya korporatsiya Energiya imeni S.P. Korolyova 1946–1996*, Moscow: RKK Energiya, 1996.

Siddiqi, A., *Challenge to Apollo*, Washington, D.C.: NASA, 2000.

Trofimov, V., *Osushchestvleniye mechty*, Moscow: Mashinostroyeniye/Polyot, 2001.

Yefremov, G. (ed.), *60 let samootverzhennogo truda vo imya mira*, Moscow: Oruzhiye i tekhnologii, 2004.

Yeftivyev, M., *Ognyonnye krylya: istoriya sozdaniya reaktivnoy aviatsii SSSR (1930–1946)*, Moscow: Veche, 2005.

Index